PROTAC-Mediated Protein Degradation: A Paradigm Shift in Cancer Therapeutics

Foreword

In the ever-evolving landscape of cancer therapeutics, innovation is paramount. The quest for more effective treatments has led to groundbreaking discoveries, and among them, PROteolysis-TArgeting Chimera- or PROTAC-mediated protein degradation represents frontier science. As we delve into the pages of *PROTAC-mediated Protein Degradation: A Paradigm Shift in Cancer Therapeutics*, edited by Dr. Mukesh Nandave and Dr. Priti Jain, we embark on a journey through the forefront of molecular medicine.

The ubiquitin proteasome system, a cornerstone of cellular regulation, has long fascinated scientists with its intricate machinery and profound implications for health and disease. This book begins by laying the foundation of our understanding and tracing the historical milestones and revolutionary breakthroughs that have shaped our perception of this system.

At the heart of this paradigm shift lies the principle of PROTACs—a transformative approach that harnesses the cell's own machinery to selectively degrade disease-causing proteins. With chapters dedicated to elucidating the principles, mechanisms, and structural considerations of PROTACs, the reader is equipped with a comprehensive understanding of this innovative strategy.

As with any groundbreaking technology, there are inherent advantages and challenges. This book meticulously examines these, offering insights into the novel technologies driving PROTAC design and the future trends shaping its development. Moreover, it navigates the complex landscape of global regulatory requirements and clinical development, illuminating the path from bench to bedside.

One of the most promising aspects of PROTACs is their potential for tissue-specific targeting, offering a tailored approach to cancer therapy. From the toxicological aspects to the management of specific malignancies such as prostate cancer, breast cancer, and hematological malignancies, each chapter delves into the intricacies of applying PROTACs in clinical practice.

Furthermore, this volume explores the synergies between PROTACs and emerging technologies like artificial intelligence and machine learning, offering a glimpse into the future of precision medicine. Whether in the treatment of glioma or cardiovascular diseases, the potential of PROTACs to reshape therapeutic landscapes knows no bounds.

In closing, "PROTAC-mediated protein degradation" transcends the boundaries of traditional oncology, offering a holistic perspective that encompasses not only the molecular intricacies but also the clinical implications and future directions of this transformative technology. It is my sincere belief that this book will serve as a cornerstone for researchers, clinicians, and students alike, driving us closer to the ultimate goal of conquering cancer.

Harvard Medical School, Boston, MA, USA
Center for Engineered Therapeutics,
Brigham and Women's Hospital, Boston, MA, USA
Division of Health Science and Technology,
Dana-Farber Cancer Institute and Harvard-MIT,
Cambridge, MA, USA

Shiladitya Sengupta

Preface

In the ongoing quest to develop more effective treatments for cancer, researchers are exploring innovative approaches that target the molecular mechanisms underlying the disease. One such promising avenue is the development of proteolysis-targeting chimeras or PROTACs. These molecules represent a novel class of therapeutic agents with the potential to revolutionize cancer therapy by exploiting the body's own protein degradation machinery to eliminate disease-causing proteins.

Though, traditional approaches like monoclonal antibodies and small molecule inhibitors have shown significant efficacy, they often encounter challenges such as drug resistance and limited selectivity. PROTACs tend to overcome these obstacles by harnessing the cell's natural protein degradation pathway to degrade target proteins, thereby achieving more potent and selective inhibition.

In the context of cancer therapy, PROTACs hold tremendous promise for targeting oncogenic drivers and overcoming drug resistance. PROTACs have the potential to disrupt multiple signaling pathways simultaneously, thereby exerting potent antitumor effects. Furthermore, the ability to target specific proteins within complex signaling networks offers the possibility of personalized treatment strategies tailored to the molecular profile of individual tumors.

Despite these exciting prospects, challenges remain in the development and optimization of PROTAC-based therapies. These include issues related to selectivity, pharmacokinetics, and off-target effects, which must be carefully addressed to ensure the safety and efficacy of PROTACs in clinical settings. Additionally, the identification of suitable E3 ligases and the design of optimal ligands pose significant hurdles in the rational design of PROTACs.

This book entitled *PROTAC-Mediated Protein Degradation: A Paradigm Shift in Cancer Therapeutics* is published with an intention to provide in-depth concepts of PROTACs and their role in the treatment of cancer. This book comprises 17 chapters presented logically to understand PROTACs from the fundamental to the latest development. This book begins with ubiquitin-proteasome system, its history and development, and principles of PROTACs. Further, it proceeds with the chemistry of PROTACs and the designing aspects for novel compounds. This book then progresses with technological aspects and global regulatory requirements for PROTACs and then discusses the pharmacological aspects covering the role in prostrate cancer, breast cancer, lymphocytic leukemia, and glioma. The delivery systems of PROTACs and exploration of AI/ML have also been dealt with in this book.

In conclusion, PROTACs represent a promising new approach in cancer therapy that capitalizes on the cell's natural protein degradation machinery to eliminate disease-causing proteins. The potential of PROTACs to overcome limitations associated with traditional inhibitors and to target previously undruggable proteins makes them an exciting area of research with the potential to significantly impact the treatment of cancer and other diseases.

Overall, our book aims to be a comprehensive and authoritative resource, catering to both experts in the field and those seeking to gain a deeper understanding of PROTAC technology and its transformative potential in cancer therapeutics.

New Delhi, Delhi, India Mukesh Nandave
New Delhi, Delhi, India Priti Jain

Contents

Editors and Contributors

About the Editors

Mukesh Nandave is currently serving as the Associate Dean (Research and Consultancy), Head, School of Pharmaceutical Sciences (SPS), and the Head of the Department of Pharmacology and Pharmaceutical Biotechnology at Delhi Pharmaceutical Sciences and Research University, under the Government of NCT of Delhi, India. He completed his Ph.D. in Pharmacology from AIIMS, Delhi, and pursued his postdoctoral training at the Ohio State University Medical Center in Columbus, USA. He has primarily focused his research on investigating the role of nutraceuticals, herbomineral formulations, and phytoconstituents in life style-related disorders such as hypertension, hyperlipidemia, diabetes, obesity, and cancer. He has published over 120 papers in national and international peer-reviewed journals and received several prestigious awards, including the distinguished leadership award of the International Academy of Cardiovascular Sciences and holds various professional memberships.

Priti Jain is an Assistant Professor at the Department of Pharmaceutical Chemistry, Delhi Pharmaceutical Sciences and Research University, New Delhi. She has more than 15 years of experience in industrial and academic research. Her area of research is mainly focused on in silico design and the discovery of low molecular weight heterocycles as anti-Alzheimer's agents, anti-cancer agents, and anti-HIV agents. She is involved in research on analytical method development and validation of API and dosage forms. She has published more than 40 research articles in peer-reviewed international journals and authored or co-authored numerous books and book chapters. She serves as an editor for various journals and a reviewer of many reputed journals. She is also a Bentham Brand Ambassador since 2018 and has been conferred with various awards. She is a member of many international scientific societies and organizations.

Contributors

V. A. Arun Department of Medicine, Naval Hospital Asvini Colaba, Mumbai, Maharashtra, India

Kalyani Barve Department of Pharmacology, Shobhaben Pratapbhai Patel School of Pharmacy & Technology Management, SVKM's Narsee Monjee Institute of Management Studies (NMIMS) Deemed-to-University, Mumbai, India

Shvetank Bhatt School of Health Sciences and Technology, Dr. Vishwanath Karad MIT World Peace University, Kothrud, Pune, Maharashtra, India

Sushma Dev Department of Pharmaceutical Chemistry, School of Pharmaceutical Sciences, Delhi Pharmaceutical Sciences and Research University, DPSRU, New Delhi, India

John Disouza Department of Pharmaceutics, Tatyasaheb Kore College of Pharmacy, Kolhapur, Maharashtra, India

Advait Dubey Department of Clinical Pharmacology, Advanced Centre for Treatment, Research & Education in Cancer (ACTREC), Tata Memorial Centre, Navi Mumbai, India

Homi Bhabha National Institute, Mumbai, India

Ashwini Gawade School of Health Sciences and Technology, Dr. Vishwanath Karad MIT World Peace University, Kothrud, Pune, Maharashtra,, India

Shikha Goswami Department of Pharmacology, Delhi Pharmaceutical Sciences and Research University, Delhi, India

Vikram Gota Department of Clinical Pharmacology, Advanced Centre for Treatment, Research & Education in Cancer (ACTREC), Tata Memorial Centre, Navi Mumbai, India

Homi Bhabha National Institute, Mumbai, India

Ramesh K. Goyal Department of Pharmacology, Delhi Pharmaceutical Sciences and Research University, Delhi, India

Dhruv Sanjay Jaya Gupta Department of Pharmacology, Shobhaben Pratapbhai Patel School of Pharmacy & Technology Management, SVKM's Narsee Monjee Institute of Management Studies (NMIMS) Deemed-to-University, Mumbai, India

Ashif Iqubal Medical Affairs, Pfizer Oncology, Mumbai, Maharashtra, India

Kashif Iqubal Irma Lerma Rangel College of Pharmacy, Texas A&M Health Science Center, Texas A&M University, College Station, TX, USA

Hemant R. Jadhav Department of Pharmacy, Birla Institute of Technology and Science, Pilani Campus, Pilani, Rajasthan, India

Keerti Jain Drug Delivery and Nanomedicine Research Laboratory, Department of Pharmaceutics, National Institute of Pharmaceutical Sciences and Research (NIPER) – Raebareli, Lucknow, India

Priti Jain Department of Pharmaceutical Chemistry, School of Pharmaceutical Sciences, Delhi Pharmaceutical Sciences and Research University, DPSRU, New Delhi, India

Vineet Kumar Jain Department of Pharmaceutics, Delhi Pharmaceutical Sciences and Research University (DPSRU), New Delhi, India

Hard K. Kalyani Department of Pharmaceutics, L. M. College of Pharmacy, Ahmedabad, Gujarat, India

Vikas Kamble Department of Pharmaceutics, Tatyasaheb Kore College of Pharmacy, Warananagar, Maharashtra, India

Deepshikha Pande Katare Proteomics and Translational Research Lab, Centre for Medical Biotechnology, Amity Institute of Biotechnology, Amity University, Noida, India

Aqsa Khan Department of Pharmaceutical Chemistry, School of Pharmaceutical Sciences, Delhi Pharmaceutical Sciences and Research University, DPSRU, New Delhi, India

Prashant S. Kharkar Department of Pharmaceutical Sciences and Technology, Institute of Chemical Technology, Matunga, Mumbai, India

Nidhi Khedpande Department of Pharmacology, Shobhaben Pratapbhai Patel School of Pharmacy & Technology Management, SVKM's Narsee Monjee Institute of Management Studies (NMIMS) Deemed-to-University, Mumbai, India

Kaustubh Kolekar Department of Pharmaceutics, Tatyasaheb Kore College of Pharmacy, Warananagar, Maharashtra, India

Amit Kumar Department of Pharmaceutical Chemistry, School of Pharmaceutical Sciences, Delhi Pharmaceutical Sciences and Research University, DPSRU, New Delhi, India

Parveen Kumar Oracle India Private Limited, Gurugram, India

Sachin Kumar Department of Medical Laboratory Technology, School of Allied Health Sciences and Management, Delhi Pharmaceutical Sciences and Research University (DPSRU), New Delhi, India

Popat Kumbhar Department of Pharmaceutics, Tatyasaheb Kore College of Pharmacy, Warananagar, Maharashtra, India

Vikas Maharshi Department of Pharmacology, All India Institute of Medical Sciences, Patna, India

Deepa S. Mandlik Bharati Vidyapeeth (Deemed to be University), Poona College of Pharmacy, Pune, India

Satish K. Mandlik Department of Pharmaceutics, Bharati Vidyapeeth (Deemed to be University), Poona College of Pharmacy, Pune, Maharashtra, India

Ruchi Jakhmola Mani Proteomics and Translational Research Lab, Centre for Medical Biotechnology, Amity Institute of Biotechnology, Amity University, Noida, India

Piyush Mehra Drug Delivery and Nanomedicine Research Laboratory, Department of Pharmaceutics, National Institute of Pharmaceutical Sciences and Research (NIPER) – Raebareli, Lucknow, India

Kundan Mishra Department of Clinical Hematology & BMT, Command Hospital, Lucknow, India

Priyanka Mohanty Department of Pharmacy, Birla Institute of Technology and Science, Pilani Campus, Pilani, Rajasthan, India

Sreevardhan Moode Drug Delivery and Nanomedicine Research Laboratory, Department of Pharmaceutics, National Institute of Pharmaceutical Sciences and Research (NIPER) – Raebareli, Lucknow, India

Riya Nagpal Department of Pharmaceutical Chemistry, School of Pharmaceutical Sciences, Delhi Pharmaceutical Sciences and Research University, DPSRU, New Delhi, India

Kavita Pal Department of Clinical Pharmacology, Advanced Centre for Treatment, Research & Education in Cancer (ACTREC), Tata Memorial Centre, Navi Mumbai, India

Homi Bhabha National Institute, Mumbai, India

Arzoo Pannu Department of Pharmacology, Delhi Pharmaceutical Sciences and Research University, Delhi, India

Bharat R. Patel Department of Pharmacology, L. M. College of Pharmacy, Ahmedabad, Gujarat, India

Dipti H. Patel Department of Pharmaceutics, Institute of Pharmaceutical sciences, Faculty of Pharmacy, Parul University, Vadodara, India

Jayvadan K. Patel Aavis Pharmaceuticals, Hoschton, GA, USA

Kashyap M. Patel Department of Pharmaceutics, L. M. College of Pharmacy, Ahmedabad, Gujarat, India

Manish P. Patel Department of Pharmaceutics, L. M. College of Pharmacy, Ahmedabad, Gujarat, India

Manisha Patel Drug Delivery and Nanomedicine Research Laboratory, Department of Pharmaceutics, National Institute of Pharmaceutical Sciences and Research (NIPER) – Raebareli, Lucknow, India

Shubham Patil School of Health Sciences and Technology, Dr. Vishwanath Karad MIT World Peace University, Kothrud, Pune, Maharashtra, India

Vandana B. Patravale Department of Pharmaceutical Sciences and Technology, Institute of Chemical Technology, Matunga, Mumbai, Maharashtra, India

Harvinder Popli Department of Pharmaceutics, Delhi Pharmaceutical Sciences and Research University (DPSRU), New Delhi, India

Rohini Pujari School of Health Sciences and Technology, Dr. Vishwanath Karad MIT World Peace University, Kothrud, Pune, Maharashtra, India

Abdul Qadir Department of Pharmaceutical Chemistry, School of Pharmaceutical Sciences, Delhi Pharmaceutical Sciences and Research University, DPSRU, New Delhi, India

Sukuru Chinna Reddy Drug Delivery and Nanomedicine Research Laboratory, Department of Pharmaceutics, National Institute of Pharmaceutical Sciences and Research (NIPER) – Raebareli, Lucknow, India

Ranbir Singh Saluja MET Institute of Pharmacy, Bhujbal Knowledge City, Bandra (West), Mumbai, India

Shreya Sharma School of Health Sciences and Technology, Dr. Vishwanath Karad MIT World Peace University, Kothrud, Pune, Maharashtra, India

Urvashi Soni School of Health Sciences and Technology, Dr. Vishwanath Karad MIT World Peace University, Kothrud, Pune, Maharashtra, India

Ruchi Tandon Translational Health Science and Technology Institute, NCR Biotech Cluster, Faridabad, India

Sofiya Tarannum Drug Delivery and Nanomedicine Research Laboratory, Department of Pharmaceutics, National Institute of Pharmaceutical Sciences and Research (NIPER) – Raebareli, Lucknow, India

Shikha Thakur Department of Pharmacy, Birla Institute of Technology and Science, Pilani Campus, Pilani, Rajasthan, India

Chukwuebuka Emmanuel Umeyor Department of Pharmaceutical Sciences and Technology, Institute of Chemical Technology, Matunga, Mumbai, Maharashtra, India

Department of Pharmaceutics and Pharmaceutical Technology, Faculty of Pharmaceutical Sciences, Nnamdi Azikiwe University, Awka, Anambra State, Nigeria

Madhura P. Vaidya MET Institute of Pharmacy, Bhujbal Knowledge City, Bandra (West), Mumbai, India

Abhishek Wahi Department of Pharmaceutical Chemistry, School of Pharmaceutical Sciences, Delhi Pharmaceutical Sciences and Research University, DPSRU, New Delhi, India

Sanjeev Yadav Department of Hematology, Sanjay Gandhi Postgraduate Institute of Medical Sciences, Lucknow, India

Abbreviations

Ab-TAC	Antibody-based PROTAC
AD	Alzheimer's disease
ADAs	Anti-drug antibodies
ADME	Absorption, distribution, metabolism, excretion
AE	Adverse effects
AhR	Aryl hydrocarbon receptor
AI	Artificial Intelligence
AKT	Protein kinase B
ALCOA	Attributable, Legible, Contemporaneous, Original, and Accurate
ALK	Anaplastic lymphoma kinase
ALL	Acute lymphoblastic leukemia
ALS	Amyotrophic lateral sclerosis
AML	Acute myeloid leukemia
AMP	Adenosine monophosphate
ANN	Artificial neural networks
AP	Affinity purification
APC	Anaphase-promoting complex
APC	Aptamer-PROTAC conjugates
API	Active pharmaceutical ingredient
APOA2	Apolipoprotein A-II precursor
AR	Androgen receptor
ARCC-4	Androgen receptor PROTAC degrader-4
ARL	Androgen receptor ligand
ATP	Adenosine triphosphate
B2M	Beta-2-microglobulin
BCL	B-cell lymphoma
BCL-2	B-cell lymphoma-2
BCL-XL	B-cell lymphoma extra large
BCM	Binary classification model
BET	Bromodomain and extra-terminal domain
BRAF	v-Raf murine sarcoma viral oncogene homolog B1
BRCA	BReast CAncer gene
BRD	Bromodomain protein
BRD4	Bromodomain containing protein 4

BTK	Bruton's tyrosine kinase
BUB1	Budding uninhibited by benzimidazole 1
CADD	Computer-aided drug design
CALR	Calreticulin gene
CAMs	Cell adhesion molecules
CAR-T	Chimeric antigen receptor T-cell therapy
CBR	Clinical benefit rate
CCR5	C-Chemokine receptor-5
CDK	Cyclic-dependent kinase
CDK4/6	Cyclin-dependent kinase(s) 4/6
CDK6	Cyclin-dependent kinase 6
CDKN2A/B	Cyclin-dependent kinase inhibitor 2A/B
CETSA	Cellular thermal shift assays
CHK	Checkpoint kinases
cIAP1	Cellular inhibitor of apoptosis protein 1
CLIPTAC	Click-formed proteolysis-targeting chimera
CMA	Chaperone-mediated autophagy
CMC	Chemistry manufacturing and controls
CNN	Convolutional neural networks
CNS	Central nervous system
COX-1/2	Cyclooxygenase ½
CRABP1	Cellular retinoic acid-binding protein 1
CRBN	Cereblon
CRFs	Case report forms
CRISPR	Clustered regularly interspaced short palindromic repeats
CRL	Cullin-Ring ligase
CRP	C-reactive protein
CRPC	Castration-resistant prostate cancer
CSRs	Clinical study reports
CTA	Clinical Trial Application
CUL2	Cullin-2
CVD	Cardiovascular disease
Cys	Cysteine
DAPK1	Protein kinase 1
DBPs	DNA-binding proteins
DDIs	Drug–drug interactions
DGM	Deep generative model
DL	Deep learning
DNA	Deoxyribonucleic acid
DSG	Di(N-succinimidyl) glutarate
DSS	Di(N-succinimidyl) suberate
DUBs	Deubiquitylating enzymes
E3RE	E3 recruiting element
EDGs	Electron-donating groups
EDP	Event-driven pharmacology

EGFR	Epidermal growth factor receptor
eGFR	Estimated glomerular filtration rate
EMA	European Medicines Agency
EPR	Enhanced permeability and retention effect
ER	Estrogen receptor
ERK	Extracellular signal-regulated protein kinase
ERα	Estrogen receptor alpha
EWGs	Electron-withdrawing groups
FAK	Focal adhesion kinase
FDA	Food and Drug Administration
FRET	Fluorescence resonance energy transfer
GAN	Generative adversarial networks
GBM	Glioblastoma
GCN	Graph convolutional neural networks
GCP	Good Clinical Practice
GMP	Good manufacturing practice
GPLR	G protein-linked receptors
GPU	Graphical processing units
GSCs	GB stem-like cells
HBA	Hemoglobin A
HBAs	Hydrogen bond acceptors
HBDs	Hydrogen bond donors
HCC	Hepatocellular carcinoma
HD	Huntington's disease
HDAC6	Histone deacetylase 6
HDACs	Histone deacetylases
HECT	Homologous to E6-AP carboxyl 74 terminus
HER2/ErbB2	Human epidermal growth factor receptor 2
HIF-1α	Hypoxia-inducible factor 1-alpha
HIPK2	Homeodomain-interacting protein kinase 2
HM	Hematological malignancies
HNSCC	Head and neck squamous cell carcinoma
HPK1	Hematopoietic progenitor kinase 1
hs-CRP	High-sensitivity C-reactive protein
HTS	High-throughput screening
IAPs	Inhibitor of apoptosis proteins (IAPs)
ICFs	Informed consent forms
ICH	International Council for Harmonisation of Technical Requirements for Pharmaceuticals for Human Use
ICIs	Immune checkpoint inhibitors
IDH	Isocitrate dehydrogenase
IDO	Indoleamine 2,3-dioxygenase
IGFALS	Insulin-like growth factor binding protein, acid labile subunit
IGFBP4	Insulin-like growth factor binding protein 4
IKBα	Inhibitor of nuclear factor kappa B

IKZF	IKAROS family zinc finger
IMDs	Immune-modulatory drugs
IND	Investigational new drug
IRBs	Institutional review boards
ITC	Isothermal titration calorimetry
IVIVE	*In vitro in vivo* extrapolation
JAK-STAT	Janus kinase-signal transducer and activator of transcription
KRAS	Kirsten rat sarcoma viral oncogene homologue
LSTM	Long short-term memory
LYTAC	Lysosome-targeting chimera
MAAs	Marketing Authorization Applications
mABs	Monoclonal antibodies
MAP4K1	Mitogen-activated protein kinase 1
MAPD	Model-free analysis of protein degradability
MAPK	Mitogen-activated protein kinase
MCL-1	Myeloid cell leukemia 1
mCRPC	Metastatic castration-resistant prostate cancer
MDM2	Mouse double minute 2 homolog
MET	Mesenchymal-epithelial transition factor receptor
MetAp-2	Methionine aminopeptidase-2
MI	Myocardial infarction
ML	Machine learning
MLP	Multi-layer perceptron
MMPs	Matrix metalloproteinases
mRNA	Messenger ribonucleic acid
MTD	Maximum tolerated dose
mTOR	Mammalian target of rapamycin
NAE	NEDD8-activating enzyme
NAPT	Nicotinamide phosphoribosyltransferase
NDAs	New drug applications
NDs	Neurodegenerative diseases
NEMO	NF-κB essential modulator
NF1	Neurofibromatosis type 1
NGAL	Neutrophil gelatinase-associated lipocalin
NPM1	Nucleophosmin
NRF2	Nuclear factor erythroid 2-related factor 2
NSCLC	Non-small cell lung cancer
NVOC	Nitroveratryloxycarbonyl
ODP	Occupancy-driven pharmacology
ORM1	Alpha-1-acid glycoprotein 1
PAKs	p21-activated kinases
PARP	Poly (ADP-ribose) polymerase
PARP1	Poly (ADP-ribose) polymerase
PC	Prostate cancer
PD	Parkinson's disease

PD	Pharmacodynamics
PD-1	Protein cell death protein-1
PDGFRα	Platelet-derived growth factor receptor alpha
PD-L1	Protein cell death ligand-1
PDX	Patient-derived xenograft
PEG	Polyethylene glycol
PIM	Proviral integration sites for Moloney murine leukemia virus
PK	Pharmacokinetics
PK	Protein kinase
POI	Protein of interest
PPARγ	Peroxisome proliferator-activated receptor gamma
PPi	Protein–protein interaction
PPIA	Peptidyl-prolyl isomerase A
PPPI3K	Phosphoinositide 3-kinase
PR	Progesterone receptors
PROTAC-DB	Proteolysis-targeting chimera database
PROTACs	Proteolysis-targeting chimeras
PSA	Prostate-specific antigen
PTEN	Phosphatase and tensin homolog
PTM	Post-translational modification
PTP1B	Protein-tyrosine phosphatase 1B
RA	Rheumatoid arthritis
RAF	Rapidly accelerated fibrosarcoma
RAS	Ras/Raf/MAPK pathway
RAS	Rat sarcoma virus
RBPs	Ribonucleic acid
RBR	Ring between ring
RIBOTACs	Ribonuclease-targeting chimeras
RING	Really interesting new gene
RING	Really interesting new gene finger
RIPK 1	Receptor-interacting protein kinase 1
RNF182	Ring finger protein 182
RNN	Recurrent neural networks
ROS	Reactive oxygen species
RP2D	Recommended phase 2 dose
RTKs	Receptor tyrosine kinases
RT-PROTAC	Radiotherapy-triggered PROTAC
SAEs	Serious adverse events
SARS-CoV2	Severe acute respiratory syndrome coronavirus 2
SASP	Senescence-associated secretory phenotype
SBD	Substrate binding domain
SCF	Skp1-Cullin-F-box
SELEX	Systematic evolution of ligands by exponential enrichment
SERD	Selective estrogen receptor degrader
SERDs	Selective estrogen receptor downregulators

SHP2	Src homology 2 domain-containing phosphatase 2
shRNA	Short hairpin ribonucleic acid
siRNA	Small interfering RNA
Sirt 2	Sirtuin 2
SMIs	Small molecule inhibitors
SOD1	Superoxide dismutase 1
SOPs	Standard operating procedures
STAT	Signal transducer and activator of transcription
STAT-3	Signal transducer and activator of transcription-3 system
TALEN	Transcription activator-like effector nuclease
TAMs	Tumor-associated macrophages
TBD	Target-based degradation
TC	Ternary complex
TCRs	T-cell receptors
TFs	Transcription factors
TGF-β	Transforming growth factor-β
TIS	Therapy-induced senescence
TKIs	Tyrosine kinase inhibitors
TLRs	Toll-like receptors
TMZ	Temozolomide
TNF	Tumor necrosis factor
TP	Target protein
Tp53	Tumor protein p53
TPD	Targeted protein degradation
TPSA	Topological polar surface area
TRIM	Tripartite motif
TRK	Tropomyosin receptor kinase
Ub	Ubiquitin
UGT	Uridine diphosphate (UDP)-glucuronosyltransferases
UPS	Ubiquitin proteasome system
VEGFR	Vascular endothelial growth factor receptor
VHL	Von Hippel–Lindau
VHL1	Von Hippel–Lindau ubiquitin ligase-1
WHO	World Health Organization
ZFN	Zinc finger nuclease
β-TRCP	Beta-transducin repeats-containing proteins

Understanding the Ubiquitin Proteasome System: History and Revolution

1

Abdul Qadir, Amit Kumar, Riya Nagpal, Aqsa Khan, Abhishek Wahi, and Priti Jain

Abstract

The Ubiquitin Proteasome System (UPS) stands as a cornerstone in the realm of cellular biology, orchestrating the regulated degradation of proteins essential for cellular homeostasis. This chapter embarks on a journey through the historical milestones that have shaped our understanding of UPS, tracing its roots from early observations to contemporary breakthroughs. Moreover, the chapter delves into the transformative impact of UPS research on various fields, ranging from cancer biology to neurodegenerative diseases, highlighting how deregulation of UPS contributes to pathogenesis. Also, this chapter illuminates the historical journey and revolutionary insights that have propelled our comprehension of UPS, underscoring its pivotal role in cellular physiology and disease mechanisms.

Keywords

Ubiquitin Proteasome System (UPS) · Cancer biology · Neurodegenerative diseases · Cellular physiology · Chimeric antigen receptor T-cell therapy

A. Qadir · A. Kumar · R. Nagpal · A. Khan · A. Wahi · P. Jain (✉)
Department of Pharmaceutical Chemistry, School of Pharmaceutical Sciences, Delhi Pharmaceutical Sciences and Research University, DPSRU, New Delhi, India
e-mail: pritijain@dpsru.edu.in

M. Nandave, P. Jain (eds.), *PROTAC-Mediated Protein Degradation: A Paradigm Shift in Cancer Therapeutics*, https://doi.org/10.1007/978-981-97-5077-1_1

Abbreviations

AD	Alzheimer's disease
ALS	Amyotrophic lateral sclerosis
AP	Affinity purification
APC	Anaphase-promoting complex
CAR-T	Chimeric antigen receptor T-cell therapy
CDKS	Cyclin-dependent kinases
CRBN	Cereblon
DUBs	Deubiquitylating enzymes
E6-AP	E6-associated protein
HD	Huntington's disease
HECT	Homologous to E6-AP Carboxyl 74 Terminus
MDM2	Murine double minute 2
MHC	Major histocompatibility complex
NDs	Neurodegenerative diseases
NEMO	NF-κB essential modulator
PD	Parkinson's disease
POI	Protein of interest
PPI	Protein–protein interaction
PROTAC	Proteolysis-targeting chimera
PRRs	Pattern recognition receptors
RING	Really interesting new gene
RIPK 1	Receptor-interacting protein kinase 1
SMAD	Suppressor of mothers against decapentaplegic
SOD1	Superoxide dismutase 1
TCRs	T-cell receptors
TLRs	Toll-like receptors
TNF	Tumour necrosis factor
Tp53	Tumour protein p53
Ub	Ubiquitin
UPS	Ubiquitin proteasome system
VHL	von Hippel-Lindau

1.1 Introduction

The ubiquitin Proteasome System (UPS) regulates the turnover of numerous cellular proteins. It is a primary process of intracellular protein breakdown (Wang and Maldonado 2006). It is a precise and universal method for the cell to get rid of biologically useless proteins, such as mutant, misfolded, damaged, terminally altered, or excessively accumulated proteins. With its ability to quickly start substrate-specific proteolysis, UPS can function as a molecular switch in a range of signalling routes, swiftly stopping a target protein's activity. Preventing the build-up of potentially hazardous, non-functional proteins is one of the main roles of the

UPS. Furthermore, it is now evident that the UPS is engaged in the majority of eukaryotic biological mechanisms, encompassing transcriptional regulation, intracellular signalling, and control of cell death. In all eukaryotic cells, the UPS is the principal mechanism governing controlled peptide chain breakdown, and it is increasingly evident that abnormalities in this route are the root cause of many human diseases. A large range of proteins with half-life period ranging from hours to days can be progressively degraded in concert via the highly conserved and strictly regulated UPS pathway (Kleiger and Mayor 2014; Zolk et al. 2006). Herein, we intend to provide the historical turning points that have influenced our knowledge of UPS, from the earliest observations to the most recent innovations. Additionally, the chapter explores how UPS research has revolutionised a number of domains, from neurodegenerative diseases to cancer biology, emphasising how UPS dysregulation plays a role in pathogenesis. Furthermore, this chapter highlights UPS's critical function in cellular physiology and disease mechanisms while also illuminating the approaches to enhance the understanding of UPS and target UPS for treatment of diseases.

1.2 History of UPS

Cell biologists were first taken aback by the requirement of ATP for the process of cytosolic protein degradation because it is widely known that metabolic energy is not needed for peptide bond hydrolysis. Hershko and colleagues employed a method that included biochemically fractionating reticulocytes, which are terminally developed red blood cells devoid of lysosomes, purifying the reticulocytes, identifying the constituents, and reconstructing the ATP-dependent cytosolic protein breakdown in vitro. These discoveries were supported by information from other organizations and genetic research revealing the intracellular functions of cytosolic protein breakdown in numerous biological systems. Avram Hershko, Aaron Ciechanover, and Irwin Rose were granted the 2004 Nobel Prize in chemistry for their groundbreaking research on the function of protein ubiquitination during cytosolic protein breakdown. A rabbit reticulocyte system was created by Etlinger and Goldberg (1977) to investigate ATP-dependent and non-lysosomal protein breakdown (Etlinger and Goldberg 1977). After reticulocyte was fractionated in 1978 by Ciechanover et al., two fractions—active principle of fraction (APF)-I and (APF)-II—were identified (Ciehanover et al. 1978). By combining APF-I with APF-II in 1979, Hershko et al. reconstructed protein degradation. Two fractions, APF-IIa and APF-IIb, were created by subdividing APF-II. The E1-E3 ubiquitin-conjugating enzymes were present in APF-IIb. Later research revealed that APF-IIa contained proteasomes (Hough et al. 1986). Wilkinson et al. discovered that APF-I was ubiquitin (Wilkinson et al. 1980). Conjugating enzymes and ATP were required for the formation of high molecular conjugates between ubiquitin and substrate proteins (Hershko et al. 1980). Hershko et al. discovered deubiquitinating enzyme activity that could recycle ubiquitin bound to substrate proteins (Hershko et al. 1980). The E1 enzyme was identified to activate the ubiquitin carboxyl terminus glycine (Hershko et al. 1981). According to Hershko et al., there is a connection between ubiquitination and protein degradation since reticulocytes produced more ubiquitin-protein conjugates when aberrant proteins were formed (Hershko et al. 1982). Further, A method for purification and identification of all three ubiquitin conjugating enzymes (E1, E2, and E3) was developed by Hershko et al. (1983) . The anatomic functions of the UPS were disclosed by Finley et al.'s 1984 genetic research. The absence of growth in the non-permissive temperature of the mutant mammary cell line ts85 was attributed to the E1 enzyme, indicating the potential role of ubiquitination in the advancement of the cell cycle. 1984 Ciechanover et al. At the non-permissive temperature, the ts85 cell line's ability to degrade short-lived proteins was hindered. Hershko and colleagues in 1986 discovered that an E3 ligase exhibited substrate specificity. A large molecular weight neutral protease that originated from rat liver, the pituitary gland of cows, and the archaebacterium *Thermoplasma acidophilum* was identified by numerous organizations. Since it was first believed that this protease was connected to RNA, it has been called "prosome". Notably, an allo-MHC serum was used to identify a large molecular weight complex made up of tiny subunits.

Several years later, it was found that the two subunits of the IFN-γ-inducible proteasome, β1i and β5i, were encoded by genes situated in the mammalian MHC region. In 1986, Hough et al. described a high molecular protease that broke down protein ubiquitin adducts but not untagged protein. Its proteolytic and particle characteristics led to the coining of the term proteasome. In the presence of ATP, isolated 20S proteasomes and 19S regulators were mixed to create active 26S proteasomes (Monaco and McDevitt 1984; Sijts and Kloetzel 2011).

1.3 Components of UPS

1.3.1 Ubiquitin

Ubiquitin is a tiny protein with a molecular weight of about 8.5 kDa that is made up of 76 amino acids folded into a compact globular shape. It exists in the cytoplasm either unbound or covalently attached to other proteins. All eukaryotic cells contain ubiquitin, a small, highly conserved peptide. Seven lysines are present in this molecule (K6, K11, K27, K29, K33, K48, and K63). K48 and K11 polyubiquitin chains mostly facilitate proteasomal breakdown. On the other hand, K63-related polyubiquitination, which is generally less prevalent in tumours, is generally linked to cellular signal building, transduction, and the repair of damaged cells rather than proteasomal breakdown (Huang et al. 2024; Park et al. 2020). A ubiquitin molecule is used to mark the substrate protein, which is then broken down by the 26 s proteasome. Ubiquitin attaches itself to proteins that need to be targeted by the proteasome. Targeted protein breakdown begins with ubiquitin's selective binding to specific proteins. The target protein and ubiquitin are connected via a branching isopeptide bond that forms between the lysine ε-amino group and the carboxyl-terminal glycine of ubiquitin. Polyubiquitin is created when a ε-amino group of lysine from one ubiquitin molecule forms a similar isopeptide bond with the carboxy terminus of ubiquitin.

Monoubiquitination and polyubiquitination are two ways that ubiquitin is altered. Monoubiquitination is the result of adding a single ubiquitin molecule to the lysine residue of a substrate (Swatek and Komander 2016). In the process of polyubiquitination, one ubiquitin molecule is joined with additional ubiquitin molecules to create polyubiquitin chains. The bare minimum length of a chain required to effectively target a protein for proteasomal destruction is four ubiquitin adducts. There are three distinct steps of protein ubiquitination, commonly known as ubiquitinylation: E1 (ubiquitin-activating enzyme), E2 (ubiquitin-conjugating enzyme), and E3 (ubiquitin ligase). In each phase, an enzyme is involved (Ebner et al. 2017). The ubiquitin chain is generated by the E1 and E2 enzymes and subsequently bound to proteins by the E3 enzyme.

1.3.2 E1, A Ubiquitin-Activating Enzyme

The ATP-dependent activation of ubiquitin by the ubiquitin-activating enzyme E1 starts the process of ubiquitination. E1-S ~ ubiquitin is created when ubiquitin is linked to an internal E1 Cys residue through an intermediary thiol ester between the target protein and the ubiquitin monomer from the E2 enzyme. Uba1 is the predominant version of this enzyme in yeast and humans (Schulman and Wade Harper 2009).

1.3.3 E2, A Ubiquitin-Conjugating Enzyme

Activated ubiquitin molecules are gradually added to the original ubiquitin proteins to form a polyubiquitin chain. Ubiquitin is trans-esterified to a conserved cysteine in an E2 enzyme subsequent to its activation by the E1. Yeast has 13 E2s, and vertebrates probably have a similar amount. The three-letter code ‘Ubc’ represents the genetic name of these enzymes. With the exception of Ubc9, an enzyme that conjugates SUMO, and Ubc12, an enzyme that conjugates NEDD8 and Rub1, the Ubcs have different genetic roles in ubiquitylation, albeit some of them overlap. Ubc3, or CDC34, is a crucial E2 enzyme in the Skp1, cullin, F-box (SCF) ubiquitin ligase complex (McKinnon and Tabrizi 2014).

1.3.4 E3, A Ubiquitin Ligase

The mechanism gains its specificity from the E3 ligase. The E3 ligase promotes the creation of a covalent link between the target protein and the ubiquitin monomer from the E2 enzyme by binding both the target protein and the complex E2-ubiquitin. To create a polyubiquitin chain, activated ubiquitin molecules are progressively added to the initial ubiquitin proteins. The 26S proteasome recognizes and targets proteins that are tagged with chains of four or more ubiquitins for destruction. By attaching itself to a protein target preferentially, the E3 ligase gives the process selectivity (Grice and Nathan 2016). Hundreds of different E3 ubiquitin ligases are enciphered by the human gene pool and are grouped into three primary classes, according to structural similarities: the U-box proteins, the RING-finger proteins, which are the most prevalent class of E3 enzymes, and the HECT domain proteins. Prior to delivering the ubiquitin to the substrate, HECT domain proteins create a covalent (thiolester) link with it (Garcia-Barcena et al. 2020; Nakamura 2011). When RING finger (including amino acids cysteines and histidines responsible for co-ordination of two metal atoms) E3s connect to the E2-ubiquitin complex, they facilitate direct ubiquitin transfer to the targeted protein, eliminating the need for further thiol ester formation—a process that is seen in the HECT family of E3 ligases (Jackson et al. 2000).

1.3.5 HECT Domain Proteins

The 350-amino acid, which is similar to the E6-AP C terminus and was initially identified in E6-AP, defines HECT domain proteins. Eukaryotes, including yeast and humans, have HECT domain proteins. The oncoprotein of the papillomavirus in humans recruits E6-AP, a cellular ubiquitin ligase, which degrades the p53 tumour suppressor 4. A conserved cysteine combines with ubiquitin to create a thioester in the E6 HECT domain. This intermediate is necessary for the process of ubiquitination. HECT domain proteins are typically big (90–200 kDa) with extended N-terminal domains. The N-terminus of these ubiquitin ligases binds to the substrate, whereas the C-terminal HECT domain transfers ubiquitin directly by a thioester bond to the substrate (Eldridge and O'Brien 2010; Jackson et al. 2000). It was recently demonstrated that Xsmurf1, a novel HECT domain E3, may bind to Smad1, ubiquitylate it, and control its stability. Smad1 is a signalling regulator that is regulated by the TGF-β family member BMP. It is unknown if other types of protein–protein interaction domains have the ability to bind HECT domain proteins to substrates for ubiquitylation.

1.3.6 Skp1, Cullin, F-Box (SCF) Complexes

The SCF class of ubiquitin ligases comprises at least four proteins: Skp1, Cul1, Roc1/Rbx1/Hrt1, and an F-box protein. F-box proteins are adaptors that directly bind SCF substrates. These proteins link to substrates by protein–protein interaction domains and feature an F-box motif, which is around 45 residues long. The protein Skp1, which is significant in SCF complexes but may have additional functions, requires the F-box to bind to it. Skp1 then forms an association with either the human homolog Cul1's N-terminus or its counterpart Cdc53p in budding yeast. Cul1 belongs to the cullin family, which in humans consists of at least Apc2 and cullins Cul1–Cul5. In known E3 complexes, Cul2 and Apc2 have roles. Cullins seem to recruit the E2 ubiquitin-conjugating enzyme and help organise and activate the E3 complex. It is possible that the remaining cullins will arrange ubiquitin ligase complexes in a similar way. For instance, Cul3 is involved in regulating cyclin E stability (McKinnon and Tabrizi 2014). Human Roc1/Rbx1 (Hrt1 in yeast), a protein with a RING-H2 finger domain, seems to increase ubiquitin ligase activity and facilitate the attachment of the Cul1 protein to the E2 enzyme (usually Ubc3/Cdc34 or Ubc5). One member of the class of finger proteins known as RING finger proteins, RING-H2 is composed of eight cysteine and histidine residues that aid in E2 binding and catalysis. These four components are sufficient to ubiquitylate certain substrates, based on biochemical reconstitution of SCF complexes in yeast and humans.

1.3.7 APC Ubiquitin Ligase

The APC, first multi-component ubiquitin ligase to be discovered, is required for the removal of cyclin B to allow for the escape from mitosis and for the degradation of substrates controlling the metaphase to anaphase transition. The APC includes a RING-H2 finger protein called Apc11 that is comparable to Roc1/Rbx1 and a cullin homologue named Apc2, which is similar to the SCF complex. Like the SCF, the APC binds to proteins to initiate its activation against particular substrates (Lehman 2009). The destruction box (also known as the D-box) and the KEN box are two destruction signals that have been found in substrates that the APC is aiming to destroy. All known APC substrates that use the Cdc20 adaptor, referred to as APC Cdc20 substrates, as well as certain APC Cdh1 substrates, referred to as APCCdh1, have the nine-residue D-box. The KEN box is a transposable motif consisting of seven residues that is known to target substrates to the APC Cdh1. Both D-box- and KEN-box-containing substrates, including Cdc20 itself, are susceptible to ubiquitylation by Cdh1 (Li and Zhang 2009).

1.3.8 Proteasomes

A hollow cylinder-shaped protein complex involved in UPS is called a proteasome. Adenosine triphosphate-dependent proteasomes are the primary non-lysosomal complexes accountable for the breakdown of most of the intracellular proteins. Proteases are essential for the survival of organisms and cells in eukaryotes. Proteasomes do not exist in eubacteria (Zolk et al. 2006). The ATP-dependent assembly of the 26S proteasome is facilitated by two primary subunits. There are several proteolytic sites in the 20S catalytic component, and several ATPases as well as a binding site for ubiquitin concatemers are present in the 19S regulatory component (Muratani and Tansey 2003).

The 26S proteasome catalytic activity is attributed to its four stacked rings, i.e. 20S subunit each of which has seven different subunits stacked one on top of the other. Two identical outside α rings and two inside β rings are present. Catalytic sites facing the hollow centre of the ring structure are located within the two inner beta rings. Beneath the beta rings are the two alpha rings. The 20S β subunit possesses three unique proteolytic activities: activities similar to peptidylglutamyl, achymotrypsin, and trypsin. The alpha subunits' amino terminus prevents entry to the proteolytic chamber. Therefore, the small holes on either end of the cylinder are the sole ways to access the proteasome's inner cavity. While the β rings feature several catalytic sites, the outer two α-rings are unknown in purpose. To create the 26S proteasome, the 19S regulatory components come together at each pore of the 20S subunit. The two 19S regulatory caps that cap the ends of the 20S complex are also referred to as PA700.19S assemblies provide substrate recognition and insertion tasks. A 19S regulatory subunit, located at either end of the 20S proteasome, is made up of two distinct subcomplexes: a base made up of ten distinct proteins that attaches to the 20S proteasome and a lid made up of nine different proteins that binds and recognizes polyubiquitinated proteins. The regulatory caps not only identify the substrates of the 20S proteasome but also make it easier for the target proteins to enter the 20S proteasome by unfolding the substrate and opening the catalytic channel. Because the 20S subunit by itself is inactive, the 19S subunit is also necessary for proteolytic action. The 19S regulatory particle participates in protein translocation into the catalytic 20S chamber for degradation, as well as substrate preparation and selection. Many subunits, including six ATPases, make up each 19S particle. These subunits presumably supply the energy needed for substrate unfolding, which is necessary prior to entry into the 20S chamber. In order to process ubiquitin chains prior to substrate translocation and destruction, the 19S component's outer-lid subcomplex is engaged (Mani and Gelmann 2005). After proteasome degradation, ubiquitin monomers are released or actively eliminated by

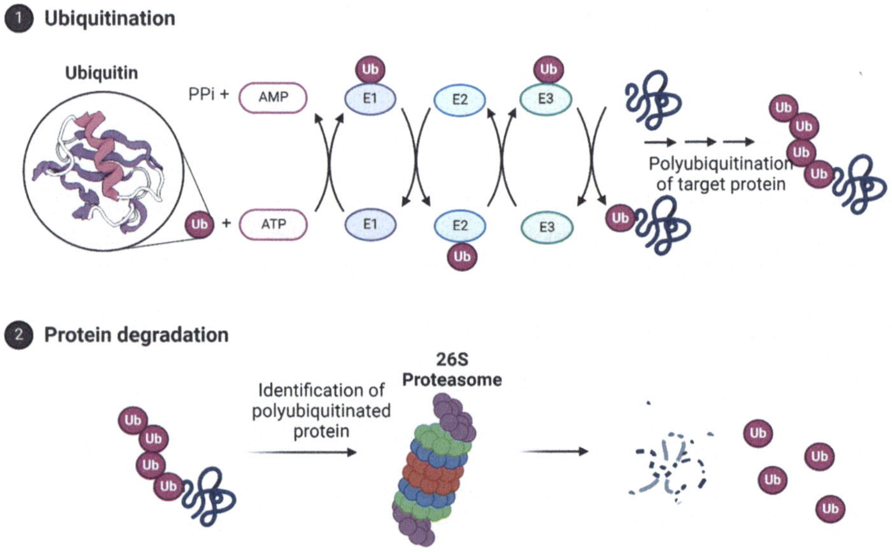

Fig. 1.1 Degradation of targeted protein through UPS

ubiquitin carboxyl-terminal hydrolases at the proteasome, where they are recycled for further usage (Fig. 1.1).

1.4 Role of UPS in Revolutionizing Cellular Biology

1.4.1 Functional Understanding

1.4.1.1 Protein Degradation

For cellular homeostasis and the control of numerous physiological functions, protein breakdown is necessary. UPS is an extremely controlled route responsible for selective protein degradation. It breaks down most of the proteins in a cell. The attachment of a polyubiquitin chain—a chain of ubiquitin molecules marks the target protein for destruction. A protein complex known as the proteasome then identifies this chain and carries the enzymatic activity necessary to break down the target protein. The enzymatic sequence of events that results in the initial attachment of the polyubiquitin chain to the target protein confers substrate specificity. Both the final protein breakdown and the transportation of polyubiquitinated proteins to the proteasome are tightly controlled processes (Bingol and Schuman 2005).

It requires three enzymatic components to attach chains of Ub onto proteins that are targeted for degradation. The preparation of Ub for conjugation is carried out by E1 which is the Ub-activating enzyme and E2, which is a Ub-carrier or conjugating protein. However, the most important enzyme in the process is E3, Ub-protein ligase, which identifies a particular protein substrate and catalyses the transfer of activated Ub to it (Lecker et al. 2006). Mechanistically, E3 enzymes can be divided into two primary types based on their RING (or RING-like) and HECT domains. The capacity of both varieties of E3 enzymes to establish specific substrate binding is similar (Ravid and Hochstrasser 2008).

There are two stages involved in the breakdown of a substrate protein: (1) covalent binding of ubiquitin to the substrate (also known as ubiquitination or ubiquitin conjugation); and (2) degradation of the polyubiquitinated substrate, which involves breaking down the polyubiquitin chain and recycling free ubiquitin (Li et al. 2022). Under stimulus, ubiquitin mediates the breakdown of protein clumps. Proteins in the endoplasmic reticulum accumulate in response to cell stimulation; these aggregates are tagged with ubiquitin when E1, E2, and E3 are present, and proteases subsequently are degenerated. Along with the hydrolysis of ATP, ubiquitin's C-terminal glycine binds to the active site of E1. Following its transfer to the E2 enzyme, the active ubiquitin forms a complex with the E2 conjugating activation. After that, this complex interacts with the specific enzyme E3, which causes the ubiquitin to go to the designated substrate (Li et al. 2022). Once the substrate protein is altered by a polyubiquitin chain including at least four ubiquitins (Ub), it can immediately bind to intrinsic Ub receptors in the 19S regulatory complex of the 26S proteasome. As an alternative, it can attach itself to adaptor proteins that include domains for both proteasome and polyubiquitin binding (Ravid and Hochstrasser 2008).

1.4.1.2 Regulatory Role in Cellular Processes

UPS plays an important role in maintaining cellular homeostasis by regulating the degradation of intracellular proteins in eukaryotic cells (Hershko and Ciechanover 1998). This sophisticated process involves binding of the ubiquitin molecules to the target proteins, preparing them for proteasomal degradation (Pickart and Cohen 2004). The regulation of the UPS is multifaceted, involving various enzymes and cofactors that govern ubiquitin conjugation, recognition, and proteasomal degradation (Chau et al. 1989; Hershko and Ciechanover 1992, 1998).

The UPS is crucial for several cellular processes, including the advancement of the cell cycle, cell differentiation, cell responsiveness, immunological response, signal transmission, and protein quality control (Ciechanover and Schwartz 1998; Hershko and Ciechanover 1998). Unregulated activity of the UPS has been linked to cancer, neurological illnesses, and immunological dysfunction (Schwartz and Ciechanover 2009). In the context of the cell cycle, the UPS regulates the quantity of cyclins, CDKs, and their inhibitors, which together control cell cycle transitions (Hershko and Ciechanover 1998). Ubiquitin-mediated degradation of cyclins and CDK inhibitors speeds up the cell cycle, ensuring accurate replication and division of genetic material (Reyes-Turcu et al. 2009).

Furthermore, the UPS plays a crucial role in cell differentiation techniques by controlling the stability of essential transcription factors and signalling molecules (Bedford et al. 2011). For example, during embryonic development, the breakdown of transcription factors like Oct4 and Sox2 is regulated by UPS, which are required for pluripotency in embryonic stem cells. Ubiquitin-dependent degradation of these molecules promotes cellular differentiation by decreasing pluripotency and activating lineage-specific gene expression pathways. In addition to its role in cell cycle and differentiation, the UPS plays a critical role in regulating cell apoptosis or programmed cell death. The timely degradation of pro-apoptotic proteins, such as caspases and Bcl-2 family members, prevents aberrant cell death and maintains tissue homeostasis (Ciechanover and Schwartz 2004; Fajerman et al. 2004). Conversely, the UPS also regulates the turnover of anti-apoptotic proteins, ensuring the efficient execution of apoptotic pathways in response to cellular stress or developmental cues (Ciechanover and Schwartz 2004).

Furthermore, the UPS is essential for the control of immunological responses., governing the turnover of key signalling molecules involved in immune cell activation and inflammation (Bedford et al. 2011). Ubiquitin-mediated degradation of signaling adaptors, such as IκB and IRAKs, controls the activation of nuclear factor-κB (NF-κB) and MAPK pathways in response to pathogens or inflammatory stimuli (Bedford et al. 2011). Furthermore, the UPS regulates the stability of immune receptors, such as TCRs and TLRs, modulating immune cell activation and cytokine production. In summary, understanding the regulatory mechanisms of the UPS holds significant therapeutic potential for the development of novel treatments targeting UPS-related diseases.

1.4.2 Disease Implications

1.4.2.1 Neurodegenerative Diseases

NDs are a distinctive group of disorders that are represented by the progressive dysfunction and loss of neurons in the central CNS (Lamptey et al. 2022; Pathak et al. 2021). Alzheimer's disease (AD), Parkinson's disease (PD), Huntington's disease (HD), and amyotrophic lateral sclerosis (ALS) are among the neurodestructive disease disorders that cause great difficulties for patients and healthcare systems around the world. Over the past decade, extensive research has been carried out on the involvement of the ubiquitin-proteasome system in the pathogenesis of NDs, putting light on potential therapeutic strategies targeting UPS dysfunction (Türker et al. 2021).

UPS involves the tagging of proteins with ubiquitin molecules, marking them for degradation by the proteasome, a large protein complex responsible for protein breakdown. Dysregulation of the UPS can lead to the accumulation of misfolded or aggregated proteins, contributing to cellular toxicity and dysfunction—a distinctive feature of many NDs. The UPS has been implicated in the removal of tau protein clumps, which are composed of six highly soluble proteins, and β-amyloid (Aβ) plaques in AD. According to earlier research, weakened UPS activity impairs the breakdown of tau and β-amyloid proteins, which causes a build-up of these proteins in the brain and ensuing neurotoxicity (Ciechanover and Kwon 2015). Similarly, the aggregation of α-synuclein aggregate that forms Lewy bodies, a typical neuropathological hallmark of PD, has been linked to defective UPS activity. It is suggested that compromised UPS activity results in an inability to effectively clear α-synuclein aggregates, which then accumulate and spread throughout the brain and cause dopaminergic neuron degeneration and motor dysfunction (Petrucelli et al. 2002). Mutant huntingtin protein is produced in HD and accumulates insolubly in neurons due to an increase of trinucleotide repeats in the huntingtin gene. When it comes to the removal of mutant huntingtin protein, the UPS is essential. On the other hand, neuronal dysfunction and the build-up of hazardous protein aggregates are caused by malfunctions in UPS-mediated protein breakdown pathways (Kumar et al. 2020). In Huntington's disease, a trinucleotide repeat expansion in the huntingtin gene leads to the production of mutant huntingtin protein, which forms insoluble aggregates within neurons. The UPS plays a crucial role in the clearance of mutant huntingtin protein. However, dysfunction in UPS-mediated protein degradation pathways results in the accumulation of toxic protein aggregates and neuronal dysfunction (Bence et al. 2001). This imbalance is partially responsible for the increasing motor and cognitive impairments seen in Huntington's disease patients. Also, amyotrophic lateral sclerosis, which is a crippling motor neuron disease, has also been connected to UPS dysfunction. Misfolded proteins that disrupt cellular proteostasis, like TDP-43 and superoxide dismutase 1 (SOD1), accumulate and accelerate motor neuron degeneration (Tai and Schuman 2008; Tashiro et al. 2012). Impaired UPS function leads to the accumulation of these dangerous protein aggregates, which speeds up the progression of the illness and results in neuronal death. Understanding the intricate relationship between UPS dysfunction and NDs is crucial for the development of targeted therapeutic interventions aimed at restoring protein homeostasis and halting disease progression.

1.4.2.2 Cancer

As discussed above, the UPS is a highly regulated pathway in cells that controls the degradation of targeted proteins related to various processes like cell cycle management, DNA repair, and programmed cell death (Hershko and Ciechanover 1998). When the UPS is dysregulated, it can contribute to the development and advancement of cancer. Changes in UPS components have been linked to the initiation and growth of tumours (Mani and Gelmann 2005). In cancer biology, the UPS can impact proteins that promote or suppress tumour formation. For instance, E3 ubiquitin ligases, such as MDM2 and SKP2, are crucial in this process by targeting tumour suppressor proteins such as p53 and p27 (Nakayama and Nakayama 2005).

While oncoprotein aggregation may result from a mutation or dysregulation of proteasome subunits, causing the initiation and spread of tumours (Adams 2004), proteasome inhibitors have emerged as attractive therapeutic medicines for the treatment of cancer because they target the UPS and cause cancer cells to undergo apoptosis. Bortezomib, which is a proteasome inhibitor, has already been approved for the treatment of Multiple Myeloma and Mantle Cell Lymphoma (Adams 2004). Furthermore, the UPS is crucial in regulating significant signalling pathways that support the development and progression of cancer. For example, the UPS controls transcription factors like NF-κB and β-catenin, as well as the stability and activity of genes involved in cell invasion, proliferation, and survival (Ciechanover and Kwon 2015). In addition, the UPS restrains the amount of proteins involved in apoptosis, consequently persuading cell survival and resistance to cell death signals in cancer cells (Salvesen and Dixit 1997). Dysfunctional degradation of proteins by the UPS has been linked to the development of drug resistance in cancer cells. Changes in UPS elements can impact the effectiveness of chemotherapy and targeted treatments by enhancing cancer cell survival and reducing the potency of anti-cancer drugs (Goldberg 2012). Besides its role in starting and advancing cancer, UPS also influences cancer prognosis and patient outcomes. Irregular expression of UPS components has been connected to prognosis and treatment resistance in types of cancers underlining the potential value of UPS-related markers, in forecasting treatment response and patient survival (Soucy et al. 2009). Moreover, tumour protein p53 (Tp53) is a transcription and has an anti-cancer property through inhibiting angiogenesis. Tumour growth requires angiogenesis to provide the necessary nutrients. Consequently, p53 inhibits the growth of tumours by contributing to the synthesis of hypoxia-inducible factors HIF1 and HIF2, which have an effect on angiogenesis by boosting the synthesis of angiogenesis inhibitors such as arresten and reducing the synthesis of factors that induce angiogenesis. HIF-1α and HIF-2α regulate p53 levels in stem cells in response to mechanical inputs; HIF-1α stabilizes p53 levels while HIF-2α decreases them (Aliabadi et al. 2021; Assadian et al. 2012). HIF-1α and HIF-2α influence the levels of p53 in stem cells in response to mechanical inputs; HIF-1α stabilizes p53 whereas HIF-2α decreases it (Das et al. 2012). The transcription factor HIF is a heterodimeric complex made up of α and β subunits.

Proline hydroxylation of α subunit is brought on by adequate oxygenation. The VHL protein binds to hydroxylated HIF-1 α which results in its ubiquitination by an E_3ligase named VHL (von Hippel-Lindau) and its rapid degradation by the proteasomes. In hypoxia or without VHL, HIF- α accumulates and activates transcription (Aliabadi et al. 2021; Mani and Gelmann 2005). Further, the constant degradation of p53 keeps its levels low in cells that are not under stress. In order to achieve this, a protein known as MDM2, or HDM2 in humans, binds to p53, stops it from working, and moves it from the nucleus to the cytoplasm. Additionally, MDM2 functions as a ubiquitin ligase, conjugating ubiquitin to p53 covalently and designating p53 for proteasome destruction (Huang et al. 2024). The creation of similar inhibitors, like PROTAC, has become the current focus of study due to the features of the UPS system. With the aid of the UPS system, this kind of inhibitor can hydrolyse proteins, tagging the problematic protein with ubiquitination to achieve target protein destruction and tumour treatment (Huang et al. 2024). All in all, targeting the UPS for cancer therapy is still an active area of research and there are continuous ongoing efforts focused on developing novel UPS inhibitors and combination therapies to improve treatment outcomes and overcome drug resistance (Bedford et al. 2011).

1.4.2.3 Inflammatory Diseases

Dysregulation of UPS has been increasingly recognized as a significant contributor to the pathogenesis of various inflammatory diseases. The UPS organizes a variety of physiological responses due to its ability to destroy numerous regulatory proteins. For example, the production of polyubiquitin chains at several levels is usually necessary for the activation of the NF-κB and MAPK pathways in response to the interaction of PRRs (Goetzke et al. 2021). The linear ubiquitin chain assembly complex (LUBAC) is responsible for assembling these polyubiquitin chains. LUBAC enhances IKK's phosphorylation of IκBα by stimulating the linear (M1-linked) ubiquitination of the IKK regulatory component NEMO (Iwai et al. 2014). Phosphorylation triggers K48-linked polyubiquitylation and proteasomal degradation, allowing NF-κB to enter the nucleus and activate genes encoding proinflammatory cytokines (e.g. TNF-α, IL-1β, and IL-6) (Alkalay et al. 1995; Çetin et al. 2021). Additional substrates for LUBAC include the receptor-interacting protein kinase 1 (RIPK1) involved in TNF signalling, which after linear ubiquitylation activates the proinflammatory NF-κB and MAPK pathways. Additionally, LUBAC-mediated linear polyubiquitylation is involved in the regulation of tumour necrosis factor (TNF)-mediated apoptosis and in canonical nuclear factor-κB (NF-κB) activation that is induced by various stimuli (Çetin et al. 2021). Furthermore, proinflammatory pathways may be activated downstream by K63-linked ubiquitin chains, a phenomenon that is inhibited by deubiquitinase A20 (Wertz et al. 2004). Also, it has been reported that cancer and autoimmune disorders are related to aberrant NF-κB activation. NF-κB is activated upon activation of the IKK complex, which is comprised of IKK1, IKK2, and NEMO (Iwai et al. 2014). Dysfunctional proteasomal activity has been linked to the abnormal build-up of proinflammatory proteins, which sustains inflammatory cascades in conditions including inflammatory bowel disease, systemic lupus erythematosus, multiple sclerosis, rheumatoid arthritis (Goetzke et al. 2021).

1.5 Recent Advancements in Understanding of UPS

1.5.1 Approaches for Investigation of UPS

The ubiquitination field is now represented in every aspect of biomedical research, from basic structural and biochemical investigations to translational and clinical research, because of the vital role that UPS plays in human disease and normal physiology. Mass spectrometry (MS) has been a potent tool for identifying and characterising protein ubiquitination in recent years. A recent review done by Sahu et al. provides an in-depth discussion on the application of mass spectrometry in understanding of ubiquitomics (Sahu et al. 2023). Other approaches for analysis of UPS include biochemical and antibody-based methods (Kliza and Husnjak 2020). Herein, transient or ectopic production of N-terminal epitope-tagged Ub variants is one of the most often used techniques. These variations can be integrated into Lys-linked Ub polymers or directly attached to the substrates as monomers. Further, affinity purification (AP) is used to enrich the Ub conjugates that have been tagged with cellular epitopes. An Ub-deficient strain of *S. cerevisiae* producing 6xHIS-Ub was used to isolate 72 Ub targets, and the original proteomic investigation found 110 ubiquitination sites in these targets (Peng et al. 2003). Using a similar method, 669 human proteins that had been changed by ubiquitin and 44 ubiquitinated peptides were found in HeLa cells (Meierhofer et al. 2008). Furthermore, biotinylation of Ub can also be used to analyse low-abundant proteins both in vitro and in vivo (Franco et al. 2011). Readers are suggested to refer to a review published by Kliza and Husnjak in 2020 for the understanding of different proteomic approaches for analysing UPS, along with their strengths and weaknesses (Kliza and Husnjak 2020).

1.5.2 UPS as a Therapeutic Target

Our understanding of proteasome biology has greatly increased as a number of approaches to target UPS with intent to discover novel drug therapies. Selecting the right target in the ubiquitin system is an important first step. This system consists of the ubiquitin molecule itself, the DUBs involved in deubiquitination, the proteasome, and the E1, E2, and E3 ubiquitinating enzymes. Furthermore, it is imperative to investigate the optimal mechanism of inhibition, be it by small molecules, peptides, antibodies, or RNA-based agents. Furthermore, it is critical to pinpoint the precise biochemical pathway to target, including whether it entails directly attaching to and impairing a certain regulatory domain or function or preventing critical protein–protein interactions that are essential for regulation. Manipulating this system has therapeutic potential, and many UPS modulators are being investigated for different medical applications. This includes E1, E2, and E3 inhibitors, proteasome inhibitors, deubiquitinating enzymes inhibitors, and many more (Aliabadi et al. 2021). One potential way to disrupt ubiquitin signalling is to target ubiquitin directly. Though not usually thought of as drugs, ubistatins are a family of compounds that have been found despite some similarities to "frequent hitters" for their capacity to attach to ubiquitin in chemical genetic screens (Verma et al. 2004). The identification of a cyclic peptide ubiquitin binder via the RaPID technology (which involves the synthesis of trillion-member macrocyclic peptide libraries, very homogenous Ub chains for the purpose of screening) is another step forward in this direction (Nawatha et al. 2019). These peptides have shown promise in causing cancer cells to undergo apoptosis by attaching themselves firmly and selectively to Lys48-linked Ub chains, which interferes with the proteasome's ability to degrade the chains. This new mode of action presents opportunities for creative treatment approaches. Immunotherapy strategies targeting the UPS have gained considerable attention, especially for cancer treatment (Yin et al. 2014). CAR T-cell therapy has shown promising results in hematologic malignancies by targeting specific UPS-regulated antigens expressed on cancer cells (Kershaw et al. 2006). Additionally, immune-checkpoint inhibitors, such as Pembrolizumab and Nivolumab, improve the anti-tumour immune responses by preventing T-cell inhibitory receptor degradation mediated by UPS. Also, gene therapy approaches targeting the UPS offer potential long-term solutions for genetic disorders associated with UPS dysfunction (Bulatov and Ciulli 2015). Adeno-associated viral vectors encoding UPS-related genes, such as ubiquitin ligases or deubiquitinases, can restore UPS function in diseased cells (D'Arcy et al. 2011). Moreover, CRISPR/Cas9-mediated genome editing enables precise modulation of UPS-related genes, offering personalized therapeutic options (Doudna and Charpentier 2014). There are still a number of obstacles in the way of developing drugs that target UPS, despite great advances. Achieving selectivity and reducing off-target effects is a significant problem, especially for medications that target ubiquitin enzymes that have numerous substrates. Furthermore, specific obstacles to the development of UPS modulators for neurodegenerative illnesses include selective protein targeting and blood-brain barrier penetration (Leestemaker and Ovaa 2017). The future research endeavours ought to concentrate on tackling these obstacles and investigating innovative treatment approaches, like targeted protein degradation technologies.

Most innovative approach targeting UPS involves targeted lyses of protein of interest (POI) by the use of PROTAC techniques in the treatment of numerous illnesses (Chen et al. 2021). The PROTACs are bivalent molecules capable of binding two ligands simultaneously through a linker: one for an E3 ligase and another for the protein of interest (POI). Unlike traditional inhibitors, PROTACs exploit the ubiquitin (Ub)-proteasome pathway, facilitating the transfer of Ub molecules from the E3 ligase to the POI, thus marking it for degradation by the 26S proteasome (Liu et al. 2022). PROTACs offer many advantages when compared to conventional small molecule inhibitors. This is mostly because of the following PROTACs features: (1) the capacity to remove "undruggable" target proteins; (2) the ability to inhibit the enzymatic and non-enzymatic activities of proteins; (3) the ability to quickly and reversibly delete protein targets; (4) the improvement of selectivity and specificity; and (5) the potential to overcome drug resistance. Since drug resistance is a major problem while treating illnesses, PROTACs technology will offer stronger drugs to counteract it. The usage of PROTACs should also be particularly beneficial due to its novel method of action, which reduces target protein levels while requiring a lower dosage of the medication than traditional ones (Toure and Crews 2016).

Though there are around 600 E3 ligases, less than ten E3s have been used as PROTACs as of yet. The Cullin RINGE3 ligase substrate recognition subunits cereblon (CRBN), Murine double minute 2 (MDM2), and von Hippel-Lindau (VHL) tumour suppressor proteins have high-affinity binders, which have made these two E3 complexes useful for PROTAC technology (Belcher et al. 2021; Zhang et al. 2020). PROTACs have been explored for the treatment of various diseases including cancer (Zhang et al. 2020) and NDs like Parkinson disease (PD) (Amirian et al. 2023), Alzheimer's disease (AD) (Inuzuka et al. 2022), Huntington's disease (HD) (Jarosińska and Rüdiger 2021), progressive supranuclear palsy, and frontotemporal dementia (Wang et al. 2023). Further chapters of this book deal with a comprehensive discussion on principles, mechanisms, and advancements of PROTACs.

1.6 Conclusion

In conclusion, the 26S proteasome tries to break down protein substrates that the highly conserved 76-amino acid protein ubiquitin covalently binds to. An isopeptide bond between ubiquitin and a substrate is formed by a cooperative set of enzymes that include an activating enzyme (E1), a conjugating enzyme (E2), and a ligase (E3). These enzymes depend on ATP for their activity. A different category of enzymes known as deubiquitylating enzymes (DUBs) is responsible for severing these bonds and is essential to ubiquitin's function. Controlling numerous cellular functions, the UPS is a significant system. The process involves five primary stages: conjugation, activation, identification of ubiquitin conjugate, ubiquitin removal, and substrate destruction mediated by proteasomes. The UPS remains a popular study issue because the human genome encodes a large number of enzymes involved in ubiquitination and deubiquitination. Finding the 26S proteasome-interacting proteins will become more crucial in the near future since it might provide insight into how the UPS enlists other proteins to control particular cellular reactions. In this context, proteomics has been explored significantly and paves the path to a better understanding of UPS. Additional knowledge of these kinds of protein interactions could result in the development of small molecule medications that block particular connections, perhaps improving conditions and processes linked to UPS. Furthermore, the discovery that certain cancer therapies benefit from proteasome inhibitors and that cancer cells are more susceptible to abnormalities in protein degradation raises the possibility that additional research could result in the development of novel therapeutic approaches to combat cancer.

References

Adams J (2004) The proteasome: a suitable antineoplastic target. Nat Rev Cancer 4(5):349–360

Aliabadi F, Sohrabi B, Mostafavi E, Pazoki-Toroudi H, Webster TJ (2021) Ubiquitin–proteasome system and the role of its inhibitors in cancer therapy. Open Biol 11(4):200390. https://doi.org/10.1098/rsob.200390

Alkalay I, Yaron A, Hatzubai A, Orian A, Ciechanover A, Ben-Neriah Y (1995) Stimulation-dependent I kappa B alpha phosphorylation marks the NF-kappa B inhibitor for degradation via the ubiquitin-proteasome pathway. Proc Natl Acad Sci 92(23):10599–10603

Amirian R, Badrbani MA, Derakhshankhah H, Izadi Z, Shahbazi M-A (2023) Targeted protein degradation for the treatment of Parkinson's disease: advances and future perspective. Biomed Pharmacother 166:115408

Assadian S, El-Assaad W, Wang XQD, Gannon PO, Barrès V, Latour M, Mes-Masson A-M, Saad F, Sado Y, Dostie J (2012) P53 inhibits angiogenesis by inducing the production of Arresten. Cancer Res 72(5):1270–1279

Bedford L, Lowe J, Dick LR, John Mayer R, Brownell JE (2011) Ubiquitin-like protein conjugation and the ubiquitin–proteasome system as drug targets. Nat Rev Drug Discov 10(1):29–46

Belcher BP, Ward CC, Nomura DK (2021) Ligandability of E3 ligases for targeted protein degradation applications. Biochemistry 62(3):588–600

Bence NF, Sampat RM, Kopito RR (2001) Impairment of the ubiquitin-proteasome system by protein aggregation. Science 292(5521):1552–1555

Bingol B, Schuman EM (2005) Synaptic protein degradation by the ubiquitin proteasome system. Curr Opin Neurobiol 15(5):536–541

Bulatov E, Ciulli A (2015) Targeting Cullin–RING E3 ubiquitin ligases for drug discovery: structure, assembly and small-molecule modulation. Biochem J 467(3):365–386. https://doi.org/10.1042/BJ20141450

Çetin G, Klafack S, Studencka-Turski M, Krüger E, Ebstein F (2021) The ubiquitin–proteasome system in immune cells. Biomol Ther 11(1):60

Chau V, Tobias JW, Bachmair A, Marriott D, Ecker DJ, Gonda DK, Varshavsky A (1989) A multiubiquitin chain is confined to specific lysine in a targeted short-lived protein. Science 243(4898):1576–1583

Chen X, Htet ZM, López-Alfonzo E, Martin A, Walters KJ (2021) Proteasome interaction with ubiquitinated substrates: from mechanisms to therapies. FEBS J 288(18):5231–5251. https://doi.org/10.1111/febs.15638

Ciechanover A, Kwon YT (2015) Degradation of misfolded proteins in neurodegenerative diseases: therapeutic targets and strategies. Exp Mol Med 47(3):e147. https://doi.org/10.1038/emm.2014.117

Ciechanover A, Schwartz AL (1998) The ubiquitin-proteasome pathway: the complexity and myriad functions of proteins death. Proc Natl Acad Sci 95(6):2727–2730

Ciechanover A, Schwartz AL (2004) The ubiquitin system: pathogenesis of human diseases and drug targeting. Biochim Biophys Acta 1695(1–3):3–17

Ciehanover A, Hod Y, Hershko A (1978) A heat-stable polypeptide component of an ATP-dependent proteolytic system from reticulocytes. Biochem Biophys Res Commun 81(4):1100–1105. https://doi.org/10.1016/0006-291x(78)91249-4

D'Arcy P, Brnjic S, Olofsson MH, Fryknäs M, Lindsten K, De Cesare M, Perego P, Sadeghi B, Hassan M, Larsson R, Linder S (2011) Inhibition of proteasome deubiquitinating activity as a new cancer therapy. Nat Med 17(12):1636–1640. https://doi.org/10.1038/nm.2536

Das B, Bayat-Mokhtari R, Tsui M, Lotfi S, Tsuchida R, Felsher DW, Yeger H (2012) HIF-2α suppresses P53 to enhance the Stemness and regenerative potential of human embryonic stem cells. Stem Cells 30(8):1685–1695

Doudna JA, Charpentier E (2014) The new frontier of genome engineering with CRISPR-Cas9. Science 346(6213):1258096. https://doi.org/10.1126/science.1258096

Ebner P, Versteeg GA, Ikeda F (2017) Ubiquitin enzymes in the regulation of immune responses. Crit Rev Biochem Mol Biol 52(4):425–460. https://doi.org/10.1080/10409238.2017.1325829
Eldridge AG, O'Brien T (2010) Therapeutic strategies within the ubiquitin proteasome system. Cell Death Differ 17(1):4–13. https://doi.org/10.1038/cdd.2009.82
Etlinger JD, Goldberg AL (1977) A soluble ATP-dependent proteolytic system responsible for the degradation of abnormal proteins in reticulocytes. Proc Natl Acad Sci USA 74(1):54–58. https://doi.org/10.1073/pnas.74.1.54
Fajerman I, Schwartz AL, Ciechanover A (2004) Degradation of the Id2 developmental regulator: targeting via N-terminal ubiquitination. Biochem Biophys Res Commun 314(2):505–512
Franco M, Seyfried NT, Brand AH, Peng J, Mayor U (2011) A novel strategy to isolate ubiquitin conjugates reveals wide role for ubiquitination during neural development *. Mol Cell Proteomics 10:5. https://doi.org/10.1074/mcp.M110.002188
Garcia-Barcena C, Osinalde N, Ramirez J, Mayor U (2020) How to inactivate human ubiquitin E3 ligases by mutation. Front Cell Dev Biol 8
Goetzke CC, Ebstein F, Kallinich T (2021) Role of proteasomes in inflammation. J Clin Med 10(8):1783
Goldberg AL (2012) Development of proteasome inhibitors as research tools and cancer drugs
Grice GL, Nathan JA (2016) The recognition of ubiquitinated proteins by the proteasome. Cell Mol Life Sci 73(18):3497–3506. https://doi.org/10.1007/s00018-016-2255-5
Hershko A, Ciechanover A (1992) The ubiquitin system for protein degradation. Annu Rev Biochem 61(1):761–807
Hershko A, Ciechanover A (1998) The ubiquitin system. Annu Rev Biochem 67:425–479. https://doi.org/10.1146/annurev.biochem.67.1.425
Hershko A, Ciechanover A, Heller H, Haas AL, Rose IA (1980) Proposed role of ATP in protein breakdown: conjugation of protein with multiple chains of the polypeptide of ATP-dependent proteolysis. Proc Natl Acad Sci USA 77(4):1783–1786. https://doi.org/10.1073/pnas.77.4.1783
Hershko A, Ciechanover A, Rose IA (1981) Identification of the active amino acid residue of the polypeptide of ATP-dependent protein breakdown. J Biol Chem 256(4):1525–1528
Hershko A, Eytan E, Ciechanover A, Haas AL (1982) Immunochemical analysis of the turnover of ubiquitin-protein conjugates in intact cells. Relationship to the breakdown of abnormal proteins. J Biol Chem 257(23):13964–13970
Hershko A, Heller H, Elias S, Ciechanover A (1983) Components of ubiquitin-protein ligase system. Resolution, affinity purification, and role in protein breakdown. J Biol Chem 258(13):8206–8214
Hough R, Pratt G, Rechsteiner M (1986) Ubiquitin-lysozyme conjugates. Identification and characterization of an ATP-dependent protease from rabbit reticulocyte lysates. J Biol Chem 261(5):2400–2408
Huang Y, Gao Y, Lin Z, Miao H (2024) Involvement of the ubiquitin-proteasome system in the regulation of the tumor microenvironment and progression. Genes Dis 101240
Inuzuka H, Liu J, Wei W, Rezaeian A-H (2022) PROTACs technology for treatment of Alzheimer's disease: advances and perspectives. Acta Mater Med 1(1):24
Iwai K, Fujita H, Sasaki Y (2014) Linear ubiquitin chains: NF-KB signalling, cell death and beyond. Nat Rev Mol Cell Biol 15(8):503–508
Jackson PK, Eldridge AG, Freed E, Furstenthal L, Hsu JY, Kaiser BK, Reimann JDR (2000) The Lore of the RINGs: substrate recognition and catalysis by ubiquitin ligases. Trends Cell Biol 10(10):429–439. https://doi.org/10.1016/S0962-8924(00)01834-1
Jarosińska OD, Rüdiger SGD (2021) Molecular strategies to target protein aggregation in Huntington's disease. Front Mol Biosci 8:769184
Kershaw MH, Westwood JA, Parker LL, Wang G, Eshhar Z, Mavroukakis SA, White DE, Wunderlich JR, Canevari S, Rogers-Freezer L, Chen CC, Yang JC, Rosenberg SA, Hwu P (2006) A phase I study on adoptive immunotherapy using gene-modified T cells for ovarian cancer. Clin Cancer Res 12(20):6106–6115. https://doi.org/10.1158/1078-0432.CCR-06-1183
Kleiger G, Mayor T (2014) Perilous journey: a tour of the ubiquitin-proteasome system. Trends Cell Biol 24(6):352–359. https://doi.org/10.1016/j.tcb.2013.12.003

Kliza K, Husnjak K (2020) Resolving the complexity of ubiquitin networks. Front Mol Biosci 7
Kumar D, Ambasta RK, Kumar P (2020) Ubiquitin biology in neurodegenerative disorders: from impairment to therapeutic strategies. Ageing Res Rev 61:101078
Lamptey RNL, Chaulagain B, Trivedi R, Gothwal A, Layek B, Singh J (2022) A review of the common neurodegenerative disorders: current therapeutic approaches and the potential role of nanotherapeutics. Int J Mol Sci 23(3):1851
Lecker SH, Goldberg AL, Mitch WE (2006) Protein degradation by the ubiquitin–proteasome pathway in Normal and disease states. J Am Soc Nephrol 17(7):1807–1819
Leestemaker Y, Ovaa H (2017) Tools to investigate the ubiquitin proteasome system. Drug Discov Today Technol 26:25–31. https://doi.org/10.1016/j.ddtec.2017.11.006
Lehman NL (2009) The ubiquitin proteasome system in neuropathology. Acta Neuropathol 118(3):329–347. https://doi.org/10.1007/s00401-009-0560-x
Li M, Zhang P (2009) The function of APC/CCdh1 in cell cycle and beyond. Cell Div 4(1):2. https://doi.org/10.1186/1747-1028-4-2
Li Y, Li S, Wu H (2022) Ubiquitination-proteasome system (UPS) and autophagy two main protein degradation machineries in response to cell stress. Cells 11(5):851
Liu Z, Mingxing H, Yang Y, Chenghao D, Zhou H, Liu C, Chen Y, Fan L, Ma H, Gong Y (2022) An overview of PROTACs: a promising drug discovery paradigm. Mol Biomed 3(1):46
Mani A, Gelmann EP (2005) The ubiquitin-proteasome pathway and its role in cancer. J Clin Oncol 23(21):4776–4789
McKinnon C, Tabrizi SJ (2014) The ubiquitin-proteasome system in neurodegeneration. Antioxid Redox Signal 21(17):2302–2321. https://doi.org/10.1089/ars.2013.5802
Meierhofer D, Wang X, Huang L, Kaiser P (2008) Quantitative analysis of global ubiquitination in HeLa cells by mass spectrometry. J Proteome Res 7(10):4566–4576
Monaco JJ, McDevitt HO (1984) H–2-linked low-molecular weight polypeptide antigens assemble into an unusual macromolecular complex. Nature 309(5971):797–799. https://doi.org/10.1038/309797a0
Muratani M, Tansey WP (2003) How the ubiquitin–proteasome system controls transcription. Nat Rev Mol Cell Biol 4(3):192–201. https://doi.org/10.1038/nrm1049
Nakamura N (2011) The role of the transmembrane RING finger proteins in cellular and organelle function. Membranes 1(4):354–393
Nakayama KI, Nakayama K (2005) Regulation of the cell cycle by SCF-type ubiquitin ligases. In: Seminars in cell & developmental biology, vol 16. Elsevier, pp 323–333
Nawatha M, Rogers JM, Bonn SM, Livneh I, Lemma B, Mali SM, Vamisetti GB, Sun H, Bercovich B, Huang Y, Ciechanover A, Fushman D, Suga H, Brik A (2019) De novo macrocyclic peptides that specifically modulate Lys48-linked ubiquitin chains. Nat Chem 11(7):644–652. https://doi.org/10.1038/s41557-019-0278-x
Park J, Cho J, Song EJ (2020) Ubiquitin–proteasome system (UPS) as a target for anticancer treatment. Arch Pharm Res 43(11):1144–1161. https://doi.org/10.1007/s12272-020-01281-8
Pathak N, Vimal SK, Tandon I, Agrawal L, Hongyi C, Bhattacharyya S (2021) Neurodegenerative disorders of Alzheimer, parkinsonism, amyotrophic lateral sclerosis and multiple sclerosis: an early diagnostic approach for precision treatment. Metab Brain Dis:1–38
Peng J, Schwartz D, Elias JE, Thoreen CC, Cheng D, Marsischky G, Roelofs J, Finley D, Gygi SP (2003) A proteomics approach to understanding protein ubiquitination. Nat Biotechnol 21(8):921–926
Petrucelli L, O'Farrell C, Lockhart PJ, Baptista M, Kehoe K, Vink L, Choi P, Wolozin B, Farrer M, Hardy J (2002) Parkin protects against the toxicity associated with mutant α-Synuclein: proteasome dysfunction selectively affects catecholaminergic neurons. Neuron 36(6):1007–1019
Pickart CM, Cohen RE (2004) Proteasomes and their kin: proteases in the machine age. Nat Rev Mol Cell Biol 5(3):177–187
Ravid T, Hochstrasser M (2008) Diversity of degradation signals in the ubiquitin–proteasome system. Nat Rev Mol Cell Biol 9(9):679–689
Reyes-Turcu FE, Ventii KH, Wilkinson KD (2009) Regulation and cellular roles of ubiquitin-specific deubiquitinating enzymes. Annu Rev Biochem 78:363–397

Sahu I, Zhu H, Buhrlage SJ, Marto JA (2023) Proteomic approaches to study ubiquitinomics. Biochim Biophys Acta Gene Regul Mech 1866(2):194940. https://doi.org/10.1016/j.bbagrm.2023.194940

Salvesen GS, Dixit VM (1997) Caspases: intracellular signaling by proteolysis. Cell 91(4):443–446

Schulman BA, Wade Harper J (2009) Ubiquitin-like protein activation by E1 enzymes: the apex for downstream signalling pathways. Nat Rev Mol Cell Biol 10(5):319–331. https://doi.org/10.1038/nrm2673

Schwartz AL, Ciechanover A (2009) Targeting proteins for destruction by the ubiquitin system: implications for human pathobiology. Annu Rev Pharmacol Toxicol 49:73–96

Sijts EJAM, Kloetzel PM (2011) The role of the proteasome in the generation of MHC class I ligands and immune responses. Cell Mol Life Sci 68(9):1491–1502. https://doi.org/10.1007/s00018-011-0657-y

Soucy TA, Smith PG, Milhollen MA, Berger AJ, Gavin JM, Adhikari S, Brownell JE, Burke KE, Cardin DP, Critchley S (2009) An inhibitor of NEDD8-activating enzyme as a new approach to treat cancer. Nature 458(7239):732–736

Swatek KN, Komander D (2016) Ubiquitin modifications. Cell Res 26(4):399–422. https://doi.org/10.1038/cr.2016.39

Tai H-C, Schuman EM (2008) Ubiquitin, the proteasome and protein degradation in neuronal function and dysfunction. Nat Rev Neurosci 9(11):826–838

Tashiro Y, Urushitani M, Inoue H, Koike M, Uchiyama Y, Komatsu M, Tanaka K, Yamazaki M, Abe M, Misawa H (2012) Motor neuron-specific disruption of proteasomes, but not autophagy, replicates amyotrophic lateral sclerosis. J Biol Chem 287(51):42984–42994

Toure M, Crews CM (2016) Small-molecule PROTACS: new approaches to protein degradation. Angew Chem Int Ed 55(6):1966–1973. https://doi.org/10.1002/anie.201507978

Türker F, Cook EK, Margolis SS (2021) The proteasome and its role in the nervous system. Cell Chem Biol 28(7):903–917

Verma R, Peters NR, D'Onofrio M, Tochtrop GP, Sakamoto KM, Varadan R, Zhang M, Coffino P, Fushman D, Deshaies RJ, King RW (2004) Ubistatins inhibit proteasome-dependent degradation by binding the ubiquitin chain. Science 306(5693):117–120. https://doi.org/10.1126/science.1100946

Wang J, Maldonado MA (2006) The ubiquitin-proteasome system and its role in inflammatory and autoimmune diseases. Cell Mol Immunol 3(4):255–261

Wang C, Zhang Y, Yang S, Xing D (2023) Recent advances of PROTACs Technology in Neurodegenerative Diseases. Arab J Chem 16(9):105015. https://doi.org/10.1016/j.arabjc.2023.105015

Wertz IE, O'rourke KM, Zhou H, Michael Eby L, Aravind SS, Ping W, Wiesmann C, Baker R, Boone DL (2004) De-ubiquitination and ubiquitin ligase domains of A20 downregulate NF-KB Signalling. Nature 430(7000):694–699

Wilkinson KD, Urban MK, Haas AL (1980) Ubiquitin is the ATP-dependent proteolysis factor I of rabbit reticulocytes. J Biol Chem 255(16):7529–7532

Yin L, Kufe T, Avigan D, Kufe D (2014) Targeting MUC1-C is synergistic with Bortezomib in downregulating TIGAR and inducing ROS-mediated myeloma cell death. Blood 123(19):2997–3006. https://doi.org/10.1182/blood-2013-11-539395

Zhang X, Linder S, Bazzaro M (2020) Drug development targeting the ubiquitin–proteasome system (UPS) for the treatment of human cancers. Cancers 12:4

Zolk O, Schenke C, Sarikas A (2006) The ubiquitin–proteasome system: focus on the heart. Cardiovasc Res 70(3):410–421. https://doi.org/10.1016/j.cardiores.2005.12.021

PROTACs: Principles and Mechanisms

2

Ranbir Singh Saluja, Madhura P. Vaidya, and Prashant S. Kharkar

Abstract

Proteolysis Targeting Chimeras (PROTACs), a groundbreaking technology that has revolutionized the landscape of targeted protein degradation and therapeutics, is a clinically validated strategy, as seen from significant number of clinical candidates at various stages of development. The present chapter focuses on the principles underlying PROTAC design, including their tripartite structure comprising a protein-of-interest (POI) ligand, an E3 ubiquitin ligase ligand, and a linker with defined chemistry, laying the groundwork for understanding the elegance of PROTACs in orchestrating targeted protein degradation, emphasizing the strategic design considerations that govern their efficacy and specificity. Next, a comprehensive exploration of the mechanisms of action of PROTACs, detailing the sequential steps from target protein recognition to ubiquitination and subsequent proteasomal degradation, is presented. The critical parameters that influence their pharmacological profiles, including the choice of E3 ligase, the nature of the linker, and the dynamics of the ternary complex formation, are highlighted with the help of apt examples from recent advancements that illustrate the iterative process of PROTAC development, from conceptualization to clinical application. A forward-looking perspective on the challenges and future directions in the field of targeted protein degradation is presented. Overall, an in-depth analysis of PROTAC technology, from its conceptual underpinnings to its application in drug discovery, with particular emphasis on the principles and

R. S. Saluja · M. P. Vaidya
MET Institute of Pharmacy, Bhujbal Knowledge City, Bandra (West), Mumbai, India
e-mail: madhurav_iop@met.edu

P. S. Kharkar (✉)
Department of Pharmaceutical Sciences and Technology, Institute of Chemical Technology, Matunga, Mumbai, India
e-mail: ps.kharkar@ictmumbai.edu.in

M. Nandave, P. Jain (eds.), *PROTAC-Mediated Protein Degradation: A Paradigm Shift in Cancer Therapeutics*, https://doi.org/10.1007/978-981-97-5077-1_2

mechanisms of PROTACs are presented, thereby conveying the transformative potential of PROTACs, setting the stage for their continued evolution and integration into the next generation of biomedical research and therapeutic development.

Keywords

PROTACs · Protein-of-interest · E3 ligase · Proteasome · Protein degradation

Abbreviations

AhR	aryl hydrocarbon receptor
cIAP1	cellular inhibitor of apoptosis protein 1
CRBN	cereblon
DCAF15	DDB1 and CUL4 associated factor 15
dTAGs	degradation tags
HECT	homologous to the E6AP carboxyl terminus
HGP	Human Genome Project
IAPs	inhibitor of apoptosis proteins
KEAP1	Kelch-like ECH-associated protein 1
LE	ligand efficiency
LipE	lipophilic efficiency
MDM2	Murine double minute 2
PD	pharmacodynamic
PEG	polyethylene glycol
PK	pharmacokinetic
POI	protein-of-interest
PROTACs	Proteolysis Targeting Chimeras
SCF	Skp1–cullin–F-box Proteins
SMIs	small molecule inhibitors
Ub	ubiquitin
UPS	ubiquitin-proteasome system
VHL	von Hippel-Lindau
XIAP	X-linked inhibitor of apoptosis protein

2.1 Introduction

The Human Genome Project (HGP) has significantly advanced drug discovery by identifying numerous proteins, many of which are considered druggable targets. This surge in identifiable targets has accelerated the development of drug ligands, particularly benefiting anticancer research. By leveraging data from the HGP, researchers have been able to target critical cellular components such as transcription factors, kinases and receptors with anticancer ligands, leading to a proliferation of small molecule inhibitors (SMIs) with antitumor properties. However, a persistent challenge with conventional dosage forms and SMIs is the development of multidrug resistance, which has prompted the exploration of alternative strategies, notably the shift from merely targeting proteins for inhibition to contemplating their complete degradation. This shift has catalyzed interest in protein degraders, among which Proteolysis Targeting Chimeras (PROTACs) have emerged as a significant class (Liu et al. 2002; Liu et al. 2022; Li et al. 2022; Qi et al. 2021). The PROTACs are bivalent molecules capable of binding two ligands simultaneously through a linker: one for an E3 ligase and another for the protein-of-interest (POI). Unlike traditional inhibitors, PROTACs exploit the ubiquitin (Ub)-proteasome pathway, facilitating the transfer of Ub molecules from the E3 ligase to the POI, thus marking it for degradation by the 26S proteasome (Liu et al. 2022; Li et al. 2022). The schematic representation of the PROTAC concept is shown in Fig. 2.1. The development of PROTACs embodies a paradigm shift in cancer therapy, (Li and Song 2020) moving from an occupancy-based model, where inhibitors bind directly to targets necessitating high systemic concentrations and risking off-target side effects, to an event-based model (Toure and Crews 2016). This innovative approach, represented by PROTACs, addresses previous limitations by inducing protein degradation rather than mere inhibition, offering a more targeted and supposedly less toxic therapeutic strategy. A large number of PROTACs are at various phases of clinical development for varied indications. Table 2.1 provides the list of representative PROTACs with their POI, indications, and the current phase of clinical development. Figure 2.2 shows the molecular structures of representative PROTACs.

The PROTAC technology was introduced at the turn of twenty-first century and the discovery and preclinical development continued with several milestones achieved in a short span of time. The history of PROTAC development is covered by excellent review articles and not discussed here (Liu et al. 2022; Xiao et al. 2022;

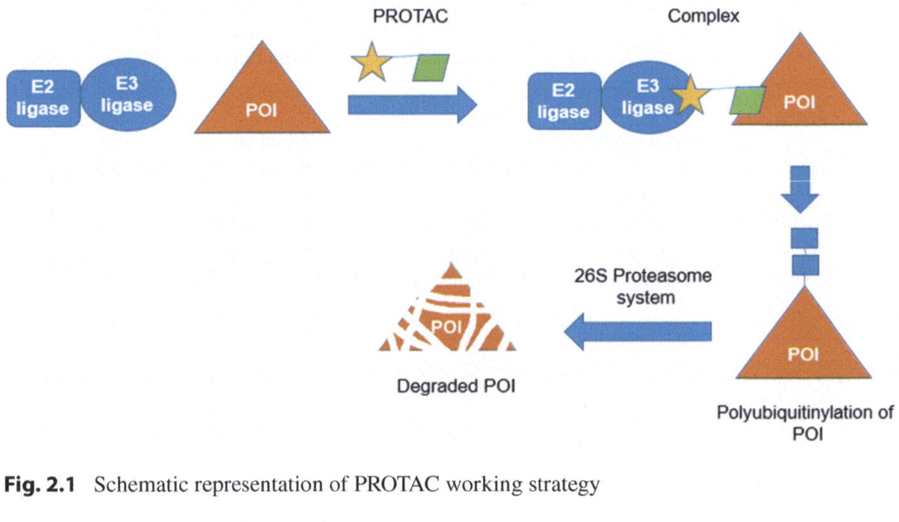

Fig. 2.1 Schematic representation of PROTAC working strategy

Table 2.1 Representative PROTACs in various phases of clinical trials/biological testing

Name (code)	Originator/ sponsor	POI	Indication	Current status
Vepdegestrant (ARV-471)	Arvinas	Estrogen receptor (ER)	$ER^+/HER2^-$ locally advanced or metastatic breast	Phase 3
Bavdegalutamide (ARV-110)	Arvinas	Androgen receptor (AR)	Metastatic castration-resistant prostate cancer (mCRPC)	Phase 2
ARV-766	Arvinas	AR	mCRPC	Phase 2
ARV-393	Arvinas	B-cell lymphoma 6 protein (BCL6)	B cell lymphomas	Preclinical
ARV-102	Arvinas	Leucine-rich repeat kinase 2 (LRRK2)	Progressive supranuclear palsy	Phase 1/1B
CC-94676 (BMS 986365)	Bristol Myers Squibb	AR	mCRPC	Phase 1
DT2216	Dialectic therapeutics	B-cell lymphoma extra-large (BCL-XL)	Liquid and solid cancers	Phase 1
BGB-16673	BeiGene	Bruton's tyrosine kinase (BTK)	Relapsed or refractory (R/R) B-cell malignancies	Phase 1
HSK29116	Liaoning Haisco pharmaceutical co.	BTK	B cell lymphomas	Phase 1
KT-474	Kymera/Sanofi	Interleukin-1 receptor-associated kinase 4 (IRAK4)	Autoimmune including AD, HS, and RA	Phase 2
KT-333	Kymera	Signal transducer and activator of transcription 3 (STAT3)	Hematologic malignancies and solid tumors	Phase 1
NX-2127	Nurix therapeutics	BTK	B cell malignancies	Phase 1b
NX-5948	Nurix therapeutics	BTK	B cell malignancies and autoimmune	Phase1a
FHD-286	Foghorn therapeutics	Brahma homologue (BRM)/ Brahma-related gene 1 (BRG1)	Relapsed/ refractory acute myeloid leukemia (AML)	Phase 1

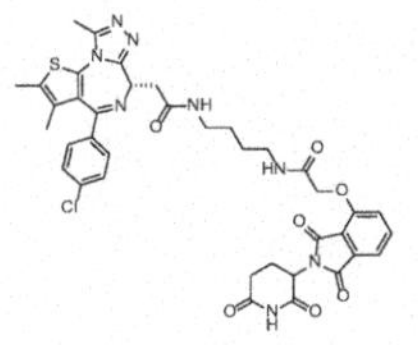

dBET1 (Bromodomain) [Cereblon]

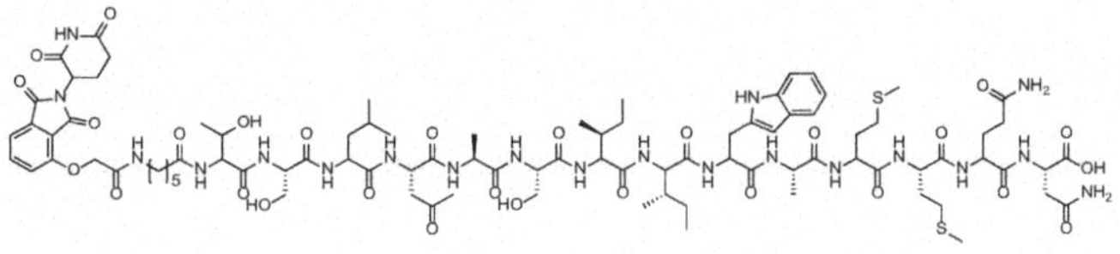

DT-6 (Transforming growth factor beta 1) [Cereblon]

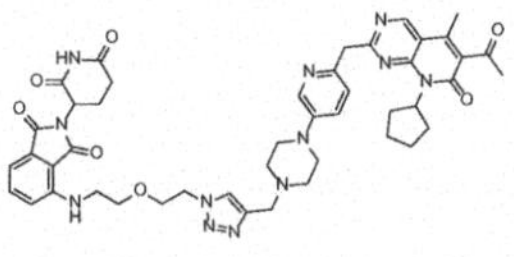

CP-10 (Cyclin-dependent kinase 6) [Cereblon]

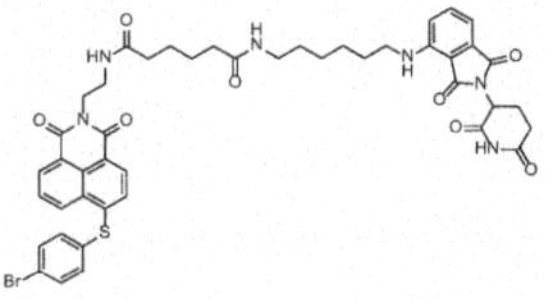

C3 (Myeloid cell leukemia-1) [Cereblon]

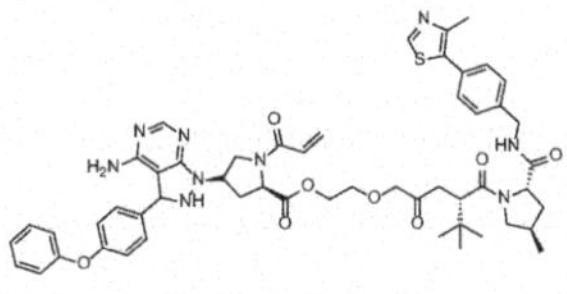

PROTAC7 (BTK)[VHL]

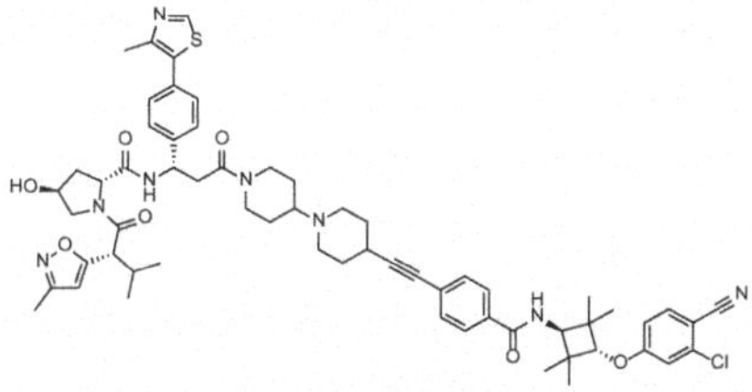

ARD-61 (AR) [VHL]

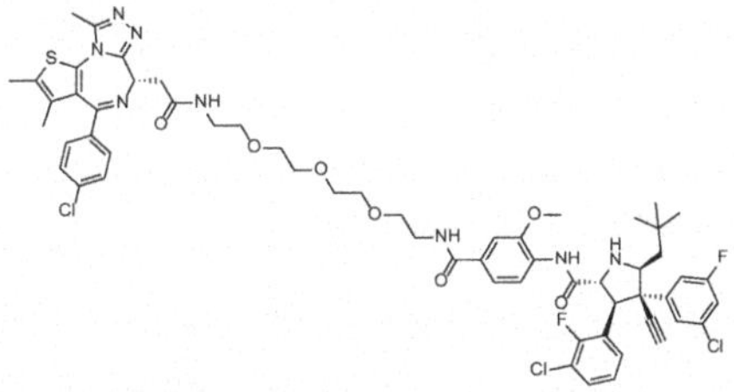

A1874 (BRD4) [MDM2]

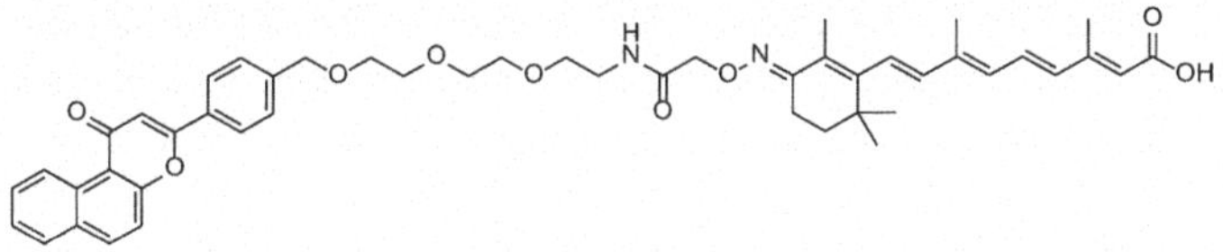

β-NF-ATRA (Cellular retinoic acid-binding protein 1) [AhR]

Fig. 2.2 Molecular structures of representative PROTACs. POI is shown in regular brackets; the corresponding E3 ligase is shown in square brackets

Targeted Protein Degradation 2024). A little over 600 E3 ligase are encoded by human genome (Yang et al. 2021). The PROTACs discovered so far uses just of few of these, e.g., SCFβ–TrCP (Ci et al. 2018), von Hippel-Lindau (VHL) (Yang et al. 2020), Murine double minute 2 (MDM2) (Li et al. 2020), inhibitor of apoptosis proteins (IAPs) (Hrdinka and Yabal 2019), and cereblon (CRBN) (Girardini et al. 2019). Several other E3 ligases are expected to be utilized for targeting by PROTACs

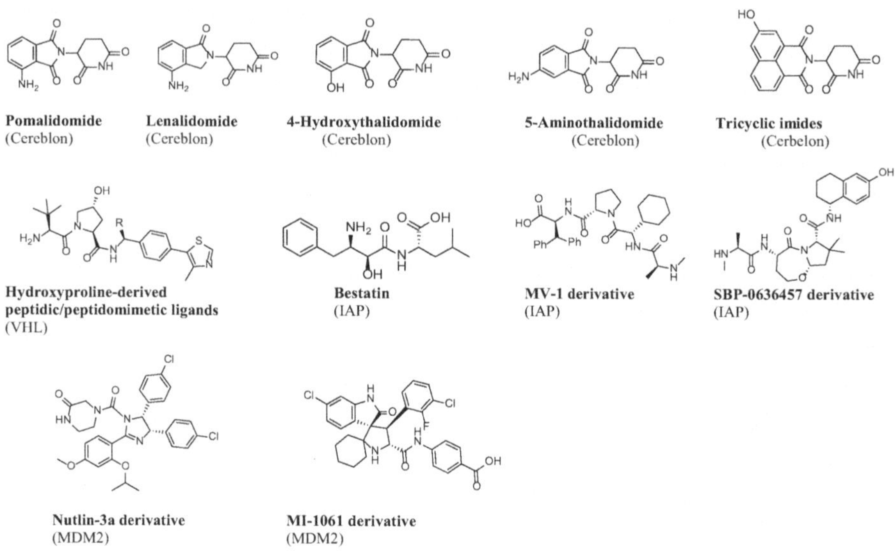

Fig. 2.3 Molecular structures of representative E3 ligase ligands used commonly in PROTACs; E3 ligase is shown in regular brackets

(Kramer and Zhang 2022). Extensive review of the discovery of E3 ligase ligands for PROTAC development appeared in the literature recently (Lee et al. 2022; Bricelj et al. 2021; Ishida and Ciulli 2021). The representative chemotypes previously used for targeting E3 ligase are shown in Fig. 2.3, along with their corresponding E3 ligase target. On similar lines, a large variety of POI ligands have been used for PROTAC development. These will be covered separately in the subsequent section(s). The interest of the scientific community in PROTACs is completely understandable due to their therapeutic potential. The present chapter is focused on the basic principles of PROTACs including the design aspects and their intricate, yet precise mechanisms of action. An overview of the challenges associated with PROTAC development, including their delivery at the target site and the pharmacokinetic (PK)-pharmacodynamic (PD) aspects.

2.2 Basic Principles of PROTACs

2.2.1 Linker Design and Optimization

The Protein Data Bank (Burley et al. 2022) contains several ternary complexes with PROTACs bound to their POI and E3 ligase components. Figure 2.4 shows a closer view of the binding pocket comprising the POI (human SMARCA2 protein) and E3 ligase (pVHL:ElonginC:ElonginB) (Farnaby et al. 2019). A recent literature reports PROTAC-DB 2.0, a repository of PROTAC structural and experimental data (Weng et al. 2023). It contains 3270 PROTACs (365 small-molecule POI binders, 1501 linkers, and 82 E3 ligands). Total number of crystal structures of the ternary

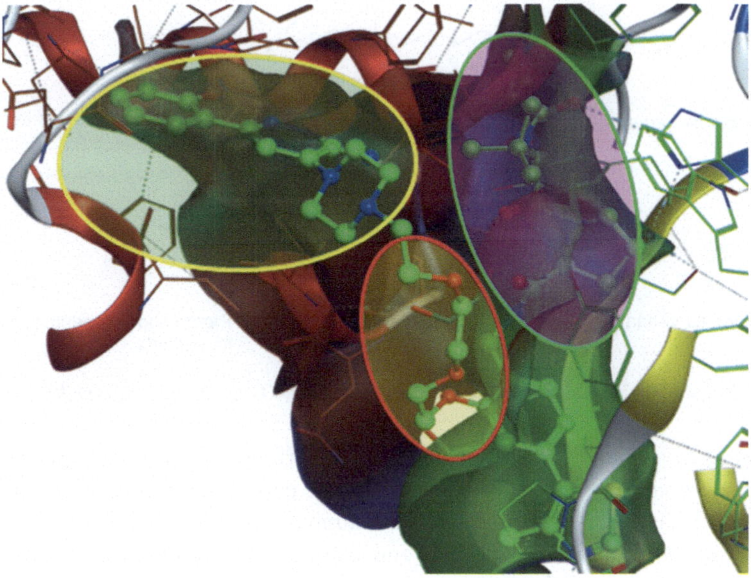

Fig. 2.4 Components of PROTAC 1 in complex with the bromodomain of human SMARCA2 (POI, shown as red surface, on the left side) and pVHL:ElonginC:ElonginB (E3 ligase, shown as green surface, on the right side) (PDB ID: 6HAY[18]). The linker is shown in the middle, while the substructure of PROTAC bound to E3 ligase is shown on the left, and the substructure bound to the POI is shown on the right side

complexes is 18 and the predicted structures are 664. Resources like these are a great boon for the rational design of PROTACs. Typically, the small-molecule ligand with higher potency at the POI is chosen as the POI binder. The E3 ligase ligand of appropriate chemistry, typically from the structurally more explored class, is chosen and the two fragments are connected via a compatible linker of appropriate length and linker chemistry. The optimization of the linker length may involve multiple rounds of design-synthesis-testing cycles.

The design of the linker connecting the POI and E3 ligase domain is crucial for the efficacy, selectivity, and pharmacokinetic properties of PROTACs. The tunable linker part also offers the ease of modulating the molecular, physicochemical, PK, and toxicity profiles. Troup et al. have recently reviewed the present strategies for PROTAC linker design (Troup et al. 2020). The key considerations in linker design include:

1. *Length and Flexibility*: The linker must be long enough to allow the two ends of the PROTAC to engage with their respective proteins, which may be spatially separated within the cell. However, excessive length can reduce the efficiency of degradation by affecting the formation of a ternary complex among the target protein, the PROTAC, and the E3 ligase. The flexibility of the linker also affects the orientation and proximity of the involved proteins, impacting the ubiquitination process.

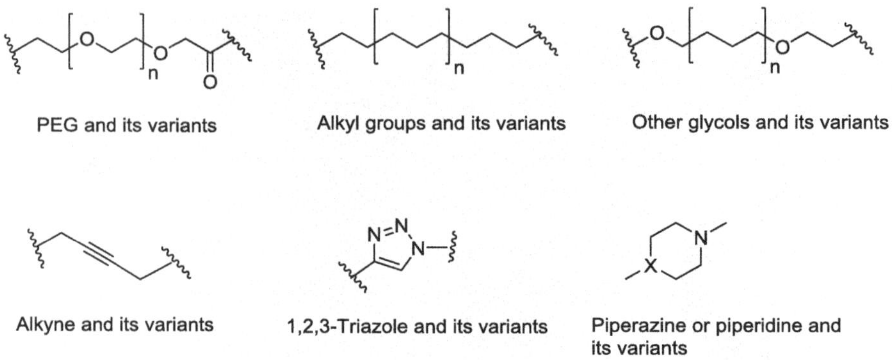

Fig. 2.5 Representative linker chemotypes explored for PROTAC design and development

2. *Composition:* Linkers can be composed of various chemical entities, including aliphatic chains, polyethylene glycol (PEG), peptides, or more rigid structures like PEG-based linkers or non-natural amino acids. The choice of materials can influence the solubility, stability, and overall pharmacokinetic profile of the PROTAC. Figure 2.5 shows a few representative linkers present in the PROTAC-DB 2.0 database. The PEG linkers are used widely for their biocompatibility, ease of synthesis, commercial availability of the appropriate building blocks, conformational flexibility, and PK-related aspects (Cecchini et al. 2021).
3. *Stability:* The chemical stability of the linker in the cellular environment is critical. Unstable linkers may lead to premature breakdown of the PROTAC, reducing its effectiveness. Stability against enzymatic degradation is particularly important to ensure that the PROTAC reaches its target intact. Goracci et al. (2020) studied the in vitro hepatic metabolism of a series of 40 PROTACs and concluded that their metabolism could not be predicted from their component ligands and the linkers played crucial role in modulating their metabolic stability.
4. *Efficacy of Ternary Complex Formation*: The optimal linker promotes efficient ternary complex formation, which is essential for effective ubiquitination and degradation of the target protein. This involves not just the physical connection between the ligands but also the proper orientation and positioning of the proteins to facilitate Ub transfer. Schwalm et al. (2023) developed cell-based assays for the measurement of PROTAC-induced degradation and the kinetics of the ternary complex formation and stability in cells.
5. *Toxicity and Off-target Effects:* The linker should not contribute to toxicity or off-target effects. Its design should minimize interactions with other proteins besides the intended E3 ligase and target protein to avoid unintended cellular consequences.

Overall, the PROTAC linker design is an active area of research, with ongoing studies aimed at understanding how different linker attributes affect the PK/PD of PROTAC molecules. Advances in computational modeling, structural biology, and biochemistry are helping to refine linker designs, leading to more potent, selective, and safe PROTACs for therapeutic use.

2.2.2 E3 Ligase Ligand Selection

The selection of the E3 ligase ligand is a critical step in the design of PROTACs, as it directly influences the ability of the PROTAC molecule to recruit an E3 Ub ligase for the targeted degradation of the POI. E3 ligase catalyze the transfer of Ub to Lys residues on target proteins, marking them for degradation by the proteasome. The choice of E3 ligase and its corresponding ligand impacts the efficacy, selectivity, and safety of PROTACs. So far, following E3 ligases, namely, von Hippel–Lindau protein (VHL), cereblon, Mouse double minute 2 homolog (MDM2), cellular inhibitor of apoptosis protein 1 (cIAP1), X-linked inhibitor of apoptosis protein (XIAP), aryl hydrocarbon receptor (AhR), Kelch-like ECH-associated protein 1 (KEAP1), and DDB1 and CUL4 associated factor 15 (DCAF15), have been used for targeting by majority of the developed PROTACs (Kramer and Zhang 2022). Despite the availability of >600 E3 ligase encoded by the human genome, only a handful of them have reported ligands. The principles dictating E3 ligase ligand selection for the PROTAC development include:

1. *E3 Ligase Abundance and Cellular Localization*: The chosen E3 ligase should be sufficiently abundant in the cells where the target protein degradation is desired. Its cellular localization is also crucial; the E3 ligase and the target protein must colocalize for efficient PROTAC-mediated degradation. For instance, nuclear proteins require E3 ligases that are present in the nucleus.
2. *Ligand Potency and Specificity*: The ligand must bind to the E3 ligase with high affinity and specificity. High-affinity binding ensures that the PROTAC can effectively recruit the E3 ligase to the target protein, while specificity minimizes off-target effects that could lead to the degradation of non-target proteins or other undesired cellular outcomes.
3. *Range of Accessible Proteins*: Different E3 ligases have different substrate specificities and are involved in the ubiquitination of distinct sets of proteins. The choice of E3 ligase ligand can thus expand or limit the range of proteins that can be targeted by PROTACs. Selecting an E3 ligase that is involved in the ubiquitination of a wide variety of proteins may offer broader applicability for PROTAC design.
4. *Structural Compatibility*: The structural compatibility between the E3 ligase ligand, the POI ligand, and the linker is crucial for the formation of an effective ternary complex. The spatial orientation and proximity enabled by the PROTAC can affect the ubiquitination efficiency of the target protein.
5. *PK/PD and Safety Profiles*: The PK/PD properties of the E3 ligase ligand can influence the overall properties of the PROTAC. Ideally, the ligand should contribute to favorable PK/PD profiles, including stability, solubility, and bioavailability of the PROTAC, as well as minimal off-target effects. The desirable safety

profile of the E3 ligase ligand is essential. The ligand should not elicit toxic responses or interfere with the normal functions of the E3 ligase beyond the intended protein degradation activity. The potential for immunogenicity or interference with critical cellular processes must be carefully evaluated.

6. *Intellectual Property (IP) Considerations*: The choice of E3 ligase ligand may be influenced by the IP considerations. Using proprietary ligands may require licensing agreements. Selection of the ligands that are not covered by the existing patents could simplify the development and commercialization process.

Given these considerations, the development of PROTACs often involves extensive screening and optimization processes to identify the most suitable E3 ligase ligands. Researchers are continually exploring new ligands and E3 ligases to expand the toolkit available for PROTAC design, aiming to increase the versatility and efficacy of this promising therapeutic approach.

2.2.3 POI Ligand Selection

The selection of POI ligand is extremely crucial for the obvious reasons of selectivity and specificity of POI recruitment for degradation (Li et al. 2020). It significantly influences the efficacy, selectivity, and therapeutic window of the final PROTAC molecule. This ligand must bind to the POI with high affinity and specificity to facilitate its ubiquitination and subsequent degradation by the proteasome. The factors considered critical for POI ligand selection include:

1. *Affinity and Specificity*: The ligand must bind to the POI with high affinity to ensure that the PROTAC can effectively recruit the target protein to the E3 ligase for ubiquitination. High specificity is also crucial to minimize off-target effects, which can lead to toxicity and side effects. Selecting a ligand that exclusively binds to the desired POI over closely related family members or other proteins is essential for achieving selective degradation. Conventionally, reversible or non-covalent ligands have been used. Nowak et al. (2023) have used POI ligands containing covalent warhead for higher potency and selectivity.
2. *Binding Site Accessibility*: The ligand must bind to a site on the POI that is accessible and conducive to forming a productive ternary complex with the E3 ligase. The orientation and proximity influenced by the binding site can affect the efficiency of ubiquitin transfer. In some cases, allosteric ligands that bind to sites other than the active site may be advantageous for targeting proteins where the active site is not amenable to ligand binding or for inducing degradation of proteins that do not have enzymatic activity. Interestingly, it is not required at all that the POI ligand target the catalytic site or the main binding site; as long as, there is a reasonable potency and selectivity of binding, the POI ligand is acceptable for incorporation in the PROTAC design.
3. *Structural Information*: Structural information about the POI, obtained through techniques such as X-ray crystallography, NMR spectroscopy or computational

methods, can greatly inform ligand selection by revealing potential binding sites, guiding the design of ligands that can fit these sites, and facilitating the rational design of PROTAC molecules that can effectively bridge the POI and E3 ligase. Several reports are based on the structure-based design of PROTACs for varied targets (Testa et al. 2020; Danishuddin et al. 2023; Drummond and Williams 2019; Bondeson et al. 2018).

4. *Ligand Efficiency and Drug-likeness*: The ligand should have favorable drug-like properties, such as good solubility, stability, and a favorable pharmacokinetic profile. It should also be efficient in its use of molecular weight and hydrophobicity to bind to the POI, as reflected by metrics such as ligand efficiency (LE) (Abad-Zapatero 2021) and lipophilic efficiency (LipE) (Johnson et al. 2018).
5. *Induced Proximity for Ubiquitination*: The chosen ligand should enable the PROTAC to bring the POI into close proximity with the E3 ligase in a manner that facilitates ubiquitination. This often requires optimization of the linker length and composition to ensure that the ternary complex is formed efficiently.
6. *Chemical and Metabolic Stability*: The ligand should be chemically stable under physiological conditions and resistant to metabolic degradation to ensure that the PROTAC remains intact and active within the cellular environment. The ligand alone and as a component of the PROTAC may show differential behavior towards metabolism, thereby affecting the metabolic stability of the final PROTAC (Cecchini et al. 2021). The chemical and metabolic liability of the final molecule may be addressed by suitable structural modifications, in any of the three components, without compromising the degradation efficiency (Chen et al. 2023).
7. *IP Considerations.* The selection of POI ligands may also be influenced by IP considerations. Novel ligands or those with favorable patent positions may be preferred to facilitate the development and commercialization of the PROTAC. There is a great opportunity to generate IP in the area of novel POI ligand discovery since one need not restrict only to the catalytic site of the POI; allosteric sites or sites not explored at all for the POI engagement can be investigated for targeting.
8. *Therapeutic Window and Toxicity*: The ligand should contribute to a favorable therapeutic window for the PROTAC, with effective degradation of the POI at concentrations that do not cause unacceptable toxicity. This involves careful evaluation of the ligand's effect on the degradation kinetics and selectivity of the PROTAC. Qin et al. (2022) have discussed key considerations for PROTAC discovery and development at length.

Overall, the selection of the POI ligand is a multifaceted process that requires careful consideration of these and other factors. Often, it involves iterative cycles of design, synthesis, and testing to identify ligands that meet these criteria and are suitable for use in PROTAC development.

2.3 Mechanism of PROTAC Action

Cellular pathways for maintaining protein homeostasis include proteasomes and lysosomes at the very end of protein turnover (Graham 2022). While proteasomes target short-lived as well as soluble misfolded proteins by ubiquitin-proteasome system (UPS), lysosome's scope include long-lived proteins, insoluble protein aggregates, macromolecular compounds, entire organelles, and microbes (few intracellular parasites and bacteria) (Ballabio and Bonifacino 2020). Proteasomes, as a part of the UPS system, contain various Ub ligases along with de-ubiquitination enzymes. The transfer of 76-mer Ub protein to POI (here any protein) via Lys isopeptide bond in a post-translational modification involves a systematic cascade in which E1 (Ub-activating) enzyme binds to Ub and transfers it to E2 (Un-conjugating) enzyme in an energy-intensive process. Subsequently, Ub ligase springs into action, whereby Ub is transferred to POI (here any protein substrate) (Fig. 2.6) (Békés et al. 2022; Zhao et al. 2022). Multiple rounds of the united action of E1-E3 enzymes yield poly-Ub POI (here any protein substrate). A total of eight poly-Ub chains on residues such as K6, K11, K27, K29, K33, K48, K63, and Met, depending on the conjugated Ub chain residue number are observed. The POIs (here any protein substrate) marked with K48-linked Ub chain are often recruited for degradation by proteasome. Surprisingly, K63-linked Ub chains are spared from proteasomal degradation (Yang et al. 2021). For description on lysosomal degradation, please see (Zhao et al. 2022) and cross-reference, therein detailed information.

The PROTACs, being bifunctional molecules, will bind to both the POI and the E3 ligase. This process marks the POI for degradation by the UPS cellular

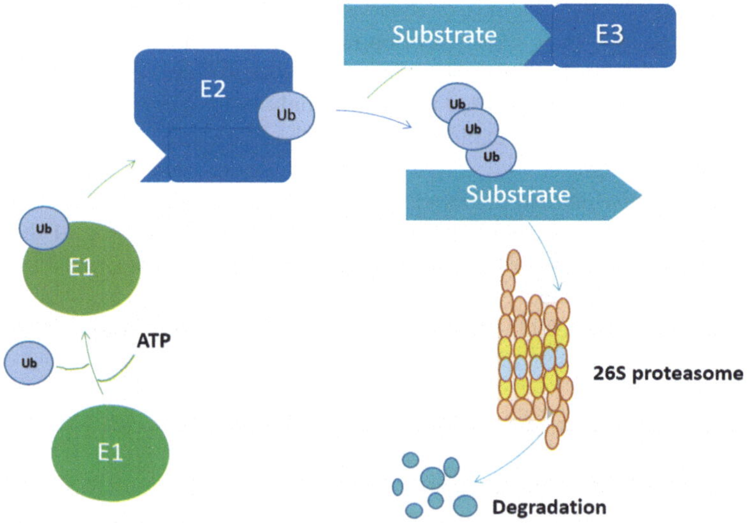

Fig. 2.6 Precise molecular events leading to POI degradation by the 26S proteasome system (Adapted from Békés et al. 2022)

machinery. A close look at the precise molecular events will shed light on the finer aspects of PROTAC mechanism of action, which will further fine-tune their design and development. Graham (2022) tried to cover these aspects in their review article. In contrast to the traditional inhibitors' occupancy-dependent mode of inhibition, the PROTACs follow event-based mechanistic mode. In the process, they play catalytic role; they are recycled following the POI degradation and continue the task at hand. The key event following the binding of the PROTAC with both the partners is POI polyubiquitylation. In the subsequent step, the polyubiquitin-containing POI is identified by the proteasome system and thus marked for degradation (Békés et al. 2022). In addition to PROTACs, a targeted protein degradation approach, several other strategies rely heavily on the UPS for targeting intracellular proteins. These include specific and non-genetic inhibitors of apoptosis protein-dependent protein erosive agents (SNIPERs) (Lin et al. 2021), molecular glues (Dong et al. 2021), trim away (Benn et al. 2022), and degradation tags (dTAGs) (Nabet et al. 2018). These approaches are not discussed here. Readers are advised to look for information on these elsewhere.

The critical component of the UPS system, i.e., E3 Ub ligases, belongs to a big family, further divided into HECT (homologous to the E6AP carboxyl terminus), U-box, RING (really interesting new gene) finger, and RBR (RING-IBR-RING) types (Yang et al. 2021). Their varied substrate specificity makes them crucial for cellular proliferation, apoptosis, DNA damage repair, immunological functions, and metabolism. Several of these E3 Ub ligases are involved in cancer progression regulation. Due to these facts, these critical enzymatic regulators have been explored as therapeutic targets. Further discussion on this aspect is beyond the scope of the present chapter. However, the E3 Ub ligase ligand selection remains critical for the success of POI degradation by the thoughtfully designed PROTACs. The first ever PROTAC designed back in 2001, recruited Skp1–cullin–F-box Proteins (SCF) E3 ligase β-TRCP to target methionine aminopeptidase-2 (MetAp-2, POI) for degradation. Several of the clinically validated PROTACs use a handful of E3 Ub ligases (Figs. 2.2 and 2.3) and target disease pathology-relevant POIs (Table 2.1).

Mares et al. (2020) studied the extended in vivo PD responses following the administration of the PROTAC targeting Receptor-Interacting Serine/Threonine Protein Kinase 2 (RIPK2) involved in inflammatory pathways. The initial PROTAC was based on VHL E3 Ub ligase. They varied the E3 ligase ligand (IAP and cereblon-based) and found that the IAP-based PROTAC was more potent (log half-maximal degradation concentration, $pDC_{50} = 9.4 \pm 0.1$) than VHL-based (pDC_{50} 8.7 ± 0.1) and cereblon-based (pDC50 8.6 ± 0.4) PROTACs. Interestingly, the IAP-based PROTAC demonstrated weaker RIPK2 potency ($IC_{50} = 10$ nM). The contrasting in vitro potency data and in cyto POI degradation data could only point to the catalytic role of the PROTAC in mediating protein degradation. Despite the encouraging in cyto degradation data, the molecule had higher microsomal clearance and poor aqueous solubility. Further structural optimization yielded a PROTAC which showed in vivo $t_{1/2}$ of 6.9 h. In summary, the in vitro potency data and in vivo PD data need to be evaluated systematically during structural optimization of PROTACs to drive further development of these candidates. Understanding the nuances of the

PROTAC mechanism of action is important for predicting the PD response following administration.

Since PROTACs work catalytically, not stoichiometrically, a single PROTAC molecule can induce the degradation of multiple target protein molecules, leading to a non-linear dose-response relationship that differs from traditional drugs. At high concentrations, PROTACs may exhibit decreased activity due to the saturation of either the target protein or E3 ligase, leading to the formation of less efficient binary complexes rather than the active ternary complex. Understanding this phenomenon is important for optimizing dosing strategies. The effect of PROTACs extends beyond their presence in the system due to the sustained depletion of the target protein. This means that the PD effects of PROTACs can last longer than their pharmacokinetic (PK) presence, influencing dosing intervals and therapeutic strategies. The activity of PROTACs can vary between tissues depending on the expression levels and activity of the target protein and the E3 ligases. This tissue-specific degradation capacity must be considered when predicting the therapeutic window and potential side effects. Identifying and validating biomarkers that can predict or indicate the PD response to PROTAC treatment is crucial for the clinical development and application of these therapeutics. These biomarkers can facilitate patient selection, dosing adjustments, and efficacy monitoring.

2.4 Challenges and Considerations

The PROTACs strategy offers several advantages, including the potential to target "undruggable" proteins, achieve more sustained protein knockdown, and reduce the likelihood of resistance mechanisms. However, the development and application of PROTACs also come with a unique set of challenges and considerations.

1. *Selectivity and Specificity.* One of the primary challenges in developing PROTACs is ensuring they have high selectivity and specificity for the POI. Since PROTACs function by recruiting E3 ligases to tag proteins for degradation, there is a risk of off-target effects where unintended proteins are also tagged and degraded (Wang et al. 2022). This can lead to toxicity and other undesirable side effects. Designing PROTACs that bind with high specificity to both the POI and the E3 ligase is crucial to minimize off-target effects.
2. *Bioavailability and Cell Permeability.* The molecular size of the PROTACs is typically larger than traditional small-molecule drugs, which can pose challenges to their bioavailability and cell permeability. Large molecules may have difficulty crossing cell membranes to reach intracellular targets. Addressing these issues often requires optimizing the PROTACs' physicochemical properties, including solubility and permeability without compromising their efficacy or specificity (Cecchini et al. 2021).
3. *Stability and Metabolism.* The stability of PROTACs in biological systems and their metabolism can significantly affect their therapeutic potential. They must be stable enough to reach their site/organ and initiate the degradation process

before being metabolized and inactivated by the drug-metabolizing enzyme systems (Liu et al. 2022). Researchers need to carefully design PROTACs to ensure an appropriate balance between stability and efficient clearance to avoid accumulation and potential toxicity.

4. *E3 Ligase Compatibility.* PROTACs rely on the recruitment of E3 ubiquitin ligases to function, which means their effectiveness is partly determined by the availability and activity of these ligases in different cell types and tissues. The choice of E3 ligase can significantly impact the potency and specificity of PROTACs, and not all ligases may be suitable for targeting specific proteins. Identifying and engineering compatible E3 ligases is a critical aspect of PROTAC development.
5. *Resistance Mechanisms.* While PROTACs offer a potential advantage in overcoming resistance mechanisms that can limit the efficacy of traditional inhibitors, resistance to PROTACs can still develop. Mechanisms such as mutations that prevent PROTAC binding, alterations in the UPS, and changes in the levels or activity of E3 ligase can all contribute to the development of resistance (Steinebach et al. 2023). Understanding and addressing these resistance mechanisms is crucial for the long-term success of PROTAC therapies.
6. *Regulatory and Safety Considerations.* The unique mechanism of action of PROTACs introduces new regulatory and safety considerations. The long-term effects of degrading target proteins, potential impacts on the ubiquitin-proteasome system, and the consequences of off-target effects require thorough investigation. Regulatory agencies may require extensive preclinical and clinical data to assess the safety and efficacy of PROTAC-based therapies.

In conclusion, while PROTACs represent a highly promising approach for targeting previously undruggable proteins and overcoming some of the limitations of traditional therapeutics, their development is accompanied by a unique set of challenges. Addressing these challenges requires innovative approaches in drug design, optimization, and testing to fully realize the therapeutic potential of PROTACs.

2.5 Conclusions

PROTACs embody a significant advancement in the field of drug discovery and development, offering a powerful tool to modulate protein function through induced degradation. The principles and mechanisms that underpin PROTAC technology showcase the potential to target a wide array of proteins, including those that have eluded traditional drug discovery efforts. By leveraging the cell's own degradation pathways, PROTACs provide a versatile and effective approach to therapeutic intervention, promising new treatments for diseases with unmet medical needs. As research in this area progresses, the continued optimization of PROTAC design and a deeper understanding of their PK/PD profiles will be essential for realizing their full therapeutic potential. The clinical approval of a late-stage clinical PROTAC is expected soon. The success in the clinic will endorse all the hard work in all these years by countless scientists and organizations. Newer tools and strategies will further expand the scope of applications of PROTACs in new avatar, e.g., Antibody-PROTAC conjugate. The multimodal structural optimization of candidate PROTACs is new to the medicinal chemist, biologist, PK/PD scientist, and other stakeholders. The future is in with a lot of surprises. Surmounting every challenge in the discovery and development stage is likely to create new opportunities and learnings. The principles involving PROTAC design, optimization, delivery, and in vivo performance need to be in sync with each other and the humane expectations.

References

Abad-Zapatero C (2021) Ligand efficiency indices for effective drug discovery: a unifying vector formulation. Expert Opin Drug Discov 16(7):763–775. https://doi.org/10.1080/17460441.2021.1884065

Ballabio A, Bonifacino JS (2020) Lysosomes as dynamic regulators of cell and organismal homeostasis. Nat Rev Mol Cell Biol 21(2):101–118. https://doi.org/10.1038/s41580-019-0185-4

Békés M, Langley DR, Crews CM (2022) PROTAC targeted protein degraders: the past is prologue. Nat Rev Drug Discov 21(3):181–200. https://doi.org/10.1038/s41573-021-00371-6

Benn JA, Mukadam AS, McEwan WA (2022) Targeted protein degradation using intracellular antibodies and its application to neurodegenerative disease. Semin Cell Dev Biol 126:138–149. https://doi.org/10.1016/j.semcdb.2021.09.012

Bondeson DP, Smith BE, Burslem GM, Buhimschi AD, Hines J, Jaime-Figueroa S, Wang J, Hamman BD, Ishchenko A, Crews CM (2018) Lessons in PROTAC design from selective degradation with a promiscuous warhead. Cell Chem Biol 25(1):78–87.e5. https://doi.org/10.1016/j.chembiol.2017.09.010

Bricelj A, Steinebach C, Kuchta R, Gütschow M, Sosič I (2021) E3 ligase ligands in successful PROTACs: an overview of syntheses and linker attachment points. Front Chem 9:707317. https://doi.org/10.3389/fchem.2021.707317

Burley SK, Berman HM, Duarte JM, Feng Z, Flatt JW, Hudson BP, Lowe R, Peisach E, Piehl DW, Rose Y, Sali A, Sekharan M, Shao C, Vallat B, Voigt M, Westbrook JD, Young JY, Zardecki C (2022) Protein Data Bank: a comprehensive review of 3d structure holdings and worldwide utilization by researchers, educators, and students. Biomol Ther 12(10):1425. https://doi.org/10.3390/biom12101425

Cecchini C, Pannilunghi S, Tardy S, Scapozza L (2021) From conception to development: investigating PROTACs features for improved cell permeability and successful protein degradation. Front Chem 9:672267. https://doi.org/10.3389/fchem.2021.672267

Chen S, Chen Z, Lu L, Zhao Y, Zhou R, Xie Q, Shu Y, Lin J, Yu X, Wang Y (2023) Discovery of novel BTK PROTACs with improved metabolic stability via linker rigidification strategy. Eur J Med Chem 255:115403. https://doi.org/10.1016/j.ejmech.2023.115403

Ci Y, Li X, Chen M, Zhong J, North BJ, Inuzuka H, He X, Li Y, Guo J, Dai X (2018) SCFβ-TRCP E3 ubiquitin ligase targets the tumor suppressor ZNRF3 for ubiquitination and degradation. Protein Cell 9(10):879–889. https://doi.org/10.1007/s13238-018-0510-2

Danishuddin, Jamal MS, Song KS, Lee KW, Kim JJ, Park YM (2023) Revolutionizing drug targeting strategies: integrating artificial intelligence and structure-based methods in PROTAC development. Pharmaceuticals (Basel) 16(12):1649. https://doi.org/10.3390/ph16121649

Dong G, Ding Y, He S, Sheng C (2021) Molecular glues for targeted protein degradation: from serendipity to rational discovery. J Med Chem 64(15):10606–10620. https://doi.org/10.1021/acs.jmedchem.1c00895

Drummond ML, Williams CI (2019) In silico modeling of PROTAC-mediated ternary complexes: validation and application. J Chem Inf Model 59(4):1634–1644. https://doi.org/10.1021/acs.jcim.8b00872

Farnaby W, Koegl M, Roy MJ, Whitworth C, Diers E, Trainor N, Zollman D, Steurer S, Karolyi-Oezguer J, Riedmueller C, Gmaschitz T, Wachter J, Dank C, Galant M, Sharps B, Rumpel K, Traxler E, Gerstberger T, Schnitzer R, Petermann O, Greb P, Weinstabl H, Bader G, Zoephel A, Weiss-Puxbaum A, Ehrenhöfer-Wölfer K, Wöhrle S, Boehmelt G, Rinnenthal J, Arnhof H, Wiechens N, Wu MY, Owen-Hughes T, Ettmayer P, Pearson M, McConnell DB, Ciulli A (2019) BAF complex vulnerabilities in cancer demonstrated via structure-based PROTAC design. Nat Chem Biol 15(7):672–680. https://doi.org/10.1038/s41589-019-0294-6

Girardini M, Maniaci C, Hughes SJ, Testa A, Ciulli A (2019) Cereblon versus VHL: hijacking E3 ligases against each other using PROTACs. Bioorg Med Chem 27(12):2466–2479. https://doi.org/10.1016/j.bmc.2019.02.048

Goracci L, Desantis J, Valeri A, Castellani B, Eleuteri M, Cruciani G (2020) Understanding the metabolism of proteolysis targeting chimeras (PROTACs): the next step toward pharmaceutical applications. J Med Chem 63(20):11615–11638. https://doi.org/10.1021/acs.jmedchem.0c00793

Graham H (2022) The mechanism of action and clinical value of PROTACs: a graphical review. Cell Signal 99:110446. https://doi.org/10.1016/j.cellsig.2022.110446

Hrdinka M, Yabal M (2019) Inhibitor of apoptosis proteins in human health and disease. Genes Immun 20(8):641–650. https://doi.org/10.1038/s41435-019-0078-8

Ishida T, Ciulli A (2021) E3 ligase ligands for PROTACs: how they were found and how to discover new ones. SLAS Discov 26(4):484–502. https://doi.org/10.1177/2472555220965528

Johnson TW, Gallego RA, Edwards MP (2018) Lipophilic efficiency as an important metric in drug design. J Med Chem 61(15):6401–6420. https://doi.org/10.1021/-acs.jmedchem.8b00077

Kramer LT, Zhang X (2022) Expanding the landscape of E3 ligases for targeted protein degradation. Curr Res Chem Biol 2:100020. https://doi.org/10.1016/j.crchbi.2022.100020

Lee J, Lee Y, Jung YM, Park JH, Yoo HS, Park J (2022) Discovery of E3 ligase ligands for target protein degradation. Molecules 27(19):6515. https://doi.org/10.3390/molecules27196515

Li X, Song Y (2020) Proteolysis-targeting chimera (PROTAC) for targeted protein degradation and cancer therapy. J Hematol Oncol 13(1):50. https://doi.org/10.1186/s13045-020-00885-3

Li W, Peng X, Lang J, Xu C (2020) Targeting mouse double minute 2: current concepts in DNA damage repair and therapeutic approaches in cancer. Front Pharmacol 11:631. https://doi.org/10.3389/fphar.2020.00631

Li R, Liu M, Yang Z, Li J, Gao Y, Tan R (2022) Proteolysis-targeting chimeras (PROTACs) in cancer therapy: present and future. Molecules 27(24):8828. https://doi.org/10.3390/molecules27248828

Lin J, Jin J, Shen Y, Zhang L, Gong G, Bian H, Chen H, Nagle DG, Wu Y, Zhang W, Luan X (2021) Emerging protein degradation strategies: expanding the scope to extracellular and membrane proteins. Theranostics 11(17):8337–8349. https://doi.org/10.7150/thno.62686

Liu Z, Hu M, Yang Y, Du C, Zhou H, Liu C, Chen Y, Fan L, Ma H, Gong Y, Xie Y (2002) An overview of PROTACs: a promising drug discovery paradigm. Mol Biomed 3(1):46. https://doi.org/10.1186/s43556-022-00112-0

Liu X, Kalogeropulou AF, Domingos S, Makukhin N, Nirujogi RS, Singh F, Shpiro N, Saalfrank A, Sammler E, Ganley IG, Moreira R, Alessi DR, Ciulli A (2022) Discovery of XL01126: a potent, fast, cooperative, selective, orally bioavailable, and blood-brain barrier penetrant PROTAC degrader of leucine-rich repeat kinase 2. J Am Chem Soc 144(37):16930–16952. https://doi.org/10.1021/jacs.2c05499. Epub 2022 Aug 25. PMID: 36007011, PMCID: PMC9501899

Mares A, Miah AH, Smith IED, Rackham M, Thawani AR, Cryan J, Haile PA, Votta BJ, Beal AM, Capriotti C, Reilly MA, Fisher DT, Zinn N, Bantscheff M, MacDonald TT, Vossenkamper A, Dace P, Churcher I, Benowitz AB, Watt G, Denyer J, Scott-Stevens P, Harling JD (2020) Extended pharmacodynamic responses observed upon PROTAC-mediated degradation of RIPK2. Commun Biol 3(1):140. https://doi.org/10.1038/s42003-020-0868-6

Nabet B, Roberts JM, Buckley DL, Paulk J, Dastjerdi S, Yang A, Leggett AL, Erb MA, Lawlor MA, Souza A, Scott TG, Vittori S, Perry JA, Qi J, Winter GE, Wong KK, Gray NS, Bradner JE (2018) The dTAG system for immediate and target-specific protein degradation. Nat Chem Biol 14(5):431–441. https://doi.org/10.1038/s41589-018-0021-8

Nowak RP, Ragosta L, Huerta F, Liu H, Ficarro SB, Cruite JT, Metivier RJ, Donovan KA, Marto JA, Fischer ES, Zerfas BL, Jones LH (2023) Development of a covalent cereblon-based PROTAC employing a fluorosulfate warhead. RSC Chem Biol 4(11):906–912. https://doi.org/10.1039/d3cb00103b

Qi SM, Dong J, Xu ZY, Cheng XD, Zhang WD, Qin JJ (2021) PROTAC: an effective targeted protein degradation strategy for cancer therapy. Front Pharmacol 7(12):692574. https://doi.org/10.3389/fphar.2021.692574

Qin L, Dai H, Wang J (2022) Key considerations in targeted protein degradation drug discovery and development. Front Chem 10:934337. https://doi.org/10.3389/fchem.2022.934337. PMID: 35978859, PMCID: PMC9376879

Schwalm MP, Krämer A, Dölle A, Weckesser J, Yu X, Jin J, Saxena K, Knapp S (2023) Tracking the PROTAC degradation pathway in living cells highlights the importance of ternary complex measurement for PROTAC optimization. Cell Chem Biol 30(7):753–765.e8. https://doi.org/10.1016/j.chembiol.2023.06.002

Steinebach C, Bricelj A, Murgai A, Sosič I, Bischof L, Ng YLD, Heim C, Maiwald S, Proj M, Voget R, Feller F, Košmrlj J, Sapozhnikova V, Schmidt A, Zuleeg MR, Lemnitzer P, Mertins P, Hansen FK, Gütschow M, Krönke J, Hartmann MD (2023) Leveraging ligand affinity and properties: discovery of novel benzamide-type cereblon binders for the design of PROTACs. J Med Chem 66(21):14513–14543. https://doi.org/10.1021/acs.jmedchem.3c00851

Targeted Protein Degradation (2024). https://www.protein-degradation.org/history/ (Accessed on 2024-02-25)

Testa A, Hughes SJ, Lucas X, Wright JE, Ciulli A (2020) Structure-based design of a macrocyclic PROTAC. Angew Chem Int Ed Engl 59(4):1727–1734. https://doi.org/10.1002/anie.201914396

Toure M, Crews CM (2016) Small-molecule PROTACS: new approaches to protein degradation. Angew Chem Int Ed Engl 55(6):1966–1973. https://doi.org/10.1002/anie.201507978

Troup RI, Fallan C, Baud MGJ (2020) Current strategies for the design of PROTAC linkers: a critical review. Explor Target Antitumor Ther 1(5):273–312. https://doi.org/10.37349/etat.2020.00018

Wang C, Zhang Y, Zhang T, Shi L, Geng Z, Xing D (2022) Proteolysis-targeting chimaeras (PROTACs) as pharmacological tools and therapeutic agents: advances and future challenges. J Enzyme Inhib Med Chem 37(1):1667–1693. https://doi.org/10.1080/14756366.2022.2076675

Weng G, Cai X, Cao D, Du H, Shen C, Deng Y, He Q, Yang B, Li D, Hou T (2023) PROTAC-DB 2.0: an updated database of PROTACs. Nucleic Acids Res 51(D1):D1367–D1372. https://doi.org/10.1093/nar/gkac946

Xiao M, Zhao J, Wang Q, Liu J, Ma L (2022) Recent advances of degradation technologies based on PROTAC mechanism. Biomol Ther 12(9):1257. https://doi.org/10.3390/biom12091257

Yang K, Wu H, Zhang Z, Leisten ED, Nie X, Liu B, Wen Z, Zhang J, Cunningham MD, Tang W (2020) Development of selective histone deacetylase 6 (hdac6) degraders recruiting Von Hippel-Lindau (VHL) E3 ubiquitin ligase. ACS Med Chem Lett 11(4):575–581. https://doi.org/10.1021/acsmedchemlett.0c00046

Yang Q, Zhao J, Chen D, Wang Y (2021) E3 ubiquitin ligases: styles, structures and functions. Mol Biomed 2(1):23. https://doi.org/10.1186/s43556-021-00043-2

Zhao L, Zhao J, Zhong K, Tong A, Jia D (2022) Targeted protein degradation: mechanisms, strategies and application. Signal Transduct Target Ther 7(1):113. https://doi.org/10.1038/s41392-022-00966-4

Structural Considerations and Chemistry of PROTACs

3

Abhishek Wahi, Hemant R. Jadhav, Shikha Thakur, Sushma Dev, Priyanka Mohanty, and Priti Jain

Abstract

The druggability of protein targets and dose-dependent side effects present challenges to the search for efficient therapies through enzyme inhibition. The druggability of protein targets and dose-dependent side effects present challenges to the search for efficient therapies through enzyme inhibition. The ubiquitin-proteasome system (UPS) has gained attention as a way to specifically break down target proteins as a result of this problem. One interesting approach is the use of target-based degradation (TBD) approaches, such as proteolysis-targeting chimeras (PROTACs). PROTACs are heterobifunctional molecules that recruit specific proteins for UPS-mediated degradation, bypassing the need to target active sites. Since their inception in 2001, PROTACs have evolved significantly, enabling the degradation of previously deemed 'undruggable' proteins. This chapter reviews the development and mechanisms of PROTACs, highlighting milestones such as the first cell-permeable PROTAC and the advent of small molecule-based E3 recruitment ligands. Notably, PROTACs have demonstrated efficacy in preclinical models, showcasing their potential in treating diseases including cancer and neurodegenerative disorders. With over 600 publications dedicated to PROTAC research, this chapter provides a comprehensive overview

A. Wahi · S. Dev · P. Jain (✉)
Department of Pharmaceutical Chemistry, School of Pharmaceutical Sciences, Delhi Pharmaceutical Sciences and Research University, DPSRU, New Delhi, India
e-mail: abhiwahi2002@gmail.com; sushma.omi@gmail.com; pritijain@dpsru.edu.in

H. R. Jadhav · S. Thakur · P. Mohanty
Department of Pharmacy, Birla Institute of Technology and Science, Pilani Campus, Pilani, Rajasthan, India
e-mail: hemantrj@pilani.bits-pilani.ac.in; p20210054@pilani.bits-pilani.ac.in; p20230072@pilani.bits-pilani.ac.in

M. Nandave, P. Jain (eds.), *PROTAC-Mediated Protein Degradation: A Paradigm Shift in Cancer Therapeutics*, https://doi.org/10.1007/978-981-97-5077-1_3

of their structural considerations and design principles, underscoring their pivotal role in contemporary drug development.

Keywords

Ubiquitin-proteasome system · Target-based degradation · Proteolysis-targeting chimeras · PROTACs · E3 ligase · Structural considerations

Abbreviations

ADPROMs	Affinity-directed proteolysis missile systems
AR	Androgen receptor
CDK	Cyclic dependent kinase
cIAP1	Cellular inhibitor of apoptosis protein 1
CRBN	Cereblon
EDGs	Electron-donating groups
EWGs	Electron-withdrawing groups
FDA	Food and Drug Administration
HBDs	Hydrogen bond donors
IMDs	Immune-modulatory drugs
MDM2	Mouse double minute 2 homolog
MetAp-2	Methionine aminopeptidase-2
PEG	Polyethylene glycol
POI	Protein of interest
PROTACs	Proteolysis-targeting chimeras
SBD	Substrate binding domain
SERDs	Selective estrogen receptor downregulators
TBD	Target-based degradation
TC	Ternary complex
TRK	Tropomyosin receptor kinase
UPS	Ubiquitin-proteasome system
VHL	Von Hippel-Lindau

3.1 Introduction

Creating molecules with high affinity that bind to and inhibit enzyme-active sites is a major focus of contemporary drug research. Although this approach has shown to be quite effective, it is constrained by the quantity of protein targets that are currently thought to be 'druggable' (Hopkins and Groom 2002). Apart from restricting drug targets to proteins possessing an active site, occupancy-driven pharmacology necessitates large drug dosages to attain sufficient IC_{90-95} concentrations for therapeutic effectiveness. This results in off-target binding and unintended side effects (Toure and Crews 2016). Moreover, various strategies to knockout the genes have been made till date (Paul et al. 2021; Tokatlian and Segura 2010) but they have suffered the disadvantage of poor pharmacokinetics. A novel approach to overcoming these challenges combines the benefits of genetic knockdown techniques with the absorption, distribution, metabolism, and excretion (ADME) properties of small molecules which involves small molecule control of intracellular protein levels via recruiting specific proteins to ubiquitin-proteasome system (UPS) for targeted degradation of desired proteins (Bondeson et al. 2018).

Investigating the possible therapeutic uses of targeting UPS has gained more attention during the past 25 years. Proteasome inhibitor approval by the US Food and Drug Administration (FDA) has shown that drugging the UPS is feasible and has prompted additional studies into this pathway's manipulation. Target-based degradation (TBD) techniques target and degrade specific proteins of interest (POI) by manipulating the UPS artificially, as an alternative to traditional small-molecule suppression. The focus of modern drug development is finding tiny molecules with higher binding affinities for target proteins. These compounds occupy the enzymatic activity site, which modulates the function of proteins (Konstantinidou et al. 2019). Numerous approaches TBD have been developed, encompassing proteolysis-targeting chimeras (PROTACs), affinity-directed proteolysis missile systems (AdPROMs), bioPROTACs, molecular glues, and selective estrogen receptor down-regulators (SERDs).

PROTACs are hetero-bifunctional molecules that consist of two ligands connected by a linker which exploit the UPS to break down a target protein associated with a certain disease as shown in Fig. 3.1. For a detailed description of the mechanism of action of PROTACs, readers are suggested to refer to Chap. 2. Since PROTACs have been developed, it is now possible to break down 'undruggable' proteins without having to take the active sites into account (Sun et al. 2019). The

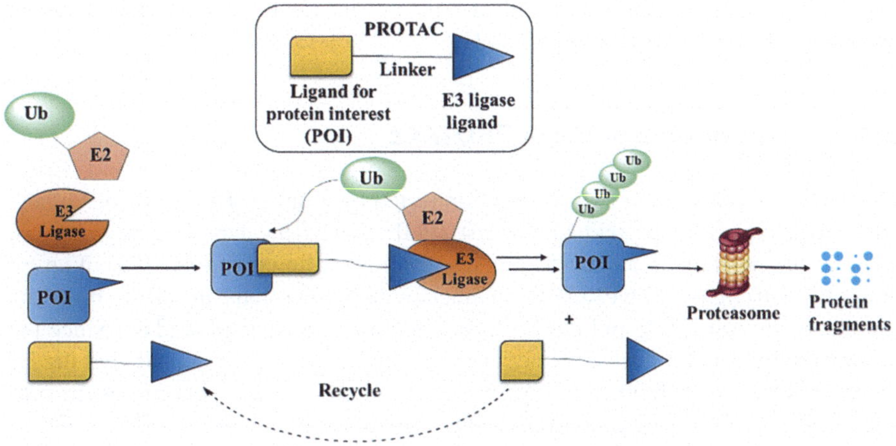

Fig. 3.1 General structure and mechanism of action of PROTACs. (POI: Protein of Interest)

first PROTAC was revealed in 2001 by the study groups of Crews and Deshaies (Sakamoto et al. 2001). It involved the Skp1-Cullin-F box-β-transducin repeat-containing protein (SCFb)-TRCP E3 to cause the breakdown of methionine amino-peptidase-2 (MetAp-2). Following the first study, PROTACs were employed in 2003 to specifically degrade the estrogen receptors and androgen receptors. Schneekloth and colleagues invented the first cell-permeable PROTAC in 2004 (Schneekloth Jr et al. 2004). It contained the E3 ligase binding peptide ALAPYIP, which activates Von Hippel-Lindau (VHL) and causes AR and FK506 binding protein 12 (FKBP12) to degrade. Simultaneously, PROTACs removed the essential for microinjection by recruiting the Von Hippel-Lindau (VHL) disease tumor-suppressor protein E3 in intact cells using a cell-penetrating peptide sequence and the HIF-1a peptide fragment (Liu et al. 2020). A shortened HIF-1a peptide segment was then included in a PROTAC intended to target the nuclear translocator of the aryl hydrocarbon receptor (ARNT). The development of small molecule-based E3 recruitment ligands marked advancement in PROTAC technology, resulting in the 2008 documentation of the first-ever fully small molecule-based PROTAC (Nalawansha and Crews 2020). The PhosphoPROTACs achieved a breakthrough in 2013 when they showed the first reports that PROTACs function in vivo by preventing tumour growth in murine models. The first small-molecule PROTAC, known as SARMnutlin, was made up of three parts: a unique substrate that binds to the androgen receptor (AR), a ligand that interacts with the E3 ligase MDM2, and a short soluble polyethylene glycol (PEG) linker that joins the two. Significant progress in the investigation of the cereblon (CRBN) E3 complex was achieved in 2010. It was determined that the main target of thalidomide immune-modulatory drugs (IMDs) is CRBN. After over 20 years of development, PROTACs have shown impressive results. More than 600 publications have been published on this topic between 2001 and 2024. At present, PROTACs play a crucial role in the realm of protein degradation and exhibit therapeutic potential across various life-threatening diseases like cancer and neurodegenerative diseases. The present book chapter overviews the structural aspects of PROTACs and provides a gist of factors considered while designing PROTACs (Hughes et al. 2021).

3.2 Structural Aspects of PROTACs

PROTACs are made up of two ligands connected by a linker. The 'warhead' ligand binds to a particular protein of interest (POI) that is the target of degradation, whereas the 'anchor' ligand binds to the substrate binding domain (SBD) of an E3 ubiquitin (Ub) ligase. The PROTAC interacts with both proteins in cells to bring the protein of interest (POI) and the E3 ligase into a ternary complex (TC). Since the ternary complex (TC) forms, the activated E2 ligase bearing loaded Ub and the E3 ligase are close to each other and the protein of interest (POI). This causes the protein of interest (POI) to acquire (poly)-ubiquitin molecules at lysine residues, designating it for 26S proteasome destruction (Yang et al. 2021). As bridge molecules between the E3 ligase and a particular protein of interest (POI), PROTACs effectively commandeer the UPS. Proteins are broken down by the UPS via substrate-specific ubiquitination and recognition. Three consecutive steps comprise the continuing process of ubiquitination: substrate-specific ligases (E3), ubiquitin-activating enzymes (E1), and ubiquitin-conjugating enzymes (E2). In an ATP-dependent mechanism, E1 forms a ubiquitin E1 thioester bond to activate free ubiquitin (Ub). Later, E1 uses trans-thioesterification to move the active Ub to E2. In the end, the target protein and the Ub-tagged E2 are brought together by the E3 ligase, which makes it easier for the target proteins to receive ubiquitin labels. The polyubiquitin chain-tagged target protein that results from these ubiquitination processes can be recycled. A significant fraction of E3 ligases is multiprotein complexes, usually made up of an adapter domain (Ad) connecting a substrate binding domain (SBD) and a substrate recognition subunit (Sc). There are estimated to be more than 600 E3 ligases in the human proteome. Each E3 ligase has a specificity towards substrates that is dictated by its molecular structure.

The requirements for ternary complex formation in PROTAC-mediated protein degradation include:

3.2.1 Affinity

A key component of PROTACs' functionality and design is their affinity for ternary complex formation, which entails the binding of an E3 ubiquitin ligase, a PROTAC molecule, and a target protein. Since longer-lasting complexes are linked to higher amounts of target protein degradation, the formation of a highly stable ternary complex is crucial to PROTACs' mode of action. The cooperativity, which is expressed as a ratio of binary to ternary complex affinities, and the individual affinities of the constituents have an impact on the affinity of the ternary complex (Bai et al. 2021). The affinity and cooperativity of ternary complex formation are studied using computational techniques and experimental techniques like surface plasmon resonance (SPR), which shed light on the behaviour and function of these degraders. An essential component of the design and effectiveness of PROTACs is the affinity of ternary complex formation, which entails the binding of a target protein, a PROTAC molecule, and an E3 ubiquitin ligase.

3.2.2 Cooperativity

Cooperativity, defined as the ratio of a PROTAC's binary to ternary complex binding affinity, is a property that affects the formation of ternary complexes in the context of PROTACs. The coefficient of cooperation, represented by α, quantifies the impact of one molecule's binding on another's binding. Higher ternary complex binding affinity as opposed to the product of the binary complex binding affinities produces positive cooperativity ($\alpha > 1$), whereas unfavourable protein-protein interactions provide negative cooperativity ($\alpha < 1$). It has been demonstrated that the ternary complex formation's affinity and cooperativity connect with PROTACs' degradation potency, making them crucial factors to take into account while creating strong degraders (Park et al. 2022; Wurz et al. 2023). To achieve target-specific degradation and to provide isoform and variant specificity, the cooperative aspect of ternary complex formation is especially crucial.

3.2.3 Stoichiometry

Target protein, PROTAC molecule, and E3 ubiquitin ligase binding often occur in a 1:1:1 stoichiometry during the formation of ternary complexes in PROTACs (Ward et al. 2023). This suggests that the ternary complex is formed by one molecule of each component.

3.2.4 Linker

In PROTACs, linker molecules are crucial for the creation of a ternary complex. Longer linkers must be optimised to guarantee effective degradation of the target protein, while shorter linkers may conflict with protein binding (Chen et al. 2023). In certain instances, the linker itself can positively influence the development of ternary complexes by establishing extra connections with one or both proteins (Bai et al. 2021). In empirical and labour-intensive design exercises, appropriate linker selection is crucial. To investigate the spatial and conformational ensemble of the ternary complex in proof-of-concept investigations, PROTACs with wide range of linkers are synthesised (Troup et al. 2020). For ternary complex formation to be successful, the linker's length and structure are essential, and computational techniques have been developed to assess a linker's fitness for ternary complex formation. The physicochemical properties of the PROTAC are also affected by the linker's composition, which must be optimized to guarantee permeability, bioavailability, and effective target protein degradation (Wurz et al. 2023).

3.2.5 Crystallisation

Crystallisation of PROTAC-mediated ternary complexes is a challenging yet crucial step in understanding their structure. The process involves over-expression and purification of the E3 ligase and target protein, forming the ternary complex, and crystallising it using techniques such as sitting drop vapor diffusion (Wijaya et al. 2023). Recent research has emphasized the significance of ternary complex crystal structures in confirming modelled structures and offering information on the selectivity profiles of protein breakdown (Hughes and Ciulli 2017). However, the potential flexibility and heterogeneity inherent to non-native complexes can hamper the crystallisation of the desired species (Wijaya et al. 2023). For successful crystallisation and ternary complex formation, appropriate linker design is essential. Despite these difficulties, PROTAC-mediated ternary complex crystallisation is necessary to learn important details about their composition and operation.

3.3 Design of PROTACs Based on Different E3 Ligases

As discussed in previous Sect. 3.2, the E3 ligases play a crucial role in the proteasome-mediated degradation of proteins. However, only a small number of the more than 600 E3 ligases that are known to exist in humans have been used in PROTACs thus far. E3 ligases that are often used to classify PROTACs include mouse double minute 2 homolog (MDM2), cereblon (CRBN), Von Hippel-Lindau (VHL), cellular inhibitor of apoptosis protein 1 (cIAP1), and other degraders. In the following subsection, a brief description of number of E3 ligase recruiters is provided.

3.3.1 MDM2 Recruiters-Based PROTACs

Though far less in number than CRBN and VHL, PROTACs based on MDM2, the primary E3 ligase targeting the tumour suppressor p53, are also reported (Anifowose et al. 2020; Han et al. 2022; Popowicz et al. 2011). To this aim, over the past 20 years, a large number of small-molecule inhibitors that target the p53-MDM2 protein-protein interactions have been found, with nutlin-3 being the first. A small number of these inhibitors are currently undergoing clinical development and may prove useful in the treatment of cancer and other related diseases. These inhibitors include HDM201, RG7388, MK-8242, and AMG-232 (Fig. 3.2).

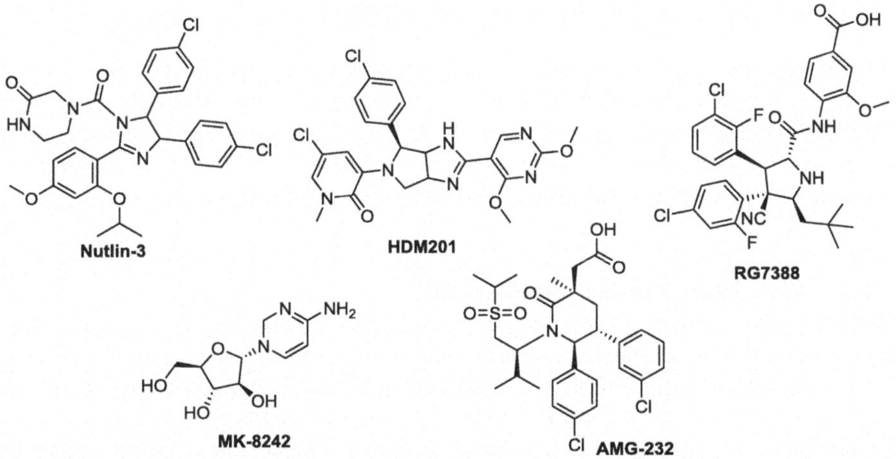

Fig. 3.2 Structure of some MDM2 inhibitors which have progressed to clinical trials for their anticancer action

Furthermore, studies have shown that significant efforts have been made to discover MDM2-based PROTACs that recruit MDM2 as E3 ligase as well as PROTACs that degrade MDM2 by recruiting CRBN as E3 ligase. Here, our focus in the following section is on PROTACs recruiting MDM2 to target other proteins of interest, i.e. our targeted protein.

These PROTACs are made using ligands called nutlins, which attach to the p53-binding pocket of MDM2 and interfere with MDM2's interaction with the transcription factor p53 without impairing MDM2's E3 ligase activity (Vassilev et al. 2004). Schneekloth and associates (2008) synthesised PROTAC that contained nutlin-3 (**1**), which is also known as MDM2 ligand and non-steroidal androgen receptor ligand (SARM) (Schneekloth et al. 2008). The androgen receptor was effectively engaged by the cell-permeable PROTAC to MDM2, whereupon the E3 ligase induced its ubiquitination and proteasomal degradation. Additionally, Crews et al. demonstrated that A1874, an MDM2-recruiting, Bromodomain 4 (BRD4)-degrading PROTAC (compound **2**), had nanomolar efficacy and could degrade the target protein by 98%. Its components include idasanutlin as an MDM2 ligand and JQ1 as a BRD4/BET inhibitor (Hines et al. 2019). Given that the PROTAC was able to both stabilise p53 and degrade BRD4, it is significant that this is the first report of a synergistic anti-proliferative action resulting from the targeted warhead and the E3 ligase ligand. The MDM2 inhibitor RG7388 was one of the PROTAC epidermal growth factor receptor (EGFR) degraders that the Ding group discovered in 2020. These degraders were based on several E3 ligase ligands. The $EGFR^{L858R/T790M}$ mutant exhibits modest degradation of compound **3** (Zhang et al. 2020). By connecting the MDM2 ligand nutlin-3 and the TrkC-targeted kinase inhibitor IY – IY, the Burgess group designed a family of TrkC-targeted kinase PROTACs in 2019 (Zhao and Burgess 2019). Compound **4** is not very effective at breaking down the TrkC protein, though. Zhao et al. reported the first Poly (ADP-ribose) polymerase 1 (PARP1)-targeting PROTAC by combining the MDM2 inhibitor nutlin-3 with the PARP1 inhibitor niraparib. Compound **5** was found following a comprehensive degradation screening in various triple-negative breast cancer cell lines. Surprisingly, compound **5** showed minimal cytotoxicity in normal cells but markedly increased PARP1 degradation and cell death in MDA-MB-231 cells with five times the potency of PARP1 inhibitors, such as niraparib (Zhao et al. 2019). The chemical structure of discussed MDM2-based PROTACs is provided in Fig. 3.3.

3.3.2 VHL Recruiter-Based PROTACs

In fact, the VHL-based mechanism predominates, most likely as a result of the targets' good, specific small molecule binders that have acceptable physicochemical properties and are well-characterized structurally and biophysically—a profile that has proven to be extremely challenging to achieve for ligands of other E3s thus far. These characteristics have allowed VHL-based PROTACs to be effectively used in the degradation of diverse target protein types linked to a range of illnesses (Setia et al. 2023).

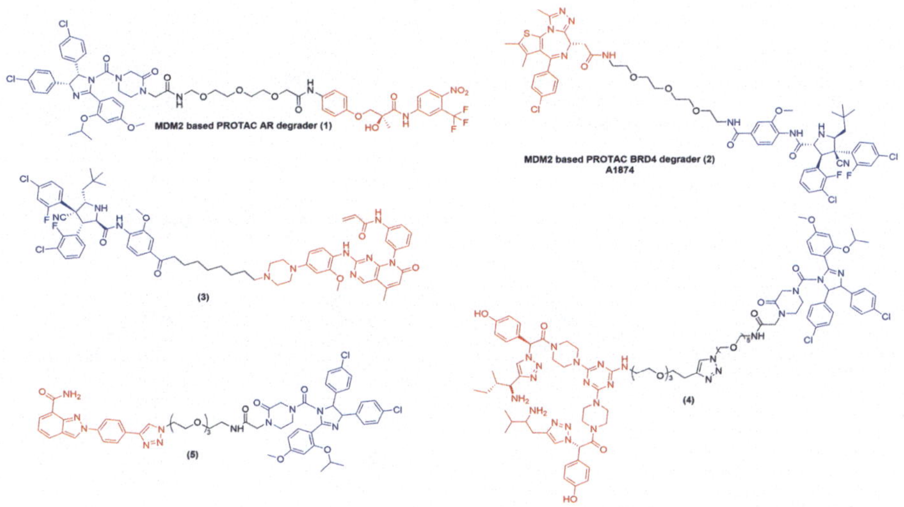

Fig. 3.3 Chemical structure of discussed MDM2-based PROTACs (red colour indicates ligand responsible for binding POI, blue represents E3 ligase ligand, and black indicates linker)

The Ciulli and Crews group made significant progress in 2012 by creating small-molecule inhibitors that break protein-protein interactions between VHL and its natural substrate HIF-1α. They used the theory that hydroxyproline (Hyp564) promotes HIF-1α binding to Ser111 and His115 on VHL in order to create the small-molecule inhibitors. Consequently, Hyp was employed as a starting material and subsequently integrated into in silico methods for the identification of hydroxyproline analogues. First, by extending the Hyp core on the left-hand side (LHS) and right-hand side (RHS), respectively, they created the first hits that contained isoxazole moieties and benzyl groups. As predicted by the *in-silico* method, including such compounds into Hyp would lead to interactions with Tyr98 and crystallographic water. Then, using a fluorescent HIF-1α peptide called FAMDEALAHyp-YIPD as a probe and a fluorescence polarisation (FP) displacement experiment, they examined their capacity to bind to VHL. Subsequently, they discovered that at larger concentrations, these substances might displace the fluorescent probe. Nevertheless, following the determination of the ligands' IC_{50} values, compound **6** (IC_{50} = 117 μM) seemed to be the most effective; thus, they aimed to maximise the benzylamine moiety situated at the RHS of 1. Later, they used electron-withdrawing groups (EWGs), electron-donating groups (EDGs), heterocyclic substitutes, and halogens in benzylamine to create a variety of VHL ligands using solid-phase synthesis. The FP test was then performed on these derivatives. Compound **7**, which possesses a heterocyclic moiety at the para position, exhibited a higher binding affinity towards VHL compared to other synthesised ligands that showed reduced affinity. Furthermore, Fig. 3.4 indicates that **7** is a more powerful derivative than 1 based on the K_d value of 5.4 μM and IC50 of 4.1 μM. The co-crystal structure of 2 then revealed important interactions with VHL. While the NH group of **7** interacts

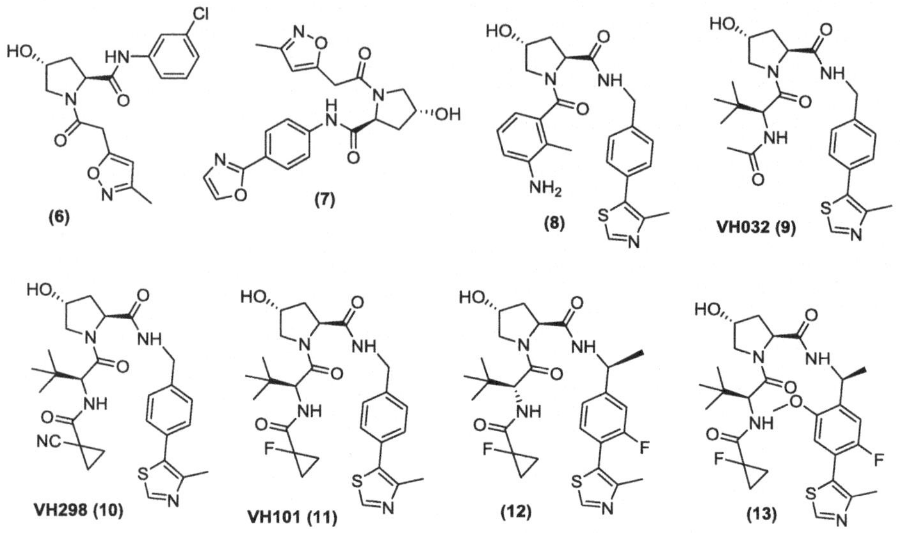

Fig. 3.4 The chemical structures of VHL-binding ligands

with the His110 carbonyl of VHL, the OH group of **7** interacts with Ser111 and His115. Furthermore, the aryl ring makes contact with Tyr98, the oxazole CH forms a hydrogen connection with Pro99, and the oxazole N forms a hydrogen link with Arg 107 (Buckley et al. 2012b). Later, the same group carried out further work on the optimisation of these derivatives, and it was found that Compound **8** ($IC_{50} = 0.9$ μM) had the highest binding potency among all other ligands found thus far. It possesses 2-methyl benzamide at the LHS and methyl thiazolyl at the RHS. Additionally, the crystal structure of **8** with the VCB complex revealed comparable interactions. Furthermore, new contacts via an H-bond were observed between the aniline in the N-terminal of 51 and the side chain of Gln96 (Buckley et al. 2012a).

A different set of VHL ligands based on an even more intricate design approach was published by Galdeano et al. (2014). Because the ligands discovered thus far have single-digit micromolar potency and low lipophilicity, scientists were compelled to refine the ligand series of pVHL/HIF-1α PPIs. In context of this, they next worked on improving the LHS and discovered ligand **9 (VH032)**, which has 4-methyl thiazole at the RHS and acetylated L-tert-leucine (t-Leu) at the LHS of the Hyp core (Galdeano et al. 2014). Remarkably, compound **9** (Kd 180 nM by ITC) demonstrated greater potency and reached a nanomolar binding affinity, which was 20 times better than other synthesised compounds in the series. Furthermore, its crystal structure demonstrated that it is the most powerful ligand since it had all of the expected interactions, as demonstrated by the HIF peptide. Notably, VH032 has been used in the bulk of VHL-recruiting PROTACs that have been released yet. The research for discovering further VHL-binding ligands was carried out a number of small molecules like VH298 (**10**), VH101 (**11**), **12, 13,** etc. have been found till date with the aim to further enhance their binding affinity, potency, and exhibit desired interactions at VHL-binding site (Soares et al. 2018; Vu et al. 2023). The chemical structures of VHL-binding ligands are provided in Fig. 3.4.

The discovery of small-molecule VHL ligands and substantial linker optimisation (discussion on linker designing is provided in Sect. 3.4) made it possible to create extremely effective PROTACs that recruit VHL as an E3 ligase. Scientists from all around the world have created a variety of VHL-based PROTACs to target a broad range of disease-causing proteins like BCR-ABL, BRD4, HDAC, CDK6, PARP-1, TBK1, and HaloTag fusion proteins, TRIM24 and tyrosine kinases (EGFR, HER2, and c-Met) (Bricelj et al. 2021; Konstantinidou et al. 2019). Using ceritinib and VHL ligand, Jong Yeon Hwang and colleagues designed the first ALK degraders in 2018 and assessed their biological activity, namely the degradation of the fusion ALK protein. Their inhibitory activity varied within ranges of 0.016–0.14 mM; nonetheless, these values were 2–20 times weaker than ceritinib (IC_{50} = 0.007 mM). In the SU-DHL-1 cell line, the representative degrader **14** with an amide linkage to ceritinib dramatically decreased the amount of NPM-ALK protein by over 90% at 1 mM. The dose-dependent degradation of another fusion protein, EML4-ALK by **14**, was also verified in the ALK-positive NSCLC cell line NCI-H3122 (Kang et al. 2018). The PROTACs approach was utilised by Shan and colleagues in 2020 to investigate the potential of antiangiogenesis drugs in vascular normalisation. Based on anti-angiogenesis drugs that have been previously reported, six new PROTACs were created. PROTACs **15** and **16** have shown significant anti-proliferative activities against human umbilical vein endothelial cells (EA.hy926) as well as strong inhibitory activity against VEGFR-2. Furthermore, they possessed the ability to precisely lower VEGFR-2 protein levels. In the meantime, HEK293 cells showed no discernible cytotoxicity from any of the PROTACs (Shan et al. 2020). For better understanding of progress and chemistry of VHL-based PROTACs, readers are suggested to refer to review articles by Setia et al. (2023) and Wang et al. (2022). The chemical structure of discussed VHL-based PROTACs is provided in Fig. 3.5.

3.3.3 IAP Recruiter-Based PROTACs

The inhibitor of apoptosis (IAP) family includes cIAP1, cIAP2, and xIAP. These members inhibit caspase activity and interfere with the control of the NF-κB signalling pathway to avoid cell death. Three baculovirus IAP repeat domains at the N-terminus charcterise IAPs. The RING domains of the majority of IAPs, such as cIAP1 and xIAP, bind to E2-binding enzymes to perform the role of E3 ubiquitination ligases (Deveraux and Reed 1999). Among IAPs, CIAP1 is overexpressed in a range of cancerous cells, and attaching to the methyl bestatin (MeBS) molecule initiates its ubiquitination activity (Che et al. 2012). In 2010, Hashimoto's group reported the PROTAC **17**, which includes all trans retinoic acid (ATRA), the endogenous ligand of retinoic acid receptors that can recruit the intracellular retinoic acid binding proteins cellular retinoic acid binding protein 1 and 2 (CRABP-1 and CRABP-2), and methyl bestatin (MeBS), which binds selectively to the BIR3

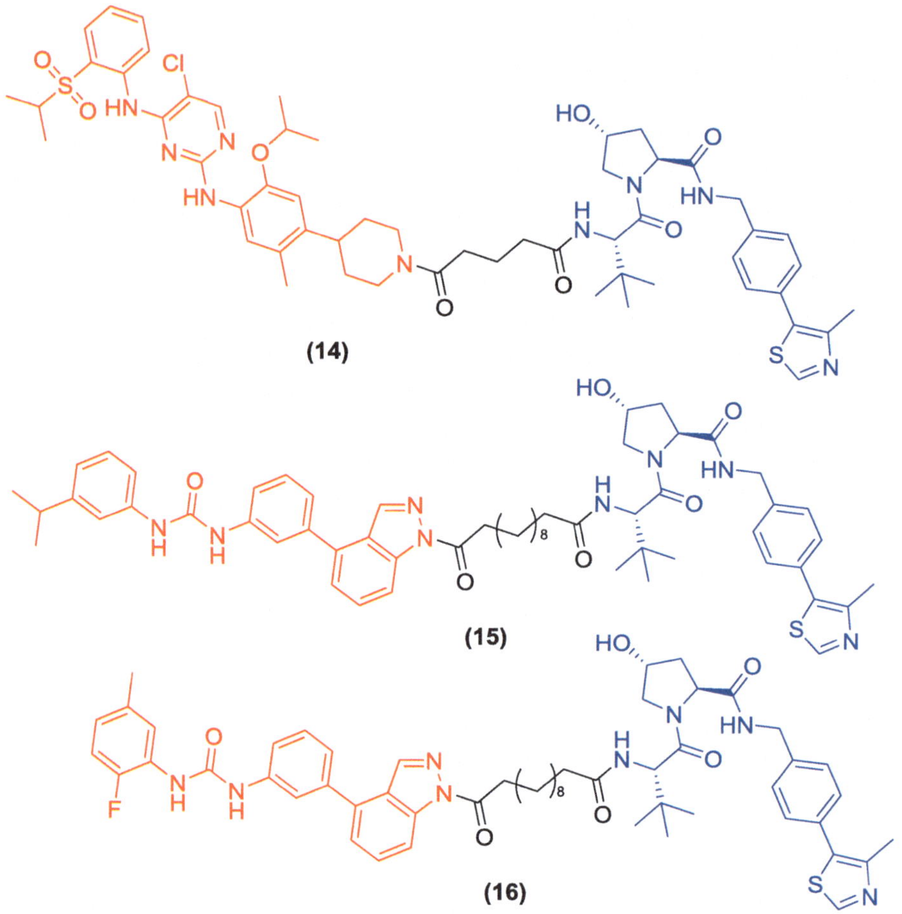

Fig. 3.5 The chemical structures of VHL-based PROTACs (red colour indicates ligand responsible for binding POI, blue represents E3 ligase ligand, and black indicates linker)

domain of cIAP1, the RING domain of which promotes auto-ubiquitination (Itoh et al. 2010). The target proteins, CRABP-1 and CRABP-2, were successfully subjected to proteasomal degradation by the cIAP1-based PROTAC. Despite this, compound **17** had limited selectivity because the ester group hydrolysed in cells with great ease. As a result, the same group produced the amide PROTAC **18**, which induced CRABP-2 to particularly degrade (Itoh et al. 2011). In another study, the Naito group reported a number of PROTACs in 2017 that contained various ER ligands and LCL161 compounds (Ohoka et al. 2017). Compound **19** was the most successful of these PROTACs, causing ER degradation with a DC_{50} of 3 nM and a D_{max} of 70%. Furthermore, the Ohoka group designed PROTACs with good ERα degradation activity in 2018. They did this by using an optimised IAP ligand (Ohoka et al. 2018). These compounds showed stronger IAP-binding affinity and ERα

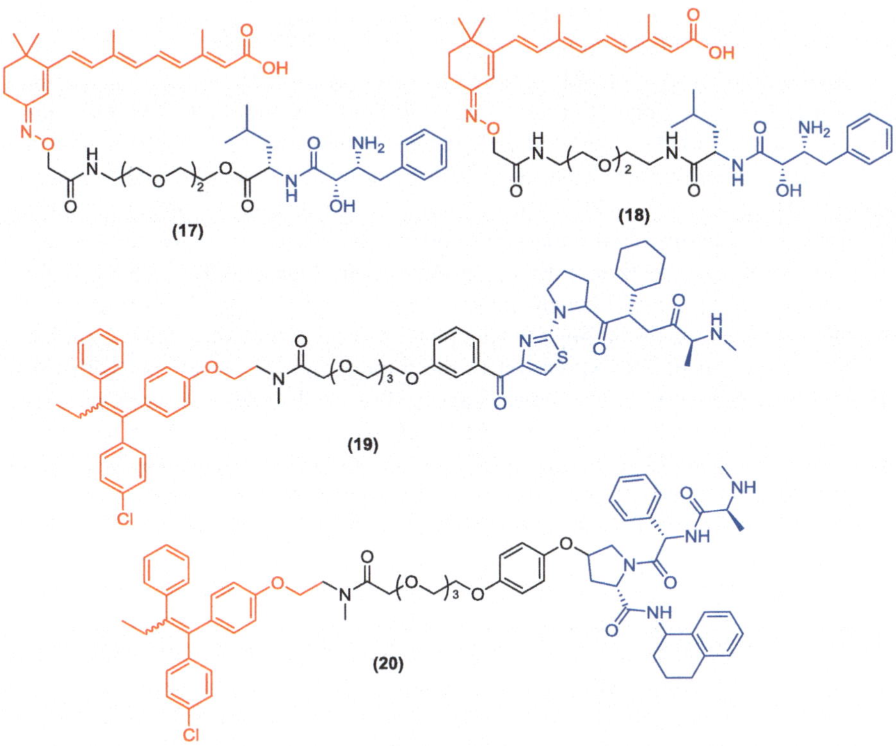

Fig. 3.6 The chemical structures of IAP-based PROTACs (red colour indicates ligand responsible for binding POI, blue represents E3 ligase ligand, and black indicates linker)

degradation efficiency than compound **19**, and they recruited xIAP more preferentially than CIAP1. Compound **20** outperformed compound **19** in terms of in-vivo anticancer efficacy in MCF-7 xenograft models based on its superior degrading activity (DC_{50} = 10 nM, D_{max} > 80%). Two further examples of protein degradation utilising hybrid small molecules called SNIPER (Specific and Non-genetic IAP-dependent Protein Eraser) are TACC3 (Ohoka et al. 2014) and BCR-ABL (Demizu et al. 2016). The chemical structure of discussed IAP-based PROTACs are provided in Fig. 3.6.

3.3.4 CRBN Recruiters-Based PROTACs

Immunoregulatory drugs (IMDs), including lenalidomide, thalidomide, and pomalidomide, attach to the E3 ligase CUL4-RBX1-DDB1-CRBN ($CRL4^{CRBN}$) subunit CRBN and facilitate the ubiquitination and destruction of transcription factors (TFs) belonging to the IKAROS family, including IKZF1 and IKZF3 (Krönke et al. 2014; Lu et al. 2014). The development of PROTACs, which use thalidomide analogues as CRBN ligands, was motivated by the molecular glue function of

IMIDs. Two CRBN-based PROTAC compounds, ARV-471 and ARV-110, are currently undergoing clinical studies.

The PROTAC ARV-471 (**21**), created by Arvinas in 2018, successfully caused ER degradation in a range of breast cancer cell lines at a dose of 11 nM. Oral treatment of compound **21** suppressed tumour development by 99% at 10 mg/kg and demonstrated a 106% inhibition rate when the dose was increased to 30 mg/kg in patient-derived xenograft (PDX) mice with estrogen receptor 1 (ER1) mutations. Compound **21** demonstrated remarkable anticancer activity when provided either in isolation or in conjunction with cyclin-dependent kinase (CDK)4/6 inhibitors. Furthermore, compound **21** has the potential to be a 'best-in-class' PROTAC targeting ER, as evidenced by findings from a phase I clinical trial showing a significant 62–90% reduction in ER expression in the tumour tissues of patients with ER+/HER2-advanced or metastatic breast cancer (Lin et al. 2020). Arvinas produced another oral active PROTAC, called ARV-110 (compound **22**), which selectively attaches to AR and subsequently facilitates the breakdown of androgen receptor (AR) (Neklesa et al. 2019). Preclinical research showed that in several prostate cancer cell lines, ARV-110 (**22**) destroyed 95%–98% of AR. In VCaP cells, compound **22** caused AR to degrade in a time-dependent manner; 4 h after injection, AR was virtually entirely broken down ($DC_{50} < 1$ nM). Compound **22** (orally, once daily [qd]) at a dose of 1 mg/kg decreased plasma PSA and more than 90% AR degradation in the VCaP xenograft mice. Notably, compound **22**'s in vivo effectiveness outperformed that of the AR inhibitor enzalutamide. Most importantly, compound **22** is now undergoing phase 1 clinical studies, and the initial results have demonstrated acceptable patient safety and tolerability characteristics. As a result, compound **22** may be developed as a potential therapeutic approach for castration resistant prostate cancer (Neklesa et al. 2019). The Crews group coupled pomalidomide with OTX015, a BRD4 inhibitor, to design ARV-825, compound **23**, a CRBN-based PROTAC molecule that efficiently induced BRD4 degradation through the proteasome pathway. Compared to the BRD4 inhibitor, compound **23** showed a greater inhibitory impact against the c-Myc gene, which is downstream of BRD4. In another research conducted, the BRD4 inhibitor JQ1 and a CRBN ligand with a shorter aliphatic chain were coupled to create the PROTAC dBET1, compound **24**. Compound **24** was found to successfully trigger BRD4 degradation at nanomolar doses, while also considerably reducing the protein levels of c-Myc, BRD2, BRD3, and BRD4. At a dosage of 50 mg/kg, it demonstrated superior in-vivo efficacy in a human leukaemia xenograft model compared to JQ1 (Lu et al. 2015). In order to break down BET proteins, the Qin group created PROTAC QCA570, compound **25**, in 2018. It has a triazole linker and a specific BRD4 (BD1) ligand. Compound **25** successfully stimulated the BET protein degradation and slowed the growth of human acute leukaemia cells at picomolar doses (Qin et al. 2018). Furthermore, compound **25** demonstrated remarkable anticancer efficacy in vivo and successfully accomplished comprehensive and durable tumour regression in a human leukaemia xenograft model. The chemical structures of discussed CRBN-based PROTACs are provided in Fig. 3.7.

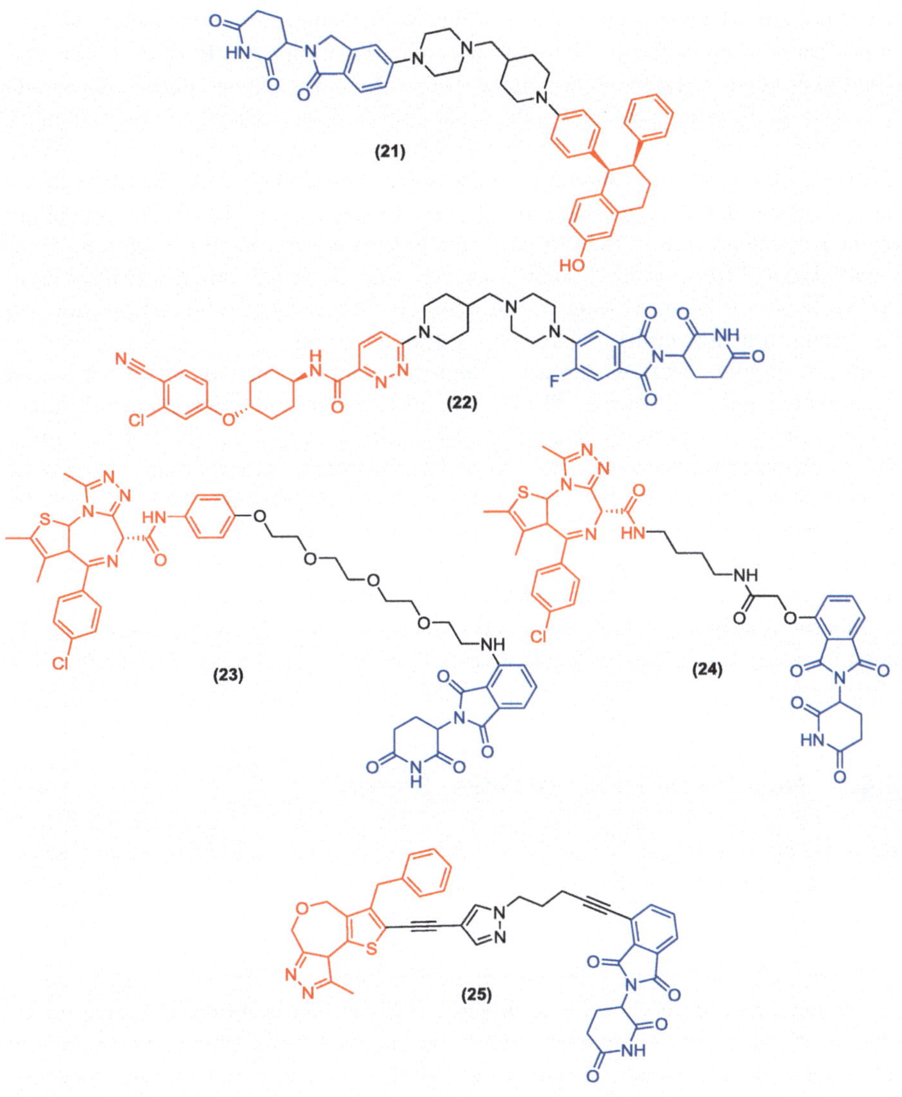

Fig. 3.7 Chemical structures of CRBN-based PROTACs (red colour indicates ligand responsible for binding POI, blue represents E3 ligase ligand, and black indicates linker)

3.4 Linker Designing

Linker design in PROTACs is a critical aspect of developing effective targeted protein degradation systems (Neklesa et al. 2017). The target protein's proximity-induced ubiquitination and subsequent destruction are facilitated by the linker, which acts as a molecular bridge between the E3 ubiquitin ligase and the ligands targeting the target protein. The composition, cleavability, linker length, and other parameters are important in influencing the effectiveness and specificity of PROTAC molecules. The linker needs to be long enough for the two ligands to attach to the appropriate locations on the target protein and the E3 ubiquitin ligase, and short enough to facilitate effective ubiquitination and target protein destruction (Chen et al. 2013). Since it can alter the molecule's binding affinity, selectivity, and pharmacokinetic characteristics, the linker design is essential to the effectiveness of PROTACs. Depending on the particular needs of the PROTAC, the linker may consist of peptides, polyethylene glycol (PEG), or alkyl chains, among other chemical moieties (Miller et al. 2023). To get the desired PROTAC properties, the linker length and composition can be optimised using empirical screening or rational design.

In developing PROTAC ternary complexes, many connector types have been documented and used, such as PEG linker, alkyl linker, and 'click chemical' linker (Poongavanam et al. 2022). The ability of the linker to assume folded conformations, which might lead to increased cell permeability, depends on its chemical makeup and flexibility. The large chemical space of possible PROTACs can be explored using computational methods, which can help in linker design and optimisation. The structure of the PROTAC-induced ternary complex, which is essential for efficient target degradation, is what these techniques attempt to model.

In structural biology, linkers are short amino acid sequences that connect multiple domains in a single protein, and their design can affect the stability and function of the protein.

3.4.1 Novel Significance of Linker Design

Drug Discovery Linker design is essential for creating novel therapeutic alternatives in a variety of domains, including oncology, neurological illnesses, and infectious diseases, because PROTACs can target hitherto 'undruggable' proteins.

Linkers are essential for tying the antibody to the chemical payload in drug delivery systems like antibody-drug conjugates (ADCs). The stability and release efficiency of the active component can be impacted by the linker chemistry and alterations chosen, which can impair the ADC's therapeutic efficacy. To ensure the effective release of the drug payload at the target location while retaining stability in circulation, for instance, several types of linkers, such as protease-sensitive peptides or acid-labile components, can be employed. To improve therapeutic effects, linker design optimisation is crucial for striking a compromise between stability in circulation and effective payload release (Su and Zhang 2021).

Computational Module PROTACs are a fast-developing field that is moving away from typical alkyl structures and towards novel linker classes to produce more potent drugs for medicinal uses. One known aspect determining degradation efficiency is the linker's length. PEG or alkyl motifs are frequently employed to systematically change the linker length (Troup et al. 2020). Using 3D data, novel models such as PROTAC-INVENT efficiently construct PROTAC linkers while taking into account the intricate architectures of target proteins and E3 ligases (Li et al. 2023). The effectiveness of PROTAC molecules is increased by using reinforcement learning to create chemically viable linker architectures and optimize molecular characteristics. Optimising PROTAC functioning and target selectivity requires adjusting the linker composition, length, and attachment locations for every ligand pair. Through particular interactions within the ternary complex, linkers can increase affinity and contribute to isoform selectivity in PROTACs (Troup et al. 2020).

Materials Science Enhancing the specificity, sensitivity, and efficacy of molecular imaging probes is made possible through the creation of intermediaries. Researchers can now construct activatable probes for molecular imaging that react to particular triggers or situations for improved imaging capabilities thanks to the successful application of stimuli-responsive linkers in probe design (Wang et al. 2024). To detect protein-protein interactions (PPIs), the use of releasable linkers in photo-affinity probes has shown efficacy. Researchers can learn more about intricate biological interactions at the molecular level by combining photo-affinity probes with releasable chemical cross-linkers (Wang et al. 2020). Through the use of targeted and responsive imaging agents, theranostic applications are made possible by the integration of stimuli-sensitive linkers in imaging probes.

3.4.2 Key Properties of Linkers

3.4.2.1 Flexibility and Length

In PROTACs, linkers' length, composition, and attachment point are crucial characteristics. Alkyl and PEG units are examples of flexible linkers, while heterocycles, alkynes, and spirocycles are examples of rigid linkers. A higher degree of freedom is offered by flexible linkers, making additional PROTAC solution-state conformations accessible. Conversely, rigid linkers can alter the physicochemical characteristics of the degraders. The permeability of a substance across a cell membrane is primarily determined by two properties: the size and polarity of the penetrating conformation(s) (Zografou-Barredo et al. 2023). The spatial distance between the E3 ligase and the ligands targeting the target protein is determined by the length of the linker. For the target protein and the E3 ligase to come together to form a ternary complex, this distance is essential. A linker that is too short may prevent the ligands from being positioned correctly, which could result in decreased binding affinity and ineffective protein breakdown. On the other hand, an excessively lengthy linker could provide steric hindrance or decreased flexibility, making it more difficult for the ligands to engage with the E3 ligase and their respective binding sites on the target protein. Flexible linkers (like alkyl and PEG units), somewhat rigid linkers, click chemistry-based linkers, and other types (like alkane, piperidine/benzene, triazole, alkynyl-alkane, and more) are the several categories into which linkers can be divided (Fig. 3.8) (Zagidullin et al. 2020).

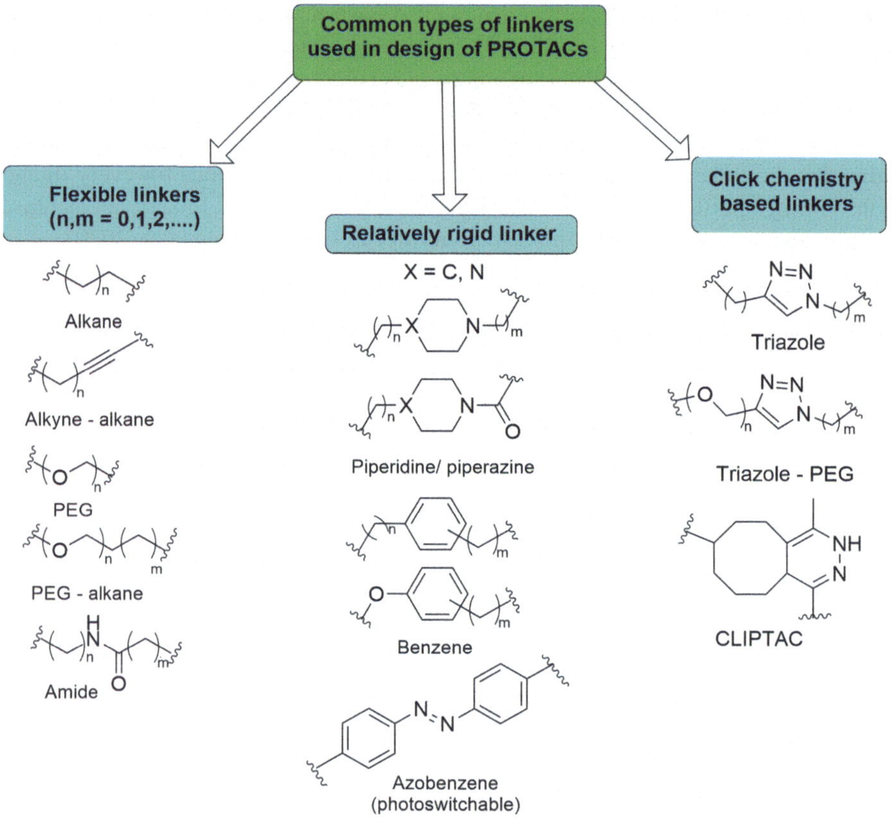

Fig. 3.8 Various types of linkers used in design of PROTACs

A successful PROTAC design depends on the linker length being as short as possible. A ternary complex may fail because of steric conflicts preventing ligands from binding to their corresponding proteins simultaneously if the linker is too short. On the other hand, if the linker is very lengthy, it can make it difficult to get the proteins close together for target ubiquitination. As a result, adjusting the linker length is essential to obtaining enhanced cell permeability and appropriate PROTAC spatial orientation (Zografou-Barredo et al. 2023). Alkyl and PEG units are examples of flexible linkers that give PROTACs additional flexibility and enable them to access a wider range of solution-state conformations. Because of its flexibility, there is a higher chance that ternary complexes involving the POI and E3 ligase will form, which will speed up POI breakdown. A flexible linker can adjust to changes in the ligands' relative positions, maximising their interactions and promoting the ternary complex's formation. Furthermore, by enabling a variety of low-energy conformations in solution, flexible linkers provide diversity in PROTAC design. Because of the conformational freedom that flexible linkers afford, PROTACs can fold in on themselves, thereby decreasing the compound's effective polar surface area (Troup et al. 2020). To increase cell permeability and oral bioavailability, this folding can conceal certain hydrogen bond donors and acceptors. Moreover, flexibility can maximize the binding affinity of the PROTAC molecule for its targets and help minimize steric conflicts. However, rigid linkers such as spirocycles, alkynes, and heterocycles can alter the physicochemical characteristics of degraders and enhance their cell permeability and water solubility. Furthermore, the majority of PROTACs exhibit poor cell membrane permeability because of their large MW. As a result, adding polar groups to linkers helps modify the PK/PD profiles of the PROTACs.

3.4.2.2 Chemical Stability

Under physiological conditions, PROTAC linkers are guaranteed to remain intact and functional due to chemical stability. The PROTAC molecule may lose its efficacy or activity if the linker is prone to breaking down. This could happen through cleavage or other chemical changes (Troup et al. 2020). To preserve the integrity and activity of the molecule and improve its pharmacokinetic profile, a chemically stable linker can withstand enzymatic degradation and metabolic transformations.

The capacity of the linker to remain stable allows it to serve as a bridge between the E3 ligase and the ligands that target the target protein, which in turn promotes the formation of the ternary complex and the eventual destruction of the protein (Wang et al. 2020). The PROTAC molecule's chemical stability guarantees that it will continue to be active over time, enabling therapeutic effects and prolonged protein degradation. The PROTAC molecule's chemical stability guarantees that it will continue to be active over time, enabling therapeutic effects and prolonged protein degradation.

3.4.2.3 Cleavable

Since many PROTAC linkers are made to be cleavable under certain circumstances, such as enzymatic cleavage by cell-resident proteases, cleavability is an essential feature of these linkers. After the ternary complex forms, cleavable linkers allow the release of the ligands that are aimed at the target protein and the E3 ligase (Sincere et al. 2023). One important way that PROTAC molecules work is by causing the target protein to degrade as a result of this release. Linkers' capacity to cleave under certain circumstances enables targeted and regulated protein degradation, improving the effectiveness and selectivity of PROTACs.

For the ligands to be released into cells at the proper time and place to cause target protein degradation, cleavable linkers in PROTACs must be designed carefully. PROTACs can effectively and selectively break down disease-causing proteins using the ubiquitin-proteasome system, the cell's natural protein degradation system, and cleavable linkers. Because it doesn't require a long time to occupy the target and allows for the degradation of several target proteins with a small number of PROTAC molecules, this method has benefits over conventional small molecule inhibitors (Troup et al. 2020).

3.4.3 Types of Linkers

Alkyl, PEG, and Extended Glycol Chains are the most prevalent linker motifs found in published degrader structures. These linkers have a number of important benefits that account for their popularity in the literature. They can be readily synthesised, are flexible, and can have their length and composition readily adjusted using a variety of reliable chemical techniques. Alkyl linkers can also provide quickly obtained preliminary SAR information regarding the ideal linker length (He et al. 2022). However, because of their high hydrophobicity, alkyl linker-based PROTACs may have trouble piercing the cell membrane. It is also possible to adjust significant physical characteristics like lipophilicity and topological polar surface area (TPSA) by utilising various PEG and alkyl motif combinations. These in turn have consequences for characteristics that influence oral absorption, such as solubility and cell permeability. In a study conducted by Liu et al., two PROTACs (compounds **26** and **27** [CG428], Fig. 3.9) that target tropomyosin receptor kinase (TRK) were synthesised (Chen et al. 2020). Despite the identical linker lengths in both PROTACS, compound **26**, which has an alkyl linker, exhibited a lower degradation rate of TRK (18% at 0.1 nM) in comparison to compound **27**, which has an oxygen atom integrated into the linker (46% at 0.1 nM). Therefore, the efficacy of the PROTAC can also be significantly influenced by the atomic composition of the linker. It has been found that degraders generated by conjugating CRBN and VHL ligands with an alkyl linker were able to reduce the amount of CRBN in HEK293T cells in a concentration-dependent manner (Kim et al. 2019). The substitution of three PEG units for a nine-atom alkyl chain resulted in only minor CRBN degradation, indicating that the addition of oxygen rather than CH_2 groups was hindering PROTAC activity, albeit the underlying mechanism was not explored. In the TC crystal structure of MZ1 (**28**) previously published, the ether oxygen atom adjacent to the amide bond of JQ1 creates a hydrogen bond contact with a BD2-specific histidine (His437) (Gadd et al. 2017). This interaction would likely vanish if the linker's PEG makeup were altered to an alkyl composition. All in all, the optimisation of linker structure frequently necessitates extensive empirical trial and error due to the complex impacts of linker composition and length on degradation efficiency.

Alkyne Chain or Aromatic Linkers (Rigid Linkers) in PROTAC design provides rigidness and occasionally permits particular interactions with target molecules. Benzene, phenyl, and naphthalene-based compounds are a few examples of aromatic linkers. Moreover, by altering the molecule's spatial orientation and binding affinity, adding aromatic rings to the linker design can improve PROTACs' overall efficacy and selectivity (Battiti et al. 2021). Furthermore, stabilisation of the ternary complex can be achieved by adding aromatic rings or alkyne chains to increase the number of π–π stacking interactions. Furthermore, rigid linkers provide conformational restriction, which may lead to enhanced physicochemical features and an increase in degradation efficacy.

Also, In PROTACs, the **triazole** motif is a frequently utilised linker that is simple to synthesis using click chemistry (Jiang et al. 2019). For example, copper-catalysed azide-alkyne cycloaddition (CuAAC) and strain-promoted azide-alkyne

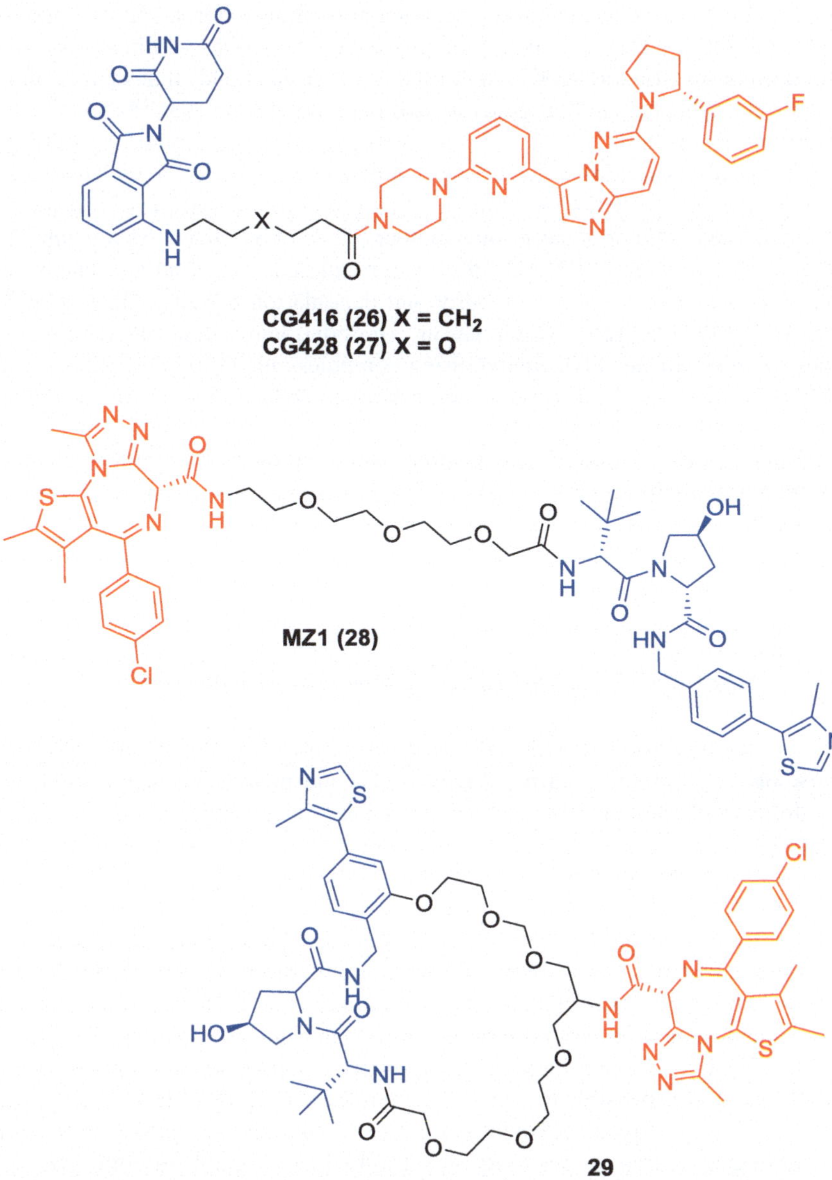

CG416 (26) X = CH_2
CG428 (27) X = O

MZ1 (28)

29

Fig. 3.9 Chemical structures of PROTACs containing flexible linker (**26**, **27**, **28**) and macrocyclic linker (**29**) [red colour indicates ligand responsible for binding POI, blue represents E3 ligase ligand and black indicates linker]

cycloaddition (SPAAC) give substituted triazole (Battiti et al. 2021). By precisely and consistently joining the various parts of PROTAC molecules, these click chemistry-based linkers improve the overall efficacy and selectivity of the molecules

(Gao et al. 2019). It is because when nitrogen-containing heterocycles like triazole are added to linkers, they may help to improve the interactions within ternary complexes because the triazole motif is a bioisostere of amide bonds that share a similar spatial orientation and can thus generate hydrogen bonds (He et al. 2022).

Macrocylic Linkers have also been investigated for the design of PROTACs. For example, after examining the ternary complex of compound **28**, which included a VHL protein and a BRD4-targeting, Ciulli's group found that the two ligands (a VHL ligand and JQ1) are near to one another (Testa et al. 2020). As a result, they suggested that locking PROTAC **28**'s conformation in the binding state would increase the ternary complex's stability and degradation activity. They created a macrocyclic PROTAC (**29**), which has an extra bond between the PEG linker and the phenol group of the VHL ligand, using computer-aided drug design. However, the PROTAC **29** showed decreased in comparison to **28** but it led to better selectivity towards BRD4-BD2.

Photoswitchable Linkers and **Linkers Formed by Self-assembly through Bioorthogonal Reactions** have also been explored to improve bioavailability, cell permeability, selectivity towards protein of interest and reduce its action towards non-target regions and thus results in reduced toxic effects (Lebraud et al. 2016; Liu et al. 2021; Pfaff et al. 2019).

3.5 Selection of Ligand Binding Protein of Interest

It is theoretically possible to build a PROTAC using any of the POI ligands. However, because of their optimal pharmacokinetic (PK) and pharmacodynamic (PD) features, drugs available in market and being used in clinical settings, drug candidates undergoing clinical trials, or highly active inhibitors are always advantageous for PROTAC design (Smith et al. 2019). Rational PROTAC design is aided by knowledge of ligand candidates' binding mechanisms (James et al. 2020). In particular, the resolution of a ligand-POI complex's crystal structure provides structural information that significantly boosts the effectiveness of PROTAC design (Gadd et al. 2017). Additionally, when ligands reversibly bind to the POI, in most cases, portions of them stretch out of the binding pocket and into the solvent. As a result, linker introduction and modification are appropriate at the solvent-exposed area of the POI ligand. It is possible to create irreversible or reversible PROTACs by altering irreversible POI ligands (Sutanto et al. 2020). Irreversible inhibitors have the potential to exhibit greater selectivity and affinity than reversible inhibitors because they can form time-dependent covalent bonds with certain amino acids.

3.6 Considerations to Improve ADME of PROTACs

PROTACs are large molecular weight heterobifunctional compounds whose physicochemical characteristics typically defy the rule of five (Cantrill et al. 2020). Thus, in order to direct structure optimisation, it is crucial to comprehend the molecular characteristics needed for a medicine like PROTAC. According to Maple's investigation, PROTACs have distinct features that set them apart from conventional small-molecule medications, making them a special kind of drug. According to a recently conducted retrospective analysis, orally active PROTACs may have estimated partition coefficient (clogP) values between 5 and 7. To achieve this, the topological polar surface area (TPSA) need not be greater than 250 Å and the number of hydrogen bond donors (HBDs) should be less than 5 (Donovan et al. 2020; Pike et al. 2020). Notably, the small quantity and size of accessible datasets have prevented a comprehensive validation of these empirical norms. Additionally, it is important to evaluate the conformation of big, flexible PROTACs in both polar and nonpolar settings. For different conformers, molecular descriptors can be calculated, such as 3D polar surface area (3D-PSA), chameleonicity indexes, number of intramolecular HBs (IMHBs), and radius of gyration (Rgyr). These measurements are helpful for forecasting cell permeability and directing the property-based design of PROTACs (Cantrill et al. 2020). Additionally, permeability and solubility are important characteristics that affect PROTACs' oral absorption and bioavailability; these aspects must to be taken into account early in the design process. Nevertheless, traditional cell-based permeability tests, such the Caco-2 cell permeability assay and the parallel artificial membrane permeability assay (PAMPA), are inappropriate for determining the permeability of PROTACs. To make these assays more appropriate for assessing degraders, they ought to be improved and adjusted. Currently, in vivo animal studies are necessary for accurate ADME or PK parameter measurements, but they are costly and time-consuming.

3.7 Conclusion

As a viable approach to tackle difficult POI targets that have shown resistance to traditional occupancy-based or tailored covalent inhibitors, PROTAC technology has gained recognition. The recognised methods for drug development are not being followed by PROTACs, and there is still much to learn about the kinetics, mechanism of action, ternary complex formation, catalytic mode of action, and notable deviations from Lipinski's rule of five of these compounds. At present, the biggest obstacle in the drug development and molecular design of PROTACs is the effective identification of drug-like candidates. The PROTAC discovery approach is still primarily empirical and requires significant, costly, and time-consuming work in synthetic and medicinal chemistry. The development of effective design techniques, the growth of novel E3 ligase ligands, deeper comprehension of the UPS and intrinsic properties of drug-like PROTACs, and a shorter discovery phase are all expected to boost PROTAC development success rates. Beyond structural biology, problems related to the pharmacokinetic features of bioactive PROTACs—specifically, cell permeability, metabolic stability, and solubility—often hamper the development of these compounds. Consequently, it is challenging to forecast a PROTAC's cell activity and offers a lot of avenues for further research for pharmacologists and medicinal chemists. In summary, advancements in understanding the mechanism of

PROTACs, figuring out ternary complex structures, finding new E3 ligases, creating effective PROTAC design tools, and fusing multidisciplinary technologies will significantly broaden the range of targets that can be drugged and degradable, advancing the development of PROTACs for actual clinical applications.

References

Anifowose A, Agbowuro AA, Yang X, Wang B (2020) Anticancer strategies by upregulating P53 through inhibition of its ubiquitination by MDM2. Med Chem Res 29:1105–1121

Bai N, Miller SA, Andrianov GV, Yates M, Kirubakaran P, Karanicolas J (2021) Rationalizing PROTAC-mediated ternary complex formation using Rosetta. J Chem Inf Model 61(3):1368–1382

Battiti FO, Zaidi SA, Katritch V, Newman AH, Bonifazi A (2021) Chiral cyclic aliphatic linkers as building blocks for selective dopamine D2 or D3 receptor agonists. J Med Chem 64(21):16088–16105

Bondeson DP, Smith BE, Burslem GM, Buhimschi AD, Hines J, Jaime-Figueroa S, Wang J, Hamman BD, Ishchenko A, Crews CM (2018) Lessons in PROTAC design from selective degradation with a promiscuous warhead. Cell Chem Biol 25(1):78–87.e5. https://doi.org/10.1016/j.chembiol.2017.09.010

Bricelj A, Steinebach C, Kuchta R, Gütschow M, Sosič I (2021) E3 ligase ligands in successful PROTACs: an overview of syntheses and linker attachment points. Front Chem 9:707317

Buckley DL, Gustafson JL, Van Molle I, Roth AG, Tae HS, Gareiss PC, Jorgensen WL, Ciulli A, Crews CM (2012a) Small-molecule inhibitors of the interaction between the E3 ligase VHL and HIF1α. Angew Chem Int Ed 51(46):11463–11467. https://doi.org/10.1002/anie.201206231

Buckley DL, Van Molle I, Gareiss PC, Tae HS, Michel J, Noblin DJ, Jorgensen WL, Ciulli A, Crews CM (2012b) Targeting the von Hippel–Lindau E3 ubiquitin ligase using small molecules to disrupt the VHL/HIF-1α interaction. J Am Chem Soc 134(10):4465–4468. https://doi.org/10.1021/ja209924v

Cantrill C, Chaturvedi P, Rynn C, Schaffland JP, Walter I, Wittwer MB (2020) Fundamental aspects of DMPK optimization of targeted protein degraders. Drug Discov Today 25(6):969–982

Che X, Yang D, Zong H, Wang J, Li X, Chen F, Chen X, Song X (2012) Nuclear CIAP1 overexpression is a tumor stage- and grade-independent predictor of poor prognosis in human bladder cancer patients. Urol Oncol 30(4):450–456. https://doi.org/10.1016/j.urolonc.2010.12.016

Chen X, Zaro JL, Shen W-C (2013) Fusion protein linkers: property, design and functionality. Adv Drug Deliv Rev 65(10):1357–1369

Chen L, Chen Y, Zhang C, Jiao B, Liang S, Tan Q, Chai H, Weihua Y, Qian Y, Yang H (2020) Discovery of first-in-class potent and selective tropomyosin receptor kinase degraders. J Med Chem 63(23):14562–14575

Chen Q, Liu C, Wang W, Meng X, Cheng X, Li X, Cai L, Linfu Luo X, He, and Huan Qu. (2023) Optimization of PROTAC ternary complex using DNA encoded library approach. ACS Chem Biol 18(1):25–33

Demizu Y, Shibata N, Hattori T, Ohoka N, Motoi H, Misawa T, Shoda T, Naito M, Kurihara M (2016) Development of BCR-ABL degradation inducers via the conjugation of an Imatinib derivative and a CIAP1 ligand. Bioorg Med Chem Lett 26(20):4865–4869

Deveraux QL, Reed JC (1999) IAP family proteins—suppressors of apoptosis. Genes Dev 13(3):239–252. https://doi.org/10.1101/gad.13.3.239

Donovan KA, Ferguson FM, Bushman JW, Eleuteri NA, Bhunia D, Ryu SS, Tan L, Shi K, Yue H, Liu X (2020) Mapping the degradable Kinome provides a resource for expedited degrader development. Cell 183(6):1714–1731

Gadd MS, Testa A, Lucas X, Chan K-H, Chen W, Lamont DJ, Zengerle M, Ciulli A (2017) Structural basis of PROTAC cooperative recognition for selective protein degradation. Nat Chem Biol 13(5):514–521

Galdeano C, Gadd MS, Soares P, Scaffidi S, Van Molle I, Birced I, Hewitt S, Dias DM, Ciulli A (2014) Structure-guided design and optimization of small molecules targeting the protein–protein interaction between the von Hippel–Lindau (VHL) E3 ubiquitin ligase and the hypoxia inducible factor (HIF) alpha subunit with in vitro nanomolar affinities. J Med Chem 57(20):8657–8663. https://doi.org/10.1021/jm5011258

Gao H, Yue W, Sun Y, Yang Y, Zhou G, Rao Y (2019) Design, synthesis, and evaluation of highly potent FAK-targeting PROTACs. ACS Med Chem Lett 11(10):1855–1862

Han X, Wei W, Sun Y (2022) PROTAC degraders with ligands recruiting MDM2 E3 ubiquitin ligase: an updated perspective. Acta Materia Medica 1(2):244

He S, Dong G, Cheng J, Ying W, Sheng C (2022) Strategies for designing proteolysis targeting chimaeras (PROTACs). Med Res Rev 42(3):1280–1342. https://doi.org/10.1002/med.21877

Hines J, Lartigue S, Dong H, Qian Y, Crews CM (2019) MDM2-recruiting PROTAC offers superior, synergistic Antiproliferative activity via simultaneous degradation of BRD4 and stabilization of P53. Cancer Res 79(1):251–262

Hopkins AL, Groom CR (2002) The druggable genome. Nat Rev Drug Discov 1(9):727–730. https://doi.org/10.1038/nrd892

Hughes SJ, Ciulli A (2017) Molecular recognition of ternary complexes: a new dimension in the structure-guided design of chemical degraders. Essays Biochem 61(5):505–516

Hughes GR, Dudey AP, Hemmings AM, Chantry A (2021) Frontiers in PROTACs. Drug Discov Today 26(10):2377–2383

Itoh Y, Ishikawa M, Naito M, Hashimoto Y (2010) Protein knockdown using methyl Bestatin–ligand hybrid molecules: design and synthesis of inducers of ubiquitination-mediated degradation of cellular retinoic acid-binding proteins. J Am Chem Soc 132(16):5820–5826

Itoh Y, Ishikawa M, Kitaguchi R, Sato S, Naito M, Hashimoto Y (2011) Development of target protein-selective degradation inducer for protein knockdown. Bioorg Med Chem 19(10):3229–3241

James KS, John WR, MichaeláCowley S (2020) PROTAC-mediated degradation of class I histone deacetylase enzymes in corepressor complexes. Chem Commun 56(32):4476–4479

Jiang X, Hao X, Jing L, Gaochan W, Kang D, Liu X, Zhan P (2019) Recent applications of click chemistry in drug discovery. Expert Opin Drug Discov 14(8):779–789

Kang CH, Lee DH, Lee CO, Du Ha J, Park CH, Hwang JY (2018) Induced protein degradation of anaplastic lymphoma kinase (ALK) by proteolysis targeting chimera (PROTAC). Biochem Biophys Res Commun 505(2):542–547. https://doi.org/10.1016/j.bbrc.2018.09.169

Kim K, Lee DH, Park S, Jo S-H, Bonsu K, Park SG, Park BC, Jeon YU, Ahn S, Kang CH (2019) Disordered region of Cereblon is required for efficient degradation by proteolysis-targeting chimera. Sci Rep 9(1):19654

Konstantinidou M, Li J, Zhang B, Wang Z, Shaabani S, Ter Brake F, Essa K, Dömling A (2019) PROTACs–a game-changing technology. Expert Opin Drug Discov 14(12):1255–1268

Krönke J, Udeshi ND, Narla A, Grauman P, Hurst SN, McConkey M, Svinkina T, Heckl D, Comer E, Li X, Ciarlo C, Hartman E, Munshi N, Schenone M, Schreiber SL, Carr SA, Ebert BL (2014) Lenalidomide causes selective degradation of IKZF1 and IKZF3 in multiple myeloma cells. Science 343(6168):301–305. https://doi.org/10.1126/science.1244851

Lebraud H, Wright DJ, Johnson CN, Heightman TD (2016) Protein degradation by in-cell self-assembly of proteolysis targeting chimeras. ACS Cent Sci 2(12):927–934

Li B, Ran T, Chen H (2023) 3D based generative PROTAC linker design with reinforcement learning. Brief Bioinform 24(5):bbad323

Lin X, Xiang H, Luo G (2020) Targeting estrogen receptor α for degradation with PROTACs: a promising approach to overcome endocrine resistance. Eur J Med Chem 206:112689

Liu J, Ma J, Liu Y, Xia J, Yuyun Li Z, Wang P, Wei W (2020) PROTACs: a novel strategy for cancer therapy. In: Seminars in cancer biology, vol 67. Elsevier, pp 171–179

Liu J, He Chen H, Kaniskan U, Xie L, Chen X, Jin J, Wei W (2021) TF-PROTACs enable targeted degradation of transcription factors. J Am Chem Soc 143(23):8902–8910

Lu G, Middleton RE, Sun H, Naniong MV, Ott CJ, Mitsiades CS, Wong K-K, Bradner JE, Kaelin WG (2014) The myeloma drug Lenalidomide promotes the Cereblon-dependent destruction of Ikaros proteins. Science 343(6168):305–309. https://doi.org/10.1126/science.1244917

Lu J, Qian Y, Altieri M, Dong H, Wang J, Raina K, Hines J, Winkler JD, Crew AP, Coleman K (2015) Hijacking the E3 ubiquitin ligase cereblon to efficiently target BRD4. Chem Biol 22(6):755–763

Miller SA, Andrianov GV, Mischley V, Wharton KA, Chen JJ, Karanicolas J (2023) Computational modeling of PROTAC ternary complexes and linker design. Inducing Targeted Protein Degradation: From Chemical Biology to Drug Discovery and Clinical Applications 151–76

Nalawansha DA, Crews CM (2020) PROTACs: an emerging therapeutic modality in precision medicine. Cell Chem Biol 27(8):998–1014

Neklesa TK, Winkler JD, Crews CM (2017) Targeted protein degradation by PROTACs. Pharmacol Ther 174:138–144

Neklesa T, Snyder LB, Willard RR, Vitale N, Pizzano J, Gordon DA, Bookbinder M, Macaluso J, Dong H, Ferraro C (2019) ARV-110: an oral androgen receptor PROTAC degrader for prostate cancer. J Clin Oncol 37(259):10–1200

Ohoka N, Nagai K, Hattori T, Okuhira K, Shibata N, Cho N, Naito M (2014) Cancer cell death induced by novel small molecules degrading the TACC3 protein via the ubiquitin–proteasome pathway. Cell Death Dis 5(11):e1513–e1513

Ohoka N, Okuhira K, Ito M, Nagai K, Shibata N, Hattori T, Ujikawa O, Shimokawa K, Sano O, Koyama R (2017) In vivo knockdown of pathogenic proteins via specific and nongenetic inhibitor of apoptosis protein (IAP)-dependent protein erasers (SNIPERs). J Biol Chem 292(11):4556–4570

Ohoka N, Morita Y, Nagai K, Shimokawa K, Ujikawa O, Fujimori I, Ito M, Hayase Y, Okuhira K, Shibata N (2018) Derivatization of inhibitor of apoptosis protein (IAP) ligands yields improved inducers of estrogen receptor α degradation. J Biol Chem 293(18):6776–6790

Park D, Izaguirre J, Coffey R, Huafeng X (2022) Modeling the effect of cooperativity in ternary complex formation and targeted protein degradation mediated by heterobifunctional degraders. ACS Bio Med Chem Au 3(1):74–86

Paul A, Anand R, Karmakar SP, Rawat S, Bairagi N, Chatterjee S (2021) Exploring gene knockout strategies to identify potential drug targets using genome-scale metabolic models. Sci Rep 11(1):213

Pfaff P, Samarasinghe KTG, Crews CM, Carreira EM (2019) Reversible spatiotemporal control of induced protein degradation by Bistable PhotoPROTACs. ACS Cent Sci 5(10):1682–1690

Pike A, Williamson B, Harlfinger S, Martin S, McGinnity DF (2020) Optimising proteolysis-targeting chimeras (PROTACs) for oral drug delivery: a drug metabolism and pharmacokinetics perspective. Drug Discov Today 25(10):1793–1800

Poongavanam V, Atilaw Y, Siegel S, Giese A, Lehmann L, Meibom D, Erdelyi M, Kihlberg J (2022) Linker-dependent folding rationalizes PROTAC cell permeability. J Med Chem 65(19):13029–13040

Popowicz GM, Dömling A, Holak TA (2011) The structure-based design of MDM2/MDMX–P53 inhibitors gets serious. Angew Chem Int Ed 50(12):2680–2688

Qin C, Yang H, Zhou B, Fernandez-Salas E, Yang C-Y, Liu L, McEachern D, Przybranowski S, Wang M, Stuckey J (2018) Discovery of QCA570 as an exceptionally potent and efficacious proteolysis targeting chimera (PROTAC) degrader of the bromodomain and extra-terminal (BET) proteins capable of inducing complete and durable tumor regression. J Med Chem 61(15):6685–6704

Sakamoto KM, Kim KB, Kumagai A, Mercurio F, Crews CM, Deshaies RJ (2001) Protacs: chimeric molecules that target proteins to the Skp1–Cullin–F box complex for ubiquitination and degradation. Proc Natl Acad Sci 98(15):8554–8559. https://doi.org/10.1073/pnas.141230798

Schneekloth Jr, John S, Fonseca FN, Koldobskiy M, Mandal A, Deshaies R, Sakamoto K, Crews CM (2004) Chemical genetic control of protein levels: selective in vivo targeted degradation. J Am Chem Soc 126(12):3748–3754

Schneekloth AR, Pucheault M, Tae HS, Crews CM (2008) Targeted intracellular protein degradation induced by a small molecule: en route to chemical proteomics. Bioorg Med Chem Lett 18(22):5904–5908

Setia N, Almuqdadi HTA, Abid M (2023) Journey of von Hippel-Lindau (VHL) E3 ligase in PROTACs design: from VHL ligands to VHL-based degraders. Eur J Med Chem 265(December 2023):116041. https://doi.org/10.1016/j.ejmech.2023.116041

Shan Y, Si R, Wang J, Zhang Q, Li J, Ma Y, Zhang J (2020) Discovery of novel anti-angiogenesis agents. Part 11: development of PROTACs based on active molecules with potency of promoting vascular normalization. Eur J Med Chem 205:112654. https://doi.org/10.1016/j.ejmech.2020.112654

Sincere NI, Anand K, Ashique S, Yang J, You C (2023) PROTACs: emerging targeted protein degradation approaches for advanced druggable strategies. Molecules 28:10

Smith BE, Wang SL, Jaime-Figueroa S, Harbin A, Wang J, Hamman BD, Crews CM (2019) Differential PROTAC substrate specificity dictated by orientation of recruited E3 ligase. Nat Commun 10(1):131

Soares P, Gadd MS, Frost J, Galdeano C, Ellis L, Epemolu O, Rocha S, Read KD, Ciulli A (2018) Group-based optimization of potent and cell-active inhibitors of the von Hippel–Lindau (VHL) E3 ubiquitin ligase: structure–activity relationships leading to the chemical probe (2S,4R)-1-((S)-2-(1-cyanocyclopropanecarboxamido)-3,3-dimethylbutanoyl)-4-Hydr. J Med Chem 61(2):599–618. https://doi.org/10.1021/acs.jmedchem.7b00675

Su D, Zhang D (2021) Linker design impacts antibody-drug conjugate pharmacokinetics and efficacy via modulating the stability and payload release efficiency. Front Pharmacol 12:687926

Sun X, Gao H, Yang Y, He M, Yue W, Song Y, Tong Y, Rao Y (2019) PROTACs: great opportunities for academia and industry. Signal Transduct Target Ther 4(1):64

Sutanto F, Konstantinidou M, Dömling A (2020) Covalent inhibitors: a rational approach to drug discovery. RSC Med Chem 11(8):876–884

Testa A, Hughes SJ, Lucas X, Wright JE, Ciulli A (2020) Structure-based design of a macrocyclic PROTAC. Angew Chem 132(4):1744–1751

Tokatlian T, Segura T (2010) SiRNA applications in nanomedicine. Wiley Interdiscip Rev Nanomed Nanobiotechnol 2(3):305–315

Toure M, Crews CM (2016) Small-molecule PROTACS: new approaches to protein degradation. Angew Chem Int Ed 55(6):1966–1973. https://doi.org/10.1002/anie.201507978

Troup RI, Fallan C, Baud MGJ (2020) Current strategies for the design of PROTAC linkers: a critical review. Explor Target Antitumor Ther 1(5):273

Vassilev LT, Vu BT, Graves B, Carvajal D, Podlaski F, Filipovic Z, Kong N, Kammlott U, Lukacs C, Klein C (2004) In vivo activation of the P53 pathway by small-molecule antagonists of MDM2. Science 303(5659):844–848

Vu LP, Diehl CJ, Casement R, Bond AG, Steinebach C, Strašek N, Bricelj A, Perdih A, Schnakenburg G, Sosič I, Ciulli A, Gütschow M (2023) Expanding the structural diversity at the phenylene core of ligands for the von Hippel–Lindau E3 ubiquitin ligase: development of highly potent hypoxia-inducible factor-1α stabilizers. J Med Chem 66(18):12776–12811. https://doi.org/10.1021/acs.jmedchem.3c00434

Wang J, Chen Q, Shan Y, Pan X, Zhang J (2020) Activity-based proteomic profiling: application of releasable linker in Photoaffinity probes. Drug Discov Today 25(1):133–140. https://doi.org/10.1016/j.drudis.2019.10.016

Wang C, Zhang Y, Wang J, Xing D (2022) VHL-based PROTACs as potential therapeutic agents: recent progress and perspectives. Eur J Med Chem 227:113906. https://doi.org/10.1016/j.ejmech.2021.113906

Wang J, Liu M, Zhang X, Wang X, Xiong M, Luo D (2024) "Stimuli-responsive linkers and their application in molecular imaging." P. 20230027 in exploration. Wiley Online Library

Ward JA, Perez-Lopez C, Mayor-Ruiz C (2023) Biophysical and computational approaches to study ternary complexes: a 'cooperative Relationship'to rationalize targeted protein degradation. Chembiochem 24(10):e202300163

Wijaya AJ, Farnaby W, Ciulli A (2023) Crystallization of VHL-based PROTAC-induced ternary complexes. Targeted Protein Degradation 241

Wurz RP, Rui H, Dellamaggiore K, Ghimire-Rijal S, Choi K, Smither K, Amegadzie A, Chen N, Li X, Banerjee A (2023) Affinity and cooperativity modulate ternary complex formation to drive targeted protein degradation. Nat Commun 14(1):4177

Yang Q, Zhao J, Chen D, Wang Y (2021) E3 ubiquitin ligases: styles, structures and functions. Mol Biomed 2(1):23. https://doi.org/10.1186/s43556-021-00043-2

Zagidullin A, Milyukov V, Rizvanov A, Bulatov E (2020) Novel approaches for the rational design of PROTAC linkers. Explor Target Antitumor Ther 1(5):381

Zhang X, Fang X, Tong L, Zhang T, Xie H, Xiaoyun L, Ren X, Ding K (2020) Design and synthesis of selective degraders of EGFRL858R/T790M mutant. Eur J Med Chem 192:112199

Zhao B, Burgess K (2019) TrkC-targeted kinase inhibitors and PROTACs. Mol Pharm 16(10):4313–4318. https://doi.org/10.1021/acs.molpharmaceut.9b00673

Zhao Q, Lan T, Shang S, Rao Y (2019) Induction of apoptosis in MDA-MB-231 breast cancer cells by a PARP1-targeting PROTAC small molecule. Chem Commun 55(3):369–372

Zografou-Barredo NA, Hallatt AJ, Goujon-Ricci J, Cano C (2023) A Beginner's guide to current synthetic linker strategies towards VHL-recruiting PROTACs. Bioorg Med Chem 88–89:117334. https://doi.org/10.1016/j.bmc.2023.117334

Advantages and Disadvantages of PROTACs

4

Rohini Pujari, Shvetank Bhatt, Urvashi Soni, Shreya Sharma, and Shubham Patil

Abstract

The present paper discusses the significant impact of Proteolysis Targeting Chimeras (PROTACs) in targeted protein degradation, presenting a paradigm shift from traditional drug inhibition to protein elimination. PROTACs offer a versatile approach to degrade "undruggable" proteins, expanding the range of potential therapeutic targets. By exploiting the ubiquitin-proteasome system, PROTACs induce the degradation of specific proteins, overcoming limitations associated with protein structure and function. The catalytic degradation function of PROTACs enables efficient and sustained protein elimination, enhancing therapeutic outcomes. Additionally, PROTACs can be combined with other modalities to synergistically target challenging proteins, such as oncogenes and proteins involved in complex interaction networks. While PROTACs hold great promise in improving treatment outcomes, ongoing research and innovation are essential to address challenges like intracellular delivery. Overall, PROTACs represent a groundbreaking approach to drug discovery, offering enhanced selectivity, specificity, and therapeutic potential in various diseases characterized by dysregulated protein expression.

Keywords

PROTACs · Undruggable · Targeted protein · Catalytic degradation · Hook effect

R. Pujari (✉) · S. Bhatt · U. Soni · S. Sharma · S. Patil
School of Health Sciences and Technology, Dr. Vishwanath Karad MIT World Peace University, Kothrud, Pune, Maharashtra, India

M. Nandave, P. Jain (eds.), *PROTAC-Mediated Protein Degradation: A Paradigm Shift in Cancer Therapeutics*, https://doi.org/10.1007/978-981-97-5077-1_4

Abbreviations

AR	Androgen receptor
BCL-2	B-Cell lymphoma-2
BRCA	Breast Cancer gene
BTK	Bruton's tyrosine kinase
DNA	Deoxyribonucleic acid
EGFR	Epidermal growth factor receptor
ER	Estrogen receptor
HBAs	Hydrogen bond acceptors
HBDs	Hydrogen bond donors
HER2	Human epidermal growth factor receptor 2
HPK1	Hematopoietic progenitor kinase 1
ICIs	Immune checkpoint inhibitors
KRAS	Kristen rat sarcoma viral oncogene homolog
MAPK	Mitogen-activated protein kinase
NSCLC	Non-small cell lung cancer
PARP	Poly (ADP-ribose) polymerase
PROTACs	Proteolysis targeting chimeras
RAF	Rapidly accelerated fibrosarcoma
STAT	Signal transducer and activator of transcription
TKIs	Tyrosine kinase inhibitors
VHL	von Hippel-Lindau

4.1 Introduction

PROTACs, or Proteolysis Targeting Chimeras, are a class of small molecules designed for targeted protein degradation. Unlike traditional drugs that inhibit the activity of specific proteins, PROTACs induce the degradation of the target proteins. This innovative approach provides a unique way to address diseases associated with the dysregulation of specific proteins (Liu et al. 2022). The basic structure of a PROTAC consists of three main components including a target protein-ligand that specifically binds to the target protein, a linker that connects the target protein-ligand to the E3 ubiquitin ligase ligand that recruits an E3 ubiquitin ligase enzyme (Cohen and Tcherpakov 2010). The overall mechanism of action involves the PROTAC molecule bringing the target protein and the E3 ubiquitin ligase into proximity (Zheng and Shabek 2017). This proximity allows the E3 ligase to ubiquitinate the target protein. Once ubiquitinated, the target protein is recognized and degraded by the proteasome, a cellular structure responsible for breaking down proteins (Bricelj et al. 2021).

The advantages of PROTACs include the potential to target a broader range of proteins, including those previously considered undruggable, and the ability to achieve a sustained and catalytic effect on protein degradation. This approach has garnered significant attention in drug development, particularly in areas such as oncology, where precise control over specific protein levels is crucial for therapeutic efficacy (Burslem et al. 2018b).

While PROTACs hold great promise, challenges remain, including achieving high specificity, optimizing the design for efficient cellular uptake, and addressing potential off-target effects. Research in the field of PROTACs is ongoing, and advancements continue to refine and expand the applications of this innovative technology in drug discovery and development (Gao et al. 2020).

4.2 Advantages of PROTRACS

4.2.1 Overcoming Drug Resistance to Cancer

Proteolysis Targeting Chimeras (PROTACs) hold significant promise in overcoming drug resistance in cancer, a challenge that often limits the efficacy of traditional cancer therapies. The unique mechanism of action of PROTACs, which induces targeted protein degradation, offers several advantages in addressing drug resistance mechanisms. Many drug resistance mechanisms in cancer involve the upregulation or activation of specific proteins that are difficult to target using traditional small molecule inhibitors. PROTACs offer a strategy to degrade these undruggable proteins, bypassing resistance mechanisms and restoring sensitivity to therapy (Buhimschi et al. 2018). Point mutations in drug targets can render them resistant to inhibition by small-molecule drugs. By inducing targeted protein degradation, PROTACs can circumvent the need for binding to specific sites on the target protein, making them less susceptible to resistance mediated by point mutations (Wispelaere et al. 2019). Drug resistance in cancer often involves the overexpression or hyperactivation of target proteins. By reducing the levels of these proteins through degradation, PROTACs can overcome resistance and restore sensitivity to therapy (Kim et al. 2022). PROTACs can be designed to target multiple proteins involved in drug resistance pathways simultaneously. By inducing the degradation of multiple proteins, PROTACs may exert synergistic effects, effectively overcoming resistance mechanisms that involve the dysregulation of multiple targets. PROTACs hold significant promise in overcoming drug resistance in cancer by offering a novel mechanism to degrade specific proteins implicated in resistance pathways (Burke et al. 2022).

PROTACs can address drug resistance in cancer in the following ways:

- ***Direct Targeting of Resistant Proteins:***
  PROTACs can directly target proteins that are overexpressed or mutated in drug-resistant cancer cells, leading to their degradation and circumventing the resistance mechanism. For example, in hormone receptor-positive breast cancer, resistance to endocrine therapies often arises due to increased expression of the estrogen receptor (ER). ARV-471, a PROTAC targeting the ER, has demonstrated efficacy in preclinical models of endocrine-resistant breast cancer by inducing ER degradation, thereby overcoming resistance to standard therapies (Lin et al. 2020).

- ***Induction of Synthetic Lethality:***
  PROTACs can induce synthetic lethality by targeting proteins that are essential for the survival of cancer cells with specific genetic alterations, thereby selectively killing drug-resistant cells. In tumors with BReast CAncer gene (BRCA) mutations, resistance to Poly (ADP-ribose) polymerase (PARP) inhibitors can occur through various mechanisms. However, targeting proteins involved in the deoxyribonucleic acid (DNA) repair pathway, such as BRCA1 or RAD51, with PROTACs could induce synthetic lethality in BRCA-mutant cells, overcoming PARP inhibitor resistance (Fu et al. 2024).
- ***Combination Therapy with Conventional Drugs:***
  PROTACs can be used in combination with conventional chemotherapy or targeted therapies to enhance their efficacy and overcome resistance mechanisms. Resistance to tyrosine kinase inhibitors (TKIs) in non-small cell lung cancer (NSCLC) often arises due to secondary mutations in the epidermal growth factor receptor (EGFR) gene. Combining TKIs with PROTACs targeting mutant EGFR variants could lead to more effective inhibition of tumor growth and overcome resistance (Marrocco and Yarden 2023).
- ***Targeting Adaptive Signaling Pathways:***
  PROTACs can target proteins involved in adaptive signaling pathways that promote resistance to therapy, thereby blocking the emergence of resistance or reversing existing resistance. Resistance to rapidly accelerated fibrosarcoma (RAF) inhibitors in BRAF-mutant melanoma often occurs through reactivation of the mitogen-activated protein kinase (MAPK) pathway via the upregulation of alternative signaling pathways. PROTACs targeting proteins upstream or downstream of RAF, such as MEK or AKT, could disrupt these adaptive signaling networks and overcome resistance (Zhong et al. 2022).
- ***Overcoming Immunotherapy Resistance:***
  PROTACs can target proteins involved in immune evasion mechanisms or tumor microenvironment modulation, thereby enhancing the efficacy of immunotherapy and overcoming resistance. Resistance to immune checkpoint inhibitors (ICIs) in melanoma can occur through various mechanisms, including upregulation of alternative immune checkpoints such as TIM-3 or LAG-3. PROTACs targeting these immune checkpoints could restore T cell-mediated cytotoxicity and overcome resistance to ICIs. Continued research and clinical development of PROTACs hold promise for improving outcomes in patients with drug-resistant cancers (Xu et al. 2020).

4.2.2 Degradation of the "Undruggable" Protein Target

Proteolysis Targeting Chimeras (PROTACs) have emerged as a promising strategy for degrading so-called "undruggable" protein targets, which are proteins that have been historically challenging to target with traditional small molecule inhibitors (Neklesa et al. 2017).

PROTACs play an important role in degrading undruggable proteins in the following ways:

- ***Expanding Target Range:***
  PROTACs can target a broader range of proteins, including those considered undruggable by conventional small molecules, such as transcription factors, scaffolding proteins, and intracellular protein-protein interaction domains. For example, BRD4, a transcriptional regulator, is involved in various cancers and is traditionally considered undruggable due to its lack of enzymatic activity. However, ARV-825, a PROTAC-targeting BRD4, has been shown to induce potent and selective degradation of BRD4, leading to anti-tumor effects in preclinical models of cancer (Wu et al. 2021).
- ***Inducing Target Protein Degradation:***
  PROTACs exploit the cell's ubiquitin-proteasome system to induce the degradation of target proteins, regardless of their draggability. This mechanism offers a new strategy to eliminate disease-associated proteins that were previously deemed undruggable. Signal transducer and activator of transcription STAT3, a transcription factor involved in oncogenesis and immune evasion, has been considered challenging to target with small molecules due to its lack of traditional binding pockets. However, PROTACs designed to recruit the E3 ligase to STAT3 have shown efficacy in degrading STAT3 and inhibiting tumor growth in preclinical cancer models (Li et al. 2023).
- ***Overcoming Binding Site Limitations:***
  PROTACs can circumvent limitations associated with targeting specific binding sites on proteins by promoting their degradation through proximity-induced ubiquitination, thus overcoming challenges posed by protein structure or function.

  Kristen rat sarcoma viral oncogene homolog (KRAS), a frequently mutated oncogene in cancer, has long been considered undruggable due to its lack of druggable binding pockets. However, recent efforts to degrade mutant KRAS using PROTACs that recruit E3 ligases to KRAS have shown promise in preclinical studies, providing a potential avenue for targeting this challenging protein (Gong et al. 2022).
- ***Addressing Protein-Protein Interactions:***
  PROTACs can disrupt protein-protein interactions by inducing the degradation of one of the interacting partners, offering a strategy to target proteins involved in complex interaction networks that were previously difficult to inhibit directly. BCL-2 family proteins, which regulate apoptosis, are challenging to target due to their intracellular interactions. However, PROTACs designed to degrade anti-apoptotic BCL-2 family members, such as Mcl-1 or Bcl-xL, have shown promise in preclinical models of cancer by disrupting the balance of pro- and anti-apoptotic proteins (Lv et al. 2021).
- ***Combination Approaches:***
  PROTACs can be used in combination with other therapeutic modalities, such as small molecule inhibitors or antibodies, to target undruggable proteins synergistically and enhance therapeutic efficacy. Human epidermal growth factor receptor 2 (HER2), a receptor tyrosine kinase overexpressed in breast cancer, can be difficult to target directly due to structural complexities. However, combining HER2-targeting antibodies with HER2-targeting PROTACs has shown synergis-

tic effects in preclinical models, leading to enhanced degradation of HER2 and improved anti-tumor activity. Thus, PROTACs offer a versatile and innovative approach to degrade undruggable protein targets, thereby expanding the scope of potential therapeutic interventions for a wide range of diseases, including cancer and other disorders characterized by dysregulated protein expression (Maneiro et al. 2020).

4.2.3 Catalytic Degradation Function

Proteolysis Targeting Chimeras (PROTACs) employ a catalytic mechanism to induce the degradation of target proteins, offering a unique advantage over traditional small molecule inhibitors. This catalytic degradation enables sustained and efficient protein elimination, leading to potent therapeutic effects (Bondeson et al. 2015). PROTACs function by recruiting an E3 ubiquitin ligase to the target protein, forming a ternary complex. This proximity facilitates the transfer of ubiquitin molecules from the E2 ubiquitin-conjugating enzyme to the target protein, leading to polyubiquitination and subsequent recognition by the proteasome for degradation. Importantly, PROTACs are not consumed in this process and can catalytically induce the degradation of multiple target protein molecules (Xiao et al. 2022). Unlike traditional small molecule inhibitors that transiently bind to their targets, PROTACs continuously recruit E3 ligases to induce the degradation of target proteins. This catalytic process allows for sustained and efficient elimination of the target protein, potentially leading to more durable therapeutic responses. The catalytic nature of PROTACs results in enhanced potency compared to traditional inhibitors, as lower concentrations of PROTACs are required to achieve effective target degradation. This increased potency can lead to improved therapeutic efficacy and reduced risk of drug resistance (Zhao et al. 2022).

Following are a few examples of catalytic degradation by PROTACs:

- ***ARV-110:*** ARV-110 is a PROTAC targeting the androgen receptor (AR) for degradation in prostate cancer cells. It has demonstrated catalytic degradation of AR, leading to tumor regression in preclinical models (Jia and Han 2023).
- ***ARV-471:*** ARV-471 is a PROTAC targeting the estrogen receptor (ER) for degradation in hormone receptor-positive breast cancer. It has shown catalytic degradation of ER and potent anti-tumor activity in preclinical studies (Bondeson et al. 2015).
- ***MT-802:*** MT-802 is a PROTAC designed to degrade Bruton's tyrosine kinase (BTK) for the treatment of B-cell malignancies. It has demonstrated catalytic degradation of BTK and potent anti-tumor effects in preclinical models (Authier et al. 2013).
- ***MT-5111:*** MT-5111 is a PROTAC targeting HER2 for degradation in HER2-positive breast cancer. It has shown catalytic degradation of HER2 and significant tumor growth inhibition in preclinical studies (Burslem et al. 2018a).

- ***HPK1-PROTAC1:*** HPK1-PROTAC1 is a PROTAC designed to degrade hematopoietic progenitor kinase 1 (HPK1) for the treatment of autoimmune diseases. It has demonstrated catalytic degradation of HPK1 and suppression of autoimmune responses in preclinical models (Zorba et al. 2018).

4.2.4 Improving Selectivity and Specificity

Proteolysis Targeting Chimeras (PROTACs) offer a unique approach to improving selectivity and specificity in drug targeting by exploiting the cell's ubiquitin-proteasome system for targeted protein degradation.

PROTACs contribute to enhanced selectivity and specificity as follows:

- ***Dual Binding Specificity:***
  PROTACs consist of two ligands connected by a linker: one that binds to the target protein and another that binds to an E3 ubiquitin ligase. This dual-binding nature allows PROTACs to selectively recruit the E3 ligase to the target protein, thereby conferring specificity to the degradation process. ARV-110, a PROTAC targeting the androgen receptor (AR) for degradation, demonstrates selectivity by recruiting the E3 ligase to AR specifically, leading to AR degradation without affecting other proteins (Neklesa et al. 2019).
- ***Enhanced Target Engagement:***
  PROTACs often exhibit enhanced target engagement compared to traditional small molecule inhibitors due to their ability to induce proximity-driven ubiquitination of the target protein. This mechanism increases the specificity of target protein degradation while minimizing off-target effects. ARV-471, a PROTAC targeting the estrogen receptor (ER), demonstrates enhanced target engagement by recruiting the E3 ligase to ER, leading to ER degradation specifically in hormone receptor-positive breast cancer cells (Grinshpun 2023).
- ***Modular Design for Specificity:***
  The modular design of PROTACs allows for fine-tuning of selectivity and specificity by optimizing the binding affinities of the target protein and E3 ligase ligands. This design flexibility enables the development of PROTACs with improved selectivity profiles.

  DT2216, a PROTAC targeting B-cell lymphoma-2 (BCL-2) for degradation, exhibits high selectivity for BCL-2 over other anti-apoptotic proteins due to the specific design of the ligands and linker (He et al. 2020).
- ***Reduction of Off-target Effects:***
  By inducing degradation of the target protein via the ubiquitin-proteasome system, PROTACs can minimize off-target effects associated with traditional inhibitors that may bind nonspecifically to other proteins. ARV-110, while designed to target the androgen receptor, has demonstrated minimal off-target effects on other proteins, contributing to its favorable safety profile in preclinical studies (Neklesa et al. 2019).

- ***Disease-Specific Targeting:***
  PROTACs can be tailored to target disease-specific proteins, allowing for precise modulation of disease pathways while sparing normal tissues. This disease-specific targeting enhances therapeutic efficacy while minimizing toxicity. MT-802, a PROTAC targeting Bruton's tyrosine kinase (BTK), shows selective degradation of BTK in B-cell malignancies, leading to potent anti-tumor effects with reduced toxicity compared to non-specific BTK inhibitors (Jaime-Figueroa et al. 2020). PROTACs thus offer a promising strategy for improving selectivity and specificity in drug targeting by leveraging the ubiquitin-proteasome system for targeted protein degradation (Arthur et al. 2020).

4.2.5 Wide Range of Targets

PROTACs offer a wide target range, allowing for the degradation of a diverse array of proteins involved in various disease pathways. This versatility makes PROTACs attractive candidates for therapeutic intervention across multiple disease indications. The ubiquitin-proteasome system is the primary intracellular protein degradation process, accounting for more than 80% of all intracellular protein breakdown. E3 ubiquitin ligase is found in a wide variety of cells. The PROTAC molecule simply needs to get the target protein near to E3 ubiquitin ligase before degrading it via proteolysome. As a result, PROTAC technology has a wide range of applications (Békés et al. 2022).

Following are a few examples demonstrating the wide target range of PROTACs:

- ***Androgen Receptor (AR):***
  PROTACs targeting the androgen receptor (AR) have been developed for the treatment of prostate cancer, aiming to degrade AR and inhibit AR signaling. ARV-110 is a PROTAC that targets AR for degradation and has shown promising anti-tumor activity in preclinical models of prostate cancer (Neklesa et al. 2019).
- ***Estrogen Receptor (ER):***
  PROTACs targeting the estrogen receptor (ER) have been developed for hormone receptor-positive breast cancer, aiming to degrade ER and inhibit ER signaling. ARV-471 is a PROTAC designed to degrade ER selectively in hormone receptor-positive breast cancer cells, demonstrating potent anti-tumor activity in preclinical studies (Hamilton et al. 2022).
- ***Bruton's Tyrosine Kinase (BTK):***
  PROTACs targeting Bruton's tyrosine kinase (BTK) have been developed for the treatment of B-cell malignancies, aiming to degrade BTK and inhibit BCR signaling. MT-802 is a PROTAC that selectively degrades BTK and exhibits potent anti-tumor effects in preclinical models of B-cell malignancies (Jaime-Figueroa et al. 2020).
- ***B-cell Lymphoma-2 (BCL-2):***
  PROTACs targeting B-cell lymphoma-2 (BCL-2) have been developed for hematologic malignancies, aiming to degrade BCL-2 and induce apoptosis. DT2216

is a PROTAC designed to degrade BCL-2 selectively, showing promising anti-tumor activity with reduced toxicity in preclinical studies (Jia et al. 2021).

- ***Hematopoietic Progenitor Kinase 1 (HPK1):***
  PROTACs targeting hematopoietic progenitor kinase 1 (HPK1) have been explored for the treatment of autoimmune diseases, aiming to degrade HPK1 and modulate immune responses. HPK1-PROTAC1 is a PROTAC that induces selective degradation of HPK1 and suppresses autoimmune responses in preclinical models (Si et al. 2020).
- ***Epidermal Growth Factor Receptor (EGFR):***
  PROTACs targeting epidermal growth factor receptors (EGFR) have been developed for the treatment of EGFR-driven cancers, aiming to degrade EGFR and inhibit downstream signaling pathways. EGFR-PROTAC1 is a PROTAC that selectively degrades EGFR, showing promising anti-tumor effects in preclinical models of EGFR-mutant cancers (Hong et al. 2022).

These examples illustrate the wide target range of PROTAC, spanning diverse disease pathways and indications.

4.2.6 Prolong the Action Time

One of the significant advantages of Proteolysis Targeting Chimeras (PROTACs) is their potential to prolong the duration of drug action by inducing sustained degradation of target proteins. This sustained effect can lead to prolonged therapeutic benefits compared to traditional small molecule inhibitors.

PROTACs contribute to prolonging drug action time in the following ways:

- ***Continuous Target Protein Degradation:***
  PROTACs continuously recruit the ubiquitin-proteasome system to degrade the target protein, resulting in sustained suppression of the target's activity. Unlike traditional inhibitors, which may have transient effects, PROTACs maintain prolonged target protein degradation, leading to a sustained therapeutic response. ARV-110 is a PROTAC targeting the androgen receptor (AR) for degradation in prostate cancer. Preclinical studies have demonstrated sustained AR degradation and tumor growth inhibition over an extended period, highlighting the prolonged drug action time of ARV-110 (Jia and Han 2023).
- ***Catalytic Degradation Mechanism:***
  PROTACs operate via a catalytic mechanism, whereby a single PROTAC molecule can induce the degradation of multiple target protein molecules. This catalytic process leads to sustained and amplified effects on target protein degradation, prolonging the duration of drug action. DT2216 is a PROTAC targeting B-cell lymphoma-2 (BCL-2) for degradation in hematologic malignancies. Preclinical studies have shown that DT2216 induces sustained BCL-2 degradation, resulting in prolonged apoptosis and tumor growth inhibition (He et al. 2020).

- ***Reduced Drug Resistance:***
  By continuously degrading the target protein, PROTACs may help reduce the emergence of drug resistance, which can occur with traditional inhibitors due to target reactivation or compensatory mechanisms. Prolonged suppression of the target protein's activity can mitigate the development of resistance over time.
  MT-802 is a PROTAC targeting Bruton's tyrosine kinase (BTK) for degradation in B-cell malignancies. Preclinical studies have demonstrated sustained BTK degradation and prolonged inhibition of BCR signaling, potentially reducing the risk of drug resistance (Buhimschi et al. 2018).
- ***Longer Half-life and Dosing Intervals:***
  The sustained pharmacodynamic effect of PROTACs may allow for longer dosing intervals and extended half-lives compared to traditional inhibitors. This prolonged drug action time can improve patient compliance and reduce the frequency of dosing.

In preclinical studies, PROTACs targeting various proteins, including kinases and transcription factors, have shown prolonged pharmacodynamic effects with dosing intervals ranging from once daily to weekly, highlighting the potential for extended drug action time (Bondeson et al. 2015; Zorba et al. 2018).

4.2.7 Novel Mode of Pharmacological Action

PROTACs represent a novel mode of pharmacological action that offers several distinct advantages over traditional small molecule inhibitors. This innovative approach harnesses the cell's own protein degradation machinery to selectively remove disease-causing proteins, leading to potent therapeutic effects. PROTACs represent a novel mode of pharmacological action that offers targeted protein degradation, catalytic activity, enhanced selectivity, and flexibility in design. Continued research and clinical development of PROTACs hold promise for expanding the therapeutic arsenal against a wide range of diseases (Wang et al. 2022).

4.2.8 Potential for Synergistic Effects

PROTACs may exhibit synergistic effects by simultaneously targeting multiple proteins involved in a disease pathway, providing a more comprehensive therapeutic approach. PROTACs hold significant potential for achieving synergistic effects when used in combination with other therapeutic modalities. By leveraging the unique mechanism of targeted protein degradation, PROTACs can complement existing treatment strategies to enhance efficacy, overcome drug resistance, and broaden the therapeutic window (Hines et al. 2019).

PROTACs can achieve synergistic effects in the following ways:

- ***Combination with Small Molecule Inhibitors:***
  PROTACs can be combined with small molecule inhibitors targeting the same pathway to achieve synergistic inhibition of the target protein. While inhibitors block the activity of the target, PROTACs induce its degradation, leading to more comprehensive pathway suppression. Combining a BRAF inhibitor with a PROTAC targeting MEK, a downstream effector of the MAPK pathway, has demonstrated synergistic inhibition of melanoma cell growth in preclinical models, overcoming resistance mechanisms associated with single-agent therapy (Zhong et al. 2022).
- ***Synergy with Immunotherapy:***
  PROTACs can enhance the efficacy of immunotherapy by modulating the tumor microenvironment and increasing tumor immunogenicity. By degrading immunosuppressive proteins or activating immune response pathways, PROTACs can synergize with immunotherapeutic agents to enhance anti-tumor immunity. Combining a PROTAC targeting PD-L1 with immune checkpoint inhibitors (ICIs) has shown synergistic effects in preclinical models of cancer. The degradation of PD-L1 by the PROTAC increases tumor immunogenicity and enhances the response to ICIs, leading to improved tumor control (Sun et al. 2018).
- ***Dual Targeting Strategy:***
  PROTACs can be designed to simultaneously target multiple proteins involved in complementary or redundant pathways, leading to synergistic inhibition of tumor growth. This dual targeting strategy allows for more effective pathway blockade and may overcome resistance mechanisms. Dual-targeting PROTACs have been developed to degrade both EGFR and HER2 in HER2-positive breast cancer. By simultaneously inhibiting two key signaling pathways, these PROTACs achieve synergistic anti-tumor effects and overcome resistance to single-agent therapies (Zhang et al. 2020).
- ***Combination with Radiotherapy or Chemotherapy:***
  PROTACs can sensitize cancer cells to radiotherapy or chemotherapy by downregulating DNA repair proteins or anti-apoptotic proteins, leading to enhanced tumor cell death. Combination therapy with PROTACs may improve treatment outcomes and reduce the risk of resistance. Combining a PROTAC targeting Mcl-1, an anti-apoptotic protein, with chemotherapy has shown synergistic effects in preclinical models of acute myeloid leukemia (AML). The PROTAC sensitizes AML cells to chemotherapy-induced apoptosis, resulting in improved therapeutic responses (Tantawy et al. 2023).
- ***Enhanced Selectivity and Reduced Toxicity:***
  PROTACs can enhance the selectivity of combination therapies by specifically targeting disease-relevant proteins, reducing off-target effects and systemic toxicity associated with non-specific treatments. This selectivity allows for higher therapeutic doses and improved efficacy. Combining a PROTAC targeting an

oncogenic driver protein with a cytotoxic chemotherapy agent has shown synergistic anti-tumor effects in preclinical cancer models (Sun et al. 2018). The PROTAC selectively degrades the oncogenic driver, while the chemotherapy agent induces non-specific tumor cell death, resulting in enhanced tumor regression with reduced systemic toxicity. PROTACs offer significant potential for achieving synergistic effects when combined with other therapeutic modalities, including small molecule inhibitors, immunotherapy, radiotherapy, or chemotherapy. Continued research and clinical development of PROTAC-based combination therapies hold promise for improving treatment outcomes across various cancer types and other diseases (Sincere et al. 2023).

4.2.9 Lower Dose Requirement

Due to their catalytic nature, PROTACs may require lower doses compared to traditional inhibitors, potentially reducing the risk of side effects associated with high drug concentrations. PROTACs offer a unique advantage in lowering the dose requirements of therapeutic agents by enhancing target engagement, promoting sustained target protein degradation, and improving therapeutic efficacy. This ability to achieve potent effects at lower doses can lead to reduced toxicity, improved safety profiles, and enhanced patient compliance (Liu et al. 2022).

PROTACs contribute to lowering dose requirements by the following mechanisms:

- ***Enhanced Target Engagement:***
  PROTACs enable enhanced target engagement by recruiting the ubiquitin-proteasome system to degrade the target protein, leading to sustained suppression of its activity. This increased target engagement allows for effective inhibition of the target at lower concentrations compared to traditional inhibitors.

  ARV-110, a PROTAC targeting the androgen receptor (AR) in prostate cancer, has demonstrated potent AR degradation and tumor growth inhibition at lower doses compared to conventional AR inhibitors in preclinical studies, suggesting enhanced target engagement (Neklesa et al. 2019).
- ***Sustained Pharmacodynamic Effects:***
  PROTACs induce sustained target protein degradation through a catalytic mechanism, where a single PROTAC molecule can degrade multiple target protein molecules. This sustained pharmacodynamic effect enables prolonged suppression of the target's activity, allowing for lower doses to achieve therapeutic efficacy.

  DT2216, a PROTAC targeting B-cell lymphoma-2 (BCL-2), has demonstrated potent and sustained BCL-2 degradation at lower doses in preclinical models of hematologic malignancies, leading to prolonged apoptosis and tumor growth inhibition (Jia et al. 2021).

- ***Reduced Off-target Effects:***
  PROTACs offer enhanced selectivity by selectively degrading the target protein while minimizing off-target effects on unrelated proteins. This reduced off-target activity allows for lower doses to be used without compromising therapeutic efficacy or increasing toxicity. ARV-471, a PROTAC targeting the estrogen receptor (ER) in hormone receptor-positive breast cancer, has shown selective ER degradation and potent anti-tumor activity at lower doses compared to non-selective ER inhibitors in preclinical studies, indicating reduced off-target effects (Hamilton et al. 2022).
- ***Minimization of Drug Resistance:***
  By inducing sustained target protein degradation, PROTACs may help minimize the development of drug resistance, which can occur with traditional inhibitors due to target reactivation or compensatory mechanisms. Lower doses of PROTACs can maintain continuous suppression of the target, reducing the likelihood of resistance. MT-802, a PROTAC targeting Bruton's tyrosine kinase (BTK) in B-cell malignancies, has demonstrated potent and sustained BTK degradation at lower doses in preclinical models, potentially reducing the risk of resistance compared to non-specific BTK inhibitors (Authier et al. 2013).
- ***Enhanced Therapeutic Index:***
  Lowering the dose requirements of therapeutic agents through PROTAC-mediated target degradation can improve the therapeutic index by widening the gap between efficacy and toxicity. This enhanced therapeutic index allows for safer and more effective treatment regimens. Various PROTACs targeting different proteins, such as kinases, transcription factors, and oncogenic drivers, have demonstrated improved therapeutic indices at lower doses in preclinical studies, suggesting their potential for clinical application (Lin et al. 2020).

  PROTACs offer a promising approach to lowering the dose requirements of therapeutic agents by enhancing target engagement, promoting sustained pharmacodynamic effects, reducing off-target effects, minimizing drug resistance, and enhancing the therapeutic index (Burslem and Crews 2020).

4.2.10 Addressing Protein Redundancy

PROTACs can address cases of protein redundancy where multiple proteins contribute to a disease phenotype by degrading multiple targets simultaneously. Protein redundancy, where multiple proteins can perform similar functions within a biological pathway or process, presents a challenge for targeted therapy. PROTACs (Proteolysis Targeting Chimeras) offer a promising strategy to address protein redundancy by enabling the simultaneous degradation of multiple target proteins (Carmony and Kim 2012).

PROTACs contribute to overcoming protein redundancy as follows:

- ***Simultaneous Degradation of Redundant Proteins:***
  PROTACs can be designed to simultaneously target multiple proteins that exhibit redundancy in function or signaling pathways. By inducing the degradation of redundant proteins, PROTACs can achieve more comprehensive pathway inhibition compared to single-target inhibitors. In cancer, multiple receptor tyrosine kinases (RTKs) may contribute to oncogenic signaling redundancy. PROTACs targeting different RTKs, such as EGFR, HER2, and HER3, have been developed to simultaneously degrade these redundant proteins, leading to synergistic antitumor effects and overcoming resistance mechanisms (Burslem et al. 2018b).
- ***Combination Therapy for Pathway Modulation:***
  PROTACs can be used in combination with other therapeutic modalities, such as small molecule inhibitors or monoclonal antibodies, to target multiple nodes within a signaling pathway affected by protein redundancy. This combination approach allows for more effective pathway modulation and inhibition of redundant proteins.

  In autoimmune diseases, multiple cytokines may contribute to disease pathogenesis through redundant signaling pathways. PROTACs targeting different cytokine receptors, combined with monoclonal antibodies against cytokines, can achieve synergistic effects by simultaneously blocking multiple nodes within the inflammatory cascade (Wang et al. 2020).
- ***Tailored Design for Protein Specificity:***
  PROTACs can be designed with specific ligands to selectively target redundant proteins while sparing non-redundant or essential proteins within the same pathway. This tailored design allows for precise modulation of protein redundancy without causing unintended effects on critical cellular functions.

  In neurodegenerative diseases, multiple proteins may contribute to pathological protein aggregation and neuronal toxicity. PROTACs can be designed to selectively degrade redundant proteins implicated in disease pathology, such as mutant forms of tau or alpha-synuclein, while sparing essential proteins involved in normal cellular function (Liu et al. 2022).
- ***Modulation of Signaling Network Dynamics:***
  PROTACs can modulate the dynamics of signaling networks by selectively degrading specific nodes or feedback regulators within redundant pathways. This modulation can reshape signaling flux and overcome compensatory mechanisms associated with protein redundancy. In cancer, compensatory feedback loops may arise in response to targeted therapy, leading to resistance and disease progression. PROTACs targeting key nodes within the feedback loops can disrupt signaling dynamics, enhance pathway inhibition, and overcome resistance to single-agent therapy (Sun et al. 2019).

PROTACs offer a versatile approach to addressing protein redundancy by enabling the simultaneous degradation of multiple target proteins, combining therapy modalities for pathway modulation, designing tailored strategies for protein specificity, and modulating signaling network dynamics (Fig. 4.1).

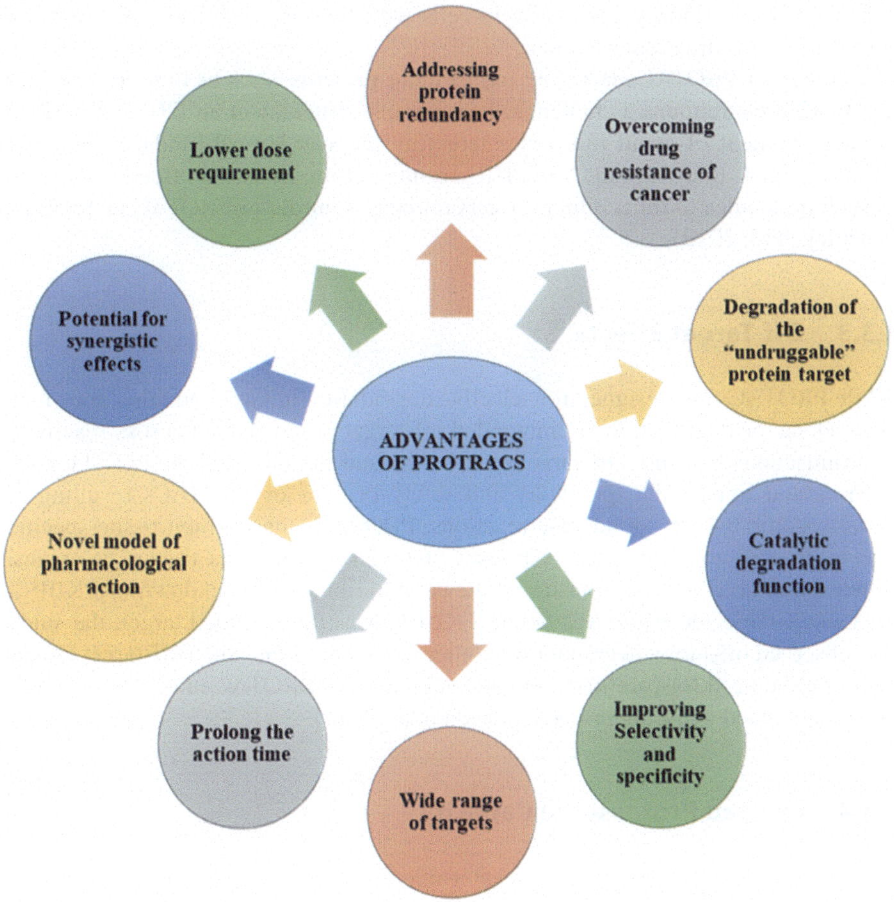

Fig. 4.1 Advantages of PROTACS

4.3 Disadvantages of PROTACs

4.3.1 Complex Design and Optimization

Designing PROTAC molecules involves a complex process that requires careful selection of ligands for both the target protein and the E3 ubiquitin ligase, along with optimizing linker length and composition. Achieving optimal binding affinities and selectivity while maintaining favorable pharmacokinetic properties can be challenging.

4.3.2 Cell Penetration and Intracellular Delivery

PROTACs need to efficiently penetrate cell membranes and reach their target proteins within the cytoplasm or nucleus. Achieving adequate intracellular delivery is crucial for their efficacy, especially in cells with barriers such as the blood-brain barrier or tumor microenvironment.

Delivery of PROTACs targeting intracellular proteins, such as transcription factors or kinases, remains a challenge. For example, Buckley et al. (2012) described the development of small molecules targeting the von Hippel-Lindau (VHL) E3 ubiquitin ligase to disrupt the VHL/HIF-1α interaction. Ensuring effective delivery of such molecules to intracellular targets remains a hurdle in PROTAC technology (Buckley et al. 2012).

4.3.3 Off-Target Effects

While PROTACs are designed for selective degradation of target proteins, they may also induce the degradation of unintended off-target proteins due to cross-reactivity or promiscuous binding. Off-target effects can lead to undesired biological consequences and impact the specificity and safety profiles of PROTACs resulting in tissue-specific toxicity or adverse reactions. PROTACs may exhibit tissue-specific effects, leading to differential responses or toxicity in various tissues or organs. Bondeson et al. (2015) demonstrated the catalytic in vivo knockdown of FKBP12 by a small-molecule PROTAC. While FKBP12 was the intended target, the study also observed off-target degradation of other related proteins. Such off-target effects highlight the need for careful design and validation of PROTAC molecules to minimize unintended consequences (Bondeson et al. 2015).

4.3.4 Limited Proteasome Capacity

The ubiquitin-proteasome system responsible for protein degradation induced by PROTACs has a finite capacity. Saturating the proteasome with excessive protein substrates may lead to cellular stress, cytotoxicity, or activation of compensatory pathways, potentially affecting cell viability and therapeutic efficacy (Liu et al. 2022).

Jaime-Figueroa et al. (2020) targeted Bruton's tyrosine kinase (BTK) using PROTACs and observed potent and sustained BTK degradation. However, excessive activation of the proteasome by PROTAC-induced protein degradation may overwhelm cellular protein degradation pathways, affecting cell viability or inducing compensatory responses (Jaime-Figueroa et al. 2020).

4.3.5 Immunogenicity

PROTACs molecules, particularly those derived from non-human proteins or containing non-natural chemical moieties, may elicit immune responses in vivo. Immunogenicity can lead to rapid clearance from circulation, reduced exposure at target sites, or adverse immune reactions in patients, impacting their pharmacokinetics and efficacy (Sincere et al. 2023).

Bondeson et al. (2015) demonstrated the in vivo efficacy of a small-molecule PROTAC in degrading target proteins. However, the immunogenicity of PROTACs may pose challenges in clinical translation, emphasizing the importance of evaluating immunogenic potential during preclinical development (Bondeson et al. 2015).

Addressing these challenges through continued research and innovation will be crucial for advancing PROTAC technology and realizing its therapeutic potential.

4.3.6 Cytotoxicity and Cellular Stress

Excessive activation of the ubiquitin-proteasome system by PROTAC-induced protein degradation may overwhelm cellular protein degradation pathways, leading to cytotoxicity or cellular stress. Arthur et al. (2020) targeted Bruton's tyrosine kinase (BTK) using PROTACs and observed potent and sustained BTK degradation. However, excessive activation of the proteasome may lead to cytotoxic effects, affecting cell viability (Authier et al. 2013).

4.3.7 Drug-Drug Interactions

PROTACs may interact with other drugs or compounds, leading to altered pharmacokinetics, efficacy, or toxicity. Co-administration of PROTACs with other medications may result in drug-drug interactions, affecting the metabolism, clearance, or activity of both the PROTAC and the co-administered drugs. Careful consideration of potential interactions is essential during drug development and clinical use (Goracci et al. 2020).

4.3.8 Poor Pharmaceutical Properties

PROTAC molecules are more complex than traditional small-molecule drugs, with more potential metabolic sites, which can affect their metabolic stability. Being bifunctional, PROTACs carry a surfeit of electronegative moieties and electropositive moieties which then serve as the superfluous reserve of hydrogen bond acceptors (HBAs) and hydrogen bond donors (HBDs) (Goracci et al. 2020). Additionally, most reported PROTACs tend to have a molecular weight greater than 700, resulting in poor permeability, low solubility, and unsatisfactory oral bioavailability. All this increases their polar surface area, making them unfavorable candidates for Lipinski's "Rule of Five". They have low bioavailability and thus low potency-typically in the micromolar range (Hofmann et al. 2024).

4.3.9 Hook Effect

The hook effect, also known as the prozone effect or high-dose hook effect, occurs in immunoassays when the concentration of analyte is too high, saturating the binding sites of both capture and detection antibodies. As a result, the signal generated by the assay may decrease or plateau at high analyte concentrations, leading to inaccurate or misleading results. The potential for the hook effect, where the efficacy of

PROTACs may decrease at higher concentrations, presents a challenge in predicting the onset of this effect and optimizing PROTAC doses to avoid it. In assays designed to measure the potency or activity of PROTACs, such as cell-based assays or biochemical assays, the hook effect can pose challenges. If the concentration of PROTACs is too high, it may saturate the binding sites on the target protein or the E3 ligase, leading to reduced or plateaued degradation kinetics. This can result in an underestimation of the true potency of the PROTACs. Dose-response studies are commonly used to determine the potency and efficacy of PROTACs by measuring the concentration-dependent degradation of the target protein (Tate and Ward 2004). However, if the PROTAC concentration exceeds the dynamic range of the assay due to the hook effect, accurate determination of the dose-response curve may be compromised. Some PROTACs may have a high affinity for the target protein or exhibit potent degradation activity at relatively low concentrations. In such cases, it can be challenging to accurately assess their activity at higher concentrations without encountering the hook effect, particularly in assays with limited dynamic range or sensitivity. Inaccurate or misleading assay results due to the hook effect can complicate data interpretation and decision-making in drug development. If the hook effect is not recognized or addressed, it may lead to erroneous conclusions regarding the potency, efficacy, or mechanism of action of the PROTAC (Liu et al. 2022) (Fig. 4.2).

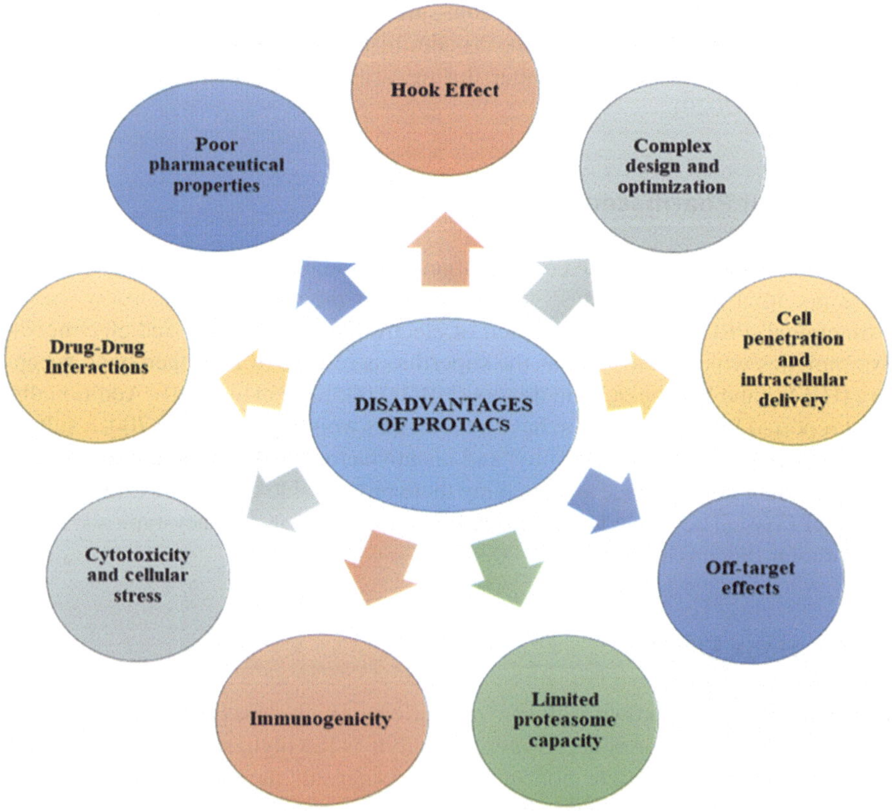

Fig. 4.2 Disadvantages of PROTACs

4.4 Conclusion

In conclusion, the development of Proteolysis Targeting Chimeras (PROTACs) represents a groundbreaking approach in the field of targeted protein degradation, offering several distinct advantages over traditional small molecule inhibitors. PROTACs provide a novel mechanism to induce selective degradation of disease-causing proteins, thereby offering the potential for enhanced therapeutic efficacy. Their ability to target proteins previously deemed "undruggable" expands the scope of druggable targets, opening up new avenues for drug discovery and development. Furthermore, PROTACs offer improved selectivity and specificity, minimizing off-target effects and reducing the risk of drug resistance compared to conventional inhibitors. Additionally, PROTACs can prolong the action time of therapeutic agents, lower dose requirements, and exhibit synergistic effects when combined with other treatment modalities. However, despite these considerable advantages, PROTACs are not without limitations. Challenges such as complex design and optimization, issues related to cell penetration and intracellular delivery, as well as potential off-target effects and adverse reactions, must be carefully addressed. Furthermore, the hook effect in assays used to evaluate PROTAC activity poses challenges in accurately assessing their potency and efficacy. Overall, while PROTACs hold immense promise as a transformative therapeutic strategy, continued research, optimization, and rigorous evaluation are essential to maximize their clinical utility and overcome existing limitations in drug development.

References

Arthur R, Valle-Argos B, Steele AJ, Packham G (2020) Development of PROTACs to address clinical limitations associated with BTK-targeted kinase inhibitors. Explor Target Antitumor Ther 1(3):131–152. https://doi.org/10.37349/etat.2020.00009

Authier S, Vargas HM, Curtis MJ, Holbrook M, Pugsley MK (2013) Safety pharmacology investigations in toxicology studies: An industry survey. J Pharmacol Toxicol Methods 68(1):44–51. https://doi.org/10.1016/j.vascn.2013.05.002

Békés M, Langley DR, Crews CM (2022) PROTAC targeted protein degraders: the past is prologue. Nat Rev Drug Discov 21(3):181–200. https://doi.org/10.1038/s41573-021-00371-6

Bondeson DP, Mares A, Smith IE, Ko E, Campos S, Miah AH, Mulholland KE, Routly N, Buckley DL, Gustafson JL, Zinn N, Grandi P, Shimamura S, Bergamini G, Faelth-Savitski M, Bantscheff M, Cox C, Gordon DA, Willard RR et al (2015) Catalytic in vivo protein knockdown by small-molecule PROTACs. Nat Chem Biol 11(8):611–617. https://doi.org/10.1038/nchembio.1858

Bricelj A, Steinebach C, Kuchta R, Gütschow M, Sosič I (2021) E3 ligase ligands in successful PROTACs: an overview of syntheses and linker attachment points [review]. Front Chem 9. https://doi.org/10.3389/fchem.2021.707317

Buckley DL, Van Molle I, Gareiss PC, Tae HS, Michel J, Noblin DJ, Jorgensen WL, Ciulli A, Crews CM (2012) Targeting the von Hippel-Lindau E3 ubiquitin ligase using small molecules to disrupt the VHL/HIF-1α interaction. J Am Chem Soc 134(10):4465–4468. https://doi.org/10.1021/ja209924v

Buhimschi AD, Armstrong HA, Toure M, Jaime-Figueroa S, Chen TL, Lehman AM, Woyach JA, Johnson AJ, Byrd JC, Crews CM (2018) Targeting the C481S Ibrutinib-resistance mutation in Bruton's tyrosine kinase using PROTAC-mediated degradation. Biochemistry 57(26):3564–3575. https://doi.org/10.1021/acs.biochem.8b00391

Burke MR, Smith AR, Zheng G (2022) Overcoming cancer drug resistance utilizing PROTAC technology [review]. Front Cell Dev Biol 10. https://doi.org/10.3389/fcell.2022.872729

Burslem GM, Crews CM (2020) Proteolysis-targeting chimeras as therapeutics and tools for biological discovery. Cell 181(1):102–114. https://doi.org/10.1016/j.cell.2019.11.031

Burslem GM, Ottis P, Jaime-Figueroa S, Morgan A, Cromm PM, Toure M, Crews CM (2018a) Efficient synthesis of immunomodulatory drug analogues enables exploration of structure-degradation relationships. ChemMedChem 13(15):1508–1512. https://doi.org/10.1002/cmdc.201800271

Burslem GM, Smith BE, Lai AC, Jaime-Figueroa S, McQuaid DC, Bondeson DP, Toure M, Dong H, Qian Y, Wang J, Crew AP, Hines J, Crews CM (2018b) The advantages of targeted protein degradation over inhibition: an RTK case study. Cell Chem Biol 25(1):67–77.e63. https://doi.org/10.1016/j.chembiol.2017.09.009

Carmony KC, Kim KB (2012) PROTAC-induced proteolytic targeting. Methods Mol Biol 832:627–638. https://doi.org/10.1007/978-1-61779-474-2_44

Cohen P, Tcherpakov M (2010) Will the ubiquitin system furnish as many drug targets as protein kinases? Cell 143(5):686–693. https://doi.org/10.1016/j.cell.2010.11.016

Fu X, Li P, Zhou Q, He R, Wang G, Zhu S, Bagheri A, Kupfer G, Pei H, Li J (2024) Mechanism of PARP inhibitor resistance and potential overcoming strategies. Genes Dis 11(1):306–320. https://doi.org/10.1016/j.gendis.2023.02.014

Gao H, Sun X, Rao Y (2020) PROTAC technology: opportunities and challenges. ACS Med Chem Lett 11(3):237–240. https://doi.org/10.1021/acsmedchemlett.9b00597

Gong RH, Chen M, Huang C, Wong HLX, Kwan HY, Bian Z (2022) Combination of artesunate and WNT974 induces KRAS protein degradation by upregulating E3 ligase ANACP2 and β-TrCP in the ubiquitin-proteasome pathway. Cell Commun Signal 20(1):34. https://doi.org/10.1186/s12964-022-00834-2

Goracci L, Desantis J, Valeri A, Castellani B, Eleuteri M, Cruciani G (2020) Understanding the metabolism of proteolysis targeting chimeras (PROTACs): the next step toward pharmaceutical applications. J Med Chem 63(20):11615–11638. https://doi.org/10.1021/acs.jmedchem.0c00793

Grinshpun A (2023) Clinician's guide to targeted estrogen receptor degradation using PROTAC in patients with estrogen receptor-positive metastatic breast cancer. Curr Opin Oncol 35(6):472–478. https://doi.org/10.1097/cco.0000000000000972

Hamilton E, Vahdat L, Han HS, Ranciato J, Gedrich R, Keung CF, Chirnomas D, Hurvitz S (2022) Abstract PD13-08: first-in-human safety and activity of ARV-471, a novel PROTAC® estrogen receptor degrader, in ER+/HER2-locally advanced or metastatic breast cancer. Cancer Res 82(4_Supplement):PD13-08-PD13-08. https://doi.org/10.1158/1538-7445.Sabcs21-pd13-08

He Y, Koch R, Budamagunta V, Zhang P, Zhang X, Khan S, Thummuri D, Ortiz YT, Zhang X, Lv D, Wiegand JS, Li W, Palmer AC, Zheng G, Weinstock DM, Zhou D (2020) DT2216—a Bcl-xL-specific degrader is highly active against Bcl-xL-dependent T cell lymphomas. J Hematol Oncol 13(1):95. https://doi.org/10.1186/s13045-020-00928-9

Hines J, Lartigue S, Dong H, Qian Y, Crews CM (2019) MDM2-recruiting PROTAC offers superior, synergistic antiproliferative activity via simultaneous degradation of BRD4 and stabilization of p53. Cancer Res 79(1):251–262. https://doi.org/10.1158/0008-5472.Can-18-2918

Hofmann N, Harms M, Mäder K (2024) ASDs of PROTACs: spray-dried solid dispersions as enabling formulations. Int J Pharm 650:123725. https://doi.org/10.1016/j.ijpharm.2023.123725

Hong D, Zhou B, Zhang B, Ren H, Zhu L, Zheng G, Ge M, Ge J (2022) Recent advances in the development of EGFR degraders: PROTACs and LYTACs. Eur J Med Chem 239:114533. https://doi.org/10.1016/j.ejmech.2022.114533

Jaime-Figueroa S, Buhimschi AD, Toure M, Hines J, Crews CM (2020) Design, synthesis and biological evaluation of proteolysis targeting chimeras (PROTACs) as a BTK degraders with improved pharmacokinetic properties. Bioorg Med Chem Lett 30(3):126877. https://doi.org/10.1016/j.bmcl.2019.126877

Jia X, Han X (2023) Targeting androgen receptor degradation with PROTACs from bench to bedside. Biomed Pharmacother 158:114112. https://doi.org/10.1016/j.biopha.2022.114112

Jia Y, Zhang Q, Zhang W, Andreeff M, Jain N, Zhang P, Zheng G, Zhou D, Konopleva M (2021) Targeting BCL-XL and BCL-2 by Protac 753B effectively eliminates AML cells and enhances efficacy of chemotherapy by targeting senescent cells. Blood 138:2230. https://doi.org/10.1182/blood-2021-147535

Kim H, Park J, Kim JM (2022) Targeted protein degradation to overcome resistance in cancer therapies: PROTAC and N-Degron pathway. Biomedicines 10:9. https://doi.org/10.3390/biomedicines10092100

Li X-M, Zhao Z-Y, Yu X, Xia Q-D, Zhou P, Wang S-G, Wu H-L, Hu J (2023) Exploiting E3 ubiquitin ligases to reeducate the tumor microenvironment for cancer therapy. Exp Hematol Oncol 12(1):34. https://doi.org/10.1186/s40164-023-00394-2

Lin X, Xiang H, Luo G (2020) Targeting estrogen receptor α for degradation with PROTACs: a promising approach to overcome endocrine resistance. Eur J Med Chem 206:112689. https://doi.org/10.1016/j.ejmech.2020.112689

Liu Z, Hu M, Yang Y, Du C, Zhou H, Liu C, Chen Y, Fan L, Ma H, Gong Y, Xie Y (2022) An overview of PROTACs: a promising drug discovery paradigm. Mol Biomed 3(1):46. https://doi.org/10.1186/s43556-022-00112-0

Lv D, Pal P, Liu X, Jia Y, Thummuri D, Zhang P, Hu W, Pei J, Zhang Q, Zhou S, Khan S, Zhang X, Hua N, Yang Q, Arango S, Zhang W, Nayak D, Olsen SK, Weintraub ST et al (2021) Development of a BCL-xL and BCL-2 dual degrader with improved anti-leukemic activity. Nat Commun 12(1):6896. https://doi.org/10.1038/s41467-021-27210-x

Maneiro MA, Forte N, Shchepinova MM, Kounde CS, Chudasama V, Baker JR, Tate EW (2020) Antibody-PROTAC conjugates enable HER2-dependent targeted protein degradation of BRD4. ACS Chem Biol 15(6):1306–1312. https://doi.org/10.1021/acschembio.0c00285

Marrocco I, Yarden Y (2023) Resistance of lung cancer to EGFR-specific kinase inhibitors: activation of bypass pathways and endogenous Mutators. Cancers (Basel) 15:20. https://doi.org/10.3390/cancers15205009

Neklesa TK, Winkler JD, Crews CM (2017) Targeted protein degradation by PROTACs. Pharmacol Ther 174:138–144. https://doi.org/10.1016/j.pharmthera.2017.02.027

Neklesa T, Snyder LB, Willard RR, Vitale N, Pizzano J, Gordon DA et al (2019) ARV-110: An Oral Androgen Receptor PROTAC Degrader for Prostate Cancer. J Clin Oncol 37, 259. https://doi.org/10.1200/jco.2019.37.7_suppl.259

Si J, Shi X, Sun S, Zou B, Li Y, An D, Lin X, Gao Y, Long F, Pang B, Liu X, Liu T, Chi W, Chen L, Dimitrov DS, Sun Y, Du X, Yin W, Gao G et al (2020) Hematopoietic progenitor Kinase1 (HPK1) mediates T cell dysfunction and is a druggable target for T cell-based immunotherapies. Cancer Cell 38(4):551–566.e511. https://doi.org/10.1016/j.ccell.2020.08.001

Sincere NI, Anand K, Ashique S, Yang J, You C (2023) PROTACs: emerging targeted protein degradation approaches for advanced Druggable strategies. Molecules 28(10):4014. https://www.mdpi.com/1420-3049/28/10/4014

Sun Y, Zhao X, Ding N, Gao H, Wu Y, Yang Y, Zhao M, Hwang J, Song Y, Liu W, Rao Y (2018) PROTAC-induced BTK degradation as a novel therapy for mutated BTK C481S induced ibrutinib-resistant B-cell malignancies. Cell Res 28(7):779–781. https://doi.org/10.1038/s41422-018-0055-1

Sun X, Gao H, Yang Y, He M, Wu Y, Song Y, Tong Y, Rao Y (2019) PROTACs: great opportunities for academia and industry. Signal Transduct Target Ther 4(1):64. https://doi.org/10.1038/s41392-019-0101-6

Tantawy SI, Timofeeva N, Sarkar A, Gandhi V (2023) Targeting MCL-1 protein to treat cancer: opportunities and challenges. Front Oncol 13:1226289. https://doi.org/10.3389/fonc.2023.1226289

Tate J, Ward G (2004) Interferences in immunoassay. Clin Biochem Rev 25(2):105–120

Wang Y, Jiang X, Feng F, Liu W, Sun H (2020) Degradation of proteins by PROTACs and other strategies. Acta Pharm Sin B 10(2):207–238. https://doi.org/10.1016/j.apsb.2019.08.001

Wang C, Zhang Y, Zhang T, Shi L, Geng Z, Xing D (2022) Proteolysis-targeting chimaeras (PROTACs) as pharmacological tools and therapeutic agents: advances and future challenges. J Enzyme Inhib Med Chem 37(1):1667–1693. https://doi.org/10.1080/14756366.2022.2076675

Wispelaere DM, Du G, Donovan KA, Zhang T, Eleuteri NA, Yuan JC, Kalabathula J, Nowak RP, Fischer ES, Gray NS, Yang PL (2019) Small molecule degraders of the hepatitis C virus protease reduce susceptibility to resistance mutations. Nat Commun 10(1):3468. https://doi.org/10.1038/s41467-019-11429-w

Wu S, Jiang Y, Hong Y, Chu X, Zhang Z, Tao Y, Fan Z, Bai Z, Li X, Chen Y, Li Z, Ding X, Lv H, Du X, Lim SL, Zhang Y, Huang S, Lu J, Pan J, Hu S (2021) BRD4 PROTAC degrader ARV-825 inhibits T-cell acute lymphoblastic leukemia by targeting 'Undruggable' Myc-pathway genes. Cancer Cell Int 21(1):230. https://doi.org/10.1186/s12935-021-01908-w

Xiao M, Zhao J, Wang Q, Liu J, Ma L (2022) Recent advances of degradation technologies based on PROTAC mechanism. Biomol Ther 12:9. https://doi.org/10.3390/biom12091257

Xu J, Brosseau JP, Shi H (2020) Targeted degradation of immune checkpoint proteins: emerging strategies for cancer immunotherapy. Oncogene 39. https://doi.org/10.1038/s41388-020-01491-w

Zhang X, Xu F, Tong L, Zhang T, Xie H, Lu X, Ren X, Ding K (2020) Design and synthesis of selective degraders of EGFR(L858R/T790M) mutant. Eur J Med Chem 192:112199. https://doi.org/10.1016/j.ejmech.2020.112199

Zhao L, Zhao J, Zhong K, Tong A, Jia D (2022) Targeted protein degradation: mechanisms, strategies and application. Signal Transduct Target Ther 7(1):113. https://doi.org/10.1038/s41392-022-00966-4

Zheng N, Shabek N (2017) Ubiquitin ligases: structure, function, and regulation. Annu Rev Biochem 86:129–157. https://doi.org/10.1146/annurev-biochem-060815-014922

Zhong J, Yan W, Wang C, Liu W, Lin X, Zou Z, Sun W, Chen Y (2022) BRAF inhibitor resistance in melanoma: mechanisms and alternative therapeutic strategies. Curr Treat Options in Oncol 23(11):1503–1521. https://doi.org/10.1007/s11864-022-01006-7

Zorba A, Nguyen C, Xu Y, Starr J, Borzilleri K, Smith J, Zhu H, Farley KA, Ding W, Schiemer J, Feng X, Chang JS, Uccello DP, Young JA, Garcia-Irrizary CN, Czabaniuk L, Schuff B, Oliver R, Montgomery J et al (2018) Delineating the role of cooperativity in the design of potent PROTACs for BTK. Proc Natl Acad Sci USA 115(31):E7285–e7292. https://doi.org/10.1073/pnas.1803662115

Novel Technologies in PROTAC Design

5

Keerti Jain, Sukuru Chinna Reddy, Sreevardhan Moode, Piyush Mehra, Sofiya Tarannum, Manisha Patel, Vineet Kumar Jain, and Harvinder Popli

Abstract

Proteolysis Targeting Chimeras (PROTACs) mediated degradation of targeted protein or protein of interest (POI) is emerging as a fascinating approach for drug discovery and exploring new therapeutic modalities. PROTACs work by hijacking the activity of E3 ubiquitin ligases responsible for ubiquitination of POI, which subsequently leads to degradation of POI by 26S proteasome. This hijacking mechanism of PROTACs is being explored to degrade different POIs involved in various types of diseases. In this chapter, we have discussed different technologies being explored for the design of efficient and potent PROTACs for various severe and critical disorders including cancer.

Keywords

Protein · Design · Target · Cancer · Mechanism

Abbreviations

ABPP	Activity-based protein profiling
AhR	Aryl hydrocarbon receptor

K. Jain (✉) · S. C. Reddy · S. Moode · P. Mehra · S. Tarannum · M. Patel
Drug Delivery and Nanomedicine Research Laboratory, Department of Pharmaceutics, National Institute of Pharmaceutical Education and Research (NIPER)-Raebareli, Lucknow, UP, India
e-mail: keertijain.02@niperraebareli.edu.in

V. K. Jain · H. Popli
Department of Pharmaceutics, Delhi Pharmaceutical Sciences and Research University (DPSRU), New Delhi, India

M. Nandave, P. Jain (eds.), *PROTAC-Mediated Protein Degradation: A Paradigm Shift in Cancer Therapeutics*, https://doi.org/10.1007/978-981-97-5077-1_5

BCL	B-cell lymphoma
BRD4	Bromodomain-containing protein 4
CADD	Computer-aided drug design
cIAP1	Cellular inhibitor of apoptosis protein 1
DSG	Di(*N*-succinimidyl) glutarate
DSS	Di(*N*-succinimidyl) suberate
E3RE	E3 recruiting element
FAK	Focal adhesion kinase
HCC	Hepatocellular carcinoma
MDM2	Mouse double minute 2 homolog
POI	Protein of interest
PROTACs	PROteolysis TArgeting Chimeras
RBR	RING-in-Between-RING
Sirt 2	Sirtuin 2
TPD	Target protein degradation
VHL	von Hippel-Lindau protein

5.1 Introduction

Proteolysis Targeting Chimeras (PROTACs) are defined as heterobifunctional molecules, which generally consist of three components i.e. first, a ligand having affinity for target protein or protein of interest (POI); second, E3 recruiting element (E3RE), which is a ligand that can recruit an E3 ubiquitin ligase; and third, a linker, which connects these two ligands (Paiva and Crews 2019; Troup et al. 2020). Bromodomain-containing protein 4 (BRD4) degrader PROTAC formed by rational structure-based design is shown in Fig. 5.1. The composition and length of linker are crucial for the physicochemical properties and bioactivity of PROTACs. With the continuous evolution of PROTAC-based therapeutics, various sophisticated functional linkers are now being explored to design PROTAC in place of conventional linkers like alkyl and polyethylene glycol (Troup et al. 2020; Zagidullin et al. 2020).

Degradation of POI by PROTAC, which was first reported in 2001, has emerged as a revolutionary path for drug discovery and redefining the traditional approach of discovering new therapeutic agents. Target protein degradation (TPD) by PROTAC is emerging as a rapidly growing technology and promising therapeutic modality having applications in chemical biology, medicinal chemistry, and drug discovery (Guedeney et al. 2023). Zhang et al. investigated the relationship between the prognosis of hepatocellular carcinoma (HCC) and expression of focal adhesion kinase (FAK) as well as effect of FAK degradation by FAK PROTAC on HCC metastasis. From the results of this study, scientists observed that FAK PROTAC degraded FAK, which led to effective suppression of HCC metastasis (Zhang et al. 2024).

PROTAC technology has fascinated scientists to develop novel therapeutic templates in recent decades. Heterobifunctional molecules, PROTACs, hijack the ubiquitin-proteasome system to degrade POI. In contrast to small molecules, PROTACs act by inhibiting the complete biological function of the POI by binding

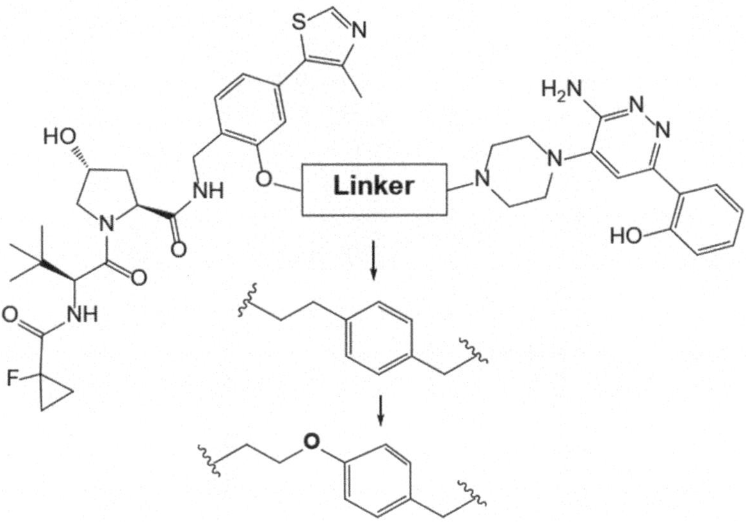

Fig. 5.1 Rational structure-based design of BRD4 degrader PROTAC. (Reproduced with permission from Zagidullin et al. 2020)

and inducing subsequent degradation. Further, PROTACs are relatively easy to handle, which makes them different from nuclear proteins, transcription factors, and other scaffolding proteins that are generally difficult to handle (Liu et al. 2022). Figure 5.2 shows schematically different milestones achieved in PROTAC technology in the last two decades. However, despite the advantages, some challenges are also associated with PROTAC technology, including rational design of an efficient PROTACs and its optimization, which is a challenging task for researchers working in this area.

5.2 Mechanism of POI Degradation by PROTAC

The mechanism of PROTAC-mediated ubiquitination of POI followed by degradation of POI by 26S proteasome is depicted graphically in Fig. 5.3. Briefly, protein substrate is conjugated with ubiquitin through an enzymatic cascade, where a thioester-bonded E1 and ubiquitin conjugate is formed by an E1 activating enzyme, which primes ubiquitin via an ATP-dependent mechanism. Then an E2 ~ ubiquitin conjugate is formed by a transthiolation reaction with an E2 conjugating enzyme. In the last step, a putative E3 ligase mediates and catalyses the transfer of ubiquitin to a specific substrate protein through a covalent or non-covalent mechanism dependent on the type of E3 ligases. RING/U-box family and the active-site cysteine-containing HECT and RING-in-Between-RING (RBR) families are the three major families of E3 ligases. Finally, PROTACs hijack the ligase activity for ubiquitination of POI by recruiting E3 to POI and inducing the formation of ternary complex (POI:PROTAC:E3 ligase) for ubiquitination followed by degradation mediated by

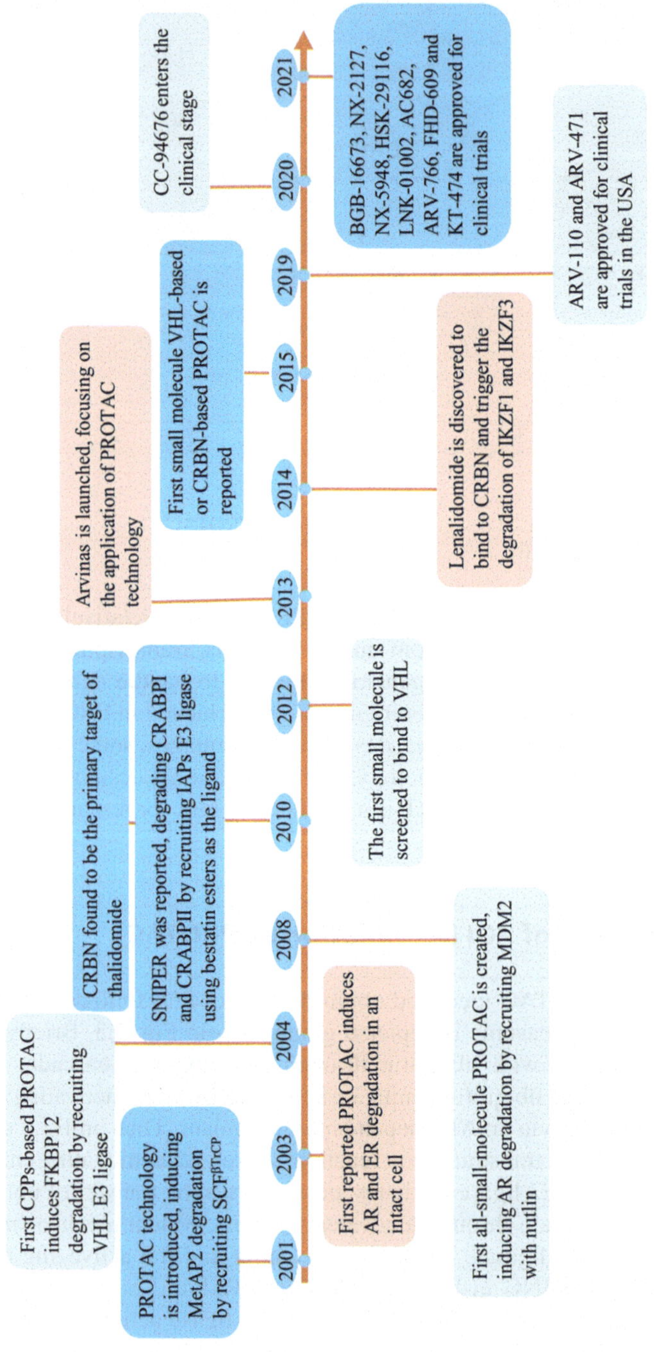

Fig. 5.2 Milestones in the development of PROTAC technology. (Adapted with permission from Liu et al. 2022)

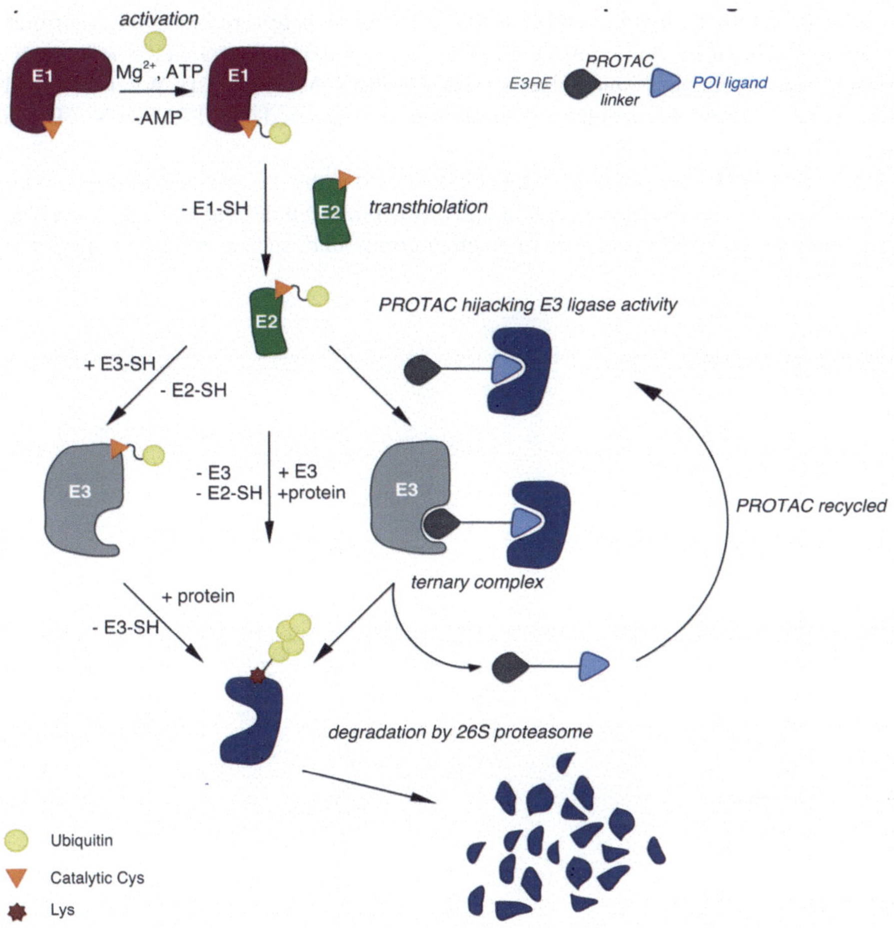

Fig. 5.3 Diagrammatic presentation of POI ubiquitination mediated by PROTAC using ubiquitination enzymatic cascade, followed by degradation of POI by 26S proteasome. (Adapted with permission from Paiva and Crews 2019)

26S proteasome. PROTACs can facilitate ubiquitination and subsequent degradation of multiple POI equivalents like a catalyst, as PROTAC is not degraded in this reaction (Paiva and Crews 2019; Liu et al. 2022; Guedeney et al. 2023).

5.3 Design of PROTAC Linkers

PROTACs are a type of compounds that modulate the POI through degradation. They usually comprise three components: one ligand that binds to the target protein or POI, another ligand that binds to an E3 ubiquitin ligase, and a linker that joins the two ligands. The ligand that binds to the target protein is often known as the 'warhead,' while the ligand that binds to the E3 ligase is referred to as the 'recruiter or anchor.' The linker that connects these two components plays a vital role in determining the effectiveness and selectivity of the PROTAC (Fig. 5.4). When POI and E3 create a ternary complex, degradation is triggered by PROTACs (Troup et al. 2020).

Most PROTAC compounds have been designed using a combination of a known ligand that binds to the target protein and a few ligands that bind to E3 ligases. The intermediate linker often consists of an ether compound, such as PEG, or a modified

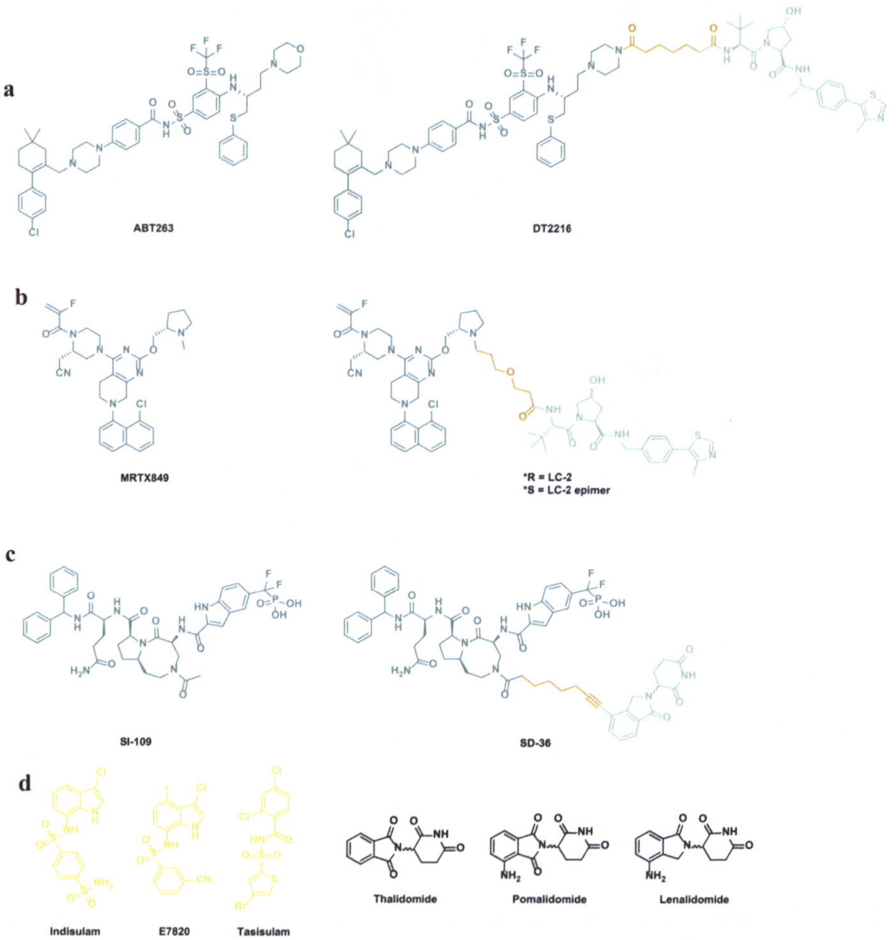

Fig. 5.4 Structures of ligands of POI (in yellow colour) with their corresponding PROTACs (in grey colour) where linkers are shown in yellow colour and the part targeting the E3 ubiquitin ligase is shown in green colour. [**a**: ABT263 and DT2216, Bcl-xL PROTAC; **b** MRTX849 and its corresponding PROTAC targeting KRASG12C; **c** SI-109 and SD-36, STAT3 degrader; and **d** "IMiD" class of small molecules and aryl-sulfonamides that act as molecular glues.] (Adapted with permission from Guedeney et al. 2023)

hydrocarbon chain. It is connected to a POI by an amide, ether, or amine bond. The linker's length and composition significantly influence the physicochemical characteristics and bioactivity of PROTACs. The PROTAC area is fast changing and is presently shifting from simpler alkyl and polyethylene glycol linkers to more advanced functional linkers. This has the potential to discover a plethora of new PROTAC medicines with improved biological activity for therapeutic purposes. To enhance stability, bioavailability, cell membrane permeability, and appropriate spatial orientation between the target POI and the E3 ubiquitin ligase, it is necessary to adjust the chemical nature, length, hydrophilicity, and stiffness of the linker characteristics for each PROTAC. Therefore, it is crucial to carefully consider the composition, length, and attachment sites while designing rational linkers to produce highly effective PROTAC compounds (Troup et al. 2020; Liu et al. 2022; Paiva and Crews 2019). A few general parameters that are required to be considered for successful designing of PROTAC linker are given below:

- **Warhead Selection**: Identify a small-molecule ligand that binds to the target protein of interest with high affinity and specificity. The ligand should have suitable physicochemical properties to engage the target protein effectively. Consider the binding mode and kinetics of the ligand-target interaction for optimal degradation efficiency.
- **Recruiter or Anchor Selection**: Choose an E3 ubiquitin ligase ligand that can recruit the ubiquitin-proteasome system for targeted protein degradation. Ensure that the recruiter has minimal off-target effects and can form a stable complex with the E3 ligase.
- **Linker Design**: Design a linker that connects the warhead and recruiter components. The linker should be of appropriate length and flexibility to allow for efficient ternary complex formation between the PROTAC, target protein, and E3 ligase. Consider linker stability, solubility, and cell permeability when designing the linker. Incorporate appropriate functional groups for conjugation to the warhead and recruiter ligands while maintaining stability during cellular uptake and intracellular processing.
- **Spacer Optimization**: Introduce a spacer between the warhead and linker to prevent steric hindrance and optimize binding affinity. The spacer can be rigid or flexible depending on the distance and orientation required for effective engagement of the target protein and E3 ligase. Spacer length and composition can influence the binding affinity and selectivity of the PROTAC (Troup et al. 2020; Liu et al. 2022).

5.4 Elements of PROTAC Linkers Design

PROTACs are heterobifunctional compounds composed of three elements. (1) PROTAC interact with the specific protein or POI through the binding of ligands. This interaction leads to greater strengthening of the tertiary complex and ultimately facilitates breakdown of the target protein's structure. (2) A compound that binds to

Table 5.1 Structure of some common linkers used to design PROTACs (reproduced with permission from Zagidullin et al. 2020)

Flexible linkers (n, m = 0, 1, 2, 3, 4, etc.)	Relatively rigid linkers	Click chemistry based linkers

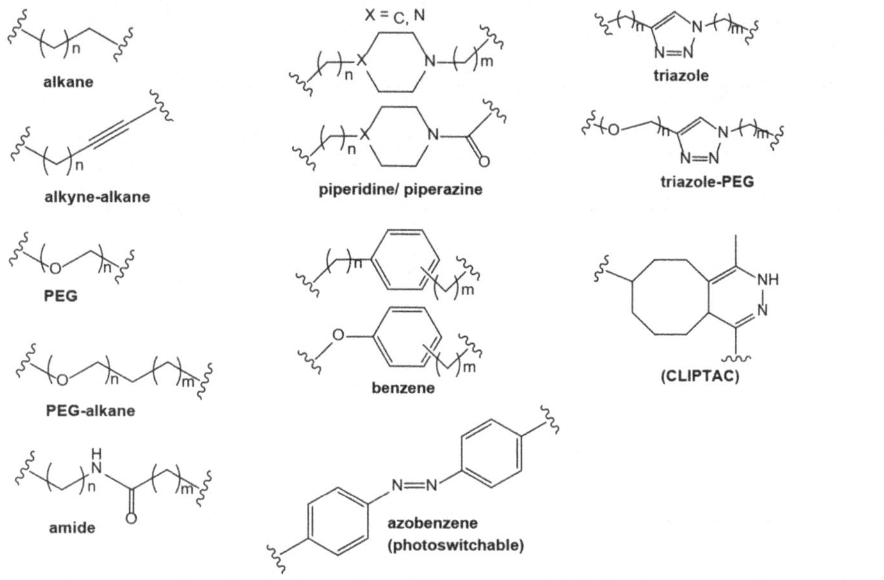

the ubiquitin ligase molecules to promote the attraction of an E3 ubiquitin ligase. (3) A useful connector to attach the target protein-ligand and ubiquitin ligase ligand in the proper arrangement. The target protein's ligand and the E3 ligase ligand are joined by the linker ease formation of ternary complex, which in turn causes the target protein to be ubiquitinated and degraded (Paiva and Crews 2019; Zagidullin et al. 2020). Due to their versatility and rapid synthesis, Alkyl, PEG, and glycol chains are frequently utilized in de novo PROTAC coupler designs. The majority were found to be alkyl and PEG, whereas alkynes, triazoles, piperazine, and piperidine add to the overall range of possibilities (Zoppi et al. 2019; Maple et al. 2019). Structure of some common linkers used in designing of PROTACs is shown in Table 5.1.

5.4.1 Alkyl and PEG Chains

Cyrus et al. reported that alkyl linkers such as di(*N*-succinimidyl) glutarate (DSG) and di(*N*-succinimidyl) suberate (DSS) facilitate PROTAC assembly synthesis of ER degraders (Cyrus et al. 2011). Lapatinib-based PROTAC is made possible by PEG linkers, which are demonstrated by substances such as pomalidomide (Troup et al. 2020). Furthermore, prolonged glycol chains provide versatility, as illustrated in Arvinas' TBK1-targeting PROTACs, where linkers varying from 7 to 29 atoms in length demonstrated sub micromolar degradation potency. The bridging components make the ability to modify PROTAC characteristics easier, thereby improving the efficiency of those molecules in breaking down specific protein molecules (Burslem and Crews 2017; Crew et al. 2018).

5.4.2 Triazole and "Click Chemistry"-Based Linkers

Several pioneering investigations have demonstrated the application of triazole moieties in PROTAC creation, facilitated by click chemistry. Wurz et al. demonstrated triazole click chemistry's efficacy in producing bromodomain-containing protein 4 (BRD4)-targeting PROTACs with different linker lengths. The study emphasized the crucial function of triazole click chemistry in controlling the strength of degradation (Wurz et al. 2018). Crew et al. provided evidence of the versatility of triazole click chemistry in the setting of enhancing anchor-linker-warhead compositions, through the generation of a library of putative Poly (ADP-ribose) polymerase (PARP1) degraders (Crew et al. 2018). Furthermore, Schiedel et al. cleverly employed triazoles to augment the water solubility of Sirtuin 2 (Sirt2) degraders, demonstrating their ability to regulate essential physical characteristics for the advancement of pharmaceuticals (Schiedel et al. 2018).

5.5 Advanced PROTAC Linkers

5.5.1 Photo Sensitive Linker

A unique method for effectively breaking down the protein via von Hippel-Lindau (VHL)-facilitated mechanisms using light PROTACs was demonstrated by Pfaff et al. (2019). Using *ortho*-F4-azobenzene in the linker region, scientists employed a novel strategy to precisely control the activation and deactivation processes. Additionally, they introduced a remarkable development known as the Azo-PROTAC. This system showed exceptional selectivity in the degradation of particular proteins by means of regulated structural alterations brought on by exposure to UV light. Treatment strategies could be advanced by this unique technology since it permits specific control over internal processes.

5.5.2 Piperazine/Piperidines

Piperazine and piperidine moieties have been used to create PROTACs like SJF620 and SJF521. SJF620 includes a piperidine moiety, whereas SJF521 contains a piperazine moiety. These PROTACs were intended to preferentially degrade the SMARCA2 and SMARCA4 proteins by activating the E3 ubiquitin ligase VHL protein, which induces proteasomal degradation. To facilitate favourable interactions and promote the development of stable ternary associations with the E3 ubiquitin ligase, which leads to the degradation of the target proteins, the piperazine and piperidine components were part of the ligand structures that target the bromodomains of SMARCA2/4 (Schiedel et al. 2018).

5.5.3 Alkynes

A variety of libraries with different linker lengths and compositions have been produced using alkyne conjugation to one ligand and azide conjugation to the other (Wurz et al. 2018). Through careful experimentation, a range of PROTACs were successfully created using different linker lengths to PEG units. The generated PROTACs had varying degradation potency depending on linker length, highlighting the need of linker optimization for each ligand pair. Moreover, tetrazine and alkyne precursor molecules were used for in-cells self-assembly of PROTAC using click chemistry were termed as CLIPTACs (Troup et al. 2020; Pfaff et al. 2019). This illustrates the adaptability of alkyne functionality in promoting PROTAC assembly with enhanced solubility and cell permeability.

5.6 Effect of Linker Length on Design of PROTAC

PROTACs have emerged as a promising class of therapeutics for TPD within cells, offering advantages over traditional inhibition-based approaches. The length of this linker plays a pivotal role in dictating the efficacy, selectivity, and pharmacokinetic properties of PROTAC molecules. Linker length profoundly influences the spatial relationship between the target protein and the E3 ligase. An optimal linker length is essential for bringing these two entities into proximity while allowing sufficient flexibility for productive protein-protein interactions. Too short a linker may hinder effective binding, while excessively long linkers could impede the necessary molecular recognition events. Thus, the linker length must be finely tuned to promote efficient target engagement and subsequent ubiquitination (Troup et al. 2020; Liu et al. 2022; Paiva and Crews 2019). Furthermore, the selectivity of PROTACs towards their intended targets is heavily influenced by linker length.

Different target proteins may necessitate varying linker lengths to achieve optimal degradation. By modulating the linker length, researchers can fine-tune the specificity of PROTACs, minimizing off-target effects and enhancing therapeutic efficacy. Moreover, linker length can impact the pharmacokinetic properties of PROTACs, including their stability, solubility, and clearance rates. Optimal linker length contributes to favourable pharmacokinetic profiles, ensuring adequate exposure to the target protein while minimizing systemic toxicity (Pfaff et al. 2019; Farnaby et al. 2019; Lebraud et al. 2016).

Additionally, linker length modulates the cellular response to PROTAC treatment. The kinetics and extent of target protein degradation, as well as downstream signalling pathways, can be influenced by the length of the linker region (Troup et al. 2020; Paiva and Crews 2019; Zagidullin et al. 2020; Zoppi et al. 2019; Maple et al. 2019). In summary, linker length is a critical parameter in the design of PROTAC molecules, with far-reaching implications for their efficacy, selectivity, pharmacokinetics, and structural integrity. By carefully adjusting the linker length, researchers can optimize PROTACs for specific target proteins, enhancing their therapeutic potential while minimizing off-target effects and systemic toxicity.

5.7 Computer-Assisted PROTAC Design

PROTACs are typically created by combining two components, with one fragment binding to the POI and the other fragment binding to the E3-ligase (E3-binder or ligand). This simultaneous binding to both proteins is necessary for the functioning of PROTACs. In order to adhere to steric constraints, the two fragments are typically not directly linked by a single covalent bond, but rather by an extra fragment of variable size known as the linker. After the formation of the ternary complex, this linker serves the purpose of not only ensuring the proximity of the two proteins but also enhancing the stability of the ternary complex by engaging in positive interactions with both proteins as shown in Fig. 5.5 (Troup et al. 2020; Tunjic et al. 2023). The linker facilitates the close contact between the target protein and the E3 ligase, resulting in a movement of ubiquitin from the E3 ligase to the POI. Ubiquitination tags induce the proteasome to break down the target protein (Pettersson and Crews 2019).

Computer-aided drug design (CADD) provides several techniques and uses to facilitate the development of conventional drugs with small molecular size. PROTACs have distinct characteristics compared to conventional drugs, including their mechanism of action, size, and other physicochemical properties. Applying CADD to PROTACs is a complex process that requires the development and validation of specific algorithms and tools. Existing CADD technologies must be

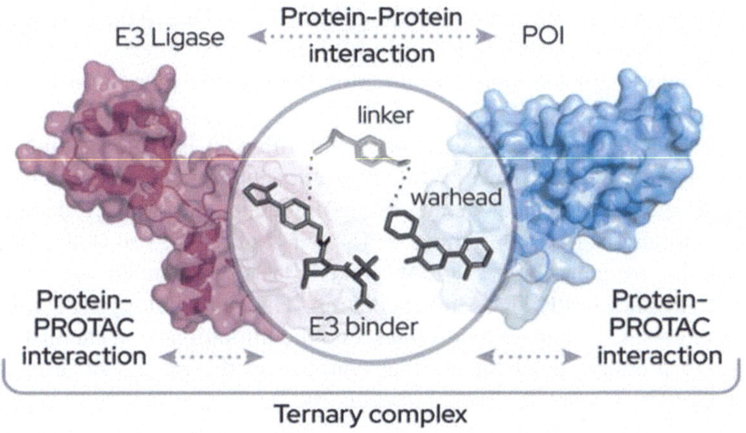

Fig. 5.5 Ternary complex of PROTAC. (Adapted with permission from Tunjic et al. 2023)

modified to accommodate the unique characteristics of PROTACs. Computational techniques can aid researchers in exploring the intricacies of PROTACs' atomic-level functioning, hence enabling them to develop strategies for creating these molecules (Tunjic et al. 2023). The overall framework of PROTACs strategy encompasses the selection of molecular elements, the prediction and assessment of the molecule's effectiveness, and an evaluation of its pharmacological characteristics in relation to its suitability as a drug.

Currently, selecting the warhead to target the POIs is not a difficult task. The primary determinant for a warhead is its strong attraction to the target protein. In contrast to the conventional method, which mostly relies on affinity, compounds undergo thorough screening for factors including selectivity to minimize the potential for off-target effects. Notably, most of the warheads used in clinical PROTACs make use of already established inhibitors of target proteins (Neklesa et al. 2019). An important advantage of this method is the possibility of achieving therapeutic effects by blocking disease-causing proteins, even in cases when PROTACs fail to degrade the targeted proteins as intended.

An important challenge in the design of PROTACs is choosing the right linker. Research has indicated that the composition and length of the linker have a substantial influence on the effectiveness of the targeted degradation process (Roy et al. 2019). An analysis of linkers used in the construction of PROTACs shows that simple PEG or alkyl chains are most used (Troup et al. 2020). Significantly, the development of linkers has primarily relied on empirical methods, driven by practical knowledge rather than established concepts. Currently, there is no conclusive knowledge on the specific characteristics or lengths of linker substrates that could potentially improve the degradation process. As an illustration, Zorba et al. previously developed a range of PROTACs including linkers of different lengths to induce degradation of the BTK protein. According to their design, PROTACs with linker lengths of approximately 20 atoms showed improved effectiveness (Zorba et al. 2018). However, previous research on PROTAC design have shown that specific PROTAC molecules provide the most effectiveness when the length of the linker is limited to only three atoms (Wang et al. 2019). Due to the absence of universally applicable rules, the choice of linkers in PROTACs design requires a customized strategy that takes into consideration the specific characteristics of every scenario.

When designing the linker's generation algorithm, it is typically important to prioritize innovation and efficacy. While most molecular generation algorithms are effective for overall molecular generation assignments and produce datasets with property distributions like the training set (Segler et al. 2018; You et al. 2018), they are not suitable for generating PROTAC linkers due to the limited size and lack of diversity in the current PROTAC linkers dataset. The methodology for synthesizing PROTAC linkers has the potential to transcend the limitations of a chemist's understanding and produce novel linkers that conform to the criteria of drug-likeness (Imrie et al. 2020; Guo et al. 2023; Tan et al. 2022).

5.8 Ubiquitin Ligases in PROTAC Design

PROTACs have successfully induced degradation of more than 60 distinct proteins. Notably, certain candidates, including B-cell lymphoma (BCL)-XL degraders and androgen receptor (AR) degraders, have advanced to the clinical trial stage. The majority of the reported PROTACs utilize E3 ligase cereblon (CRBN) or VHL for recruitment, resulting in the degradation of more than 30 proteins via CRBN and more than 20 proteins using VHL. An underlying drawback of solely depending on CRBN or VHL for PROTAC development is the occurrence of drug resistance. Another hurdle in the production of PROTACs is the requirement for the chimeric small molecule to establish a ternary complex with the E3 ligase and its neo-substrate protein. This complex must endure for a sufficient duration to allow the E3 ligase to ubiquitinate its substrate. While CRBN and VHL have demonstrated the ability to ubiquitinate a diverse range of target proteins, certain protein-ligase combinations have a limited duration and therefore do not undergo ubiquitination (Bricelj et al. 2021; Gadd et al. 2017; Kramer and Zhang 2022; Mullard 2021). To assure the degradation of these proteins by PROTACs, it is crucial to find different E3 ligases that can be hijacked. E3 ligases exhibit varied expression patterns, and their levels can differ depending on the tissue type, tumor type, cellular compartment, or cell state. Hence, the recruitment of E3 ligases with a precise expression profile could potentially result in the creation of highly specialized PROTACs for the advancement of precision medicine. The following section discusses the development of some of the potential E3 ligases.

5.8.1 Mouse Double Minute 2 Homolog (MDM2)

MDM2 functions as an E3 ligase that specifically identifies and breaks down the tumor suppressor protein p53. The nutlin-based PROTAC A1874 can effectively recruit MDM2 to degrade BRD4 while still maintaining nutlin's ability to stabilize p53. This results in a synergistic effect between BRD4 degradation and p53 stabilization (Kramer and Zhang 2022).

5.8.2 cIAP1

The E3 ligase cellular inhibitor of apoptosis protein 1 (cIAP1), often referred to as BIRC2, hinders apoptosis by ubiquitinating its substrate RIP1. Its excessive expression has been associated with several forms of cancer. Methyl bestatin (MeBS), a ligand, inhibits the association between cIAP1 and RIP1. Additionally, it triggers the process of cIAP1 auto-ubiquitination and subsequent destruction (Sekine et al. 2008; Kramer and Zhang 2022).

5.8.3 Aryl Hydrocarbon Receptor (AhR)

AhR has a dual role as both an E3 ligase and a transcription factor, regulating various cellular responses (Ohtake et al. 2007; Kramer and Zhang 2022).

5.8.4 Kelch-Like ECH-Associated Protein-1 (KEAP1)

KEAP1 functions as an E3 ligase involved in the oxidative stress response. It controls the ubiquitination and subsequent degradation of NRF2, a transcription factor that plays a crucial role in anti-oxidative responses (Kramer and Zhang 2022).

5.8.5 DCAF15

DCAF15 is an E3 ligase with limited knowledge about its functions. However, it has been recently shown that DCAF15 plays a crucial role in inhibiting the ability of natural killer cells to eliminate cancer cells. The anticancer sulfonamides cause the dimerization of DCAF15 and the RNA splicing factor protein RBM39. This dimerization process results in the ubiquitination and subsequent destruction of RBM39 (Han et al. 2017).

5.8.6 RNF114

Nimbolide, a terpene natural substance, is recognized for its anticancer properties. The activity-based protein profiling (ABPP) technique was employed to identify the E3 ligase RNF114 as the specific target of nimbolide. Nimbolide was utilized to enlist RNF114 in the process of degrading BRD4 (Spradlin et al. 2019).

5.8.7 FEM1B

Previously, it was demonstrated that the E3 ligase FEM1B degrades its substrate FNIP1 as a component of the reductive stress response. A targeted ABPP methodology was employed to discover covalent ligands for FEM1B. The chloroacetamide compound forms a covalent bond with FEM1B (Henning et al. 2022; Kramer and Zhang 2022).

5.8.8 DCAF11

DCAF11 functions as an E3 ligase and is responsible for the degradation of p21, a cell cycle checkpoint protein that is known to be dysregulated in osteosarcoma. A different PROTAC was created that links the DCAF11 ligand to the androgen receptor ligand. This PROTAC was then demonstrated to cause the degradation of the androgen receptor (AR) in the 22Rv1 prostate cancer cell line (Zhang et al. 2021).

5.9 Design of E3 Ligase Ligands for PROTAC

The design of ligands for E3 ligase follows a similar approach to the selection of warheads targeting POIs, where binding affinity is a key factor in both cases. Nevertheless, it is crucial to emphasize a distinctive characteristic; while the aim of POIs is to achieve destruction, E3 ligases must maintain their biological functionality, which includes the transfer of ubiquitin and re-entry into the degradation cycle. Therefore, once ligands with a strong attraction are identified, it is crucial to confirm that these ligands do not undermine the inherent functions of the E3 ligase.

The initial PROTACs were composed of a peptidic ligand that targeted an E3 ligase. The restricted capacity of peptide moieties to pass across cell membranes, their ease of synthesis, and their tendency to degrade in biological systems have prompted researchers to focus on developing nonpeptidic E3 ligase ligands that are more similar to traditional drugs. Through the utilization of these small-molecule E3 ligase ligands, the scientific community has extensively shown that PROTACs may effectively cause the degradation of many proteins, even at sub nanomolar concentrations (Ishida and Ciulli 2021).

In 2012, Crews and Ciulli created novel peptidomimetic pVHL ligands by utilizing the hydroxyproline central segment. This part plays a crucial role in binding interactions with Ser111 and His115 in the hydroxyproline binding pocket on pVHL. Using computational methods, researchers developed potential ligands and determined cocrystal structures. They then used rational medicinal chemistry techniques to optimize these ligands, resulting in the creation of a small-molecule inhibitor that targets the pVHL-HIF1α protein-protein interaction. This inhibitor has dissociation constants in the single-digit micromolar range. Subsequently, second-generation ligands were created employing structure-based pharmaceutical design techniques. This involved utilizing cocrystal structures and isothermal titration calorimetry (ITC) to systematically enhance interactions on the left-hand side of the VHL ligand. Through their optimization work, researchers successfully obtained potent VHL binders that exhibit nanomolar binding affinity to VHL. This led to the discovery of the VHL ligand VH032. VH101, VH298, MZ1, PROTAC_ RIPK2, PROTAC_ ERRα, ARD-266, VZ185, ACBI1, thalidomide, dBET1, dFKBP12, QCA570, PROTAC6, ZNL-02-096, and d9A-2 represent a selection of ligase ligands that have been developed so far (Ishida and Ciulli 2021; Galdeano et al. 2014).

5.10 Clinical Considerations/Translation of PROTAC Design

PROTAC is characterized by three essential structure aspects that form its design, a ligand that exhibits specific binding affinity towards POI, a ligand that interacts with the ubiquitin E3 ligase, commonly referred to as E3 ligase, and the linker, functions as the intermediary that establishes a connection between these two components (Ocaña and Pandiella 2020). Over the past two decades, PROTACs have become significant in addressing disease-related proteins. Notably, there are several PROTACs that have progressed to clinical trials, with a primary focus on cancer treatment. Besides cancer, PROTACs show the greatest potential in the treatment of various diseases such as neurodegenerative diseases and viral infections.

PROTACs, which were initially developed to address drug resistance, are currently being developed for their potential in antiviral applications. Traditional antiviral medications facing problems due to the emergence of resistance results in decrease in the efficacy of clinical treatment. Wispelaere et al. designed a compound DGY-08-097, degrader of HCV NS3/4A protease, showed promising efficacy in cellular models, suggesting it a novel approach to combat viral infection (de Wispelaere et al. 2019).

Neurodegenerative disorders like Alzheimer's, Huntington's, and Parkinson's primarily affect the elderly, resulting in cognitive impairment. Misfolded protein aggregation such as tau, β-amyloid, and α-synuclein are the major reasons for these neurodegenerative disorders (Hyun and Shin 2021). Regulating proteins such as tau which is associated with Alzheimer's poses a challenge due to the absence of active pockets. Accumulation of tau is a characteristic feature of neurodegenerative disorders like Alzheimer's. Lu et al. developed a peptide-based PROTAC with a Keap1 E3 ligase ligand. This PROTAC demonstrated a strong binding affinity for both Tau and Keap1, resulting in a significant reduction in Tau levels (Sonwani et al. 2024; Lu et al. 2018). Summary of PROTACs that are either under or approaching clinical trial is given in Table 5.2.

5.11 Delivery Systems for PROTAC

In the field of drug delivery, nanoparticles had a considerable impact because of enhanced stability and solubility, overcoming biological barriers, and prolonging the duration of drug circulation (Hui et al. 2019; Jain and Jain 2023; Jain et al. 2024; Sahu et al. 2024). Significant progress has been achieved in the field of heterobifunctional PROTACs, particularly in the domain of PROTAC-based nanoparticles. These advancements demonstrated the potential to accelerate the transition from laboratory experimentation to clinical implementation (Chen et al. 2022). Nanoparticles in the field of drug delivery are classified into two primary categories, organic and inorganic nanoparticles, that have been engineered and formulated for PROTAC delivery. There are two major subcategories of organic nanoparticles, namely polymeric and lipidic nanoparticles. The aforementioned classification scheme highlights the utilization of various materials in the pursuit of novel drug

Table 5.2 Summary of PROTACs that are either under or approaching clinical trials (adapted with permission from Chen et al. 2022)

Names	Targets	E3 ligases	Ligands for E3 ligases	Indications	Status
ARV-471	ER	CRBN	Lenalidomide derivative	Breast cancer	Phase II
ARV-110	AR	CRBN	Thalidomide derivative	Prostate cancer	Phase II
CFT8634	BRD9	CRBN	Thalidomide analogs	Synovial sarcoma	Phase I/II
ARV-766	AR	VHL	VH032	Prostate cancer	Phase I
AR-LDD	AR	CRBN	Undisclosed	Prostate cancer	Phase I
GT20029	AR	Undisclosed	Undisclosed	Androgenetic alopecia and acne	Phase I
NX-2127	BTK	CRBN	Undisclosed	B-cell malignancies	Phase I
HSK29116	BTK	Undisclosed	Undisclosed	B-cell malignancies	Phase I
BGB-16673	BTK	Undisclosed	Undisclosed	B-cell malignancies	Phase I
DT-2216	BCL-XL	VHL	VH032	Liquid and solid tumors	Phase I
FHD-609	BRD9	Undisclosed	Undisclosed	Synovial sarcoma	Phase I
LNK01002	RAS GTPase	Undisclosed	Undisclosed	Acute myeloid leukemia	Phase I
KT-474	IRAK4	Undisclosed	Undisclosed	Atopic dermatitis and hidradenitis suppurativa	Phase I
KT-413	IRAK4	CRBN	Undisclosed	B-cell lymphomas	Phase I
NX-5948	BTK	CRBN	Undisclosed	B-cell malignancies and autoimmune diseases	Phase I
KT-333	STAT3	Undisclosed	Undisclosed	Liquid and solid tumors	Phase I
CG001419	TRK	CRBN	Undisclosed	Cancer and other diseases	IND-e
CFT8919	EGFR	CRBN	Undisclosed	Non-small cell lung cancer	IND-e

IND-e, in investigational new drug (IND)-enabling preclinical studies

delivery options, each presenting distinct benefits and customized functionality. Nanoparticle designs such as polymeric structures with molecular precision and liposomal formulation with lipid-based elegance, offer targeted and effective drug delivery techniques. These diverse compositions are influencing the future of therapeutic interventions (Jain et al. 2010, 2012; Bagre et al. 2013; Moon et al. 2023; Pathak and Jain 2023). The role of delivery systems in advancing PROTACs to clinical translation is graphically shown in Fig. 5.6.

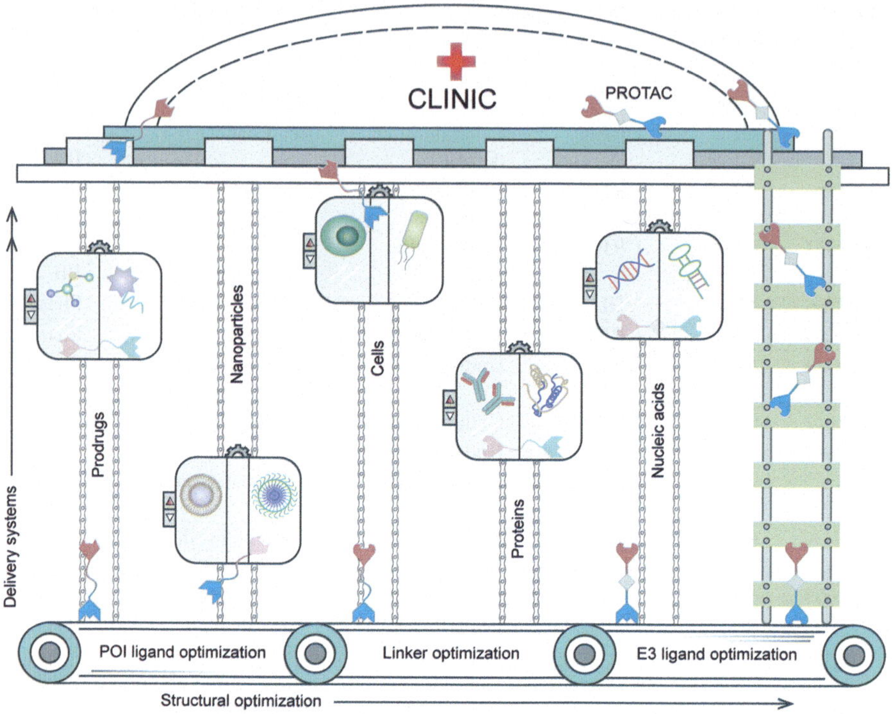

Fig. 5.6 Graphical presentation of structural optimization and delivery systems to advance PROTACs technology to clinical translation. (Adapted with permission from Chen et al. 2022)

5.11.1 Lipid-Based Nanocarriers

Lipid-based nanocarriers are ideal options for transporting drugs due to their remarkable properties such as biocompatibility, degradability, low toxicity, immune compatibility, structural versatility, and large-scale production potential. Lipid-based nanoparticles have gained global recognition and are currently being utilized in the treatment of a wide variety of illnesses. Lipid-based nanocarriers, liposomes and modified liposomes, which are primarily composed of phospholipids, can form both unilamellar and multilamellar vesicular structures. These liposomes are composed of the lipid bilayer that surrounds an aqueous cavity, which encapsulates both the hydrophilic and hydrophobic drug molecules (Jain et al. 2014a, b; Yonezawa et al. 2020; Pardhi et al. 2022; Joyson et al. 2023; Patel et al. 2023).

5.11.2 Polymeric Nanoparticles

Polymeric nanoparticles, with their versatile structures and properties derived from both natural and synthetic sources, are prime examples of novel drug delivery systems. The inherent biocompatibility, simple fabrication methods, and configurable physicochemical properties make them suitable hosts for a wide range of therapeutic medicines from small to macromolecules (Jain et al. 2014a, b, 2021; Begines et al. 2020; Chen et al. 2022; Ojha et al. 2022; Singh et al. 2015). These nanoparticles encapsulate drugs through either physical or chemical interaction, while the surface of these nanoparticles can be easily modified to impart additional functionalities, this has led to many drug delivery vehicles being developed and investigated over a decade due to their properties such as high biocompatibility, physiological stability and multifunctional drug delivery (El-Say and El-Sawy 2017; Jain 2019; Jain et al. 2020; Suthar et al. 2022).

In a study synthetic polymer PLGA-PEG was used to prepare the nanoparticles for the delivery of BRD4 targeting PROTAC, ARV-825 called ARV-NP exhibited excellent physical stability and precise and significant anticancer effects in laboratory tests. These nanoparticles displayed an optimal size for targeting tumors passively and had an extended circulation time due to their unique surface properties. The PLGA-PEG nanoparticles released ARV-825 within the cancer cells leading to the degradation of BRD4 and subsequent inhibition of undruggable c-Myc transcription (Saraswat et al. 2020).

5.11.3 Inorganic Nanoparticles

Because of their ability to take on a wide variety of sizes, shapes, and geometries, nanoparticles that are generated from inorganic materials like gold, iron, silica, and quantum dots hold a greater deal of promise as flexible drug delivery vehicles. To convey PROTACs, these inorganic platforms that are biocompatible and long-lasting are appropriate vehicles since they successfully avoid the intrinsic limits of PROTAC (Liang et al. 2014; Jain and Zhong 2022; Patel et al. 2024; Sonam et al. 2024; Bajwa et al. 2016).

In the research carried out by Yan et al. peptide PROTACs with ligands targeting MDMX protein and VHL were attached to gold (I) via thiol gold interaction to generate self-assembled nanoparticles called nano-MP@PSIs for tumor-targeted MDMX-degrading PROTAC administration. Nano-MP@PSIs were produced to release free PROTACs in glutathione overexpressing tumor microenvironments. After an intravenous injection to mice xenografted with pancreatic cancer, Nano-MP@PSIs showed sustained bloodstream circulation and selective release of PROTAC within the tumor. The released PROTAC degraded the MDMX, consequently reactivating p53 and p73, reducing anticancer resistance and causing cancer cell death (Yan et al. 2021).

5.12 Recent Progress and Future Challenges

In recent years, there has been an exponential rise in the prevalence of PROTAC, due to its capacity for the exploration and progression of novel therapies. This innovative method was first introduced by Craig M. Crews in 2001, and since then it has undergone nearly two decades of development, resulting in notable accomplishments. One of the notable accomplishments is the creation of the first orally available small-molecule PROTAC, ARV-110. This innovative molecule is specially designed to target the androgen receptor for the treatment of metastatic castration-resistant prostate cancer and is currently undergoing clinical trials under the name NCT03888612 known as Arvinas. The progress made in this area generated great excitement among researchers and improvement in the field of PROTAC (Zeng et al. 2021). The following are the recent progress made in the PROTAC, for targeting diverse proteins:

- PROTAC for Targeting Protein Kinases
- Protein kinase inhibitors have captured significant interest and demonstrated remarkable success in the field of cancer treatment (Roskoski 2020). In the realm of PROTAC-mediated targeting of protein kinases, these innovative molecules offer a versatile approach for degrading key proteins involved in the various signaling pathways. By harnessing the power of PROTAC technology, researchers can design molecules capable of selectively degrading the specific protein targets implicated in diseases such as cancer. This targeted degradation can lead to potential therapeutic effects, overcoming drug resistance mechanisms (Lai et al. 2016; Gabizon et al. 2020).
- PROTAC for Targeting Nuclear Receptors
- In addition to playing critical roles in the regulation of a wide variety of physiological activities, nuclear receptors like androgen and estrogen receptors, which are intricately related to the process of carcinogenesis, particularly in the case of breast and prostate cancer. The androgen receptors play a crucial role in the development of prostate cancer. However, the rise of androgen receptor gene amplification, mutations, and alternative splicing renders some drugs ineffective. To overcome this challenge PROTAC-based degraders are introduced, among them, one of the notable compounds is ARD-69, which exhibits its potency exceeding 100-fold in the suppression of gene transcription mediated by androgen receptor, the androgen receptor protein in prostate cancer is rapidly degraded resulting in the notable decrease even at the low doses, showing it a potential approach to fight against prostate cancer (Han et al. 2019).

Ubiquitin-proteasome system and PROTAC-mediated degradation of POI is shown graphically in Fig. 5.7 along with few major challenges associated with PROTACs technology.

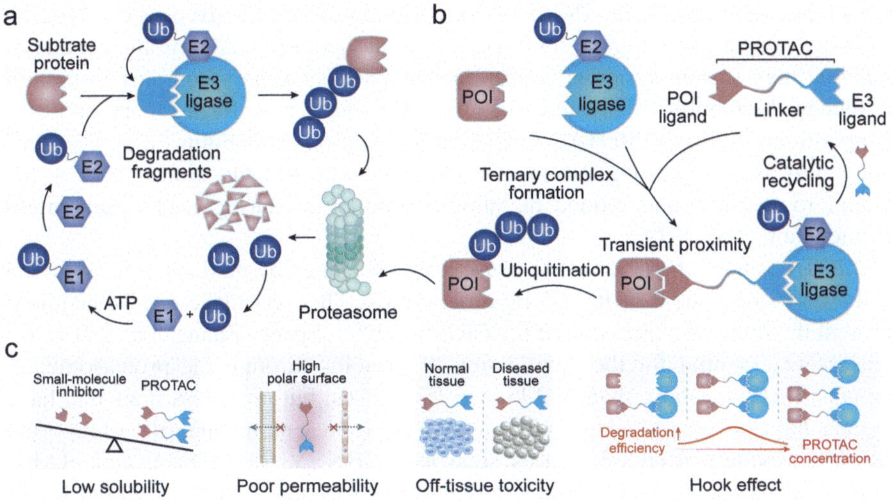

Fig. 5.7 Graphical representation of (**a**) ubiquitin-proteasome system; (**b**) PROTAC-mediated degradation of POI and (**c**) major challenges associated with PROTAC technology. (Adapted with permission from Chen et al. 2022)

5.13 Challenges and Future Aspects

PROTAC research has made noteworthy progress in the field of drug development over the past two decades, characterized by notable achievements. The introduction of the first and second orally bioavailable PROTACs into clinical trials has significantly advanced the field. However similar to every pioneering domain, the development of PROTAC is overflowed with formidable obstacles and compelling prospects, enticing researchers and pharmaceutical visionaries alike to traverse this unexplored realm with ingenuity.

At present, most PROTAC molecules exceed the traditional limits set by Lipinski's "rule of five" (Edmondson et al. 2019). The advancement of PROTACs as potential agents poses unique obstacles. Further investigation is required to better comprehend how PROTACs can cross the cell membrane, which is an obstacle associated with the PROTAC. This includes exploring their absorption, distribution, metabolism, excretion, and toxicity profiles in both theoretical and practical settings. Overcoming the challenge of enhancing cell uptake and bioavailability to maintain the optimal concentrations of PROTAC for pharmacological efficacy, as well as obtaining molecules with favorable physicochemical attributes, is still a significant task that needs to be addressed (Zeng et al. 2021).

A distinct challenge that is specific to the PROTACs is the phenomenon referred to as the "hook" effect. At elevated concentrations of PROTACs, cellular degradation of target protein in vitro is impeded. Potentially attributed to the formation of competing POI-PROTAC or E3 ligase-PROTAC binary complexes, rather than the desired POI-PROTAC-E3 ligase ternary complex (Gadd et al. 2017).

Another challenge facing PROTAC for optimal in vivo effectiveness involves the necessity of a ternary complex for degradation. In vivo, PROTACs produce metabolites that may sustain systemic concentrations but lose ternary complex formation efficiency. Metabolites that bind to POI or E3 ligase may form binary complex competitively limiting PROTAC-mediated degradation. Metabolites may also provide safety issues, limiting therapeutic advantage and to address these issues, it's crucial to monitor and reduce metabolite production during lead optimization (Edmondson et al. 2019).

PROTACs have the potential to cause drug resistance by modifying the genome of the core component of the E3 ligase complex. This resistance can be acquired through the reduced expression of the CRBN or CUL2 gene (Zhang et al. 2019). E3 ligases are essential for the degradation of proteins through the proteasome. In humans, there are more than 600 known E3 ligases, but only, less than 1%, have ligands that can be targeted by small molecules. At present, the majority of reported PROTACs widely used E3 ligases such as CRBN, VHL, MDM2, and cIAP1. Increasing the range of applicable E3 ubiquitin ligases poses a significant hurdle for PROTAC technology. Identifying more precise E3 ligases and ligands within specific cells or tissues is a prominent scientific inquiry (Zeng et al. 2021).

5.14 Conclusion

Overall, PROTAC technology has made a notable advancement in drug development, showing promising progress in clinical trials. However, there are still challenges that need to be addressed, including Lipinski's rule of five, understanding the hook effect, and optimizing in vivo efficacy, cell permeability, and stability. In addition to these, oral bioavailability and drug resistance are also an ongoing challenge for PROTACs. Despite the challenges, ongoing innovation has the potential to revolutionize therapeutic applications.

Acknowledgments The authors (KJ, SCR, SM, PM, ST, MP) are grateful to the Department of Pharmaceuticals, Ministry of Chemicals and Fertilizers, and the Governments of India, for providing the facilities to write this manuscript. The NIPER-R communication number for this manuscript is NIPER-R Communication/583…..

Conflicts of Interest The authors declare no conflicts of interest related to this manuscript.

References

Bagre AP, Jain K, Jain NK (2013) Alginate coated chitosan core shell nanoparticles for oral delivery of enoxaparin: in vitro and in vivo assessment. Int J Pharm 456(1):31–40

Bajwa N, Kumar Mehra N, Jain K, Kumar JN (2016) Targeted anticancer drug delivery through anthracycline antibiotic bearing functionalized quantum dots. Artif Cells Nanomed Biotechnol 44(7):1774–1782

Begines B, Ortiz T, Pérez-Aranda M, Martínez G, Merinero M, Argüelles-Arias F, Alcudia A (2020) Polymeric nanoparticles for drug delivery: recent developments and future prospects. Nanomaterials 10(7):1403

Bricelj A, Steinebach C, Kuchta R, Gütschow M, Sosič I (2021) E3 ligase ligands in successful PROTACs: an overview of syntheses and linker attachment points. Front Chem 9:707317

Burslem GM, Crews CM (2017) Small-molecule modulation of protein homeostasis. Chem Rev 117(17):11269–11301

Chen Y, Tandon I, Heelan W, Wang Y, Tang W, Hu Q (2022) Proteolysis-targeting chimera (PROTAC) delivery system: advancing protein degraders towards clinical translation. Chem Soc Rev 51(13):5330–5350

Crew AP, Raina K, Dong H, Qian Y, Wang J, Vigil D, Serebrenik YV, Hamman BD, Morgan A, Ferraro C, Siu K, Neklesa TK, Winkler JD, Coleman KG, Crews CM (2018) Identification and characterization of Von Hippel-Lindau-recruiting proteolysis targeting chimeras (PROTACs) of TANK-binding kinase 1. J Med Chem 61(2):583–598

Cyrus K, Wehenkel M, Choi EY, Han HJ, Lee H, Swanson H, Kim KB (2011) Impact of linker length on the activity of PROTACs. Mol Biosyst 7(2):359–364

de Wispelaere M, Du G, Donovan KA, Zhang T, Eleuteri NA, Yuan JC, Kalabathula J, Nowak RP, Fischer ES, Gray NS, Yang PL (2019) Small molecule degraders of the hepatitis C virus protease reduce susceptibility to resistance mutations. Nat Commun 10(1):3468

Edmondson SD, Yang B, Fallan C (2019) Proteolysis targeting chimeras (PROTACs) in 'beyond rule-of-five' chemical space: recent progress and future challenges. Bioorg Med Chem Lett 29(13):1555–1564

El-Say KM, El-Sawy HS (2017) Polymeric nanoparticles: promising platform for drug delivery. Int J Pharm 528(1–2):675–691

Farnaby W, Koegl M, Roy MJ, Whitworth C, Diers E, Trainor N, Zollman D, Steurer S, Karolyi-Oezguer J, Riedmueller C, Gmaschitz T, Wachter J, Dank C, Galant M, Sharps B, Rumpel K, Traxler E, Gerstberger T, Schnitzer R, Petermann O, Greb P, Weinstabl H, Bader G, Zoephel A, Weiss-Puxbaum A, Ehrenhöfer-Wölfer K, Wöhrle S, Boehmelt G, Rinnenthal J, Arnhof H, Wiechens N, Wu MY, Owen-Hughes T, Ettmayer P, Pearson M, McConnell DB, Ciulli A (2019) BAF complex vulnerabilities in cancer demonstrated via structure-based PROTAC design. Nat Chem Biol 15(7):672–680

Gabizon R, Shraga A, Gehrtz P, Livnah E, Shorer Y, Gurwicz N, Avram L, Unger T, Aharoni H, Albeck S, Brandis A, Shulman Z, Katz BZ, Herishanu Y, London N (2020) Efficient targeted degradation via reversible and irreversible covalent PROTACs. J Am Chem Soc 142(27):11734–11742

Gadd MS, Testa A, Lucas X, Chan KH, Chen W, Lamont DJ, Zengerle M, Ciulli A (2017) Structural basis of PROTAC cooperative recognition for selective protein degradation. Nat Chem Biol 13(5):514–521

Galdeano C, Gadd MS, Soares P, Scaffidi S, Van Molle I, Birced I, Hewitt S, Dias DM, Ciulli A (2014) Structure-guided design and optimization of small molecules targeting the protein-protein interaction between the von Hippel-Lindau (VHL) E3 ubiquitin ligase and the hypoxia inducible factor (HIF) alpha subunit with in vitro nanomolar affinities. J Med Chem 57(20):8657–8663

Guedeney N, Cornu M, Schwalen F, Kieffer C, Voisin-Chiret AS (2023) PROTAC technology: a new drug design for chemical biology with many challenges in drug discovery. Drug Discov Today 28(1):103395

Guo J, Knuth F, Margreitter C, Janet JP, Papadopoulos K, Engkvist O, Patronov A (2023) Link-INVENT: generative linker design with reinforcement learning. Digit Discov 2:392–408

Han T, Goralski M, Gaskill N, Capota E, Kim J, Ting TC, Xie Y, Williams NS, Nijhawan D (2017) Anticancer sulfonamides target splicing by inducing RBM39 degradation via recruitment to DCAF15. Science 356(6336):eaal3755

Han X, Wang C, Qin C, Xiang W, Fernandez-Salas E, Yang CY, Wang M, Zhao L, Xu T, Chinnaswamy K, Delproposto J, Stuckey J, Wang S (2019) Discovery of ARD-69 as a highly potent proteolysis targeting chimera (PROTAC) degrader of androgen receptor (AR) for the treatment of prostate cancer. J Med Chem 62(2):941–964

Henning NJ, Manford AG, Spradlin JN, Brittain SM, Zhang E, McKenna JM, Tallarico JA, Schirle M, Rape M, Nomura DK (2022) Discovery of a covalent FEM1B recruiter for targeted protein degradation applications. J Am Chem Soc 144(2):701–708

Hui Y, Yi X, Hou F, Wibowo D, Zhang F, Zhao D, Gao H, Zhao CX (2019) Role of nanoparticle mechanical properties in cancer drug delivery. ACS Nano 13(7):7410–7424

Hyun S, Shin D (2021) Chemical-mediated targeted protein degradation in neurodegenerative diseases. Life 11(7):607

Imrie F, Bradley AR, van der Schaar M, Deane CM (2020) Deep generative models for 3D linker design. J Chem Inf Model 60(4):1983–1995

Ishida T, Ciulli A (2021) E3 ligase ligands for PROTACs: how they were found and how to discover new ones. SLAS Discov 26(4):484–502

Jain K (2019) Nanohybrids of dendrimers and carbon nanotubes: a benefaction or forfeit in drug delivery? Nanosci Nanotechnol Asia 9(1):21–29. https://doi.org/10.2174/2210681208666171204163622

Jain K, Jain NK (2023) Multifunctional and targeted theranostic nanomedicines. Springer Nature, Berlin, pp 1–17. https://doi.org/10.1007/978-981-99-0538-6. ISBN: 9789819905386

Jain K, Zhong J (2022) Theranostic applications of nanomaterials. Curr Pharm Des 28(2):77. https://doi.org/10.2174/138161282802211223150153

Jain K, Kesharwani P, Gupta U, Jain NK (2010) Dendrimer toxicity: let's meet the challenge. Int J Pharm 394(1–2):122–142

Jain K, Kesharwani P, Gupta U, Jain NK (2012) A review of glycosylated carriers for drug delivery. Biomaterials 33(16):4166–4186

Jain K, Gupta U, Jain NK (2014a) Dendronized nanoconjugates of lysine and folate for treatment of cancer. Eur J Pharm Biopharm 87(3):500–509

Jain V, Gupta A, Pawar VK, Asthana S, Jaiswal AK, Dube A, Chourasia MK (2014b) Chitosan-assisted immunotherapy for intervention of experimental leishmaniasis via amphotericin B-loaded solid lipid nanoparticles. Appl Biochem Biotechnol 174(4):1309–1330

Jain K, Mehra NK, Jain VK, Jain NK (2020) IPN dendrimers in drug delivery. In: Jana S, Jana S (eds) Interpenetrating polymer network: biomedical applications. Springer, Singapore. https://doi.org/10.1007/978-981-15-0283-5_6

Jain K, Patel AS, Pardhi VP, Flora SJS (2021) Nanotechnology in wastewater management: a new paradigm towards wastewater treatment. Molecules 26(6):1797

Jain VK, Jain K, Popli H (2024) Conjugates of amphotericin B to resolve challenges associated with its delivery. Expert Opin Drug Deliv 21(2):187–210

Joyson N, Pathak A, Jain K (2023) One platform comparison of polymeric and lipidic nanoparticles for the delivery of amphotericin B. AAPS PharmSciTech 24(8):226

Kramer LT, Zhang X (2022) Expanding the landscape of E3 ligases for targeted protein degradation. Curr Res Chem Biol 2(100020):1–5

Lai AC, Toure M, Hellerschmied D, Salami J, Jaime-Figueroa S, Ko E, Hines J, Crews CM (2016) Modular PROTAC design for the degradation of oncogenic BCR-ABL. Angew Chem Int Ed 55(2):807–810

Lebraud H, Wright DJ, Johnson CN, Heightman TD (2016) Protein degradation by in-cell self-assembly of proteolysis targeting chimeras. ACS Cent Sci 2(12):927–934

Liang R, Wei M, Evans DG, Duan X (2014) Inorganic nanomaterials for bioimaging, targeted drug delivery and therapeutics. Chem Commun 50(91):14071–14081

Liu Z, Hu M, Yang Y, Du C, Zhou H, Liu C, Chen Y, Fan L, Ma H, Gong Y, Xie Y (2022) An overview of PROTACs: a promising drug discovery paradigm. Mol Biomed 3(1):46

Lu M, Liu T, Jiao Q, Ji J, Tao M, Liu Y, You Q, Jiang Z (2018) Discovery of a Keap1-dependent peptide PROTAC to knockdown Tau by ubiquitination-proteasome degradation pathway. Eur J Med Chem 146:251–259

Maple HJ, Clayden N, Baron A, Stacey C, Felix R (2019) Developing degraders: principles and perspectives on design and chemical space. Med Chem Commun 10(10):1755–1764

Moon Y, Jeon SI, Shim MK, Kim K (2023) Cancer-specific delivery of proteolysis-targeting chimeras (PROTACs) and their application to cancer immunotherapy. Pharmaceutics 15(2):411

Mullard A (2021) Targeted protein degraders crowd into the clinic. Nat Rev Drug Discov 20(4):247–250

Neklesa T, Snyder LB, Willard RR, Vitale N, Pizzano J, Gordon DA, Bookbinder M, Macaluso J, Dong H, Ferraro C, Wang G, Wang J, Crews CM, Houston J, Crew AP, Taylor I (2019) ARV-110: an oral androgen receptor PROTAC degrader for prostate cancer. J Clin Oncol 37:259. https://doi.org/10.1200/JCO.2019.37.7_suppl.259

Ocaña A, Pandiella A (2020) Proteolysis targeting chimeras (PROTACs) in cancer therapy. J Exp Clin Cancer Res 39(1):189

Ohtake F, Baba A, Takada I, Okada M, Iwasaki K, Miki H, Takahashi S, Kouzmenko A, Nohara K, Chiba T, Fujii-Kuriyama Y, Kato S (2007) Dioxin receptor is a ligand-dependent E3 ubiquitin ligase. Nature 446(7135):562–566

Ojha B, Jain VK, Gupta S, Talegaonkar S, Jain K (2022) Nanoemulgel: a promising novel formulation for treatment of skin ailments. Polym Bull 79:4441–4465

Paiva SL, Crews CM (2019) Targeted protein degradation: elements of PROTAC design. Curr Opin Chem Biol 50:111–119

Pardhi VP, Suthar T, Sharma A, Jain K (2022) Bedaquiline fumarate microemulsion: formulation optimization, rheological characterization and in vitro studies. Nanomedicine (Lond) 17(21):1529–1546

Patel P, Kumar K, Jain VK, Popli H, Yadav AK, Jain K (2023) Nanotheranostics for diagnosis and treatment of breast cancer. Curr Pharm Des 29(10):732–747

Patel P, Pathak A, Jain K (2024) Novel ligand decorated theranostic zein nanoparticles coloaded with paclitaxel and carbon quantum dots: formulation and optimization. Nanomedicine (Lond) 19(5):367–382

Pathak A, Jain K (2023) Ch. 13: Dendrimer–drug conjugates. In: Polymer-drug conjugates: linker chemistry, protocols and applications. Elsevier, Amsterdam, pp 315–345. https://doi.org/10.1016/B978-0-323-91663-9.00005

Pettersson M, Crews CM (2019) PROteolysis TArgeting Chimeras (PROTACs)—past, present and future. Drug Discov Today Technol 31:15–27

Pfaff P, Samarasinghe KTG, Crews CM, Carreira EM (2019) Reversible spatiotemporal control of induced protein degradation by bistable photoPROTACs. ACS Cent Sci 5(10):1682–1690

Roskoski R (2020) Properties of FDA-approved small molecule protein kinase inhibitors: a 2020 update. Pharmacol Res 152:104609

Roy MJ, Winkler S, Hughes SJ, Whitworth C, Galant M, Farnaby W, Rumpel K, Ciulli A (2019) SPR-measured dissociation kinetics of PROTAC ternary complexes influence target degradation rate. ACS Chem Biol 14(3):361–368

Sahu RC, Suthar T, Kumar D, Singh P, Datusalia AK, Jain K (2024) Novel ligand conjugated poly(propylene imine) dendrimers for brain targeted delivery of tacrine hydrochloride. J Drug Deliv Sci Technol 92(105336):1–10

Saraswat A, Patki M, Fu Y, Barot S, Dukhande VV, Patel K (2020) Nanoformulation of PROteolysis TArgeting chimera targeting 'undruggable' c-Myc for the treatment of pancreatic cancer. Nanomedicine 15(18):1761–1777

Schiedel M, Herp D, Hammelmann S, Swyter S, Lehotzky A, Robaa D, Oláh J, Ovádi J, Sippl W, Jung M (2018) Chemically induced degradation of sirtuin 2 (Sirt2) by a proteolysis targeting chimera (PROTAC) based on sirtuin rearranging ligands (SirReals). J Med Chem 61(2):482–491

Segler MHS, Kogej T, Tyrchan C, Waller MP (2018) Generating focused molecule libraries for drug discovery with recurrent neural networks. ACS Cent Sci 4(1):120–131

Sekine K, Takubo K, Kikuchi R, Nishimoto M, Kitagawa M, Abe F, Nishikawa K, Tsuruo T, Naito M (2008) Small molecules destabilize cIAP1 by activating auto-ubiquitylation. J Biol Chem 283(14):8961–8968

Singh J, Jain K, Mehra NK, Jain NK (2015) Effect of anticancer drug on delivery potential of PPI dendrimers. J Colloid Sci Biotechnol 4(2):133–140
Sonam S, Patel P, Jain K (2024) PAMAM dendrimer and carbon quantum dots complexes as theranostic nanocarrier: synthesis, optimization and photophysical characterization. Opt Laser Technol 175(110761):1–12
Sonwani A, Pathak A, Jain K (2024) Nanocarriers-mediated nose-to-brain drug delivery: a novel approach for the management of Alzheimer's disease. J Drug Deliv Sci Technol 98(105855):1–16
Spradlin JN, Hu X, Ward CC, Brittain SM, Jones MD, Ou L, To M, Proudfoot A, Ornelas E, Woldegiorgis M, Olzmann JA, Bussiere DE, Thomas JR, Tallarico JA, McKenna JM, Schirle M, Maimone TJ, Nomura DK (2019) Harnessing the anti-cancer natural product nimbolide for targeted protein degradation. Nat Chem Biol 15(7):747–755
Suthar T, Jain VK, Popli H, Jain K (2022) 12—Nanoemulsions as effective carriers for targeting brain tumors. In: Nanocarriers for drug-targeting brain tumors: micro and nano technologies. Elsevier, Amsterdam, pp 347–363. https://doi.org/10.1016/B978-0-323-90773-6.00008-7
Tan Y, Dai L, Huang W, Guo Y, Zheng S, Lei J, Chen H, Yang Y (2022) DRlinker: deep reinforcement learning for optimization in fragment linking design. J Chem Inf Model 62(23):5907–5917
Troup RI, Fallan C, Baud MGJ (2020) Current strategies for the design of PROTAC linkers: a critical review. Explor Target Antitumor Ther 1(5):273–312
Tunjic TM, Weber N, Brunsteiner M (2023) Computer aided drug design in the development of proteolysis targeting chimeras. Comput Struct Biotechnol J 21:2058–2067
Wang B, Wu S, Liu J, Yang K, Xie H, Tang W (2019) Development of selective small molecule MDM2 degraders based on nutlin. Eur J Med Chem 176:476–491
Wurz RP, Dellamaggiore K, Dou H, Javier N, Lo MC, McCarter JD, Mohl D, Sastri C, Lipford JR, Cee VJ (2018) A "click chemistry platform" for the rapid synthesis of bispecific molecules for inducing protein degradation. J Med Chem 61(2):453–461
Yan S, Yan J, Liu D, Li X, Kang Q, You W, Zhang J, Wang L, Tian Z, Lu W, Liu W, He W (2021) A nano-predator of pathological MDMX construct by clearable supramolecular gold(I)-thiol-peptide complexes achieves safe and potent anti-tumor activity. Theranostics 11(14):6833–6846
Yonezawa S, Koide H, Asai T (2020) Recent advances in siRNA delivery mediated by lipid-based nanoparticles. Adv Drug Deliv Rev 154–155:64–78
You J, Ying R, Ren X, Hamilton WL, Leskovec J (2018) GraphRNN: generating realistic graphs with deep auto-regressive models. arXiv:1802.08773
Zagidullin A, Milyukov V, Rizvanov A, Bulatov E (2020) Novel approaches for the rational design of PROTAC linkers. Explor Target Antitumor Ther 1(5):381–390
Zeng S, Huang W, Zheng X, Cheng L, Zhang Z, Wang J, Shen Z (2021) Proteolysis targeting chimera (PROTAC) in drug discovery paradigm: recent progress and future challenges. Eur J Med Chem 210:112981
Zhang L, Riley-Gillis B, Vijay P, Shen Y (2019) Acquired resistance to BET-PROTACs (proteolysis-targeting chimeras) caused by genomic alterations in core components of E3 ligase complexes. Mol Cancer Ther 18(7):1302–1311
Zhang X, Luukkonen LM, Eissler CL, Crowley VM, Yamashita Y, Schafroth MA, Kikuchi S, Weinstein DS, Symons KT, Nordin BE, Rodriguez JL, Wucherpfennig TG, Bauer LG, Dix MM, Stamos D, Kinsella TM, Simon GM, Baltgalvis KA, Cravatt BF (2021) DCAF11 supports targeted protein degradation by electrophilic proteolysis-targeting chimeras. J Am Chem Soc 143(13):5141–5149. https://doi.org/10.1021/jacs.1c00990
Zhang X, Li S, Song M, Chen Y, Chang L, Liu Z, Dai H, Wang Y, Yang G, Jiang Y, Lu Y (2024) Degradation of FAK-targeting by proteolytic targeting chimera technology to inhibit the metastasis of hepatocellular carcinoma. Oncol Res 32(4):679–690
Zoppi V, Hughes SJ, Maniaci C, Testa A, Gmaschitz T, Wieshofer C, Koegl M, Riching KM, Daniels DL, Spallarossa A, Ciulli A (2019) Iterative design and optimization of initially inactive proteolysis targeting chimeras (PROTACs) identify VZ185 as a potent, fast, and selective von Hippel-Lindau (VHL) based dual degrader probe of BRD9 and BRD7. J Med Chem 62(2):699–726

Zorba A, Nguyen C, Xu Y, Starr J, Borzilleri K, Smith J, Zhu H, Farley KA, Ding W, Schiemer J, Feng X, Chang JS, Uccello DP, Young JA, Garcia-Irrizary CN, Czabaniuk L, Schuff B, Oliver R, Montgomery J, Hayward MM, Coe J, Chen J, Niosi M, Luthra S, Shah JC, El-Kattan A, Qiu X, West GM, Noe MC, Shanmugasundaram V, Gilbert AM, Brown MF, Calabrese MF (2018) Delineating the role of cooperativity in the design of potent PROTACs for BTK. Proc Natl Acad Sci U S A 115(31):E7285–E7292

Future of Trends in the Design and Development of PROTAC

6

Popat Kumbhar, Kaustubh Kolekar, Vikas Kamble, Chukwuebuka Emmanuel Umeyor, John Disouza, and Vandana B. Patravale

Abstract

Proteolysis-targeting chimeras (PROTACs), which precisely breakdown proteins of interest, have emerged as one of the most commonly employed tools in recent years for drug development. In a variety of illnesses, including malignancies, PROTAC has shown potential as a targeted therapeutic option. However, PROTACs are allied with a range of challenges. To overcome the many hurdles, choosing the right components such as the protein of interest (POI), linker, and E3 ligase ligand as well as design strategies are essential. This chapter discusses the significance of PROTACs, major trends in the future development of PROTAC, and the scope of PROTAC beyond oncology. Additionally, designs and development of PROTAC technologies and PROTACs in clinical trials are briefed.

P. Kumbhar · K. Kolekar · V. Kamble · J. Disouza
Department of Pharmaceutics, Tatyasaheb Kore College of Pharmacy, Warananagar, Maharashtra, India
e-mail: pskumbhar@tkcpwarana.ac.in; jidisouza@tkcpwarana.ac.in

C. E. Umeyor
Department of Pharmaceutical Sciences and Technology, Institute of Chemical Technology, Mumbai, Maharashtra, India

Department of Pharmaceutics and Pharmaceutical Technology, Faculty of Pharmaceutical Sciences, Nnamdi Azikiwe University, Awka, Anambra State, Nigeria
e-mail: ec.umeyor@unizik.edu.ng

V. B. Patravale (✉)
Department of Pharmaceutical Sciences and Technology, Institute of Chemical Technology, Mumbai, Maharashtra, India
e-mail: vb.patravale@ictmumbai.edu.in

M. Nandave, P. Jain (eds.), *PROTAC-Mediated Protein Degradation: A Paradigm Shift in Cancer Therapeutics*, https://doi.org/10.1007/978-981-97-5077-1_6

Keywords

PROTACs · Protein degradation · Protein of interest · E3 ligase · Linkers · Clinical trials

Abbreviations

ARCC-4	Androgen receptor PROTAC degrader-4
BCL-XL	B-cell lymphoma-extra-large
BTK	Bruton's tyrosine kinase
cIAPs	Cellular inhibitor of apoptosis
CLIPTAC	Click-formed proteolysis targeting chimera
CNS	Central nervous system
CRBN	Cereblon
CRISPR	Clustered regularly interspaced short palindromic repeats
DBPs	DNA-binding proteins
DNA	Deoxyribonucleic acid
ERK	Extracellular signal-regulated protein kinases
FAK	Focal adhesion kinase
FDA	Food and Drug Administration
GPLR	G protein-linked receptors
HER2	Human epidermal growth factor receptor 2
IAP	Inhibitor of apoptosis protein
MAP4K1	Mitogen-activated protein kinase 1
MDM2	Murine double minute 2
NVOC	Nitroveratryloxycarbonyl
PD	Pharmacodynamics
PD-1	Programmed cell death protein-1
PD-L1	Programmed cell death ligand-1
PK	Pharmacokinetics
POI	Protein of interest
PROTAC-DB	Proteolysis-targeting chimeras database
PROTACs	Proteolysis-targeting chimeras
RAS	Rat sarcoma virus
RBPs	RNA-binding proteins
RNA	Ribonucleic acid
RNF182	Ring finger protein 182
ROS	Reactive oxidation species
RT-PROTAC	Radiotherapy-triggered PROTAC
STAT-3	Signal transducer and activator of transcription-3
TALEN	Transcription activator-like effector nuclease
TP	Target protein

TRIM	Tripartite motif
VHL	von-Hippel Lindau
ZFN	Zinc-finger nuclease

6.1 Introduction

Proteolysis-targeting chimeras (PROTACs) are tiny, heterobifunctional molecules made up of a protein of interest (POI) ligands, a linker, and an E3 ligase ligand. POI is also recognized as a target protein (TP). The linker's two ends bind to the POI ligands and the E3 ligase ligand to form a ternary complex (Fig. 6.1). E3 ligase ligands have more than 600 known E3 ubiquitin ligases but only a few of them like cereblon (CRBN), von-Hippel Lindau (VHL), inhibitor of apoptosis protein (IAP), and murine double minute 2 (MDM2) are widely used in the development of PROTACs. The POI ligands target primarily cancer, neurodegenerative, and auto-inflammatory disease proteins by acting as either agonists or inhibitors of these proteins (Pei et al. 2019; Zou et al. 2019). Linkers are selected based on their chemical properties, length, hydrophilicity, and rigidity parameters. The binding leads to the recruitment of the E3 ligase to the POI, resulting in its ubiquitination and subsequent breakdown by the proteasome. Unlike the majority of small-molecule inhibitors, antagonists, and blockers, clinically employed for a long time, PROTAC offers distinct benefits. PROTAC requires no direct action on the active site, in contrast to conventional drugs. When it binds to the target protein, it has the power to make it break down. PROTACs are designed to specifically target different types of protein moieties for destruction (Pei et al. 2019; Zou et al. 2019; Pettersson and Crews

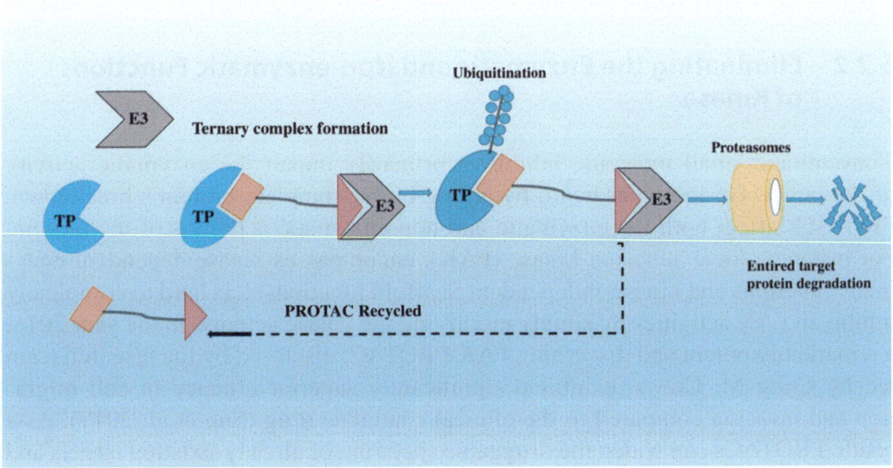

Fig. 6.1 The process of PROTAC generation and protein degradation (TP: target protein, E3: E3 ligase, PROTAC: proteolysis-targeting chimera)

2019; Dale et al. 2021). Furthermore, they can be used for enhancing specificity and selectivity toward targeted proteins, improving therapeutic potential, and minimizing drug resistance (Liu et al. 2022a, b). Along with that, a few challenges PROTACs have to face like development and complexity challenges, adverse and toxic effects, and permeability with high molecular weight compounds (Zou et al. 2019; Pettersson and Crews 2019; Dale et al. 2021). In this chapter, the significance of PROTAC development and major trends in the development of PROTAC are discussed. Further, the scope of PROTAC beyond oncology and different designs and development of PROTAC are provided.

6.2 Significance of PROTAC's Development

6.2.1 Overcoming Drug Resistance of Cancer

In the realm of cancer treatment, alongside the conventional use of chemotherapy, there has been rapid progress in the development of kinase inhibitors over the last two decades (Liu et al. 2022a, b). While these inhibitors prove highly efficacious in treating cancer, people often develop resistance to the drugs, leading to cancer recurrence. In response to this challenge, PROTACs have emerged as a more advantageous option for combating drug resistance to various cancers by effectively degrading the entire target protein. For instance, androgen receptor PROTAC Degrader, ARCC-4, which targets the androgen receptor, has demonstrated the capability to overcome drug resistance in prostate cancer associated with enzalutamide (Wu et al. 2015). Similarly, PROTAC Degrader of Bruton's tyrosine kinase (BTK) mutants L18I, has exhibited efficacy in overcoming the series of clinically relevant BTK mutants-C481S alteration (Salami et al. 2018).

6.2.2 Eliminating the Enzymatic and Non-enzymatic Functions of Kinase

Conventional small molecule inhibitors primarily impair the enzymatic activity of the target. On the other hand, by aiding in the complete protein's breakdown, PROTACs affect both the enzymatic and non-enzymatic activities of the protein. For instance, focal adhesion kinase (FAK), encompasses kinase-dependent enzymatic functions and kinase-independent Scaffold functions. It is hard to completely inhibit all FAK activities by simply modifying the kinase activity. In the year 2018, a remarkably potent and discerning FAK PROTAC, disclosed by the research team led by Craig M. Crews, exhibited significantly superior efficacy in cell migration and invasion compared to the clinical candidate drug (Sun et al. 2019). As a result, PROTACs can widen the druggable spectrum of already-existing targets and modulate proteins that are tough to control through conventional small-molecule inhibitors.

6.2.3 Degrade the "Undruggable" Protein Target

Nowadays, conventional drug discovery techniques can only target nearly 20–25% of established target proteins, comprising kinases, G protein-linked receptors (GPLR), nuclear receptors, and iron channels (Cromm et al. 2018; Wells and McClendon 2007). Proteins lacking catalytic activity or featuring catalytically autonomous functions are still considered "undruggable" targets. Despite the attractiveness of the signal transducer and activator of transcription-3 (STAT3) as a therapeutic target due to its involvement in multiple signaling pathways, the absence of an overtly druggable site on STAT3's surface hampers the development of STAT3 inhibitors. Consequently, the Food and Drug Administration (FDA) has yet to approve any directly targeted drugs against STAT3. Wang's group created a potent PROTAC that targets STAT3 exclusively and has significant biological effects both in vitro and in vivo (Kim et al. 2014). This effective instance underscores the pivotal usefulness of the PROTAC technique, particularly in the realm of "undruggable" targets like Kirsten-rat sarcoma viral oncogene homolog (K-Ras), a challenging tumor target that can be activated by numerous mutations including G12A, G12C, G12D, G12S, G12V, G13C, and G13D in clinical settings (Bai et al. 2019).

6.2.4 Quick and Reversible Chemical Silencing Approach In Vivo

Conventional approaches for genetic protein suppression, including transcription activator-like effector nuclease (TALEN) and zinc-finger nuclease (ZFN), or clustered regularly interspaced short palindromic repeats (CRISPR)-Cas9, typically involve extended timelines, irreversible mechanisms, and substantial costs, posing significant challenges for research, particularly in non-human primate studies. Moreover, these genetic animal models occasionally yield phenotypic misconceptions attributable to possible gene mutation. Notably, traditional genetic approaches are unsuitable for investigating the in vivo functionality of genes that are lethal to embryos. In contrast to DNA-based protein knockout methods, PROTACs directly reduce target proteins without intervening at the genomic level, making them well-suited for exploring the functions of embryonic-lethal proteins in mature organisms. Furthermore, PROTACs offer precise spatiotemporal control, enabling targeted protein reduction at specific intervals and target protein recovery upon end of drug treatment. PROTACs are a novel, quick, and repeatable chemical cut-down method that effectively supplements current genetic techniques (Jinesh et al. 2018).

6.3 Major Trends in the Future Development of PROTAC

PROTAC technology has essentially attained an advanced point in recent years. It is proposed that the next major achievements in the PROTAC realm will focus on four distinct domains involving finding the most promising targets for protein degradation, expanding the therapeutic uses of E3 ligases, broadening the clinical treatment space outside of oncology, and developing new PROTAC approaches (Békés et al. 2022).

6.3.1 Identifying Optimal Targets for Protein Degradation

Clinically validated targets, such as the androgen and estrogen receptors selected by Arvinas, comprise the first group of proteolytic agents in the clinical phase. Specific products aimed at these targets have shown moderate success, confirming the therapeutic potential of PROTACs. Nevertheless, the true promise of this treatment paradigm lies in addressing challenging or traditionally undruggable targets not effectively targeted by existing treatment modalities. PROTAC targets exhibit diverse characteristics, which encompass (a) proteins that are expressed excessively, mutated, aggregate, express in isoforms, or pinpoint inside cells and cause gain-of-function disease, (b) having an attaching region that is open for E3 ligase, (c) preferably with an unstructured area that can reach the proteasome (Békés et al. 2022).

6.3.2 Expansion of Available E3 Ligases

The human genome contains over 600 encoded E3 ubiquitin ligases, yet only a limited number, such as VHL and CRBN, are currently employed in the design of PROTACs. Expanding the repertoire of E3 ubiquitin ligases applicable to PROTAC technology stands out as a significant challenge confronting PROTAC development. A distinctive advantage of PROTAC molecules lies in their ability to be judiciously designed based on the target. This involves the selection of molecules and linkers that effectively recruit the ligase. The modular nature of PROTACs allows for the substitution of the E3 ligase binding module, catering to diverse targets. This adaptability enables the recruitment of specific E3 ligases, optimizing therapeutic efficacy. Additional E3 ligases can be identified through the following methods (Békés et al. 2022).

6.3.2.1 Ubiquitous Ligase

An effective and valuable strategy involves the identification of widely applicable and prevalent ligases, akin to CRBN and VHL. These ligases are flexible enough to be used in a wide range of therapeutic indications since they can be coupled with any sort of protein. Despite the existence of numerous ligases, current PROTAC molecules predominantly utilize a limited selection, information about which is accessible through the online proteolysis-targeting chimeras database (PROTAC-DB). Intriguingly, employing VHL or CRBN for the same target protein may yield distinct degradation efficiencies. For instance, in the development of protein degraders targeting KRAS G12C, previous investigations revealed the feasibility of utilizing the PROTAC system, with the degradation outcome being influenced by the choice between recruiting VHL or CRBN as the E3 ligase. Ligases suitable for PROTAC-mediated protein degradation can be pinpointed by assessing the structural traits of these ligases. The ability to anticipate the three-dimensional structure of proteins from amino acid sequences, as demonstrated in previous studies, may modify the ligase selection criteria used in the creation of protein degradation treatments (Békés et al. 2022).

6.3.2.2 Tissue and Cell-Specific E3 Ligases

A distinct approach that emphasizes essential ligase characteristics such as tissue and cell preference, tumor development, and tumor need to prepare the way for the development of domain-driven protein degradation treatments. Previous research has pinpointed several E3 ligases expressed in distinct tissues, exemplified by ring finger protein 182 (RNF182) and tripartite motif (TRIM9), specifically expressed in the central nervous system (CNS). This specificity holds potential significance for targeting neuronal diseases. Targeted degradation in the CNS can circumvent systemic off-target effects and mitigate toxic side effects. Interestingly, certain ligases exhibit "reverse specificity," characterized by low expression levels in particular tissues and cells. This attribute may also create opportunities for the development of protein degraders. A notable instance is the low expression of VHL in platelets. Consequently, the PROTAC molecule DT2216, recruiting VHL, can avoid platelet damage while degrading b-cell lymphoma-extra-large (BCL-XL), thereby reducing platelet-induced toxicity and enhancing its therapeutic index. Currently, DT2216 is undergoing Phase 1 clinical development (Békés et al. 2022).

6.3.2.3 E3 Ligases Enriched in Diseased Tissues or Cells

A novel frontier in targeted protein degradation comprises PROTAC molecules customized to target tumor cells precisely. This is attainable through focusing on E3 ligases specific or enriched in tumors. Generally, E3 ligases abundant in tumors often align with tumor-dependent E3 ligases. This alignment aids in identifying E3 ligases crucial for various cancer cell types, offering the advantage that tumor cells find it challenging to develop resistance to PROTAC molecules via ligase mutations. Exploring, investigating, and developing tumor-specific E3 ligases for protein degradation constitute just one strategy to attain tumor-specific PROTAC molecules. Another approach capitalizes on the specificity of antibodies for tumors by conjugating degradation molecules to tumor-specific antibodies. Research conducted by Genentech and its partners has demonstrated the tumor-targeting specificity of antibody-conjugated PROTAC molecules. However, it is worth noting that antibody-conjugated PROTACs lose the oral characteristics inherent in small-molecule PROTACs (Békés et al. 2022).

6.3.3 Scope of PROTAC Beyond Oncology

Until now, the primary focus of protein degrader research has been in the field of oncology. However, given the versatility of protein degraders to target any chosen protein, their application extends far beyond oncology. Currently, protein degraders have progressively proven promise in a variety of applications, especially neurodegenerative illnesses, marking a breakthrough in the inflammation/immunology domain.

Following immune checkpoint inhibitors' accomplishment, the creation of small-molecule medications to boost anticancer immune responses has emerged as a critical field in drug research. PROTAC molecules, operating as small molecule drugs, can activate immune cells and replicate the effects of PD-1/PD-L1 targeted therapy, emerging as a potential "first-in-class" therapy. Notably, recent preclinical studies with PROTAC molecules targeting mitogen-activated protein kinase 1 (MAP4K1) have demonstrated promising activity. While small molecule inhibitors can serve a similar purpose, PROTAC molecules may offer better tolerability and lower toxic side effects. Because PROTAC molecules lack an active site, they have the unique ability to degrade proteins that are not affected by conventional small-molecule inhibitors. This attribute renders proteins that accumulate in various neurodegenerative diseases, potential targets for PROTACs (Békés et al. 2022).

6.3.4 Develop Innovative PROTAC Modes

In addition to PROTAC, several innovative technologies for protein degradation have emerged, significantly broadening the spectrum of targets accessible through this methodology. PROTAC techniques that are considered advanced, including hybrid PROTACs and biology-based PROTACs, have the potential to expand the range of targets that can be reached by PROTACs.

Bio-PROTACs employing biological products such as peptides, fusion proteins, and oligonucleotides serve as ligands for target recognition. In research, PROTACs based on fusion proteins or peptides have successfully degraded cancer-related targets like human epidermal growth factor receptor 2 (HER2), MYC, and KRAS. Oligonucleotides-based PROTACs offer the ability to target transcription factors and RNA-binding proteins, known for their involvement in various cancers, obesity, cardiovascular issues, and neurological diseases due to mutations.

While these bio-PROTACs hold promise as a new treatment modality, overcoming hurdles related to their delivery and administration is imperative. These therapies might necessitate a delivery mechanism akin to gene therapy, utilizing precisely regulated promoters to control expression levels within cells. Advancements in nanoparticle-based delivery technologies and viral vector-based gene delivery systems will play a pivotal role in the progress of this therapeutic model (Békés et al. 2022).

6.4 Designs and Development of PROTACs

6.4.1 Peptide-Based PROTACs

This sort of PROTAC incorporates peptide ligands for POI, imitating the sequences noticed in native proteins binding to the POI. For instance, signal transducer and activator of transcription 3 (STAT3), a crucial transcription factor linked to oncology inception and development (Johnson et al. 2018). SI-109 is a peptide-based PROTAC that was created using SD-36, a peptide that was obtained through the gp130 protein's STAT3-binding domain. This innovation resulted in robust STAT3 degradation, effectively suppressing blood cancer and lymph cancer both in vitro and in vivo (Bai et al. 2019). Benefits of this kind of PROTACs encompass improved target selectivity, chemical production, and binding capacity. Nevertheless, their capacity to tolerate digestion and membrane permeability is limited (Jin et al. 2020a, b). Consequently, peptide-based PROTACs are typically administered intravenously and are particularly advisable for treating different types of proteins or addressing hematological diseases.

6.4.2 Small Molecule-Based PROTACs

FDA-approved anticancer inhibitors serve as a viable approach to constructing ligands for the POI in small molecule-based PROTACs. A case in point is the prevalence of the BRAFV600E mutation in melanoma and colorectal cancer, which activates oncogenic extracellular signal-regulated protein kinases (ERK) signaling independent activation of RAS (Poulikakos and Rosen 2011). Posternak et al. employed the small-molecule BRAFV600E inhibitor BI882370 as the POI ligand and pomalidomide as the E3 ligand for CRBN. The resulting PROTAC P4B effectively facilitated BRAF degradation, thereby inhibiting melanoma and colon cancer harboring the BRAF mutation (Posternak et al. 2020). In contrast to peptide-based PROTACs, those based on small molecules exhibit enhanced cell penetration and aversion to digestion. This characteristic allows for diverse modes of administration and an expanded scope of target options. It is essential to note that small-molecule ligands often lack target specificity (Bai et al. 2019), emphasizing the need for careful consideration when selecting or optimizing small-molecule ligands for the POI.

6.4.3 Nucleotide-Based PROTACs

These PROTACs utilize oligonucleotides as ligands for the POI. Human malignancies exhibit up-regulation or excessive stimulation of certain RNA-binding proteins (RBPs) such as Lin28, IGF2BPs, and Musashi-1/2, as well as DNA-binding proteins (DBPs) like NF-κB, c-myc, and STAT3, which foster tumorigenesis and cancer progression. Nevertheless, since these sorts of proteins lack suitable binding sites, the difficulty arises in creating medications that specifically target RBPs or DBPs. For example, by interacting with the let-7 precursor (pre-let-7), an especially shielded RBP, Lin28 drives carcinogenesis by impeding the synthesis of let-7, a tumor-suppressive microRNA (Posternak et al. 2020). Still, Lin28 fails to possess precise targetable regions that may be employed in small-molecule therapy. Ghidini et al., created RNA-PROTAC, a nucleotide-based PROTAC, by employing Lin28-binding oligoribonucleotides made from pre-let-7 as the POI ligand. This novel

PROTAC demonstrated exceptional minimally toxic and highly selective Lin28 breakdown in leukemia cells (Ghidini et al. 2021). Additionally, Samarasinghe et al., presented TRAFTAC, a DNA-based PROTAC that identifies the transcription factor NF-κB for targeted breakdown by employing DNA sequences as the POI ligand (Samarasinghe et al. 2021). These nucleotide-based PROTACs propose an innovative way to combat cancer alongside extending the ideal basis for PROTACs.

6.4.4 Folate-Conjugated PROTACs

Healthy cells possess minimal or no expression of the folate receptor, whereas tumor cells involving the lung, ovary, and breast have substantial amounts of folate receptor-α, making them specialised targets for delivering medications. These cells have an affinity to transport folate toward cells. Given this, many approaches have been utilized in cancer-specific drug delivery systems and some are already in clinical trials. This advice applies a folate-conjugated strategy to target particular cancer cells with PROTAC administration, thereby preventing detrimental side effects on healthy cells. For instance, Liu and co-workers fabricated folate-conjugated PROTAC and investigated the targeted degradation of protein in both cancerous and normal cells. They found that the folate-conjugated PROTACs can deactivate proteins at folate expression (folate-ARV-771, folate-MS432, and folate-MS99) in cancer cells. This confirms that the targeted degradation of POIs in cancer cells is achieved by using folate-conjugated PROTACs (Liu et al. 2022a, b).

6.4.5 Hypoxia-Activated PROTACs

Solid tumors are frequently linked to an abnormal circulatory system, which impairs medicine delivery and causes a lack of oxygen or insufficient supply of nutrients. Hypoxia causes aggressiveness, metastasis, and resistance to chemotherapy and radiotherapy at tumor sites. In the interim, it provides a range of hypoxia-activated prodrugs, targets for medications, and more for the management of various sorts of tumors. For example, Cheng and the research team successfully fabricated tumor hypoxia-activated proteolysis targeting chimeras (ha-PROTACs) by incorporating the hypoxia-activated leaving group, into the cereblon (CRBN) E3 ligand of an epidermal growth factor receptor 19 deletions (EGFRDel19)-based PROTAC. They observed that EGFRDel19 was efficiently and selectively deactivated at the tumor location by ha-PROTACs. The heightened potency in inhibiting cell viability and migration, along with promoting cell apoptosis in tumor hypoxia was also noticed with PROTAC. Moreover, PROTAC successfully released the active moiety. This work validates the application of a hypoxia-activated CRBN E3 ligase ligand in ha-PROTACs to boost the selectivity of PROTACs (Liu et al. 2021).

6.4.6 Radiotherapy-Triggered PROTAC (RT-PROTAC)

Conventional PROTACs are considerably targeting undruggable proteins, but somehow they exhibit restricted use in clinical practice owing to their evident systemic toxicity attributable to untargeted degradation of protein. Therefore, RT-PROTACs are employed for cancer therapy. This tactic uses X-ray radiation to destroy DNA, induce tumor necrosis, and cause cell apoptosis. Additionally, they promote specific chemical reactions. For the past few decades, X-ray radiation has been exploited to regulate the release of drugs from chemically modified cages by initiating a chemical reaction. The medication is then released at a specified time and area to work on the tumor site. In light of this, Yang et al. developed RT-PROTAC to precisely regulate medication activation and protein breakdown caused by radiation therapy. For formulating RT-PROTAC, they directly incorporated X-rays into a bromodomain (BRD)-targeting PROTAC. The in vitro and in vivo study demonstrated that the X-ray radiation actives prodrugs and degrades BRD targeting protein resulting in a synergistic antitumor potency with radiotherapy. Additionally, an in vivo study revealed the controlled degradation of proteins and the reduction of PROTACs' systemic toxicity by achieving a formulation that targets a particular pathway (Cheng et al. 2023).

6.4.7 Reactive Oxidation Species (ROS)-Responsive PROTAC

ROS is a list of oxidants that contains superoxide radicals, hydrogen peroxide (H_2O_2), and hydroxyl radicals. ROS are considerably more prevalent in cancer cells than in normal cells. As a result of this, the ROS-responsive PROTAC has a greater affinity for cancer cells, increasing the cancer cells' selectivity while lessening its toxicity to healthy cells. Given this, Liu and co-workers successfully prepared ROS-responsive PROTACs containing CRBN E3 ligase for specific suppression of BRD3 along with aryl boronic acid. They noticed that the PROTAC containing aryl boronic acid and H_2O_2 rapidly changed into a transitional compound derived from phenols along with the elimination of 1,6-benzyl, releasing the active PROTAC at the tumor site whereas the analog of the boronic group containing glutarimide was unsuccessful to release the drug payload. In addition, compared to typical healthy cells, higher inhibition of specific targeted proteins was observed (Yang et al. 2022).

6.5 New Avenues in PROTAC Design and Development Technologies

6.5.1 PhotoPROTAC

High spatiotemporal precision light finds widespread usage in biological research and illness therapy by means of the optically controlled development or release of active small-molecule modulators (Szymanski et al. 2013). Under the influence of light, some molecular components, including azobenzene, can alter either irreversibly or reversibly, affecting the molecules' spatial configuration in addition to their physical, chemical, and biological traits. This idea enthused the invention of PhotoPROTAC, which employs photoswitches (Hull et al. 2018) or photocages (Silva et al. 2019) to control PROTAC performance spatiotemporally.

By adding a photoswitchable moiety like azobenzene to the linker or E3 ligand, photoswitchable PROTACs impose reversible optical regulation of the breakdown of proteins. The PROTAC is incapable of creating a stable ternary complex as it preserves an inactive conformation when exposed to insufficient light. When the PROTAC is subjected to light with the specified wavelength, it transforms into the active conformation and forms an operational ternary complex which results in the target degradation (Fig. 6.2). For instance, Reynders et al. constructed PROTACs containing an azobenzene portion on the linker that hinder acute lymphoblastic leukemia cells and destruct BET class proteins under 390 nm UV light. Interestingly, target decomposition might be blocked by turning the PROTAC into an inactive conformation at a suitable light exposure (for example, 500 nm for azobenzene-containing PROTAC) (Jin et al. 2020a, b; Pfaff et al. 2019; Reynders et al. 2020).

Photolabile inhibiting groups, notably the nitroveratryloxycarbonyl (NVOC), are integrated into photocaged PROTACs and trigger irreversible protein breakdown. The photocaging domain on the E3 ligand precludes PROTAC and the E3 ligase from interacting when there is no light stimulation. The photocaging group splits from PROTAC when exposed to light, which makes it simpler for the POI-PROTAC-E3 ternary complex to develop (Liu et al. 2020; Naro et al. 2020; Kounde et al. 2020; Xue et al. 2019). As an illustration, Liu et al. constructed photocaged PROTACs through the use of NVOC on the CRBN ligand. These opto-PROTACs, when subjected to 365 nm UV irradiation, effectively degraded IKZF1/3, BRDs, or

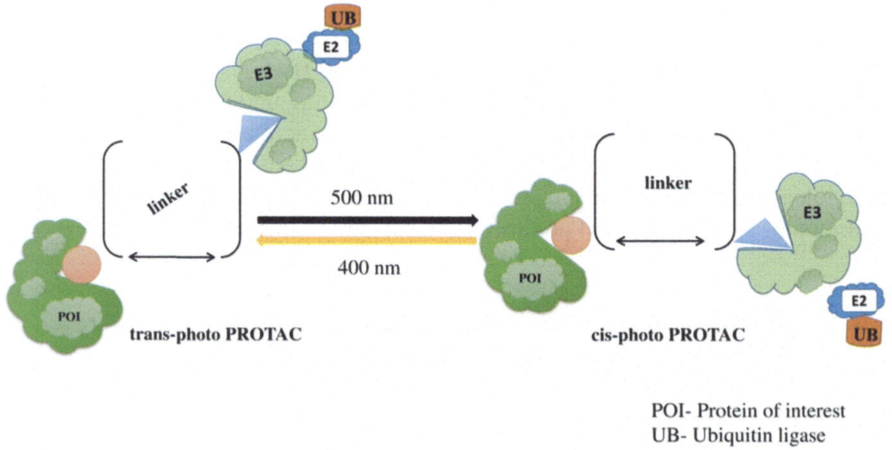

Fig. 6.2 Light-sensitive transformation of PhotoPROTAC (POI: protein of interest, UB: ubiquitin)

the ALK fusion protein utilizing corresponding POI ligands, thereby suppressing malignant cell proliferation in an optically controlled manner (Liu et al. 2020).

6.5.2 CLIPTAC

PROTACs typically possess a substantial molecular weight, imposing limitations on their solubility, pharmacokinetics, and bioavailability. Addressing the challenge of reducing the molecular weight of PROTACs is crucial. Chemical reactions known as "click chemistry" have the advantages of gentle reaction conditions, substantial yields, selectivity, and wide utility (Kolb et al. 2001). Click chemistry has grown into a basic technique for covalently remodeling macromolecules in physiological settings over time. It is especially well-suited for creating conjugated frameworks inside cells from two tiny precursors. The process of formation of CLIPTAC via click chemistry is depicted in Fig. 6.3 (Ji et al. 2019; Smeenk et al. 2021). Lebraud et al. designed a tetrazine-tagged E3 and a *trans*-cyclooctene-tagged POI ligand as precursors, taking cues from the ideas of click chemistry. Tetrazine and trans-cyclooctene united to generate integrated PROTACs, or CLIPTACs, inside of cells via the click reaction resulting in a covalent six-membered ring motif. Notably, these CLIPTACs successfully degraded oncogenic BRD4 or ERK1/2 (Lebraud et al. 2016). As a result, CLIPTAC has emerged as an appealing strategy for mitigating the molecular weight of PROTACs.

6.5.3 Tag-Based PROTAC

It necessitates a lot of effort and time to make PROTAC initially. It entails chemical synthesis, molecular design, and animal or cell-based testing (Cecchini et al. 2021). For PROTAC research to succeed, an apt E3 ligase/E3 ligand system must be chosen. The entire spectrum of POI-E3 ligase interactions is yet unknown, nevertheless, considering the human proteome contains over 600 E3 ligases. Consequently, tag-based PROTAC systems involve the fusion of tag and POI resulting tag-POI

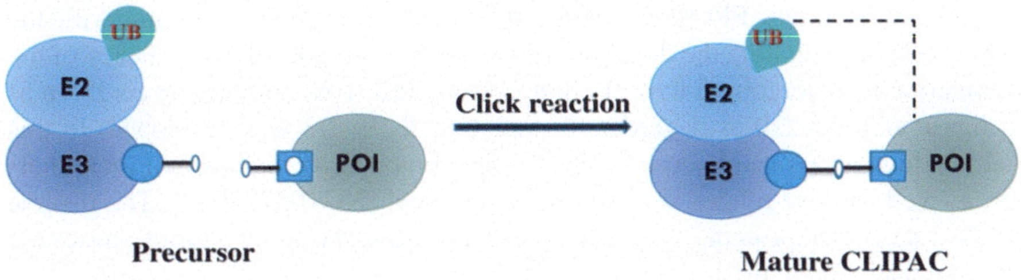

Fig. 6.3 CLIPTAC formation via click chemistry (POI: protein of interest, UB: ubiquitin)

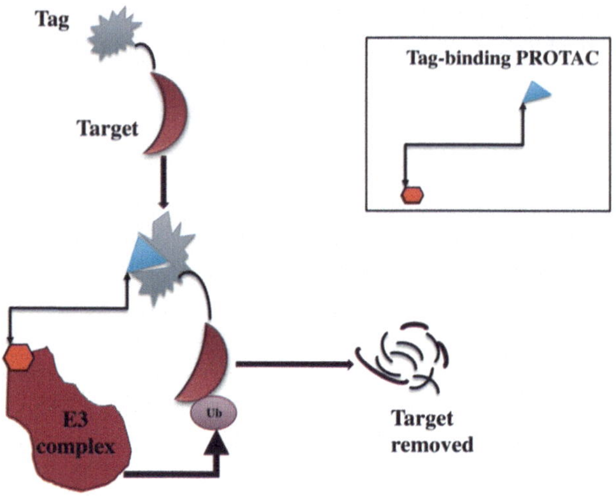

Fig. 6.4 Process of protein degradation using tag-based PROTAC through E3 complex (E3: E3 ligase, Ub: ubiquitin)

protein complex in cells. The potential E3 ligase and tag of the tag-POI protein are brought in through administering a universal PROTAC molecule. The tag-POI protein abundance facilitates assurance of the potential E3 ligase's ability to begin POI breakdown (Fig. 6.4). Two notable tag-based PROTACs are dTAG and Halo-PROTAC (Buckley et al. 2015; Tomoshige et al. 2015; England et al. 2015; Nabet et al. 2018). These tag-based PROTACs cannot be considered suited as therapies for the management of disease, even if they demonstrate potential as molecular tools for determining whether the recommended E3 ligase/E3 ligand system is appropriate for PROTAC.

6.6 PROTACs in Clinical Trials

Several PROTACs are available in clinical trials for a variety of indications including multiple myeloma, acute myeloid leukemia, non-small cell lung carcinoma, etc. CC-220 monotherapy PROTAC for the treatment of multiple myeloma with the objective of investigating the dose, safety, tolerability, drug level, and efficacy of the formulation is undergoing Phase II clinical trials. PROTAC comprising cereblon as E3 ligase and CC-220 as degrader for targeting IKZF 1/3 was developed. In this trial, the highest allowable dose of CC-220 in combination therapy and monotherapy, as well as dose-related toxicity, was examined [NCT02773030]. The diverse PROTACs targeted protein degraders undergoing clinical trials at various phases are summarized in Table 6.1.

Table 6.1 Diverse PROTACs targeted protein degraders undergoing clinical trials

Clinical trial no.	Phase	Indications	Target	Degrader	E3 ligase
NCT02773030	Phase II	Multiple myeloma	IKZF 1/3	CC-220	Cereblon
NCT03989414	Phase II	Multiple myeloma	IKZF 1/3	CC-92480	Cereblon
NCT02848001 NCT04336982	Phase II	Acute myeloid leukemia	GSPT 1	CC-90009	Cereblon
NCT04072952	Phase II	Breast cancer	Estrogen receptor	ARV-471	Cereblon
NCT03930953 NCT04434196	Phase I	Non-Hodgkin lymphoma, chronic myeloid leukemia	IKZF 1/3	CC-99282	Cereblon
NCT04756726	Phase I	Multiple myeloma	IKZF 1/3	CFT7455	Cereblon
NCT03891953	Phase I	Non-small cell lung carcinoma	Helios	DKY709	Cereblon
NCT05080842	Phase I	Breast cancer	Estrogen receptor	AC682	Cereblon
NCT04428788	Phase I	Prostate cancer	Androgen receptor	CC-94676	Cereblon
NCT04886622	Phase I	Liquid and solid tumors	BCL-xL	DT2216	von-Hippel Lindau

6.7 Challenges in Future of PROTAC Technology in Drug Development

The future of PROTAC technology in drug development is allied with range of challenges. As of now, only a single instance of PROTAC targeting an "undruggable" target has been documented (Bai et al. 2019); additional instances are requisite to validate the potential benefits of PROTAC in addressing "undruggable" targets in the future. The process of reinforcing protein-protein connections via small molecule modulators of E3 ligases is illustrated by the notion of "molecular glue" encountered in nature. Considering the auxin (plant hormone) that links to the ligase SCF-TIR1, prompting the Aux/IAA proteins to be activated and the breakdown process to begin. Additionally, certain small molecules capable of inducing the desired protein decomposition through the "molecular glue" mode of action have been documented (Thiel et al. 2012; Chamberlain et al. 2019). Moreover, recent findings suggest that certain PROTACs may accomplish desired protein decomposition by employing a mechanism that involves "molecular glue" or solely through "molecular glue" (Yang et al. 2019). Distinguishing between these two mechanisms and determining how to effectively integrate them present challenges for future research. Additionally, due to its catalytic mode of action, traditional approaches are inadequate for accurately assessing the pharmacokinetics (PK) and pharmacodynamics

(PD) attributes of PROTACs. Consequently, there is an urgent need for further investigations to establish evaluation systems for the PK and PD of PROTACs. Another challenge lies in swiftly and effectively screening for ligands targeting the proteins essential for PROTACs, particularly those involved in protein-protein interactions. Knowing the specificity, degrading activity, and possible off-target consequences which vary based on the targets, cell lines, and animal models alongside the reasonable layout of PROTACs is yet unexplored. A small number of the approximately 600 E3 ubiquitin ligases that the human genome encodes including VHL, CRBN, cellular inhibitor of apoptosis (cIAPs), and MDM2 are used in PROTAC design. An additional hurdle in this sector is to broaden the range of E3 ubiquitin ligases (Gao et al. 2020).

6.8 Conclusion

The PROTACs are small heterobifunctional ternary complex comprised of POI, E3 ligase ligand, and linker. In recent years, PROTAC has demonstrated considerable improvements with respect to its catalytic mechanism-based solubility, permeability, stability, and affinity. PROTAC technology has the potential to mitigate drug resistance, enhance cell migration efficacy, increase undrugability, etc. However, there are a couple of problems with PROTACs in the context of the technology used in drug development pertaining to protein targeting, such as undruggable, molecular glue, pharmacokinetics, pharmacodynamics, challenges with selectively screening ligand targeting during protein-protein interaction, and expedition E3 ubiquitin ligase. These challenges are overcome by utilizing optimal target protein degrader, expansion of E3 ligase, clinical trials, and fabrication of newer PROTAC design techniques.

References

Bai L, Zhou H, Xu R, Zhao Y, Chinnaswamy K, McEachern D, Wang S (2019) A potent and selective small-molecule degrader of STAT3 achieves complete tumor regression in vivo. Cancer Cell 36(5):498–511

Békés M, Langley DR, Crews CM (2022) PROTAC targeted protein degraders: the past is prologue. Nat Rev Drug Discov 21(3):181–200

Buckley DL, Raina K, Darricarrere N, Hines J, Gustafson JL, Smith IE, Crews CM (2015) HaloPROTACs: use of small molecule PROTACs to induce degradation of HaloTag fusion proteins. ACS Chem Biol 10(8):1831–1837

Cecchini C, Pannilunghi S, Tardy S, Scapozza L (2021) From conception to development: investigating PROTACs features for improved cell permeability and successful protein degradation. Front Chem 9:672267

Chamberlain PP, D'Agostino LA, Ellis JM, Hansen JD, Matyskiela ME, McDonald JJ et al (2019) Evolution of cereblon-mediated protein degradation as a therapeutic modality. ACS Med Chem Lett 10(12):1592–1602

Cheng W, Li S, Han S, Miao R, Wang S, Liu C, Zhang X (2023) Design, synthesis and biological evaluation of the tumor hypoxia-activated PROTACs bearing caged CRBN E3 ligase ligands. Bioorg Med Chem 82:117237

Cromm PM, Samarasinghe KT, Hines J, Crews CM (2018) Addressing kinase-independent functions of Fak via PROTAC-mediated degradation. J Am Chem Soc 140(49):17019–17026
Dale B, Cheng M, Park KS, Kaniskan HÜ, Xiong Y, Jin J (2021) Advancing targeted protein degradation for cancer therapy. Nat Rev Cancer 21(10):638–654
England CG, Luo H, Cai W (2015) HaloTag technology: a versatile platform for biomedical applications. Bioconjug Chem 26(6):975–986
Gao H, Sun X, Rao Y (2020) PROTAC technology: opportunities and challenges. ACS Med Chem Lett 11(3):237–240
Ghidini A, Cléry A, Halloy F, Allain FH, Hall J (2021) RNA-PROTACs: degraders of RNA-binding proteins. Angew Chem 133(6):3200–3206
Hull K, Morstein J, Trauner D (2018) In vivo photopharmacology. Chem Rev 118(21):10710–10747
Ji X, Pan Z, Yu B, De La Cruz LK, Zheng Y, Ke B, Wang B (2019) Click and release: bioorthogonal approaches to "on-demand" activation of prodrugs. Chem Soc Rev 48(4):1077–1094
Jin YH, Lu MC, Wang Y, Shan WX, Wang XY, You QD, Jiang ZY (2020a) Azo-PROTAC: novel light-controlled small-molecule tool for protein knockdown. J Med Chem 63(9):4644–4654
Jin J, Wu Y, Chen J, Shen Y, Zhang L, Zhang H, Luan X (2020b) The peptide PROTAC modality: a novel strategy for targeted protein ubiquitination. Theranostics 10(22):10141
Jinesh GG, Sambandam V, Vijayaraghavan S, Balaji K, Mukherjee S (2018) Molecular genetics and cellular events of K-Ras-driven tumorigenesis. Oncogene 37(7):839–846
Johnson DE, O'Keefe RA, Grandis JR (2018) Targeting the IL-6/JAK/STAT3 signalling axis in cancer. Nat Rev Clin Oncol 15(4):234–248
Kim J, Kim H, Park SB (2014) Privileged structures: efficient chemical "navigators" toward unexplored biologically relevant chemical spaces. J Am Chem Soc 136(42):14629–14638
Kolb HC, Finn MG, Sharpless KB (2001) Click chemistry: diverse chemical function from a few good reactions. Angew Chem Int Ed 40(11):2004–2021
Kounde CS, Shchepinova MM, Saunders CN, Muelbaier M, Rackham MD, Harling JD, Tate EW (2020) A caged E3 ligase ligand for PROTAC-mediated protein degradation with light. Chem Commun 56(41):5532–5535
Lebraud H, Wright DJ, Johnson CN, Heightman TD (2016) Protein degradation by in-cell self-assembly of proteolysis targeting chimeras. ACS Cent Sci 2(12):927–934
Liu J, Chen H, Ma L, He Z, Wang D, Liu Y, Wei W (2020) Light-induced control of protein destruction by opto-PROTAC. Sci Adv 6(8):eaay5154
Liu J, Chen H, Liu Y, Shen Y, Meng F, Kaniskan HU, Wei W (2021) Cancer selective target degradation by folate-caged PROTACs. J Am Chem Soc 143(19):7380–7387
Liu Z, Hu M, Yang Y, Du C, Zhou H, Liu C, Xie Y (2022a) An overview of PROTACs: a promising drug discovery paradigm. Mol Biomed 3(1):46
Liu H, Ren C, Sun R, Wang H, Zhan Y, Yang X, Chen H (2022b) Reactive oxygen species-responsive pre-PROTAC for tumor-specific protein degradation. Chem Commun 58(72):10072–10075
Nabet B, Roberts JM, Buckley DL, Paulk J, Dastjerdi S, Yang A, Bradner JE (2018) The dTAG system for immediate and target-specific protein degradation. Nat Chem Biol 14(5):431–441
Naro Y, Darrah K, Deiters A (2020) Optical control of small molecule-induced protein degradation. J Am Chem Soc 142(5):2193–2197
Pei H, Peng Y, Zhao Q, Chen Y (2019) Small molecule PROTACs: an emerging technology for targeted therapy in drug discovery. RSC Adv 9(30):16967–16976
Pettersson M, Crews CM (2019) Proteolysis targeting chimeras (PROTACs)—past, present and future. Drug Discov Today Technol 31:15–27
Pfaff P, Samarasinghe KT, Crews CM, Carreira EM (2019) Reversible spatiotemporal control of induced protein degradation by bistable PhotoPROTACs. ACS Cent Sci 5(10):1682–1690
Posternak G, Tang X, Maisonneuve P, Jin T, Lavoie H, Daou S, Sicheri F (2020) Functional characterization of a PROTAC directed against BRAF mutant V600E. Nat Chem Biol 16(11):1170–1178
Poulikakos PI, Rosen N (2011) Mutant BRAF melanomas—dependence and resistance. Cancer Cell 19(1):11–15

Reynders M, Matsuura BS, Bérouti M, Simoneschi D, Marzio A, Pagano M, Trauner D (2020) PHOTACs enable optical control of protein degradation. Sci Adv 6(8):eaay5064

Salami J, Alabi S, Willard RR, Vitale NJ, Wang J, Dong H, Crews CM (2018) Androgen receptor degradation by the proteolysis-targeting chimera ARCC-4 outperforms enzalutamide in cellular models of prostate cancer drug resistance. Commun Biol 1(1):1–9

Samarasinghe KT, Jaime-Figueroa S, Burgess M, Nalawansha DA, Dai K, Hu Z et al (2021) Targeted degradation of transcription factors by TRAFTACs: transcription factor targeting chimeras. Cell Chem Biol 28(5):648–661

Silva JM, Silva E, Reis RL (2019) Light-triggered release of photocaged therapeutics—where are we now? J Control Release 298:154–176

Smeenk ML, Agramunt J, Bonger KM (2021) Recent developments in bioorthogonal chemistry and the orthogonality within. Curr Opin Chem Biol 60:79–88

Sun Y, Ding N, Song Y, Yang Z, Liu W, Zhu J, Rao Y (2019) Degradation of Bruton's tyrosine kinase mutants by PROTACs for potential treatment of ibrutinib-resistant non-Hodgkin lymphomas. Leukemia 33(8):2105–2110

Szymanski W, Beierle JM, Kistemaker HA, Velema WA, Feringa BL (2013) Reversible photocontrol of biological systems by the incorporation of molecular photoswitches. Chem Rev 113(8):6114–6178

Thiel P, Kaiser M, Ottmann C (2012) Small-molecule stabilization of protein–protein interactions: an underestimated concept in drug discovery? Angew Chem Int Ed 51(9):2012–2018

Tomoshige S, Naito M, Hashimoto Y, Ishikawa M (2015) Degradation of HaloTag-fused nuclear proteins using bestatin-HaloTag ligand hybrid molecules. Org Biomol Chem 13(38):9746–9750

Wells JA, McClendon CL (2007) Reaching for high-hanging fruit in drug discovery at protein–protein interfaces. Nature 450(7172):1001–1009

Wu P, Nielsen TE, Clausen MH (2015) FDA-approved small-molecule kinase inhibitors. Trends Pharmacol Sci 36(7):422–439

Xue G, Wang K, Zhou D, Zhong H, Pan Z (2019) Light-induced protein degradation with photocaged PROTACs. J Am Chem Soc 141(46):18370–18374

Yang J, Li Y, Aguilar A, Liu Z, Yang CY, Wang S (2019) Simple structural modifications converting a bona fide MDM2 PROTAC degrader into a molecular glue molecule: a cautionary tale in the design of PROTAC degraders. J Med Chem 62(21):9471–9487

Yang C, Yang Y, Li Y, Ni Q, Li J (2022) Radiotherapy-triggered proteolysis targeting chimera prodrug activation in tumors. J Am Chem Soc 145(1):385–391

Zou Y, Ma D, Wang Y (2019) The PROTAC technology in drug development. Cell Biochem Funct 37(1):21–30

Global Regulatory Requirements Applicable for PROTACs

7

Rohini Pujari, Urvashi Soni, and Ashwini Gawade

Abstract

The present chapter highlights the pivotal role of global regulatory guidelines in overseeing the development, evaluation, and commercialization of PROTACs, a novel class of therapeutic agents with unique mechanisms of action. It emphasizes the importance of adhering to Good Clinical Practice (GCP), Chemistry Manufacturing and Controls (CMC) regulations, and Good Manufacturing Practice (GMP) standards to ensure the quality, safety, and consistency of PROTACs. By following these guidelines, developers can demonstrate the ethical conduct, efficacy, and quality of PROTACs, ultimately supporting their successful regulatory approval and commercialization. The chapter underscores the significance of regulatory compliance in advancing innovative therapeutic approaches for the benefit of patients worldwide.

Keywords

PROTACs · Good clinical practice · Chemistry manufacturing and controls · Good manufacturing practice · Preclinical development · Clinical development

Abbreviations

ADME	Absorption, distribution, metabolism, excretion
AEs	Adverse events
ALCOA	Attributable, legible, contemporaneous, original, accurate
API	Active pharmaceutical ingredient

R. Pujari (✉) · U. Soni · A. Gawade
School of Health Sciences and Technology, Dr. Vishwanath Karad MIT World Peace University, Kothrud, Pune, Maharashtra, India

M. Nandave, P. Jain (eds.), *PROTAC-Mediated Protein Degradation: A Paradigm Shift in Cancer Therapeutics*, https://doi.org/10.1007/978-981-97-5077-1_7

CMC	Chemistry manufacturing and controls
CRFs	Case report forms
CSRs	Clinical study reports
CTA	Clinical trial application
DDIs	Drug-drug interactions
EMA	European Medicines Agency
FDA	Food and Drug Administration
GCP	Good clinical practice
GMP	Good manufacturing practice
ICFs	Informed consent forms
ICH	International Council for Harmonisation of Technical Requirements for Pharmaceuticals for Human Use
IND	Investigational new drug
IRBs	Institutional Review Boards
MAAs	Marketing authorization applications
NDAs	New drug applications
PD	Pharmacodynamics
PK	Pharmacokinetics
PROTACs	Proteolysis-targeting chimeras
SAEs	Serious adverse events
SOPs	Standard operating procedures

7.1 Introduction

Global regulatory guidelines play a pivotal role in guiding the development, evaluation, and commercialization of new drug molecules, ensuring that they meet stringent standards for safety, efficacy, and quality before reaching patients. Compliance with these guidelines is essential for successful drug development and market access, ultimately benefiting public health and patient care (Chakraborty and Yadav 2018).

Regulatory guidelines set standards for the safety of new drug molecules. This includes requirements for preclinical studies to assess toxicity and safety pharmacology, as well as clinical trials to evaluate safety and efficacy in humans. By adhering to these guidelines, developers minimize the risks associated with new drug molecules, ensuring patient safety. Regulatory guidelines provide frameworks for conducting well-controlled clinical trials to assess the efficacy of new drug molecules. This involves defining appropriate endpoints, study designs, and statistical analyses to demonstrate therapeutic benefit. Ensuring that clinical trials are conducted in accordance with these guidelines helps establish the efficacy of new drugs, providing evidence of their effectiveness in treating the targeted disease (Mahapatra et al. 2014). Regulatory guidelines specify requirements for the manufacturing, quality control, and distribution of drug products. Adherence to Good Manufacturing Practice (GMP) ensures that drug molecules are consistently produced to high-quality standards, minimizing variability and ensuring product integrity. Compliance with these guidelines is essential to guarantee the quality, purity, and potency of drugs reaching the market (Chakraborty and Yadav 2018).

Global regulatory guidelines provide a harmonized framework for drug development, facilitating the exchange of scientific information and data across international borders. This helps streamline the drug development process by providing clear expectations and requirements for regulatory submissions, reducing duplication of efforts, and accelerating the time to market for new drugs. Regulatory approval is a prerequisite for market access and commercialization of new drug molecules. Compliance with regulatory guidelines is necessary to obtain marketing authorization from regulatory authorities, allowing developers to bring their products to market and make them available to patients in need. Without regulatory approval, new drug molecules cannot be marketed or distributed to patients (Kumar et al. 2023).

Adherence to regulatory guidelines helps build public confidence in the safety, efficacy, and quality of new drug molecules. By demonstrating compliance with rigorous regulatory standards, developers instill trust in healthcare professionals, patients, and regulatory agencies, fostering acceptance and adoption of new therapies.

PROTACs, or Proteolysis-Targeting Chimeras, are a class of small molecules designed to degrade specific target proteins within cells. They represent an innovative and revolutionary approach to drug development with the potential to address a wide range of diseases, including cancer, neurodegenerative disorders, and autoimmune conditions (Liu et al. 2022).

Global regulatory guidelines are essential for the development evaluation and commercialization of the PROTACs. Global regulatory guidelines ensure consistency and harmonization in the development, evaluation, and approval of PROTACs across different regions and jurisdictions. This reduces duplication of efforts, streamlines the regulatory process, and facilitates global drug development and access (Chakraborty and Yadav 2018).

PROTACs are a novel class of therapeutic agents with unique mechanisms of action (Burslem and Crews 2020). Regulatory guidelines provide frameworks for assessing the safety of PROTACs in preclinical studies and clinical trials, helping to identify and mitigate potential risks associated with their use. Guidelines outline requirements for designing and conducting clinical trials to evaluate the efficacy of PROTACs in treating specific diseases or conditions. This includes defining appropriate endpoints, study populations, and statistical analyses to demonstrate therapeutic benefit. Regulatory guidelines establish standards for the manufacturing, quality control, and distribution of PROTACs. Adherence to Good Manufacturing Practice (GMP) ensures that PROTACs are consistently produced to high-quality standards, minimizing variability and ensuring product integrity (Franco et al. 2023).

Regulatory approval is necessary for PROTACs to be marketed and made available to patients. Global regulatory guidelines assure healthcare professionals, patients, and regulatory authorities regarding the safety, efficacy, and quality of PROTAC-based therapies, fostering public confidence and acceptance. Global regulatory guidelines encourage international collaboration and information exchange among regulators, industry stakeholders, and researchers. This facilitates knowledge sharing, innovation, and the development of best practices for PROTAC research and development (Kumar et al. 2023).

Guidelines outline requirements for post-marketing surveillance and pharmacovigilance to monitor the safety and efficacy of approved PROTACs in real-world

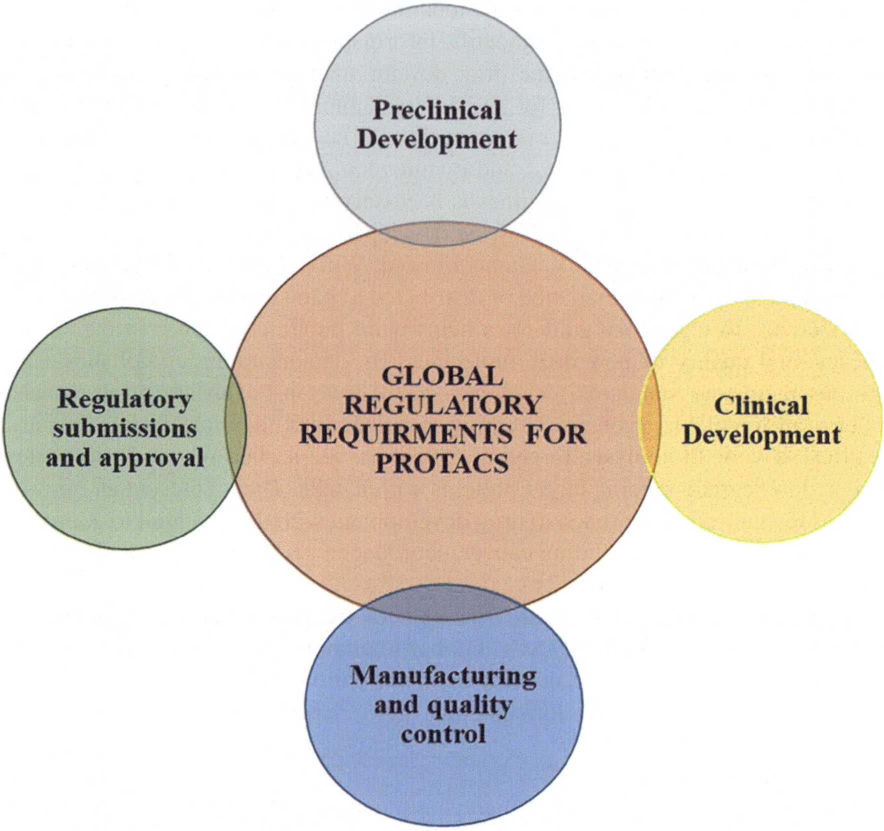

Fig. 7.1 Global regulatory requirements of PROTACs

settings. This helps ensure ongoing evaluation of benefit-risk profiles and the timely identification of any emerging safety concerns. Compliance with these guidelines is crucial for advancing PROTAC-based therapies and improving patient outcomes worldwide (Mahapatra et al. 2014).

This chapter covers the overall aspects of global regulatory requirements applicable to PROTACs for ensuring the safe, effective, and high-quality development, evaluation, and commercialization of these innovative therapeutic agents ensuring that they meet stringent standards for safety, efficacy, and quality (Fig. 7.1).

7.2 Global Regulatory Requirements Applicable for PROTACs

Regulatory requirements for the development and preclinical and clinical trials of PROTACs are governed by various agencies worldwide. Following are the global regulatory requirements of the PROTACs:

7.2.1 Preclinical Development

Preclinical studies are used to assess the safety profile and efficacy of PROTACs in animal models. This includes acute and chronic toxicity studies, genotoxicity assessments, and safety pharmacology evaluations along with the pharmacokinetic and pharmacodynamic profile of the PROTACs (Steinmetz and Spack 2009).

7.2.1.1 Safety Pharmacology and Toxicology Studies

Safety pharmacology and toxicology studies are critical components of the development and commercialization process for any new therapeutic agent, including PROTACs. The safety pharmacology and toxicology assessments are conducted early in the development process to identify potential safety concerns and inform compound design. Toxicology studies are designed according to regulatory guidelines (e.g., ICH S9 for nonclinical evaluation of anticancer pharmaceuticals). These include acute, subacute, and chronic toxicity studies as appropriate. Appropriate doses are chosen for toxicology studies based on pharmacokinetic data, target engagement, and efficacy. Use a range of doses to assess dose-response relationships and determine the therapeutic index. For conducting preclinical studies, relevant animal species are selected for toxicology studies based on pharmacological activity, metabolic pathways, and potential toxicities. Typically, rodents (rats and mice) and non-rodents (e.g., dogs, non-human primates) are used (Mohs and Greig 2017).

PROTACs are administered via routes intended for clinical use such as oral, intravenous, or other routes depending on the intended clinical application. After the administration of PROTACs, the animals are closely monitored throughout the study for signs of toxicity, including clinical signs, body weight changes, food consumption, and clinical pathology parameters (Begley and Ellis 2012).

Specialized assessments are also conducted for cardiovascular, respiratory, and central nervous system evaluations, to assess potential off-target effects. Also, the genotoxic potential of PROTACs is evaluated using appropriate in vitro and in vivo assays (e.g., Ames test, chromosomal aberration assay). The carcinogenic potential of PROTACs is assessed in long-term carcinogenicity studies, if warranted based on the duration of treatment and the target population (Andrade et al. 2016).

The effects of PROTACs on vital organ systems (e.g., cardiovascular, respiratory, central nervous system) are assessed using standardized safety pharmacology studies. The PROTACs are also evaluated for the potential effects of PROTACs on reproductive function and fetal development in reproductive and developmental toxicity studies. Immunotoxicity assessment is done to evaluate the immunotoxic potential of PROTACs using appropriate assays to evaluate immune function and potential immunomodulatory effects (Pugsley et al. 2008).

Along with this, the metabolism and pharmacokinetic profile of PROTACs are characterized in preclinical species to understand potential species differences and inform clinical development.

The safety pharmacology and toxicology studies of PROTACs are conducted in relevant disease models to assess safety and efficacy in a more clinically relevant context. All the studies are conducted in accordance with regulatory guidelines (e.g., FDA, EMA, ICH) and ensure data quality, integrity, and compliance with Good Laboratory Practice (GLP) regulations (Guth et al. 2019).

The obtained safety pharmacology and toxicology studies with other nonclinical and clinical data are compiled to comprehensively assess the safety profile of PROTACs and support regulatory submissions. Continual evaluation is done to evaluate safety pharmacology and toxicology data throughout the development process to identify and address any emerging safety concerns (Engwall et al. 2022).

7.2.1.2 Pharmacokinetics and Pharmacodynamics Studies

Pharmacokinetics (PK) and Pharmacodynamics (PD) studies are essential components of the development and commercialization process for PROTACs. These studies help characterize the absorption, distribution, metabolism, and excretion (ADME) of PROTACs, as well as their pharmacological effects on the target protein(s) and downstream biological pathways. PK/PD studies are designed to address the key questions related to the compound's PK properties (e.g., clearance, half-life, bioavailability) and PD effects (e.g., target engagement, downstream signaling) (Amouzadeh and Vargas 2013). PK/PD studies are conducted in accordance with regulatory guidelines (e.g., FDA, EMA, ICH) and ensure data quality, integrity, and compliance with Good Laboratory Practice (GLP) regulations. PK/PD studies are initiated in the early stages of the development process to understand the compound's behavior in vivo and guide compound optimization.

For conducting these studies, appropriate routes of administration (e.g., oral, intravenous, subcutaneous) are selected based on the intended clinical use and formulation, and appropriate doses are selected that cover a range of exposures relevant to the anticipated therapeutic window and clinical dosing regimens (Authier et al. 2013).

The PK/PD studies at the initial stages are conducted in the appropriate animal models (e.g., rodents, non-human primates) based on species relevance and translational potential to humans. Blood or tissue samples at multiple time points are collected following dosing to capture the PK profile (e.g., concentration-time curve) and PD effects (e.g., target engagement, biomarker modulation) using sensitive and specific analytical methods (e.g., LC-MS/MS, ELISA) to quantitate PROTAC concentrations in biological matrices and assess PD endpoints (Bhatt et al. 2019).

The PK parameters (e.g., area under the curve, clearance, volume of distribution) are calculated using non-compartmental or compartmental analysis methods to characterize the compound's PK profile, and the PD biomarkers (e.g., target protein degradation, downstream pathway modulation) are identified and validated to assess target engagement and pharmacological effects (Engwall et al. 2022).

Pharmacokinetic-pharmacodynamic modeling approaches are utilized to integrate PK and PD data, elucidate concentration-effect relationships, and optimize dosing regimens, which also gives the basis to evaluate the potential drug-drug interactions (DDIs) between PROTACs and concomitant medications using in vitro and in vivo studies, focusing on metabolic enzymes and transporters. The data also help in assessing the metabolic pathways and clearance mechanisms of PROTACs to identify major metabolites and potential liabilities (e.g., drug-drug interactions, toxic metabolites) (Caruso et al. 2014).

The comparability of PK/PD profiles is evaluated between preclinical species and humans to predict human PK/PD relationships and inform clinical study design. Also, dose-ranging studies are conducted to establish dose-response relationships for efficacy and safety endpoints, informing dose selection for subsequent clinical trials (Parkinson et al. 2013).

PK/PD data is evaluated continuously to refine study designs as needed to address emerging questions and optimize the development trajectory of PROTACs. PK/PD data with other nonclinical and clinical data is integrated to support decision-making throughout the drug development process and regulatory submissions (Caruso et al. 2014).

7.3 Clinical Development

Clinical trials for PROTACs follow a phased approach, starting with Phase 1 trials in healthy volunteers to assess safety, PK/PD, and dose-ranging. Subsequent phases evaluate efficacy and safety in patient population.

7.3.1 Clinical Trial Design

Clinical trial design for the development of PROTACs follows established principles and guidelines to ensure the generation of robust data regarding safety, efficacy, and pharmacokinetics. While there may not be specific guidelines tailored exclusively for PROTACs, developers typically adhere to general principles outlined by regulatory authorities such as the International Council for Harmonisation of Technical Requirements for Pharmaceuticals for Human Use (ICH), the U.S. Food and Drug Administration (FDA), and the European Medicines Agency (EMA) (Kandi and Vadakedath 2023).

Following are the key considerations for clinical trial design in the clinical development of PROTACs:

- ***Phase-based Approach***: Clinical development typically progresses through several phases (Phase 1 to Phase 3) to assess safety, efficacy, and dosage optimization. Phase 1 trials focus on safety and pharmacokinetics, Phase 2 trials evaluate efficacy and dose-response relationships, and Phase 3 trials confirm efficacy and safety in larger patient populations (Sverdlov et al. 2021).
- ***Study Population***: The selection of appropriate study populations is crucial to demonstrate the safety and efficacy of PROTACs in the target patient population. Inclusion and exclusion criteria should be clearly defined to ensure enrollment of eligible patients and minimize confounding factors (Orloff and Stanski 2011).
- ***Endpoints and Outcome Measures***: Clinical trials should include clinically meaningful endpoints and outcome measures to assess the efficacy and safety of

PROTACs. Endpoints may include measures such as progression-free survival, overall survival, objective response rate, and biomarker assessments (Sverdlov et al. 2021).

- ***Dosing and Administration***: Protocols should specify the dosing regimen and route of administration for PROTACs based on preclinical data and pharmacokinetic/pharmacodynamic (PK/PD) considerations. Dose escalation strategies may be employed in Phase 1 trials to determine the maximum tolerated dose and establish the recommended Phase 2 dose (Kandi and Vadakedath 2023).
- ***Control Groups***: Comparator groups, such as placebo or standard of care, may be included in clinical trials to evaluate the relative efficacy and safety of PROTACs compared to existing treatments or placebo. Randomization and blinding procedures help minimize bias and ensure the validity of trial results (Fruhner et al. 2017).
- ***Trial Duration and Follow-up***: Protocols should define the duration of treatment and follow-up assessments to adequately evaluate the long-term safety and efficacy of PROTACs. Follow-up periods may vary depending on the disease indication and study objectives (Sverdlov et al. 2021).
- ***Safety Monitoring and Reporting***: Clinical trials should incorporate robust safety monitoring procedures to detect and assess adverse events associated with PROTAC treatment. Adverse event reporting, safety assessments, and monitoring of laboratory parameters are essential components of safety evaluation (Mohs and Greig 2017).
- ***Statistical Considerations***: Protocols should include statistical methods for sample size calculation, data analysis, and interpretation of trial results. Adequate statistical power is essential to detect treatment effects and draw meaningful conclusions from clinical trial data (Rajadhyaksha 2010).
- ***Ethical Considerations***: Clinical trial protocols must be reviewed and approved by independent ethics committees or institutional review boards (IRBs) to ensure that the rights, safety, and well-being of trial participants are protected. Informed consent should be obtained from all participants before enrollment in the trial (Sverdlov et al. 2021).
- ***Regulatory Compliance***: Clinical trials should be conducted in accordance with regulatory requirements and guidelines issued by relevant authorities. Compliance with Good Clinical Practice (GCP) standards ensures the ethical conduct, integrity, and quality of clinical trials.

By adhering to these guidelines, developers can design and conduct clinical trials that generate robust data on the safety, efficacy, and pharmacokinetics of PROTACs, ultimately supporting their successful development and regulatory approval for clinical use (Sverdlov et al. 2021).

7.3.2 Good Clinical Practice (GCP)

Good Clinical Practice (GCP) guidelines are essential for ensuring the ethical conduct, integrity, and quality of clinical trials, including those involving PROTACs. While there are no specific GCP guidelines exclusively tailored for PROTACs, developers conducting clinical trials involving PROTACs are expected to adhere to established GCP principles outlined by regulatory authorities such as the International Council for Harmonisation of Technical Requirements for Pharmaceuticals for Human Use (ICH), the U.S. Food and Drug Administration (FDA), and the European Medicines Agency (EMA) (Vijayananthan and Nawawi 2008).

Following are the key GCP principles relevant to the clinical development of PROTACs:

- Protocols for clinical trials involving PROTACs must be reviewed and approved by an independent ethics committee or institutional review board (IRB). Participants must provide informed consent before enrolling in the trial, and their rights, safety, and well-being must be protected throughout the study (Otte et al. 2005).
- Clinical trials must be conducted in compliance with the approved protocol and relevant regulatory requirements. Adequate measures should be in place to ensure data accuracy, integrity, and confidentiality. Monitoring of the trial should be conducted regularly to verify adherence to the protocol, assess the progress of the study, and ensure participant safety (Devine et al. 2008).
- Investigators conducting clinical trials involving PROTACs must be qualified and experienced in accordance with GCP principles. They are responsible for ensuring that the trial is conducted in compliance with the protocol, relevant regulations, and ethical principles. Investigators must also ensure the safety and well-being of trial participants and accurately record and report data (Vijayananthan and Nawawi 2008).
- Accurate and complete records of all data generated during the clinical trial must be maintained. This includes source documents, case report forms (CRFs), and other trial-related documents. Data should be attributable, legible, contemporaneous, original, and accurate (ALCOA) to ensure integrity and traceability (Sather et al. 2003).
- Adverse events (AEs) and serious adverse events (SAEs) occurring during the trial must be promptly reported to the appropriate regulatory authorities and ethics committees. Safety data should be monitored continuously throughout the trial to identify and mitigate potential risks associated with PROTACs (Otte et al. 2005).
- Adequate quality assurance and quality control measures should be implemented to ensure the reliability and validity of trial results. This includes regular audits, inspections, and oversight of trial conduct, data collection, and reporting processes (Devine et al. 2008).
- Detailed documentation of clinical trial protocols, informed consent forms, investigator brochures, and safety reports are submitted to regulatory authorities for review and approval (Vijayananthan and Nawawi 2008).

By adhering to GCP guidelines, developers can ensure that clinical trials involving PROTACs are conducted ethically, with the highest standards of quality, integrity, and participant safety. Compliance with GCP principles is essential for obtaining regulatory approval and advancing PROTAC-based therapies through the clinical development process.

7.4 Manufacturing and Quality Control

Manufacturing and quality control guidelines are essential for ensuring the consistency, safety, efficacy, and regulatory compliance of PROTACs throughout the development process. Adherence to these guidelines is critical for advancing PROTAC-based therapies through preclinical and clinical development and ultimately bringing them to market for the benefit of patients.

7.4.1 Good Manufacturing Practice (GMP) Regulations

Manufacturing facilities must adhere to GMP regulations to ensure consistent production of high-quality PROTACs. Good Manufacturing Practice (GMP) guidelines are critical in the development of PROTACs to ensure that they are manufactured consistently, safely, and in accordance with regulatory standards (Gouveia et al. 2015).

Following are the GMP regulations applicable for the development of PROTACs:

- ***Facilities and Equipment***: GMP requires that facilities and equipment used in the production of PROTACs meet specific standards to ensure cleanliness, suitability, and proper functioning. Facilities should be designed to prevent contamination and facilitate efficient workflow, while equipment should be regularly calibrated and maintained to ensure accuracy and reliability (Petrelli et al. 2019).
- ***Raw Materials***: GMP mandates stringent controls over the sourcing, testing, and handling of raw materials used in PROTAC synthesis. All raw materials must be of appropriate quality, adequately characterized, and sourced from reliable suppliers. Testing procedures should be in place to verify the identity, purity, and quality of raw materials before use in PROTAC manufacturing.
- ***Process Controls***: GMP requires the implementation of robust process controls to ensure consistency and reproducibility in PROTAC manufacturing. This includes establishing and validating manufacturing processes, documenting standard operating procedures (SOPs), and conducting in-process controls to monitor critical parameters and ensure product quality at each stage of production.
- ***Quality Control Testing***: GMP emphasizes the importance of quality control testing throughout the manufacturing process to verify the identity, purity, potency, and stability of PROTACs. Analytical methods should be validated for accuracy, precision, specificity, and sensitivity, and testing should be conducted in compliance with approved specifications and protocols.
- ***Documentation and Record-keeping***: GMP requires comprehensive documentation and record-keeping practices to ensure traceability, accountability, and compliance with regulatory requirements. This includes maintaining detailed records of manufacturing activities, testing results, deviations, and corrective

actions taken to address any issues that arise during PROTAC production (Chaudhri et al. 2015).
- ***Personnel Training and Qualifications***: GMP stipulates that personnel involved in PROTAC manufacturing must be adequately trained and qualified to perform their duties effectively and in compliance with GMP standards. Training programs should cover GMP principles, SOPs, safety protocols, and any specific requirements related to PROTAC manufacturing.
- ***Quality Assurance Oversight***: GMP mandates the establishment of a quality assurance unit responsible for overseeing compliance with GMP guidelines and ensuring the integrity and quality of PROTAC manufacturing processes. This includes conducting internal audits, evaluating deviations and non-conformances, and implementing corrective and preventive actions as needed (Patel and Chotai 2011).
- ***Packaging and Labeling***: GMP guidelines extend to the packaging and labeling of PROTACs to ensure product integrity, identification, and compliance with regulatory requirements. Packaging materials should be suitable for their intended use and provide adequate protection against contamination, while labeling should accurately reflect the contents, dosage form, strength, and usage instructions of the product (Gouveia et al. 2015).

By adhering to GMP guidelines throughout the development of PROTACs, manufacturers can ensure that these innovative therapeutic agents are produced to the highest standards of quality, safety, and consistency, ultimately supporting their successful clinical development and commercialization.

7.4.2 Chemistry, Manufacturing, and Controls (CMC)

Chemistry, Manufacturing, and Controls (CMC) regulations play a crucial role in the development of PROTACs. CMC sections of regulatory submissions provide comprehensive information on the chemical composition, manufacturing process, and quality controls of PROTACs. These regulations ensure that the manufacturing processes, analytical methods, and controls are well-defined, consistent, and compliant with regulatory standards (Barrett 2022).

The chemistry section provides detailed information on the synthesis, characterization, and specifications of the active pharmaceutical ingredient (API) of the PROTAC. It includes the chemical structure, synthetic route, purity profile, and physicochemical properties of the drug substance. It also describes the formulation and composition of the final drug product, including excipients, stabilizers, and any other components used in the formulation of the PROTAC (Popkin et al. 2022).

The manufacturing section outlines the manufacturing process for the synthesis of the PROTAC, including the selection and sourcing of raw materials, reaction conditions, purification steps, and quality control measures implemented throughout the manufacturing process. CMC regulations require detailed descriptions of the manufacturing facilities and equipment used in the production of PROTACs. This includes information on facility design, cleanliness, and compliance with Good

Manufacturing Practice (GMP) standards. Developers must provide evidence of process validation to demonstrate the reproducibility, reliability, and robustness of the manufacturing process. This involves conducting validation studies to ensure that the process consistently produces PROTACs of the desired quality and purity (Barber et al. 2017).

CMC regulations require the maintenance of comprehensive batch records documenting all manufacturing activities, including raw material usage, in-process controls, deviations, and quality control testing results for each batch of PROTACs produced (Popkin et al. 2022).

The controls section of CMC regulations mandates the establishment of specifications for the drug substance and drug product, including criteria for identity, purity, potency, and stability. These specifications ensure that PROTACs meet quality standards and are suitable for clinical use. Developers must validate analytical methods used for testing the identity, purity, potency, and stability of PROTACs. This includes methods such as chromatography, spectroscopy, and mass spectrometry, which are used to assess the quality of both the drug substance and drug product. CMC regulations require stability studies to evaluate the long-term and accelerated stability of PROTACs under various storage conditions. These studies assess the degradation kinetics, shelf life, and storage requirements of PROTACs to ensure product quality and safety over time (Barrett 2022).

Developers must compile all CMC data and information into regulatory submissions, such as Investigational New Drug (IND) applications, Marketing Authorization Applications (MAAs), or New Drug Applications (NDAs), depending on the stage of development and intended market. Regulatory authorities review the CMC data to assess the quality, consistency, and reliability of PROTAC manufacturing processes and controls, ensuring compliance with regulatory standards before granting approval for clinical trials or marketing authorization (Algorri et al. 2020).

By adhering to CMC regulations, developers can ensure that the manufacturing processes, analytical methods, and controls for PROTACs are well-defined, robust, and compliant with regulatory standards, ultimately supporting their successful development and commercialization as safe and effective therapeutic agents (Farid et al. 2020).

7.5 Regulatory Submissions and Approval

Regulatory submissions for the clinical development of PROTACs follow standard procedures similar to other investigational drugs.

Following is the overview of the typical regulatory submissions involved:

7.5.1 Investigational New Drug (IND) Application (United States) or Clinical Trial Application (CTA) (European Union)

Before initiating clinical trials with PROTACs, developers must submit an IND application to the U.S. Food and Drug Administration (FDA) or a CTA to the relevant regulatory authorities in the European Union (e.g., European Medicines Agency, national competent authorities). The submission includes comprehensive data from preclinical studies, chemistry, manufacturing, and controls (CMC) information, as well as the proposed clinical trial protocol, investigator brochure, and any other relevant documentation. Regulatory authorities review the IND or CTA to ensure that the proposed clinical trial is scientifically sound, ethically conducted, and meets regulatory requirements. The approval allows the clinical trial to proceed (Holbein 2009; Lambers et al. 2008).

7.5.2 Clinical Trial Protocol and Informed Consent Forms

Detailed protocols outlining the objectives, design, methodology, statistical considerations, and endpoints of the clinical trial are submitted to regulatory authorities for review and approval.

Informed consent forms (ICFs) outlining the trial's purpose, procedures, risks, and benefits are provided to participants, and the approved version is submitted to regulatory authorities (Gupta 2013).

7.5.3 Safety Reporting

Developers are required to submit safety reports to regulatory authorities throughout the clinical trial, including expedited reporting of serious adverse events (SAEs) and periodic safety updates. These safety reports provide information on the safety profile of the investigational PROTAC and contribute to ongoing risk assessment and management (Samara et al. 2023).

7.5.4 Clinical Study Reports (CSRs)

Following completion of the clinical trial, developers prepare CSRs that summarize and analyze the trial's results, including efficacy and safety data. CSRs are submitted to regulatory authorities as part of marketing authorization applications (e.g., New Drug Application [NDA] in the U.S., Marketing Authorization Application [MAA] in the EU) or for other regulatory purposes (Miller et al. 2019).

7.5.5 Marketing Authorization Application (MAA) or New Drug Application (NDA)

Once clinical development is complete and sufficient data have been gathered to support the safety, efficacy, and quality of the PROTAC, developers submit marketing authorization applications to regulatory authorities. These applications include comprehensive data on preclinical studies, clinical trials, CMC information, pharmacovigilance plans, and proposed labeling. Regulatory authorities review the application and, if satisfied with the data, grant marketing authorization, allowing the PROTAC to be marketed and distributed for clinical use (Murthy and Murthy 2023).

7.5.6 Post-marketing Surveillance

After approval, developers continue to monitor the safety and efficacy of the PROTAC through post-marketing surveillance activities, including pharmacovigilance programs and ongoing clinical studies. Regulatory submissions for PROTACs must adhere to the specific requirements and guidelines of each regulatory authority, ensuring that the development and clinical testing of PROTACs are conducted in compliance with applicable regulations and standards (Raj et al. 2019).

7.6 Conclusion

In conclusion, the global regulatory requirements for PROTACs emphasize the critical importance of adhering to established guidelines and principles throughout the development, clinical trials, and commercialization of these innovative therapeutic agents. By following Good Clinical Practice (GCP) standards, Chemistry, Manufacturing, and Controls (CMC) regulations, and Good Manufacturing Practice (GMP) guidelines, developers can ensure the ethical conduct, quality, and consistency of PROTACs, ultimately supporting their successful regulatory approval and commercialization. The document underscores the significance of robust data generation, safety reporting, and post-marketing surveillance in demonstrating the safety, efficacy, and quality of PROTACs, highlighting the essential role of regulatory compliance in advancing novel therapeutic approaches for patient benefit.

References

Algorri M, Cauchon NS, Abernathy MJ (2020) Transitioning chemistry, manufacturing, and controls content with a structured data management solution: streamlining regulatory submissions. J Pharm Sci 109(4):1427–1438. https://doi.org/10.1016/j.xphs.2020.01.020

Amouzadeh HR, Vargas HM (2013) Safety pharmacology assessment of biopharmaceuticals. In: Vogel HG, Maas J, Hock FJ, Mayer D (eds) Drug discovery and evaluation: safety and pharmacokinetic assays. Springer, Berlin, pp 555–560. https://doi.org/10.1007/978-3-642-25240-2_18

Andrade EL, Bento AF, Cavalli J, Oliveira SK, Schwanke RC, Siqueira JM, Freitas CS, Marcon R, Calixto JB (2016) Non-clinical studies in the process of new drug development—part II: good laboratory practice, metabolism, pharmacokinetics, safety and dose translation to clinical studies. Braz J Med Biol Res 49(12):e5646. https://doi.org/10.1590/1414-431x20165646

Authier S, Vargas HM, Curtis MJ, Holbrook M, Pugsley MK (2013) Safety pharmacology investigations in toxicology studies: an industry survey. J Pharmacol Toxicol Methods 68(1):44–51. https://doi.org/10.1016/j.vascn.2013.05.002

Barber C, Antonucci V, Baumann JC, Brown R, Covey-Crump E, Elder D, Elliott E, Fennell JW, Gallou F, Ide ND, Jordine G, Kallemeyn JM, Lauwers D, Looker AR, Lovelle LE, McLaughlin

M, Molzahn R, Ott M, Schils D et al (2017) A consortium-driven framework to guide the implementation of ICH M7 Option 4 control strategies. Regul Toxicol Pharmacol 90:22–28. https://doi.org/10.1016/j.yrtph.2017.08.008

Barrett JS (2022) Chemistry, manufacturing, and controls. In: Fundamentals of drug development. Wiley, New York, pp 301–318. https://doi.org/10.1002/9781119913276.ch16

Begley CG, Ellis LM (2012) Drug development: raise standards for preclinical cancer research. Nature 483(7391):531–533. https://doi.org/10.1038/483531a

Bhatt S, Northcott C, Wisialowski T, Li D, Steidl-Nichols J (2019) Preclinical to clinical translation of hemodynamic effects in cardiovascular safety pharmacology studies. Toxicol Sci 169(1):272–279. https://doi.org/10.1093/toxsci/kfz035

Burslem GM, Crews CM (2020) Proteolysis-targeting chimeras as therapeutics and tools for biological discovery. Cell 181(1):102–114. https://doi.org/10.1016/j.cell.2019.11.031

Caruso A, Frances N, Meille C, Greiter-Wilke A, Hillebrecht A, Lavé T (2014) Translational PK/PD modeling for cardiovascular safety assessment of drug candidates: methods and examples in drug development. J Pharmacol Toxicol Methods 70(1):73–85. https://doi.org/10.1016/j.vascn.2014.05.004

Chakraborty K, Yadav K (2018) Drug approval process in US, Europe and India and its regulatory requirements: a review. Int J Drug Regul Aff 6(3):31–39. https://doi.org/10.22270/ijdra.v6i3.266

Chaudhri V, Yadav V, Verma PK, Singh A (2015) A review on good manufacturing practice (GMP) for medicinal products. J Pharm Bioallied Sci 7(2):87–96

Devine S, Dagher RN, Weiss KD, Santana VM (2008) Good clinical practice and the conduct of clinical studies in pediatric oncology. Pediatr Clin N Am 55(1):187–209, xi–xii. https://doi.org/10.1016/j.pcl.2007.10.008

Engwall MJ, Sutherland W, Vargas HM (2022) Safety pharmacology evaluation of biopharmaceuticals. In: Hock FJ, Gralinski MR, Pugsley MK (eds) Drug discovery and evaluation: safety and pharmacokinetic assays. Springer International Publishing, Berlin, pp 1–16. https://doi.org/10.1007/978-3-030-73317-9_18-1

Farid SS, Baron M, Stamatis C, Nie W, Coffman J (2020) Benchmarking biopharmaceutical process development and manufacturing cost contributions to R&D. MAbs 12(1):1754999. https://doi.org/10.1080/19420862.2020.1754999

Franco P, Jain R, Rosenkrands-Lange E, Hey C, Koban MU (2023) Regulatory pathways supporting expedited drug development and approval in ICH member countries. Ther Innov Regul Sci 57(3):484–514. https://doi.org/10.1007/s43441-022-00480-3

Fruhner K, Hartmann G, Sudhop T (2017) Analysis of integrated clinical trial protocols in early phases of medicinal product development. Eur J Clin Pharmacol 73(12):1565–1577. https://doi.org/10.1007/s00228-017-2335-y

Gouveia BG, Rijo P, Gonçalo TS, Reis CP (2015) Good manufacturing practices for medicinal products for human use. J Pharm Bioallied Sci 7(2):87–96. https://doi.org/10.4103/0975-7406.154424

Gupta UC (2013) Informed consent in clinical research: revisiting few concepts and areas. Perspect Clin Res 4(1):26–32. https://doi.org/10.4103/2229-3485.106373

Guth BD, Grobler AF, Frazier KS, Greiter-Wilke A, Herzyk D, Hough TA, Khan AA, Markert M, Smith JD, Svenson KL, Wells S, Pugsley MK (2019) Drug safety Africa: an overview of safety pharmacology & toxicology in South Africa. J Pharmacol Toxicol Methods 98:106579. https://doi.org/10.1016/j.vascn.2019.106579

Holbein ME (2009) Understanding FDA regulatory requirements for investigational new drug applications for sponsor-investigators. J Investig Med 57(6):688–694. https://doi.org/10.2310/JIM.0b013e3181afdb26

Kandi V, Vadakedath S (2023) Clinical trials and clinical research: a comprehensive review. Cureus 15(2):e35077. https://doi.org/10.7759/cureus.35077

Kumar AP, Renu GM, Gowda S, Hu T, Kumar V, Hu Y (2023) Current regulatory requirements and regulation on submission of ANDA in India comparison with Singapore. World J Adv Res Rev 20:404–409. https://doi.org/10.30574/wjarr.2023.20.1.2056

Lambers HHJ, Dobre D, Hillege HL, Grobbee DE, de Zeeuw D (2008) Does the European clinical trials directive really improve clinical trial approval time? Br J Clin Pharmacol 66(4):546–550. https://doi.org/10.1111/j.1365-2125.2008.03246.x

Liu Z, Hu M, Yang Y, Du C, Zhou H, Liu C, Chen Y, Fan L, Ma H, Gong Y, Xie Y (2022) An overview of PROTACs: a promising drug discovery paradigm. Mol Biomed 3(1):46. https://doi.org/10.1186/s43556-022-00112-0

Mahapatra AK, Sameeraja NH, Murthy PN (2014) Drug approval process—in United States of America, European Union and India: a review. Appl Clin Res Clin Trials Regul Aff 1(1):13–22. https://doi.org/10.2174/2213476X01666140123234357

Miller J, Ross JS, Wilenzick M, Mello MM (2019) Sharing of clinical trial data and results reporting practices among large pharmaceutical companies: cross sectional descriptive study and pilot of a tool to improve company practices. BMJ 366:l4217. https://doi.org/10.1136/bmj.l4217

Mohs RC, Greig NH (2017) Drug discovery and development: role of basic biological research. Alzheimers Dement (N Y) 3(4):651–657. https://doi.org/10.1016/j.trci.2017.10.005

Murthy S, Murthy S (2023) Chapter 81—New drug application process. In: Eltorai AEM, Liu T, Chand R, Kalva SP (eds) Translational interventional radiology. Academic, New York, pp 401–403. https://doi.org/10.1016/B978-0-12-823026-8.00081-X

Orloff JJ, Stanski D (2011) Innovative approaches to clinical development and trial design. Ann Ist Super Sanita 47(1):8–13. https://doi.org/10.4415/ann_11_01_03

Otte A, Maier-Lenz H, Dierckx RA (2005) Good clinical practice: historical background and key aspects. Nucl Med Commun 26(7):563–574. https://doi.org/10.1097/01.mnm.0000168408.03133.e3

Parkinson J, Visser SA, Jarvis P, Pollard C, Valentin JP, Yates JW, Ewart L (2013) Translational pharmacokinetic-pharmacodynamic modeling of QTc effects in dog and human. J Pharmacol Toxicol Methods 68(3):357–366. https://doi.org/10.1016/j.vascn.2013.03.007

Patel K, Chotai N (2011) Documentation and records: harmonized GMP requirements. J Young Pharm 3(2):138–150. https://doi.org/10.4103/0975-1483.80303

Petrelli F, Caraffa A, Scuri S, Grappasonni I, Magrini E, Cocchini A (2019) The requirements for manufacturing highly active or sensitising drugs comparing good manufacturing practices. Acta Biomed 90(2):288–299. https://doi.org/10.23750/abm.v90i2.8340

Popkin ME, Goese M, Wilkinson D, Finnie S, Flanagan T, Campa C, Clinch A, Teasdale A, Lennard A, Cook G, Mohan G, Osborne MD (2022) Chemistry manufacturing and controls development, industry reflections on manufacture, and supply of pandemic therapies and vaccines. AAPS J 24(6):101. https://doi.org/10.1208/s12248-022-00751-9

Pugsley MK, Authier S, Curtis MJ (2008) Principles of safety pharmacology. Br J Pharmacol 154(7):1382–1399. https://doi.org/10.1038/bjp.2008.280

Raj N, Fernandes S, Charyulu NR, Dubey A, Ravi GS, Hebbar S (2019) Postmarket surveillance: a review on key aspects and measures on the effective functioning in the context of the United Kingdom and Canada. Ther Adv Drug Saf 10:2042098619865413. https://doi.org/10.1177/2042098619865413

Rajadhyaksha V (2010) Conducting feasibilities in clinical trials: an investment to ensure a good study. Perspect Clin Res 1(3):106–109

Samara C, Garcia A, Henry C, Vallotton L, Cariolato L, Desmeules J, Pinçon A (2023) Safety surveillance during drug development: comparative evaluation of existing regulations. Adv Ther 40(5):2147–2185. https://doi.org/10.1007/s12325-023-02492-3

Sather MR, Raisch DW, Haakenson CM, Buckelew JM, Feussner JR (2003) Promoting good clinical practices in the conduct of clinical trials: experiences in the department of veterans affairs cooperative studies program. Control Clin Trials 24(5):570–584. https://doi.org/10.1016/S0197-2456(03)00074-6

Steinmetz KL, Spack EG (2009) The basics of preclinical drug development for neurodegenerative disease indications. BMC Neurol 9 Suppl 1(Suppl 1):S2. https://doi.org/10.1186/1471-2377-9-s1-s2

Sverdlov O, Ryeznik Y, Wong WK (2021) Opportunity for efficiency in clinical development: an overview of adaptive clinical trial designs and innovative machine learning tools, with examples from the cardiovascular field. Contemp Clin Trials 105:106397. https://doi.org/10.1016/j.cct.2021.106397

Vijayananthan A, Nawawi O (2008) The importance of good clinical practice guidelines and its role in clinical trials. Biomed Imaging Interv J 4(1):e5. https://doi.org/10.2349/biij.4.1.e5

Tissue-Specific Targeting Strategies with PROTAC Technology

8

Ruchi Jakhmola Mani and Deepshikha Pande Katare

Abstract

This chapter provides an in-depth exploration of PROTAC (Proteolysis Targeting Chimeras) technology, a novel approach to drug discovery that harnesses the ubiquitin-proteasome system to selectively degrade disease-causing proteins. The mechanisms of action of PROTACs, including the recruitment of E3 ubiquitin ligases to target proteins, are elucidated, emphasizing the importance of target tissue detectivity in achieving therapeutic specificity. Factors influencing target tissue detectivity, such as ligand binding specificity, cellular localization of target proteins, stability, pharmacokinetics, and cell signaling pathways, are discussed in detail. Methods for assessing target tissue detectivity, including in vitro assays and in vivo models, provide valuable tools for evaluating PROTAC efficacy and selectivity. Case studies highlight successful examples of PROTAC-mediated protein degradation in various disease settings, underscoring the potential of this technology in drug discovery and therapeutics. Despite challenges and limitations, including the need for optimization of PROTAC design, pharmacokinetics, and drug delivery, PROTACs offer promising implications for overcoming drug resistance, expanding the druggable target space, and improving patient outcomes.

Keywords

Drug discovery · Proteolysis targeting chimeras (PROTAC) · Targeted protein degradation · Target tissue detectivity · Ubiquitin-proteasome system

R. J. Mani · D. P. Katare (✉)
Proteomics and Translational Research Lab, Centre for Medical Biotechnology, Amity Institute of Biotechnology, Amity University, Noida, India
e-mail: rjakhmola@amity.edu; dpkatare@amity.edu

M. Nandave, P. Jain (eds.), *PROTAC-Mediated Protein Degradation: A Paradigm Shift in Cancer Therapeutics*, https://doi.org/10.1007/978-981-97-5077-1_8

Abbreviations

AD	Alzheimer's disease
AR	Androgen receptor
CETSA	Cellular thermal shift assays
FRET	Fluorescence resonance energy transfer
PDX	Patient-derived xenograft
PROTACs	Proteolysis targeting chimeras
RA	Rheumatoid arthritis
UPS	Ubiquitin-proteasome system

8.1 Introduction

8.1.1 Overview of PROTAC

PROTAC, an acronym for Proteolysis Targeting Chimeras, represents an innovative approach to drug discovery and development. These molecules harness the cellular machinery to selectively degrade target proteins, offering a promising strategy for tackling diseases with challenging therapeutic targets. The core principle of PROTAC involves linking a ligand for the target protein to a ligand for an E3 ubiquitin ligase, forming a ternary complex that leads to ubiquitination and subsequent proteasomal degradation of the target protein. By inducing the degradation of disease-causing proteins rather than simply inhibiting their activity, PROTACs offer several potential advantages, including increased selectivity, reduced drug resistance, and the ability to target previously undruggable proteins (Bondeson et al. 2015).

8.1.2 Historical Background

The concept of targeted protein degradation through small molecules traces its roots back to the early 2000s when scientists began exploring ways to exploit the ubiquitin-proteasome system for therapeutic purposes. The development of PROTAC technology can be attributed to the seminal work of Prof. Craig Crews and colleagues at Yale University in the mid-2000s. Their groundbreaking research demonstrated the feasibility of using bifunctional molecules to recruit E3 ligases to specific target proteins, leading to their degradation (Bondeson et al. 2015; Crews and Pearson 2012). This landmark study laid the foundation for subsequent advancements in PROTAC design and application.

8.2 Importance of Target Tissue Detectivity

The concept of target tissue detectivity is central to the success of PROTAC technology in drug discovery and development. Traditional small-molecule inhibitors often lack the desired selectivity, leading to off-target effects and adverse reactions. In contrast, PROTACs offer the potential for enhanced selectivity by exploiting the natural substrate specificity of E3 ligases (Winter et al. 2015). This selectivity is particularly critical in the context of target tissue detectivity, where the goal is to selectively degrade disease-associated proteins in specific tissues or cell types while minimizing systemic effects.

Target tissue detectivity is essential for several reasons:

1. Enhanced Therapeutic Efficacy: By selectively targeting disease-causing proteins in affected tissues, PROTACs can achieve higher concentrations at the site of action, thereby enhancing therapeutic efficacy while minimizing systemic toxicity.
2. Minimization of Off-Target Effects: Selective degradation of target proteins in specific tissues reduces the risk of off-target effects associated with traditional small-molecule inhibitors, improving the safety profile of PROTAC-based therapies.
3. Overcoming Drug Resistance: Target tissue detectivity enables PROTACs to bypass mechanisms of drug resistance by selectively targeting disease-associated proteins in resistant cell populations or tissues, potentially overcoming resistance mechanisms that limit the effectiveness of traditional inhibitors.
4. Expansion of Druggable Targets: PROTAC technology has the potential to expand the druggable target space by enabling the selective degradation of proteins that are traditionally considered "undruggable" using conventional small-molecule inhibitors.

8.3 Mechanism of Action of PROTAC

8.3.1 Ubiquitin-Proteasome System

The ubiquitin-proteasome system (UPS) is a crucial cellular pathway responsible for the targeted degradation of proteins. This system maintains cellular homeostasis by regulating the levels of key proteins involved in various cellular processes, including cell cycle control, signal transduction, and protein quality control. The UPS consists of three main components: ubiquitin, E1 ubiquitin-activating enzymes, E2 ubiquitin-conjugating enzymes, and E3 ubiquitin ligases. The process of protein degradation via the UPS begins with the covalent attachment of ubiquitin molecules to target proteins, marking them for recognition and subsequent degradation by the proteasome.

The ubiquitination process involves the sequential action of E1, E2, and E3 enzymes. E1 enzymes activate ubiquitin in an ATP-dependent manner, transferring it to E2 enzymes. Subsequently, E3 ligases facilitate the transfer of ubiquitin from E2 enzymes to the target protein, leading to the formation of a polyubiquitin chain. Once polyubiquitinated, the target protein is recognized and degraded by the 26S proteasome, a large multi-subunit protease complex responsible for protein degradation in eukaryotic cells (Churcher 2018; Burslem et al. 2018). The proteasome

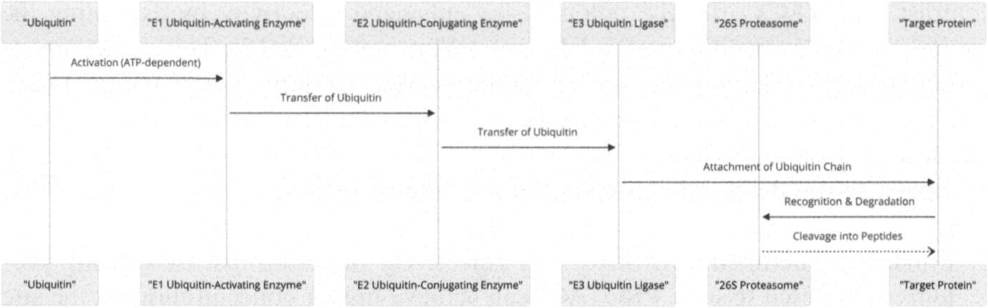

Fig. 8.1 Diagram illustrating the ubiquitin-proteasome system

cleaves the polyubiquitin chain and degrades the target protein into small peptides, which can then be recycled for cellular processes (Fig. 8.1).

8.3.2 Principle of PROTAC-Mediated Protein Degradation

PROTACs, or Proteolysis Targeting Chimeras, represent a novel approach for inducing the degradation of specific target proteins within the cell. The core principle of PROTAC-mediated protein degradation involves the recruitment of target proteins to the ubiquitin-proteasome system for proteasomal degradation. PROTAC molecules consist of three key components: a ligand for the target protein, a ligand for an E3 ubiquitin ligase, and a linker connecting the two ligands.

The mechanism of action of PROTACs can be summarized as follows:

1. Target Engagement: The PROTAC molecule binds to the target protein of interest through its specific ligand moiety, forming a ternary complex.
2. Recruitment of E3 Ligase: Simultaneously, the PROTAC molecule interacts with an E3 ubiquitin ligase through its ligand moiety for the E3 ligase.
3. Formation of Ternary Complex: The binding of the PROTAC molecule to both the target protein and the E3 ligase brings them into proximity, facilitating the transfer of ubiquitin molecules from the E2 enzyme to the target protein.
4. Ubiquitination and Degradation: The polyubiquitinated target protein is recognized by the 26S proteasome and subsequently degraded into small peptides.
5. Recycling of Components: Following degradation, ubiquitin molecules are released and recycled, while the PROTAC molecule can undergo additional rounds of target engagement and degradation (Fig. 8.2).

The ability of PROTACs to induce the degradation of target proteins offers several advantages over traditional small-molecule inhibitors (Toure and Crews 2016). By promoting the elimination of disease-causing proteins rather than simply inhibiting their activity, PROTACs can achieve potent and durable therapeutic effects with potentially lower doses and reduced risk of drug resistance.

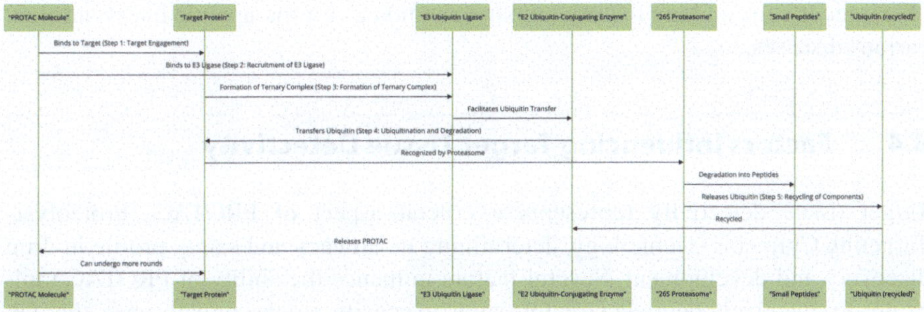

Fig. 8.2 The mechanism of action of PROTACs

8.3.3 Selective Targeting Mechanisms

Selective targeting is a critical aspect of PROTAC technology, as it determines the specificity of protein degradation and minimizes off-target effects. Several mechanisms contribute to the selective targeting of proteins by PROTACs:

1. Ligand Binding Specificity: The specificity of PROTAC-mediated protein degradation is primarily determined by the affinity and selectivity of the ligands for the target protein and the E3 ligase. High-affinity interactions between the PROTAC molecule and its target proteins ensure selective engagement and degradation while minimizing interactions with non-target proteins (Sakamoto et al. 2001).
2. Cellular Localization of Target Proteins: The subcellular localization of target proteins can influence their accessibility to PROTAC molecules and E3 ligases. Proteins localized to specific cellular compartments may be more or less susceptible to PROTAC-mediated degradation depending on the accessibility of the target and E3 ligase (Bondeson et al. 2015).
3. Stability and Pharmacokinetics of PROTAC Molecules: The stability and pharmacokinetic properties of PROTAC molecules play a crucial role in determining their selectivity and efficacy. Optimizing the pharmacokinetic profile of PROTACs can enhance their tissue distribution and reduce systemic exposure, thereby improving selectivity for target tissues (Buckley et al. 2016).
4. Influence of Cell Signaling Pathways: Cell signaling pathways can modulate the activity and expression of target proteins and E3 ligases, thereby influencing the efficacy and selectivity of PROTAC-mediated protein degradation. Understanding the regulatory mechanisms governing target protein expression and degradation can help optimize the design of selective PROTAC molecules (Cromm and Crews 2017).

Overall, the selective targeting mechanisms employed by PROTACs contribute to their ability to induce the degradation of specific target proteins while minimizing off-target effects, making them promising candidates for therapeutic intervention in various diseases.

8.4 Factors Influencing Target Tissue Detectivity

Target tissue detectivity represents a crucial aspect of PROTAC (Proteolysis Targeting Chimeras) technology, determining its efficacy and safety profile in drug discovery and development. Several factors influence the ability of PROTAC molecules to selectively degrade target proteins in specific tissues or cell types, thereby minimizing off-target effects and maximizing therapeutic benefits. This section explores the key factors influencing target tissue detectivity in PROTAC-mediated protein degradation (Churcher 2018), including ligand binding specificity, cellular localization of target proteins, stability and pharmacokinetics of PROTAC molecules, and the influence of cell signaling pathways.

8.4.1 Ligand Binding Specificity

The ligand binding specificity of PROTAC molecules plays a fundamental role in determining their ability to selectively recruit target proteins for degradation while minimizing interactions with off-target proteins. Ligand binding specificity is governed by the affinity and selectivity of the small-molecule ligands for both the target protein and the E3 ubiquitin ligase. High affinity and specificity toward the target protein ensure efficient recruitment to the PROTAC ternary complex, facilitating ubiquitination and subsequent proteasomal degradation (Lai et al. 2021). Conversely, nonspecific interactions with off-target proteins may lead to unwanted side effects and reduce the therapeutic window of PROTAC-based therapies.

Several strategies have been employed to enhance ligand binding specificity in PROTAC design:

1. Rational Design of Ligands: Structure-activity relationship studies and computational modeling techniques are utilized to design ligands with optimal binding affinity and selectivity for the target protein and E3 ligase.
2. Chemical Modifications: Introduction of specific chemical modifications, such as stereochemical changes or addition of functional groups, can improve ligand specificity by enhancing interactions with the target protein while reducing nonspecific binding to off-target proteins.
3. Fragment-Based Design: Fragment-based drug discovery approaches enable the identification of small-molecule fragments that bind to distinct regions of the target protein and E3 ligase, which can be combined to create high-affinity PROTAC molecules with enhanced specificity.

Optimizing ligand binding specificity is essential for achieving target tissue detectivity and maximizing the therapeutic potential of PROTAC-based therapies.

8.4.2 Cellular Localization of Target Proteins

The cellular localization of target proteins plays a critical role in determining the accessibility of PROTAC molecules and their ability to induce protein degradation in specific tissues or subcellular compartments. Target proteins may exhibit distinct subcellular localization patterns, including cytoplasmic, nuclear, membrane-bound, or organelle-associated localization, which can influence the efficacy and selectivity of PROTAC-mediated degradation. The subcellular localization of target proteins is influenced by various factors, including protein structure, post-translational modifications, and protein-protein interactions. Subcellular localization determines the accessibility of target proteins to PROTAC molecules and E3 ubiquitin ligases, as well as the efficiency of ubiquitination and proteasomal degradation. Additionally, the subcellular distribution of target proteins may vary across different cell types or tissues, further influencing the selectivity of PROTAC-mediated protein degradation.

Strategies to enhance target tissue detectivity based on cellular localization include:

1. Subcellular Targeting Sequences: Incorporation of subcellular targeting sequences or motifs into PROTAC molecules can direct their localization to specific cellular compartments, thereby increasing the efficiency of target protein degradation in the desired tissue or organelle.
2. Protein Engineering: Site-directed mutagenesis or protein engineering approaches can be employed to modify the subcellular localization signals of target proteins, facilitating their relocalization to specific cellular compartments where PROTAC molecules are more effective.

Understanding the cellular localization of target proteins is essential for optimizing the design and application of PROTAC molecules to achieve target tissue detectivity and minimize off-target effects (Zhang et al. 2021).

8.5 Stability and Pharmacokinetics of PROTAC Molecules

The stability and pharmacokinetic properties of PROTAC molecules influence their biodistribution, systemic exposure, and duration of action, ultimately impacting their efficacy and safety profile in vivo. PROTAC molecules must exhibit adequate stability in biological fluids and tissues to ensure prolonged circulation and effective target engagement. Additionally, favorable pharmacokinetic properties, including appropriate absorption, distribution, metabolism, and excretion (ADME) characteristics, are essential for achieving optimal therapeutic outcomes and minimizing potential toxicities (Paiva et al. 2021).

Several factors contribute to the stability and pharmacokinetics of PROTAC molecules:

1. Metabolic Stability: PROTAC molecules should resist degradation by endogenous metabolic enzymes, such as cytochrome P450 (CYP) enzymes, to maintain their integrity and activity in vivo. Chemical modifications, such as incorporation of metabolically stable linkers or modification of susceptible functional groups, can enhance metabolic stability and prolong the half-life of PROTAC molecules.
2. Plasma Protein Binding: Interactions with plasma proteins may affect the distribution, clearance, and systemic exposure of PROTAC molecules. High plasma protein binding can limit the free fraction of PROTAC available for target engagement and subsequent protein degradation. Optimization of physicochemical properties, such as lipophilicity and molecular size, can modulate plasma protein binding and improve the pharmacokinetic profile of PROTAC molecules.
3. Routes of Administration: The route of administration can significantly impact the pharmacokinetics and tissue distribution of PROTAC molecules. Different administration routes, including oral, intravenous, subcutaneous, and intraperitoneal routes, may result in distinct absorption, distribution, and elimination kinetics, influencing the overall efficacy and tolerability of PROTAC-based therapies (Krönke et al. 2014).
4. Clearance Pathways: Clearance pathways, such as hepatic metabolism, renal excretion, and biliary excretion, play a crucial role in the elimination of PROTAC molecules from the body. Understanding the predominant clearance pathways of PROTACs is essential for optimizing their pharmacokinetic properties and minimizing the risk of accumulation and toxicity.

Strategies to optimize the stability and pharmacokinetics of PROTAC molecules include:

- Rational Drug Design: Structure-activity relationship studies and computational modeling techniques can be employed to identify chemical modifications that improve metabolic stability, reduce plasma protein binding, and enhance overall pharmacokinetic properties.
- Prodrug Strategies: Prodrug formulations can be designed to improve the bioavailability and tissue distribution of PROTAC molecules by enhancing their solubility, stability, and membrane permeability.
- Formulation Optimization: Formulation approaches, such as nanoparticle encapsulation or lipid-based delivery systems, can be utilized to enhance the stability, solubility, and tissue-specific targeting of PROTAC molecules (Crews et al. 2017).

Optimizing the stability and pharmacokinetic properties of PROTAC molecules is essential for achieving target tissue detectivity and maximizing the therapeutic potential of PROTAC-based therapies.

8.5.1 Influence of Cell Signaling Pathways

Cell signaling pathways play a crucial role in regulating protein expression, stability, and localization, thereby influencing the efficacy of PROTAC-mediated protein degradation. Several cellular signaling pathways have been implicated in modulating PROTAC activity, including:

1. Ubiquitin-Proteasome System: The core machinery responsible for PROTAC-mediated protein degradation, the ubiquitin-proteasome system, is tightly regulated by various cellular signaling pathways. Modulation of signaling pathways involved in ubiquitin ligase activity or proteasomal function can affect the efficiency of PROTAC-induced protein degradation.
2. Protein Kinases and Phosphorylation: Protein kinases play key roles in regulating protein stability and localization through phosphorylation-dependent mechanisms. Modulation of protein kinase activity or phosphorylation status can influence the accessibility of target proteins to PROTAC molecules and subsequent degradation.
3. Protein-Protein Interactions: Cell signaling pathways often involve protein-protein interactions that can impact PROTAC-mediated protein degradation. Disruption of protein-protein interactions by PROTACs or alterations in protein complex formation can modulate target protein stability and degradation (Zeng et al. 2021).

8.6 Methods for Assessing Target Tissue Detectivity

8.6.1 In Vitro Assays

8.6.1.1 Cellular Thermal Shift Assays (CETSA)

Cellular Thermal Shift Assays (CETSA) have emerged as powerful tools for assessing target tissue detectivity of PROTAC molecules in vitro. CETSA exploits the principle that ligand binding stabilizes target proteins against thermal denaturation, resulting in a shift in their thermal melting profile. In the context of PROTACs, CETSA allows researchers to evaluate the engagement of target proteins by measuring changes in their thermal stability upon PROTAC treatment (Burslem et al. 2018). This technique offers several advantages, including its simplicity, versatility, and compatibility with various cell types and target proteins. By providing quantitative data on target engagement and tissue selectivity, CETSA facilitates the optimization of PROTAC molecules for enhanced therapeutic efficacy and reduced off-target effects.

8.6.1.2 Fluorescence-Based Assays

Fluorescence-based assays represent another valuable approach for assessing target tissue detectivity of PROTAC molecules in vitro. These assays leverage fluorescent tags or labels to monitor the interaction between PROTACs, target proteins, and E3 ligases in real time (Lai and Craig 2017). For example, fluorescence polarization assays can be used to measure changes in the rotational diffusion of fluorescently labeled ligands upon binding to their target proteins. Similarly, fluorescence resonance energy transfer (FRET) assays enable the detection of protein-protein interactions between target proteins and E3 ligases. By providing insights into the kinetics and specificity of PROTAC-mediated protein degradation, fluorescence-based assays offer valuable information for guiding the design and optimization of PROTAC molecules for clinical applications.

8.6.2 In Vivo Models

8.6.2.1 Animal Studies

Animal studies represent a critical component of preclinical evaluation for assessing the target tissue detectivity of PROTAC molecules in vivo. These studies involve the administration of PROTACs in animal models, followed by the analysis of target protein degradation and tissue distribution (Zorba et al. 2018). Various animal models, including mice, rats, and non-human primates, can be utilized to investigate the pharmacokinetics, efficacy, and safety of PROTAC-based therapies. By using genetically engineered animals or disease models, researchers can assess the selectivity and therapeutic potential of PROTACs in relevant physiological contexts. Animal studies provide valuable insights into the in vivo behavior of PROTAC molecules, including their tissue distribution, metabolic stability, and potential off-target effects (Ottis et al. 2016).

8.6.2.2 Patient-Derived Xenograft Models

Patient-derived xenograft (PDX) models offer a translational platform for evaluating the target tissue detectivity of PROTAC molecules in vivo. These models involve the engraftment of patient-derived tumor tissues into immunodeficient mice, allowing researchers to assess the response of human tumors to PROTAC treatment (Bondeson et al. 2015). PDX models recapitulate the heterogeneity and complexity of human tumors, providing valuable insights into PROTAC efficacy, selectivity, and resistance mechanisms. By using PDX models derived from different patient populations or tumor types, researchers can evaluate the therapeutic potential of PROTACs in specific clinical contexts. PDX studies enable the assessment of target protein degradation, tumor regression, and biomarker responses, facilitating the translation of PROTAC-based therapies into clinical trials (Raina et al. 2014).

8.7 Case Studies and Applications

8.7.1 Examples of Successful Target Tissue Detectivity

The application of PROTAC technology has yielded promising results in various disease settings, demonstrating its potential for achieving target tissue detectivity and therapeutic efficacy. Several case studies highlight successful examples of

PROTAC-mediated protein degradation in specific tissues or cell types. One notable example is the development of ARV-110, a PROTAC molecule targeting the androgen receptor (AR) for degradation in prostate cancer cells. ARV-110 selectively degrades AR in prostate cancer cells while sparing AR in non-cancerous tissues, demonstrating exquisite target tissue detectivity (Winter et al. 2016). Preclinical studies have shown that ARV-110 effectively inhibits the growth of AR-dependent prostate cancer xenografts in vivo, leading to tumor regression and prolonged survival in animal models. These findings highlight the potential of PROTACs for achieving tissue-specific protein degradation and therapeutic benefits in cancer.

Another compelling example is the use of PROTACs to target disease-associated proteins in neurodegenerative disorders such as Alzheimer's disease (AD). Researchers have developed PROTAC molecules capable of degrading tau protein, a key pathological hallmark of AD, in neuronal cells. These tau-targeting PROTACs selectively induce tau degradation in neurons while sparing other cell types, offering a promising approach for mitigating tau pathology in AD. Preclinical studies have shown that tau-targeting PROTACs reduce tau aggregation and neurotoxicity in cellular and animal models of AD, providing evidence of their potential therapeutic utility in neurodegenerative diseases (Winter et al. 2016; Crews et al. 2017).

Furthermore, PROTAC technology has been applied to target proteins involved in inflammatory and autoimmune disorders, such as rheumatoid arthritis (RA) and psoriasis. For example, PROTACs targeting specific kinases implicated in inflammatory signaling pathways have been developed to selectively degrade disease-associated proteins in immune cells. Preclinical studies have demonstrated that these kinase-targeting PROTACs effectively inhibit inflammatory cytokine production and suppress autoimmune responses in animal models of RA and psoriasis. These findings suggest that PROTAC-mediated protein degradation holds promise for the development of novel therapies for autoimmune and inflammatory diseases.

8.8 Challenges and Limitations in Achieving Selectivity

While PROTAC technology holds great promise for achieving target tissue detectivity and therapeutic specificity, several challenges and limitations must be addressed to maximize its clinical utility. One major challenge is the design of PROTAC molecules with optimal selectivity and potency. Achieving selectivity requires careful consideration of the binding affinities of both the target protein ligand and the E3 ubiquitin ligase ligand, as well as their relative expression levels in different tissues. Moreover, off-target effects may arise due to promiscuous binding of PROTACs to unintended proteins, leading to undesired degradation and potential toxicity. Thus, rational design strategies and screening approaches are needed to minimize off-target effects and enhance the selectivity of PROTAC molecules.

Another challenge is the identification of suitable E3 ubiquitin ligases for targeting specific proteins of interest. While several E3 ligases have been validated for use in PROTAC-mediated protein degradation, not all target proteins may be efficiently degraded by existing E3 ligase ligands. Furthermore, the expression levels and subcellular localization of E3 ligases may vary across different tissues, influencing the efficacy and selectivity of PROTACs. Thus, efforts to identify novel E3 ligases and optimize their binding properties are essential for expanding the scope of PROTAC technology and overcoming limitations in achieving selectivity (Steinebach et al. 2020).

Additionally, the pharmacokinetic properties of PROTAC molecules, including their stability, biodistribution, and metabolic fate, pose significant challenges for achieving target tissue detectivity. PROTACs must exhibit favorable pharmacokinetic profiles to ensure sufficient exposure and selective degradation of target proteins in vivo. However, factors such as poor oral bioavailability, rapid clearance, and off-target tissue distribution may limit the efficacy and selectivity of PROTAC-based therapies. Thus, the optimization of PROTAC pharmacokinetics through medicinal chemistry approaches and formulation strategies is crucial for overcoming these challenges and improving target tissue detectivity. Furthermore, the complex interplay between target proteins and cellular signaling pathways can influence the selectivity and efficacy of PROTAC-mediated protein degradation. Crosstalk between signaling pathways, feedback loops, and compensatory mechanisms may impact the response of target proteins to PROTAC treatment, leading to variability in target tissue detectivity and therapeutic outcomes. Therefore, a comprehensive understanding of the molecular mechanisms underlying PROTAC action and the signaling networks regulating target protein stability is essential for addressing challenges and optimizing selectivity in PROTAC-based therapies (Wang et al. 2020).

Therefore, while PROTAC technology offers exciting opportunities for achieving target tissue detectivity and therapeutic specificity, addressing challenges and limitations in design, target identification, pharmacokinetics, and signaling pathways is essential for realizing its full potential in clinical applications.

8.9 Future Directions and Opportunities for Improvement

Despite the challenges and limitations associated with PROTAC technology, ongoing research efforts are focused on addressing these issues and advancing the field toward clinical translation. Several future directions and opportunities for improvement hold promise for enhancing target tissue detectivity and therapeutic specificity of PROTAC-based therapies.

One key area of focus is the development of novel ligands for targeting E3 ubiquitin ligases with enhanced selectivity and potency. By expanding the repertoire of E3 ligase ligands, researchers can overcome limitations in target protein degradation and achieve improved selectivity in PROTAC-mediated protein degradation. High-throughput screening approaches, structure-based design strategies, and computational modeling techniques are being employed to identify and optimize E3 ligase ligands with desirable pharmacological properties.

Another promising avenue for improving target tissue detectivity is the incorporation of tissue-specific targeting strategies into PROTAC design. By engineering PROTAC molecules with tissue-specific ligands or delivery systems, researchers can enhance the accumulation and selective degradation of target proteins in specific tissues or cell types. Nanoparticle-based delivery systems, antibody-drug conjugates, and cell-penetrating peptides are being explored as potential strategies for achieving tissue-specific targeting and enhancing therapeutic efficacy of PROTAC-based therapies (Winter et al. 2016).

Furthermore, advancements in chemical biology and proteomics are enabling researchers to gain deeper insights into the molecular mechanisms underlying PROTAC action and target tissue detectivity. By elucidating the determinants of PROTAC efficacy, selectivity, and resistance, researchers can inform rational design strategies and optimize PROTAC molecules for enhanced therapeutic outcomes. Moreover, the development of innovative assay technologies and biomarker approaches for assessing target tissue detectivity in preclinical and clinical settings is critical for accelerating the translation of PROTAC-based therapies into the clinic. Additionally, efforts to overcome challenges associated with pharmacokinetics and drug delivery are essential for maximizing the clinical potential of PROTAC technology. By optimizing the pharmacokinetic properties of PROTAC molecules through medicinal chemistry approaches, formulation strategies, and prodrug design, researchers can enhance tissue penetration, stability, and systemic exposure, thereby improving target tissue detectivity and therapeutic efficacy. Moreover, the development of innovative drug delivery systems and administration routes, such as oral formulations, sustained-release formulations, and targeted delivery approaches, can further enhance the selectivity and efficacy of PROTAC-based therapies while minimizing off-target effects and toxicity (Ottis et al. 2016; Wang et al. 2020).

Therefore, while challenges and limitations exist, PROTAC technology holds tremendous promise for achieving target tissue detectivity and therapeutic specificity in a wide range of disease settings.

8.10 Conclusions

The exploration of PROTAC technology has revealed its potential to revolutionize drug discovery and therapeutic interventions. The mechanism of action of PROTACs, leveraging the ubiquitin-proteasome system for targeted protein degradation, offers a novel approach to address diseases with challenging therapeutic targets. Through bifunctional molecules that recruit E3 ubiquitin ligases to specific proteins, PROTACs can selectively degrade disease-causing proteins, demonstrating the importance of target tissue detectivity. Factors influencing target tissue detectivity, such as ligand binding specificity, cellular localization of target proteins, stability, pharmacokinetics, and cell signaling pathways, play critical roles in the success of PROTAC-based therapies. Methods for assessing target tissue detectivity, including in vitro assays such as CETSA and fluorescence-based assays, as well as in vivo models like animal studies and patient-derived xenograft models, provide valuable tools for evaluating PROTAC efficacy and selectivity. Despite challenges and limitations, successful case studies demonstrate the potential of PROTACs in achieving target tissue detectivity and therapeutic specificity in various disease settings, including cancer, neurodegenerative disorders, and autoimmune diseases.

However, further research is needed to optimize PROTAC design, pharmacokinetics, and drug delivery, as well as to elucidate the molecular mechanisms underlying PROTAC action.

The implications of PROTAC technology for drug discovery and therapeutics are vast and multifaceted. Firstly, PROTACs offer a promising approach for targeting previously undruggable proteins, expanding the scope of druggable targets and enabling the development of therapies for diseases with limited treatment options. By selectively degrading disease-associated proteins while sparing healthy tissues, PROTACs have the potential to improve therapeutic efficacy and minimize off-target effects, thereby enhancing patient outcomes and quality of life. Additionally, PROTAC technology may overcome drug resistance mechanisms by targeting proteins through a different mechanism of action, providing a solution to challenges associated with traditional small-molecule inhibitors. Furthermore, the development of tissue-specific PROTACs and innovative drug delivery systems holds promise for enhancing target tissue detectivity and improving the clinical utility of PROTAC-based therapies. Overall, the continued exploration and refinement of PROTAC technology have significant implications for advancing drug discovery and therapeutic interventions, paving the way for the development of next-generation treatments for a wide range of diseases.

References

Bondeson DP et al (2015) Catalytic in vivo protein knockdown by small-molecule PROTACs. Nat Chem Biol 11(8):611–617

Buckley DL et al (2016) Small-molecule inhibitors of the orphan nuclear receptor SHP2 (PTPN11). J Med Chem 59(5):1862–1875

Burslem GM et al (2018) The advantages of targeted protein degradation over inhibition: an RTK case study. Cell Chem Biol 25(1):67–77

Churcher I (2018) Protac-induced protein degradation in drug discovery: breaking the rules or just making new ones? J Med Chem 61(2):444–452

Crews CM, Pearson NL (2012) Targeted degradation: a prototype for small-molecule drug discovery. Drug Discov Today Technol 9(1):e3–e10

Cromm PM, Crews CM (2017) Targeted protein degradation: from chemical biology to drug discovery. Cell Chem Biol 24(9):1181–1190

Krönke J et al (2014) Lenalidomide causes selective degradation of IKZF1 and IKZF3 in multiple myeloma cells. Science 343(6168):301–305

Lai AC et al (2017) Induced protein degradation: an emerging drug discovery paradigm. Nature reviews Drug discovery 16(2):101–114.

Lai AC, Craig MC (2021) Cereblon modulator iberdomide induces degradation of the transcription factors Ikaros and Aiolos: immunomodulation in healthy volunteers and relevance to systemic lupus erythematosus. Ann Rheum Dis 80(5):557–567

Ottis P et al (2016) Assessment of cellular E3 ubiquitin ligase activity and substrate ubiquitination by immunoprecipitation assays. Methods Mol Biol 1449:79–89

Paiva S-L et al (2021) Discovery and preclinical characterization of PROTAC BCL-XL degraders for the treatment of hematologic malignancies. Cancers 13(2):333

Raina K et al (2014) Targeted protein destabilization reveals an estrogen-mediated ER stress response. Nat Chem Biol 10(12):957–962

Sakamoto KM et al (2001) Development of protacs to target cancer-causing proteins for ubiquitination and degradation. Mol Cell Proteomics 1(4):506–518

Steinebach C et al (2020) Targeted degradation of endogenous proteins by nanobodies. Nat Commun 11(1):1–10

Toure M, Crews CM (2016) Small-molecule PROTACS: new approaches to protein degradation. Angew Chem Int Ed 55(6):1966–1973

Wang Y et al (2020) De novo targeted degradation of BRD9 by PROTACs. Chem Commun 56(33):4527–4530

Winter GE et al (2015) Phthalimide conjugation as a strategy for in vivo target protein degradation. Science 348(6241):1376–1381

Winter GE et al (2016) Drug development: PROTACs show target binding induced degradation of proteins. ChemMedChem 11(14):1531–1534

Zeng M et al (2021) Targeted protein degradation by small molecules: opportunities and challenges. J Med Chem 64(23):16948

Zhang X et al (2021) Discovery of amino acid conjugates of E3 ligase ligands that elicit degradation of estrogen receptor α. ACS Med Chem Lett 12(1):72–79

Zorba A et al (2018) Delineating the role of cooperativity in the design of potent PROTACs for BTK. Proc Natl Acad Sci 115(31):E7285–E7292

Clinical Development of PROTACs

9

Advait Dubey, Kavita Pal, and Vikram Gota

Abstract

PROTACs have gained significant attention in recent years due to their capability to target and degrade disease-causing proteins. It is looked upon as a therapeutic that would address resistance mechanisms thus expanding treatment opportunities. The deployment of PROTACs as therapeutic modalities depends upon their translational efficacy from being confined as a lab success to patient utility. Results of the clinical trials done recently heighten the anticipation of integration of PROTACs into treatment regimens. In this chapter, we delve into the cutting-edge realm of targeted protein degradation, specifically focusing on PROTACs. Our exploration encompasses a multifaceted analysis, beginning with pharmacology of PROTACs and examination of biomarkers crucial for monitoring the efficacy and response to PROTAC therapies. Moving forward, we present an overview of ongoing clinical trials, shedding light on their trial designs and the potential these trials hold for the future of anticancer treatments. Moreover, our discussion encompasses dosing strategies for PROTACs, emphasizing the approaches employed in optimizing therapeutic outcomes. Addressing safety concerns and considerations is paramount in our examination, offering a comprehensive understanding of the challenges and advancements in ensuring the well-being of patients undergoing PROTAC-based therapies. Lastly, we explore the frontier of combining PROTACs with other anticancer molecules.

A. Dubey · K. Pal · V. Gota (✉)
Department of Clinical Pharmacology, Advanced Centre for Treatment, Research & Education in Cancer (ACTREC), Tata Memorial Centre, Navi Mumbai, India

Homi Bhabha National Institute, Mumbai, India
e-mail: kpal@actrec.gov.in; vgota@actrec.gov.in

M. Nandave, P. Jain (eds.), *PROTAC-Mediated Protein Degradation: A Paradigm Shift in Cancer Therapeutics*, https://doi.org/10.1007/978-981-97-5077-1_9

Keywords

Targeted protein degradation · Ubiquitin protease system · ARV-471 · ARV-110 · ADME · Recommended phase 2 dose · Clinical benefit rate · Maximum tolerated dose · Pharmacokinetics · Therapy-induced senescence

Abbreviations

ADME	Absorption, distribution, metabolism and excretion
AE	Adverse effects
AML	Acute myeloid leukemia
AR	Androgen receptor
BRD4	Bromodomain containing protein 4
BTK	Burton's tyrosine kinase
CBR	Clinical benefit rate
CCR5	C-chemokine receptor-5
CRBN	Cereblon
EDP	Event driven pharmacology
ER	Estrogen receptor
ERK	Extracellular signal-regulated kinase
HBA	Hemoglobin A
IVIVE	In vitro in vivo extrapolation
mCRPC	Metastatic castration-resistant prostate cancer
MTD	Maximum tolerated dose
ODP	Occupancy driven pharmacology
PEG	Polyethylene glycol
PK	Pharmacokinetics
POI	Protein of interest
PSA	Prostate specific antigen
PTM	Post-translational modification
RP2D	Recommended phase 2 dose
SASP	Senescence associated secretory phenotype
SERD	Selective estrogen receptor degrader
TIS	Therapy-induced senescence
TPSA	Topological polar surface area
UGT	Uridine diphosphate (UDP)-glucuronosyltransferases
UPS	Ubiquitin protease system
VEGFR	Vascular endothelial growth factor receptor

9.1 Introduction

In the realm of cancer treatment, traditional approaches like surgery, chemotherapy, and radiotherapy are commonly favoured. However, these methods come with drawbacks, including pronounced side effects, toxicities, morbidities, and enduring adverse effects. Such consequences significantly affect the quality of life for cancer patients both during and after treatment. With advancements in treatment strategies, targeted cancer therapies were introduced, which focus on cancer-related biomolecules, such as oncoproteins, and disrupt their cellular processes leading to oncogenesis within cancer tissues. Targeted therapies have made significant strides in the field of cancer treatment, emerging as a potent plan of action against cancer. Notably, successful development of small molecular inhibitors and monoclonal bodies has progressed for target proteins that are over-expressed or hyperactivated in cancer (Padma 2015). Nevertheless, researchers continue to explore more effective and specific strategies to address challenges such as off-target effects, limited therapeutic benefits, and drug resistance associated with these targeted therapies. Moreover, the challenge posed by the undruggable proteome, which constitutes over 85% restricts the options for cancer treatment (Neklesa et al. 2017). On another front, tumour heterogeneity and the emergence of drug resistance to existing therapeutics stand as significant obstacles to achieving effective cancer treatment. Proteolysis targeting chimeras (PROTACs) in clinical trials are seen as a solution to the existing limitations of the available therapies against cancer and other diseases (He et al. 2022). The initiation of clinical trials for ARV-471 in 2019 marked a significant milestone as the first proteolysis targeting chimeras (PROTAC) to be approved for testing in individuals with breast cancer. This development heralded a new era in therapeutics, showcasing the potential for targeted protein degradation to be explored in human patients. PROTACs are heterobifunctional molecules comprising of a linker, a ligand for binding to the target protein, and E3 ubiquitin ligase binding ligand. PROTACs inhibit the function of their target proteins by pinpointing towards degradation by the ubiquitin protease system (UPS). These components can be finely adjusted to enhance the pharmacokinetic characteristics of PROTACs, thereby refining their effectiveness as therapeutic agents. The molecular weight of PROTACs typically falls within the range of 900–1100 Da, with the number of rotatable bonds ranging from 20 to 25 (Atilaw et al. 2021). The type of linker employed significantly influences cell permeability, impacting the bioavailability of PROTACs.

One limitation of PROTACs lies in the scarcity of ligands that bind to E3 ligases beyond those extensively utilized. This aspect presents an opportunity for improvement in the preclinical setting. Clinical development of optimized PROTAC is achieved when the ligand binding to the target protein, the physicochemical properties of the linker, and the ligase utilized for degradation collectively enhance the therapeutic effects of PROTACs by improving permeability, half-life, and clearance. The following chapter would expound on the various facets of clinical development of PROTACs. We would also explore various pharmacological considerations and dosing strategies that would maximize the ADME properties of PROTACs.

Along with that we would also bring into light various safety concerns of PROTACs and synergistic treatment modalities that can be used with PROTACs.

9.2 Pharmacology and Biomarkers

9.2.1 Pharmacology of PROTACs

The effectiveness of PROTACs as a therapeutic approach lies in their ability to specifically degrade targeted proteins, selectively pinpoint malignant or diseased cells, and employ a catalytic mechanism of action that acts as a countermeasure against resistance development. The success of this therapy is largely influenced by the pharmacological model through which PROTACs operate in combating cancer or other disorders. The two primary pharmacological concepts are Occupancy Driven Pharmacology (ODP) and event driven pharmacology (EDP). In ODP, the inhibitor must bind to the target protein, blocking its functions. While the inhibitor is bound, the protein's function remains inhibited, requiring sustained and potentially high concentrations of the inhibitor drug. This prolonged exposure increases the risk of off-target toxicities and development of drug resistance. Mutations leading to alterations in the catalytic binding site of proteins can diminish the drug efficacy or lead to its rapid dissociation from the protein. On the contrary, PROTACs operate under the event driven pharmacology model. In this approach, the ligand doesn't need to occupy the binding site to affect protein function. Instead, PROTACs induce protein degradation by positioning the targeted protein in proximity to ubiquitin ligase, that is by inducing selective intracellular proteolysis. Figure 9.1 depicts the mechanism of action of PROTACs. This catalytic mechanism enables PROTACs to function through multiple cycles against the targeted proteins, making them effective even at sub-stoichiometric concentrations. Even if molecules cannot independently initiate degradation, their ability to bind to the target protein can still be valuable in constructing a PROTAC, thereby increasing their therapeutic significance (Koravović et al. 2022). Moreover, PROTACs prove effective against mutated versions of target proteins, making them preferred treatments for addressing cancer, a condition prone to drug resistance and mutations.

9.2.2 Role of Biomarkers in PROTAC Therapies

According to the Biomarkers, Endpoints, and other Tools (BEST) glossary, "a biomarker is a quantifiably measurable characteristic used as an indicator for various biological processes, including normal functions, disease pathways, or responses to exposures and interventions, such as therapeutic treatments". BEST classifies biomarkers into seven categories: (1) Diagnostic, (2) Susceptibility/risk, (3) Monitoring, (4) Prognostic, (5) Predictive, (6) Pharmacodynamic/response, and (7) Safety. In the context of PROTACs, evaluating the levels of target proteins during therapy can function as a biomarker for assessing therapeutic efficacy. Common biomarkers for

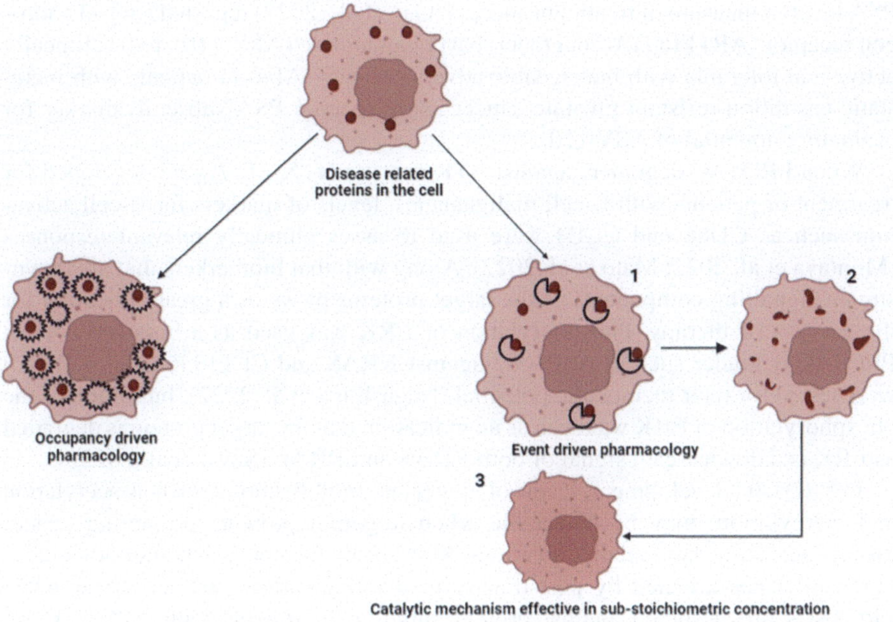

Fig. 9.1 Mechanism of action of PROTAC through event driven pharmacology. The action of inhibitor drugs through occupancy driven model of pharmacology is constant but event driven model is dynamic as depicted in the figure, wherein, (1) PROTACs being bound to target proteins in sub-stoichiometric condition in the cell; (2) the protein is entirely degraded giving rise to many small fragments; (3) after clearance of the fragments, the cell is healthy without any disease proteins

measuring PROTAC function include cell proliferation rates, apoptosis rates, and invasion capacity. Additionally, examining downstream targets such as substrate phosphorylation or transcription factor activity can provide insights into whether a PROTAC is effectively engaging and inducing the degradation of the target protein.

Biomarkers play a crucial role in monitoring not only the on-target effects but also the off-target toxicity of PROTACs. By observing the levels of target proteins in tumour tissue or blood samples, researchers can gain valuable information about the overall impact of PROTACs. Pharmacodynamic markers become essential in determining the appropriate dosage and dosing frequency, emphasizing the need for thorough testing of these markers even in the preclinical stages of development. This comprehensive approach involving biomarkers contributes to a better understanding of PROTAC functionality and safety in therapeutic applications.

It is noteworthy that a myriad of molecular indicators exhibit specificity to distinct cancer types and can be crucial for monitoring the efficacy and response to PROTAC therapies. These biomarkers can be explored as a marker for efficient activity of PROTACs. For example, Prostate Specific Antigen (PSA) is a critical biomarker for evaluating therapy efficacy in prostate cancer (Lilja et al. 2008). Increased PSA levels in blood are indicative of prostate cancer, making reduced

PSA levels a measure of treatment success. Gao et al. (2022) reported a novel androgen receptor (AR) PROTAC degrader, bavdegalutamide (ARV-110), to be clinically active and tolerable with manageable adverse effects (AEs) in patients with metastatic castration-resistant prostate cancer using specific PSA subset as marker for evaluating potential of ARV-110.

When PROTAC degraders against BTK such as NTX-2127 were developed for treatment of patients with B-cell malignancies, levels of markers for B-cell activation such as CD86 and CCL4 were used to asses clinically relevant responses (Montaya et al. 2022; Mato et al. 2022). Along with that biomarkers that are downstream signalling components of the target proteins prove as a great indication for the success of therapy. Phosphorylation of ERK was used as a biomarker when PROTAC degrader such as ASP3082 against KRAS and CFT1946 against BRAF were tested for their therapeutic potential (Nagashima et al. 2022). Inhibition of the phosphorylation of ERK was seen as an indication that the target protein is degraded as ERK is a downstream signal of both KRAS and BRAF (Sowa et al. 2022).

In PROTAC development, levels of epigenetic modifications, such as acetylation and methylation, may be detectable when targeting proteins promoting cancer through post-translational modifications. Monitoring the upregulation or downregulation of genes affected by post-translational modifications (PTMs) occur when PROTACs are designed against proteins binding to regions with PTMs. These altered levels of expression of genes can also be explored as marker for activity of PROTACs. In addition, detection of fragments of the targeted protein upon action of PROTACs would help us to determine the therapy effects along with other evidence like tumour growth inhibition or reduction. It would be a robust biomarker as it would give us an idea about the therapy in a much direct manner than other markers.

9.3 Ongoing Clinical Trials of PROTACs

1. *Bavdegalutamide* (*ARV-110*): PROTAC developed against prostate cancer that targets the androgen receptor. A critical role in prostate cancer initiation and progression is played by androgen signalling. Androgen receptor is a key component of androgen signalling which regulate gene transcriptional processes via nuclear binding to androgen response elements on multiple target genes. Disruption of androgen signalling by degradation of the androgen receptor is the mechanism by which ARV-110 acts against metastatic castration-resistant prostate cancer (mCRPC).

   In ARV-110, the binding to the AR ligand binding domain is through a cyclohexyl moiety which then upon binding employs E3 ubiquitin ligase that contains cereblon through immunomodulatory based moiety that leads to polyubiquitination of the Androgen receptor.

   In prostate cancer cell lines VCap (vertebral cancer of prostate) and LNCap (lymph node carcinoma of the prostate), ARV-110 showed degradation of the AR receptor with a DC_{50} of 1 nM and when treated with 10 nM of ARV-110 for 8 h, AR degradation with D_{Max} of 85% was seen (Snyder et al. 2021).

ARV-110 was also able to degrade AR receptors having several mutations such as T878A, H875V, F877L, M896V which were resistant to abiraterone and enzalutamide which are hormone based therapeutic agents. ARV-110 showed greater induction of apoptosis, more inhibition of cell proliferation and greater inhibition of Prostate specific antigen (PSA) synthesis when compared with enzalutamide. The combination of ARV-110 with abiraterone greatly inhibited the tumour than each agent alone in a preclinical study on castrated mouse model (Neklesa et al. 2019).

Preclinical trials of ARV-110 were done using castrated and even in non-castrated mouse models in whom tumour was induced by injecting VCap tumour cells. Oral administration of ARV-110 lead to tumour reduction more than enzalutamide. In mouse models that were resistant to abiraterone and enzalutamide, ARV-110 was tested and it was found that once daily administration of 3 mg/kg ARV-110 reduced tumour growth by 60% and 10 mg/kg daily reduced tumour growth by 70%. Also, when tested in a patient-derived xenograft, 100% tumour inhibition with 90% decrease in PSA levels was due to treatment with ARV-110. (Chirnomas et al. 2023).

The ongoing clinical trials of PROTACs along with the study designs are shown in Table 9.1. PROTAC Bavdegalutamide (NCT03888612) became the first PROTAC to enter clinical trials in year 2019. During clinical trials, the study design used was phase-1 dose escalation design. The trial enrolled patients with mCRPC who had undergone two prior therapies and progressed on their most recent therapy. A total of 22 patients were initially enrolled. Among these patients were 77% who had received abiraterone and enzalutamide therapy earlier and 77% had previously undergone chemotherapy. Such patients were administered dosages of 35–700 mg once daily or 140–420 mg twice daily orally (Petrylak et al. 2020). After 6 weeks of ARV-110 treatment, reduced AR proteins levels were observed in patient's tumour tissue and 2 patients had clinically relevant response as decreased levels of serum PSA of 74% and 97% (Chirnomas et al. 2023). This was the clinical proof of action of ARV-110. Exposure-activity relationship was observed in bavdegalutamide and with further analysis of the patient cohort it was concluded that 420 mg total daily dose would be efficacious. Predictions from enzalutamide-resistant preclinical models were also compared and thus 420 mg once daily was selected as recommended phase 2 dose (RP2D) based on safety, efficacy and pharmacokinetic studies (Gao et al. 2022). Treatment-related adverse effects were seen in patients in phase-1 clinical trials (Chirnomas et al. 2023).

When patients (N = 138) of both phase 1 and 2 were treated with RP2D 420 mg total daily intake, TRAEs were observed in >20% patients and adverse effects were grade 1 or 2 and no series TRAEs were observed in patients (Nausea-48%; fatigue-36%; vomiting-26%; Decreased appetite-25% and Diarrhoea-20%).

2. *ARV-471* (*Vepdegestrant*): Breast cancer is the second most common cancer in women. ARV-471 is a CRBN-based PROTAC targeting estrogen receptor (ER) and is in clinical trials for treating ER+ breast cancer patients. Treatment with

Table 9.1 Ongoing clinical trials of PROTACs

Protein degrader	Trial phase	Estimated/ actual sample size	Patient population	Trial design
Bavdegalutamide (ARV-110)	I/II	250	mCRPC	Dose escalation and cohort expansion, open label clinical trial **NCT03888612**
	Ib	40	Metastatic prostrate cancer with rising PSA levels on abiraterone	Open label, single group assessment, clinical trial **NCT05177042**
CC-94676	I	250	mCRPC	Multi-centre, open label, dose finding study **NCT04428788**
AC176	I	36	mCRPC	Open label, first in human, multi-centre, dose escalation design **NCT05241613**
	I	08	mCRPC	Multi-centre, open label, dose escalation design **NCT05673109**
HP518	I	22	mCRPC	First in human, open label, dose escalation and dose expansion study **NCT05252364**
ARV-766	I/II	220	mCRPC	Open label, dose escalation and cohort expansion trial **NCT05067140**
AC682	I	30	Advanced-stage ER+ HER− breast cancer	First in human, open label, dose escalation clinical trial **NCT05080842**
	I	06	Advanced-stage ER+ HER− breast cancer	Open label, dose escalation study **NCT05489679**
NX-2127	I	160	B-cell malignancies	Multi-centre, open label and dose escalation study **NCT04830137**
NX-5948	I	226	B-cell malignancies	Multi-centre, open label, dose escalation and cohort expansion **NCT05131022**

(continued)

Table 9.1 (continued)

Protein degrader	Trial phase	Estimated/ actual sample size	Patient population	Trial design
BGB-16673	I/II	291	B-cell malignancies	Open label, dose escalation and expansion studies **NCT05006716**
	I	116	B-cell malignancies	Open label, dose escalation and expansion study **NCT05294731**
HSK29116	I	156	B-cell malignancies	Multi-centre, open label, dose escalation study **NCT04861779**
CFT8634	I/II	50	Advanced-stage synovial sarcoma or SMARC1-null tumours	First in human, non-randomized, open label, multi-centre study **NCT05355753**
FHD-609	I	104	Advanced-stage synovial sarcoma or SMARCB1-loss tumours	Multicentre, open label, dose escalation and dose expansion study **NCT04965753**
DT2216	I	20	Relapsed and/or refractory malignancies	Open label, dose escalation and cohort expansion studies **NCT04886622**
KT-413	I	80	B-cell non-Hodgkin lymphoma	Multi-centre, open label, dose escalation and expansion studies **NCT05233033**
KT-333	I	120	Blood cancers and solid tumours	Multi-centre, open label, dose escalation and expansion studies **NCT05225584**
ASP3082	I	356	Advanced-stage $KRAS^{G12D}$-mutant solid tumours	Open label, dose escalation, non-randomized study **NCT05382559**
CFT1946	I/II	102	Advanced-stage $BRAF^{V600}$-mutant solid tumours	Open label, multi-centre, dose escalation studies **NCT05668585**
ARV-471	I/II	215	Advanced-stage ER+ HER2− breast cancer	Open label, dose escalation and cohort expansion studies **NCT04072952**
ARV-471	I	06	Advanced-stage ER+ HER2− breast cancer	Open label, single group assessment **NCT05463952**

(continued)

Table 9.1 (continued)

Protein degrader	Trial phase	Estimated/ actual sample size	Patient population	Trial design
ARV-471	Ib	32	Advanced-stage ER+ HER2− breast cancer	Open label, single group assessment **NCT05501769**
ARV-471	II	152	Localized ER+ HER2− breast cancer	Open label, randomized, non-comparative proof of concept, neoadjuvant study **NCT05549505**
ARV-471	I/II	35	Advanced-stage ER+ HER2− breast cancer	Open label, umbrella study **NCT05548127**
ARV-471	I/II	35	Advanced-stage ER+ HER2− breast cancer	Prospective, open label, multi-centre study **NCT05573555**
ARV-471	III	560	Advanced-stage ER+ HER2− breast cancer	Randomized, open label, multi-centre studies **NCT05654623**

The above table contains information available at https://clinicaltrials.gov/ and is adapted from Chirnomas et al. 2023.

selective ER degraders (SERDs) like fulvestrant has shown that ER+ breast cancer can be controlled by targeting the ER receptors but even after 6 months of treatment with fulvestrant, up to 50% of the baseline levels of ER remains the same. Thus, a better ER degrader was needed to combat ER+ breast cancer. When ARV-471 was tested with cell lines of breast cancer like MCF-7 and ESR-1 mutant cell lines, it resulted in ER degradation at DC_{50} of 1–2 nM. ARV-471 also successfully degraded ER mutants with Y537S and D538G mutations. ARV-471 was preclinically tested with various xenograft models such as ESR1 mutant model, tamoxifen-resistant MCF7 xenograft model and oestradiol-dependent xenograft model of MCF7 (Flanagan et al. 2018).

In PDX mutant model of ESR1, ARV-471 was administered orally and daily dose of 10 mpk leads to inhibition of tumour growth by 99% and dose of 30 mpk leads to inhibition of tumour growth by 106%. These were much better results than fulvestrant that upon dosage of 200 mpk twice in a week showed 62% tumour growth inhibition. In orthotropic oestradiol-dependent MCF-7 xenograft model, dosage of 10 mpk lead to >90% of tumour growth and dosage of 30 mpk lead to >100% tumour inhibition and in tamoxifen-resistant xenograft model of MCF-7, once daily dosage 30 mpk lead to 65% tumour growth inhibition and 73% reduction in tumoural ER (Flanagan et al. 2018).

Clinical trials of ARV-471 (NCT04072952) were performed in patients with locally advanced or metastatic ER+ HER2− breast cancer. In phase 1, patients who were selected had to fulfil certain conditions such as they must have undergone one prior CDK 4/6 inhibitor treatment, two endocrine therapies and at most three prior chemotherapy lines. A daily oral administration of ARV-471 was given to 60 patients in doses ranging from 30 to 700 mg. Pharmacokinetic parameters proportionally increased from 30 to 500 mg doses and ER degradation of up to 89% was observed in tumour biopsy samples collected post-treatment and observed by ER immunofluorescence (Hamilton et al. 2021). The clinical benefit rate in 47 evaluable patients was 40%. Clinical benefit rate (CBR) is a value that indicates the total number or percentage of patients who achieved a complete, partial response or had stable disease for 6 months or more. This value indicates the percentage of the patient who had actually benefitted by the therapeutic intervention.

In phase II cohort expansion phase, two doses were tested in patients. They were 200 and 500 mg of ARV-471 daily. Seventy-one patients who participated had advanced-stage ER+ HER2− breast cancer among which 100% had undergone prior CDK4/6 inhibitor and 73% with prior chemotherapy, among them were 45% who had received it in metastatic setting and 79% had even received prior fulvestrant treatment. Such cohort of patients received oral ARV-471 (Hamilton et al. 2021; Chirnomas et al. 2023). The Clinical Benefit Rate in one cohort of 35 patients at dosage of 200 mg was 37% and in another cohort of 36 patients at dosage of 500 mg was 38.9%. However, the sub-group of patients having *ESR1* mutation had superior CBR of 47.4% ($N = 19$) and 54.5% ($N = 22$) at doses of 200 and 500 mg respectively.

The most common treatment-related Adverse Effect observed in the treatment cohorts was fatigue. Median progression-free survival for the cohort treated with 200 mg dosage was 3.5 months and for the same dosage *ESR1* mutated group gave results such as median progression-free survival of 5.5 months and phase III monotherapy dosage selected was 200 mg once daily of ARV-471 (Chirnomas et al. 2023).

The clinical trial outcomes of leading PROTACs, specifically ARV-110 and ARV-471, will serve as benchmark to determine progress in the field. As a new class of compounds, the PROTACs have suffered considerable attrition during the early phases of clinical development. ARV-471 and ARV-110, the initial two PROTAC drugs entering phase II clinical trials, have demonstrated robust clinical performance in initial trials. The forthcoming phase II and phase III clinical trial results will determine if these PROTACs can meet anticipated outcomes, potentially paving the way for their clinical implementation in the near future.

9.4 Dosing Strategies

The effectiveness, potential toxicity, including the risk of death, or lack of any effect experienced by patients can be influenced by the dosing strategies of a drug. The dosing strategy depends on the half-life and stability, target protein turnover rate, duration of target protein engagement, E3 ligase engagement, therapeutic window, pharmacodynamic markers and combination therapies.

9.4.1 Intermittent Dosing

Intermittent dosing is a strategy in which doses of medication are administered with periods of time in between each dose where the plasma levels of the drug concentration rise and fall between drug doses (Tirona and Kim 2017). The frequency of dosing depends on the pharmacokinetic half-life along with the drug formulation for desired therapeutic outcome (Grover and Benet 2011). The dosage and the route of administration depend on the pharmacological effects and safety profile of the therapeutic. PROTAC ARV-771 was used to treat castration-resistant prostate cancer mice implanted with 22Rv1 tumour cells to induce tumour formation. Intermittent dosing regimen of Q3D (once in 3 days) with 30 mg/kg dosage produced a 50% reduction in tumour growth, >55% BRD4 degradation and also a noticeable >90% suppression of tumour cMYC (Watt et al. 2019). For protein degraders like PROTACs, intermittent dosing will be effective as it would also give the system appropriate washout periods where the PROTAC could be eliminated from the body. This will reduce the problems like resistance development, toxicity, off-target degradation etc. thus improving the efficacy of the treatment.

9.4.2 Pulse Dosing

PROTACs follow the event driven model of pharmacology in which irreversible degradation of protein takes place. This event driven model might support the pulse dosing strategy in which pulses or bursts of doses are administered with extended period of no dosing between them. Pulse dosing must be thoroughly tried in preclinical models and effects of the time interval where no doses are administered must be established (Ibrahim et al. 2004). Pulse dosing would also give the body time for complete metabolic degradation of the PROTAC so that there is clearance of the earlier dose reducing safety and toxicity issues that might arise due to accumulation of PROTAC. Continuous protein degradation in cell might also activate apoptotic pathways leading to cell death and this will not be feasible when simple protein knockdown is enough for a therapeutic outcome. A single high dosage might help to reach efficacious concentration required for onset of action early thus reducing the time required to see desired clinical effects of therapy. This would help especially if the disease or condition needs immediate and strong therapeutic administration. Though a single high dose might also lead to issues like off-target degradation of other proteins and thus pulse dosing needs testing.

9.4.3 Sequential Dosing

In sequential dosing, a dual or multiple combination of PROTACs can be sequentially administered against various target proteins that may be proteins of the same pathway or collective proteins responsible for a diseased condition. In a study where sequential doses of chemotherapeutic agents were compared with concurrent doses, the sequential arm performed better than the concurrent arm and even managed to reduce the overall accumulation of the drugs (LeVasseur and Chia 2017). This would help to combat the condition in its entirety and also it might lead to low doses of administration of each PROTAC to achieve desired therapeutic effects. Sequential dosing would also require the optimization of multiple factors such as the dosage of each PROTAC, the timing, the order of administration and the prediction of overall synergy displayed by the combination. PROTACs named dAurA383 were used in combination with PROTAC dAurA450 and these were administered in AML cells against the kinase target protein AURKA. Both dAurA383 and dAurA450 when used in 1:1 ratio of dosage synergistically inhibited the growth and even the stemness of AML cells. Another finding from this study was when used in combination, the PROTACs show hook effect at a higher concentration (8×10^{-6} M) than they would have shown when used individually (2×10^{-6} M) in KG1A and Kasumi-1 cells. dAurA383 is a CRBN based PROTAC and preferentially degraded mitotic AURKA and dAurA450 is a cIAP based PROTAC which degraded the interphase AURKA in AML cells (Liu et al. 2022). Thus, sequential dosing of PROTACs would help to eliminate both cancer cells as well as cancer stem cells, thus making PROTAC a therapy that would solve the problem of relapse in cancer. But with many PROTACs come many factors that would need to be tested and predetermined to assure synergy as well as these PROTACs if combined together must not aggravate the off-target degradation, toxicity as the concentration to show hook effect is increased. The metabolism of PROTACs can also become a problem if the body needs to clear multiple PROTACs at one time and also the synergy between multiple PROTACs must not elicit a hyperactive immune response that would worsen the efficacy of the therapy. Thus, combinational and sequential dosing must be rigorously tested in preclinical models.

9.5 Safety Concerns with PROTACs

Despite their targeted activity, the PROTACs are not completely devoid of safety concerns. Figure 9.2 illustrates some of the important safety concerns of PROTACs.

9.5.1 Off-Target Effects

The activity and safety of PROTAC are confounded when it begins to degrade proteins other than its target protein. Mainly, the off-target effect would arise when there is a close structural similarity between the target protein and the other destructed proteins. Several studies have observed examples of off-target effects, such as (Bai et al. 2021; Bondeson et al. 2018; Yang et al. 2019). As PROTACs are created using multiple combinations of ligands and linkers, there would be some combinations in which the linker or the ligand would indiscriminately bind and degrade proteins other than the target protein. Tissue specificity is also necessary, as they might enter healthy tissues; others can target tissues such as cancer, and cause

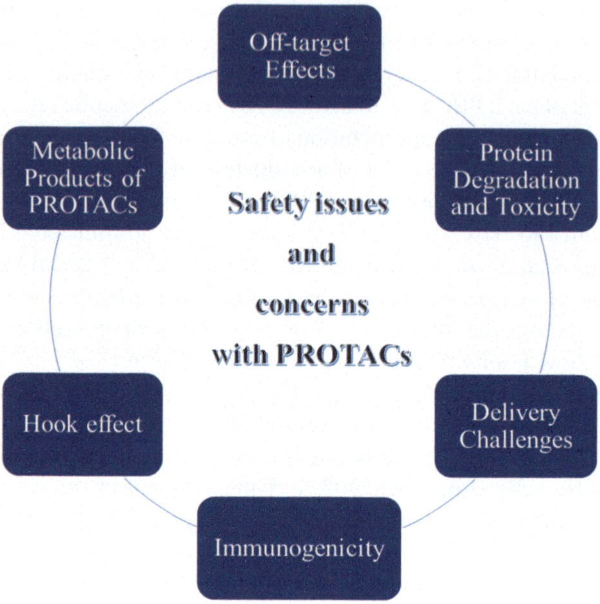

Fig. 9.2 Safety concerns with use of PROTACs

off-target degradation, destabilizing homeostasis of the normal tissues (Liu et al. 2020). This non-selectivity becomes a hurdle when we intend to develop PROTAC as a therapeutic, which might become a personalized medicine.

9.5.2 Protein Degradation and Toxicity

PROTACs, through their degradation, might bring down the protein even below necessary levels, and prolonged loss of the protein may not be well tolerated by the system, especially if the protein has a slow turnover rate. This may compromise the entire therapy by leading to various on-target toxicities especially and more critically if the protein has an important function in the system (Khan et al. 2019). Thus, the dosage and potency of the PROTAC must be well understood and optimized to prevent such unwanted safety failures. To improve tissue selectivity and cell permeability, a variety of PROTACs have been created in multiple formulations to optimize their effects (Sincere et al. 2023).

9.5.3 Delivery Challenges

Cellular penetration of PROTACs is dependent on multiple factors like size, solubility, and polarity. Reduced cell permeability of PROTACs is due to their molecular size (>800 da) and the linker used. The design of a PROTAC becomes an important factor that would increase the overall cell penetration of the PROTAC. PROTACs should not accumulate in the system and circulation creating opportunities for off-target effects and toxicities. For a PROTAC to have, various linkers of different lengths, rigidity, and hydrophilicity could be tested which would improve the design of PROTAC in areas of stability, cell membrane permeability, and compatible spatial variation between the E3 ubiquitin ligase and target POI (Zagidullin et al. 2020). Suboptimal penetration due to inefficient delivery might lead to inappropriate data about safe exposure levels and the degree of protein knockdown needed for a good therapeutic response. Delivery systems could be developed such as lipid encapsulation, increasing lipophilicity, nanoparticles, antibodies, peptides and nucleic acid-based delivery systems that might lead to the optimization of PROTAC (Chen et al. 2023).

9.5.4 Immunogenicity

Antibody-based PROTACs are also developed to improve and optimize the PROTAC design (Cotton et al. 2021). If the PROTAC elicits an immune response that is detrimental to the body, not only would the purpose of the therapy fail, but it would also lead to adverse effects aggravating the condition against which PROTAC was administered. PROTACs should be humanized such that they contain peptide sequences and ligand components that would not hyperactivate our immune response. Even if individual components are not immunogenic, the conjugation of multiple components of PROTACs together might make them immunogenic thus reducing their efficacy. The accumulation of PROTACs in circulation might induce an immune response. Thus, the design and dosage of PROTAC must be modulated to avoid an unwanted immune response.

9.5.5 Hook Effect

For PROTACs to be productive and functional, they need to form a ternary complex that comprises a target protein, a linker, and an E3 ubiquitin ligase. Such a ternary complex is formed when PROTACs are in optimal concentration, and at this concentration, efficient degradation of target proteins is carried out. At high concentrations, there is a saturation observed in PROTAC binding to the target protein and the E3 ligase resulting in binary complexes that would not result in the degradation of proteins, and the mechanism of such binary and ternary complexes is well studied (Douglass et al. 2012). This formation of binary complexes is called the 'Hook Effect'. The binary complex formation may give rise to an incorrect determination of the maximum tolerated dose (MTD) estimation in preclinical models. The metabolic implications of the binary complexes in various tissues are still to be seen, as there is tissue variability, as is the function of a PROTAC. These binary complexes may also create off-target effects when they bind a low-affinity target protein, which might not require degradation through E3 ubiquitin ligase. Safety issues of the hook

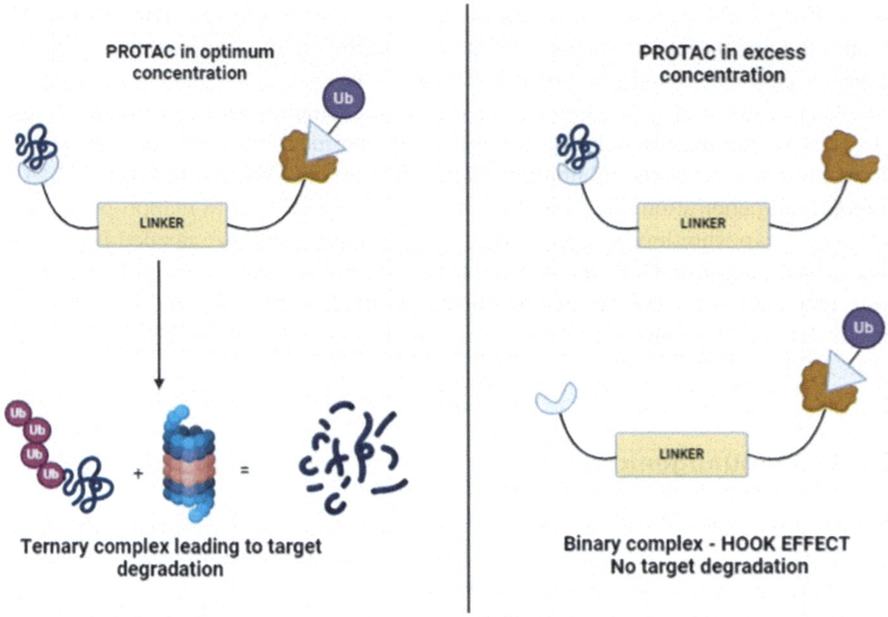

Fig. 9.3 Depiction of Hook effect. Excess concentration of PROTACs compared to the target proteins leads to the formation of binary complexes which nullify the degradation of target protein. Formation of ternary complexes consisting of target protein and E3 ubiquitin ligase bound to it leads to efficient degradation

effect must be determined thoroughly through experiments and the maximum tolerated dose (MTD) must be evaluated by studying protein degradation at lower concentrations to correct errors that might arise due to hook effect (Moreau et al. 2020). Figure 9.3 is an illustration of the hook effect.

9.5.6 Metabolism of PROTACs

Finally, metabolism of PROTAC itself in the liver and its hepatic clearance should not create any toxicity in the system. When linkers of PROTACs are metabolized, their metabolites must not interact with important proteins or other enzymes which would disrupt the normal functioning of the liver cells. Toxicity caused due to metabolites of degraded PROTACs might confound the therapeutic effects leading to overall less efficacy of the PROTACs administered.

9.6 ADME Properties of PROTACs

As PROTAC therapies are expected to pass through the clinical phase trials and ultimately reach the market, multiple companies surveyed the impact of PK properties of PROTACs. The aim of the survey was to achieve logical scientific speculations with inclusive information. In the survey, PROTACs, the targeted protein degraders (TPDs) were administered through three routes of administration, namely, oral, intravenous and subcutaneous. Chameleonicity is observed in large molecules like PROTACs where lower topological polar surface area (tPSA) is achieved by these molecules by adopting folded conformation, and thus, higher permeability was achieved than expected compared to their molecular weight alone. This property of PROTACs inspires companies to take the oral route of administration even when PROTACs defy the Lipinski rule of 5 (Ro5). They only adhere to one of the five components listed in Ro5, a number of hydrogen bond donors essential for chameleonicity.

The low solubility of PROTACs in aqueous solutions is a significant challenge in the intravenous route of administration. PROTACs with low solubility in aqueous solutions may also result in acute toxicity, thus undermining the therapeutic benefit. Therefore, optimized formulations that improve the solubility of TPDs are needed. Multiple issues clouded over the subcutaneous route of administration, and a need for linearity, challenges in formulations, insufficiency of translatability emerged. While maintaining the anonymity of the species, two companies in the survey observed lymphatic absorption of the TPDs. Generally, lymphatic absorption requires the compounds to have high HBA count, polar surface area, higher values of log *P* and higher molecular weight (Lu et al. 2015). Thus, overtly showing that further investigation is necessary.

When analyzing the tissue distribution, a paradox of higher-than-expected values in normal tissues and lower-than-expected values in tumour tissues was observed, which could be attributed to P-glycoprotein-mediated efflux. This efflux would limit the distribution of TPD into tumour cells. In the metabolization of TPDs through cytochrome, the majority of the survey responders observed P450. Another major metabolic pathway was glucuronidation by UGT enzymes, and as literature supports, these are also pathways for the metabolism of small molecule inhibitors. In the survey, amidase and esterase were pointed out as important enzymes in metabolism, and some responses even reported elimination through hydrolytic clearance. As a result, researchers have started finding suitable delivery agents.

The survey responders reported multiple routes of elimination in which biliary excretion was most frequently observed, followed by renal elimination. Some responders observed even transporter-mediated elimination and intestinal secretion, with one referring to OATP-mediated hepatic uptake. When questions related to the optimization of PROTAC design were asked in the survey, the majority of the responders noted that linker design would improve PROTAC design altogether. The responders have used allometry, in vitro-in vivo extrapolation (IVIVE) and PBPK modelling for human PK predictions. Sometimes, these strategies have been used simultaneously for predictions. This survey painted a comprehensive picture of the ADME properties of the PROTACs and also tried to string them together with the physicochemical properties of the PROTACs (Volak et al. 2023).

Linkers can significantly affect the pharmacokinetic properties of a PROTAC; linker length and linker composition can impact the PK properties of a PROTAC. The linker composition can be of many types: examples of flexible linkers are polyethene glycol (PEG), alkyl groups and amide, and examples of comparatively rigid linkers are piperidines, piperazine, benzene and linkers that are based on click chemistry are triazole, triazole-PEG etc. (Zagidullin et al. 2020). PROTACs with higher linker lengths demonstrate more degradation of the target protein. Linker length and attachment point would also lead to differential ligase recruitment, thus displaying highly selective degradation. Rigid linkers are generally considered to have better pharmacokinetic properties by increasing their solubility and potency (Cecchini et al. 2022) but yet as already reported in literature, rigid linkers also have shown an unpredictable pattern of potency at times where potency would increase in a particular cell line but decrease in other. Thus, careful experimentation and judgement are necessary during formulation (Troup et al. 2020). PROTACs that would incorporate linkers with high lipophilicity and less number of hydrogen bond donor exposed to the solvent would be responsible for their higher permeability. Adding groups or atoms to the linker motif that would decrease their tendency to be effluxed by P-glycoprotein would improve the therapeutic efficiency. An opposite impact of length is seen on metabolic stability, i.e. with increasing length, metabolic stability of a PROTAC decreases as the number of spots that can be attacked for degradation increases (Cecchini et al. 2022). Thus, a balance between linker length must be maintained, and multiple combinations must be tried. Linkers containing rigid groups were shown to improve metabolic stability significantly. Rigid and short linkers in ARV-110 and ARV-471 consist of piperidine and piperazine moieties. Thus, according to the requirement of the PROTACs, linkers can be finely tuned to produce good pharmacokinetic properties. Clinical development needs fine tuning that would result in stable PK properties and linkers are a hotspot in PROTACs where a lot of modifications can be made. Contribution of click chemistry would substantially augment the development of linkers. Wherein ARV 471 with a rigid linker has reached phase III clinical trials, and this is a good indication, but linker might show different effects against different cancer tissues and which type of linker is most suitable for human body must be determined in future clinical trials.

9.7 Combination Therapy and Synergy

The evidence of the capabilities of PROTACs to target various tumours is steadily increasing, but it is also essential to establish the scope of synergism in PROTACs. The synergism of PROTACs with other cancer treatment regimens and molecules would increase the therapeutic efficacy of the line of treatment. Many drugs used to treat malignant tumour cells promote a form of therapy-induced senescence (TIS) in vitro and in vivo (Saleh et al. 2020). Senescence is considered a tumour-repressing factor, but the evidence is mounting up that due to the reversal of phenotype, it might be the reason for relapse of the tumour, its metastasis and chemotherapy resistance. The tumour promotion is mainly due to the pro-inflammatory cytokines such as Interleukin-8, Interleukin-6, Interleukin-1β and chemokines such as CCR5 along with tissue remodelling factors known as senescence associated secretory phenotype (SASP) secreted by senescent cells. To take advantage of tumour-repressing role of TIS and then eliminate the senescent cells, an approach called "one-two punch" is applied in which senescence is first induced by use of drugs. Later, these senescent cells are killed by a senolytic agent (Fitsiou et al. 2022). PROTAC 753B exhibited senolytic properties when used against the leukaemia cell line MOLM-14. Ara-C, a chemotherapeutic agent used in treatment of AML, induced senescence in these cells, producing pro-inflammatory cytokines and chemokines. This effect was reversed by cotreatment of 753B with Ara-C. The effect was further validated on another AML cell line, Kasumi-1, and it was established that PROTAC 753B can target and eliminate senescent AML cells induced by chemotherapy. It was found that 753B targets the BCL-XL, which is a crucial anti-apoptotic protein and is over-expressed in senescence-induced cells. These results indicate that when used in combination with PROTAC 753B and chemotherapy sequentially or concomitantly, it may produce superior results than chemotherapy alone due to the clearance of senescent cells by 753B (Jia et al. 2023). Apart from chemotherapy, PROTACs can also be combined with radiotherapy to target proteins that confer radioresistance, such as epidermal growth factor receptor (EGFR), androgen receptor (AR), estrogen receptor (ER), cyclin-dependent kinases (CDKs), etc. (Zheng and Tao 2020). These proteins are mainly targeted for radiosensitization because these proteins are regulators of DNA damage repair, autophagy induction, and apoptosis inhibition, and they can weaken the radiotherapy by activating repair and replication in cancer cells, thus reducing the cytotoxicity and efficacy of radiotherapy (Ouellette et al. 2022). DNA damage and generation of free radicals such as reactive oxygen species (ROS) are the primary mechanisms by which ionizing radiations exert cytotoxic effects. When PROTACs were tested as radiosensitizers against EGFR, lapatinib-based PROTAC and gefitinib-based PROTAC selectively degraded even mutant forms of EGFR, and afatinib-based PROTAC even was successful in degrading a double mutant of EGFR (Burslem et al. 2018). PROTAC ARD-61 was used against castration-resistant prostate cancer (CRPC) cell lines, and it efficiently degraded androgen receptor (AR) and even the mutant forms of it while not affecting the cells not expressing AR at the same time. Even when PROTACs were used with dimethylcurcumin, which is an FDA-approved AR degradation enhancer in xenograft models, it showed radiosensitizing effects (Kregel et al. 2020). Radiotherapy induced PROTACs (RT-PROTAC) have been already reviewed in literature. In this novel approach a PROTAC that targets Bromodomain (BRD) is attached to a phenyl azide cage that is X-ray inducible. This approach has shown potential for antitumour activity in xenograft model of MCF-7 and it shows a promise of synergy with radiotherapy due to its inducible nature (Tamatam and Shin 2023).

PROTAC DT2216 was combined with Sotorasib to treat predominantly KRAS-mutated cancers and was tested in various KRAS-mutated cancer cell lines. Synergistic combination effects include reduced cell viability of cell lines that were only partially sensitive to Sotorasib along with increased inhibition of colony formation and apoptosis induction. When this combination of Sotorasib + DT2216 was used in mouse xenografts, significant tumour reduction with no side effects such as a decrease in mouse weight or a decrease in blood counts was seen. Thus, in cancers where monotherapies are only partially effective, such combinations with PROTACs can be used primarily against non-small cell lung cancer (NSCLC), colorectal cancer (CRC) and pancreatic cancer where KRAS mutations create a greater impact on cancer development and progression (Khan et al. 2022). When a BET (bromodomain extracellular terminal) PROTAC was conjugated with aptamer AS1411 antitumour and antiproliferative activity against MCF-7 breast cancer cells was observed which was significantly more than aptamer and PROTAC when tested individually. The combination was selective against nucleolin overexpressing tumour cells and increased BRD4 degradation in mice xenografts. Severe lung lesions were observed in MCF-7 xenografted mice when treated with PROTAC but were not observed in combinatorial treatment (He et al. 2021). Thus, pronounced effects of tumour degradation and decreased toxicity were seen in the aptamer and PROTAC combination. In another experiment, when combinations of BET PROTAC ARV-771 and ibrutinib or venetoclax were used against mino cells and Z-138 cell lines of mantle cell lymphoma, synergistic activity was seen against both ibrutinib sensitive as well as ibrutinib resistant cell lines (Sun et al. 2018).

When a panel of drugs consisting of drugs approved by FDA or PROTACs in clinical trials was tested for their activity, synergistic activity was seen with some drugs in cell lines (Lim et al. 2019) as shown in Table 9.2.

The emerging field of PROTACs holds tremendous promise in revolutionizing cancer treatment strategies. The evidence presented demonstrates the potential of PROTACs not only as standalone therapies but also in synergistic combinations with existing cancer treatment regimens. By effectively targeting and eliminating therapy-induced senescent cells, PROTACs offer a "one-two punch" approach that may enhance the therapeutic efficacy against metastasis and relapse (Fitsiou et al.

Table 9.2 PROTACs show promising synergistic activity with FDA-approved drugs

PROTAC	Drug	Activity
ARV-825	Cediranib VEGFR inhibitor	Synergistic activity against both KMS11 and KMS28BM multiple myeloma cell lines
	Cerolanib PDGFR α/β and FLT3 inhibitor	
	GSK 1904529A IGF-1R inhibitor	
	Motesanib VEGFR 1/2/3 inhibitor	
	LY3023414 Dual PI3K/mTOR inhibitor	
MZ1	Cediranib	Synergistic activity against both KMS11 and KMS28BM multiple myeloma cell lines
	Crenolanib	
	GSK 1904529A	
	Motesanib	
	LY3023414	Synergistic activity only against KMS11 but not KMS28BM

2022). Moreover, the combination of PROTACs with traditional chemotherapy and radiotherapy showcases their versatility in addressing various aspects of cancer biology, including DNA damage repair and apoptosis inhibition. The success of PROTACs in degrading specific proteins associated with radioresistance and mutation-driven cancers, such as EGFR and KRAS, suggests a potential breakthrough in treating various cancers. Overall, the findings presented underscore the multifaceted potential of PROTACs in reshaping cancer therapeutic landscapes, providing a novel and promising avenue for future research and clinical applications.

First in Human (FIH) also known as Phase 1 clinical trials are done to assess the safety as well as tolerability of new drug or treatment in Humans. These are often short-term trials usually involving small number of healthy volunteers. The highest dose that can be safely administered to humans is determined. PROTACs which successfully pass these FIH trials must be subjected to Phase 1b, open label, randomized control clinical trials in which they would be administered to patients with a target disease to determine their efficacy. Moreover, PROTACs should be administered in combination with an approved therapeutic agent against cancer/target disease to determine synergy. Efficacious positive results in these trials would speed up the process of translation of PROTACs into Clinics. Including multiple arms in the study consisting of patients given PROTACs alone, chemotherapy alone and both in combination must be tested for their efficacy, safety and then pharmacokinetic parameters. Such arms were created to test the synergistic effects of Palbociclib which is a CDK4/6 inhibitor and ARV 471 which is an ER degrading PROTAC in $ER^{+}HER2^{-}$ breast cancer patients. This synergy showed potency in preclinical trials as because of Palbociclib (Liu et al. 2018), the cell cycle is blocked during progression from G1 to S phase but estrogen eventually activates E-Cdk2 complexes through ER-mediated transcription resulting in cell growth. This pathway would be obstructed by degradation of ER by ARV-471. Thus, by synergism, there would be complete growth arrest of cancer cells thus establishing a more final therapy (NCT04072952). ARV-110, a potent AR degrader is tried in combination with abiraterone which irreversibly binds and inhibits CYP17 enzyme essential in androgen synthesis after it showed success in preclinical trials (NCT05177042) (Rehman and Rosenberg 2012). Phase 1b clinical trial results would determine if the combination of existing chemotherapeutic agents with PROTACs would result in clinically significant results. Maximum tolerated Dose (MTD) of such combinations along with sustainable dosage frequency would play an important part in clinical translation of such combinations.

9.8 Conclusion

The promising preclinical results and the transition of PROTACs into clinical trials holds promise in overcoming the lack of efficacy, resistance, and toxicity associated with standard drugs. However, the optimization of PROTAC design becomes crucial, particularly in enhancing ADME properties and minimizing toxicity related to tissue non-specificity and off-target effects. Linker design plays a pivotal role in improving pharmacokinetic properties, demanding careful consideration of length, lipophilicity, and cyclic ionizable groups to balance solubility and permeability. Given the unique pharmacology of PROTACs, the dose and frequency of administration would be critical, requiring careful consideration. Indicators of ubiquitination can possibly serve as useful pharmacodynamic biomarkers to monitor the activity of PROTACs in early-phase dose escalation trials. Late-phase trials should explore the efficacy of PROTACs used as single agents or their synergistic and complementary effect when used in combination with conventional chemotherapy and targeted therapies.

References

Atilaw Y, Poongavanam V, Nilsson CS, Nguyen D, Giese A, Meibom D, Erdelyi M, Kihlberg J (2021) Solution conformations shed light on PROTAC cell permeability. ACS Med Chem Lett 12(1):107–114

Bai N, Miller S, Andrianov G, Yates M, Kirubakaran P, Karanicolas J (2021) Rationalizing PROTAC-mediated ternary complex formation using Rosetta. J Chem Inf Model 61:1368

Bondeson DP, Smith BE, Burslem GM, Buhimschi AD, Hines J, Jaime-Figueroa S, Wang J, Hamman BD, Ishchenko A, Crews CM (2018) Lessons in PROTAC design from selective degradation with a promiscuous warhead. Cell Chem Biol 25:78–87

Burslem GM, Smith BE, La AC, Crew AP, Hines J, Crews CM, Jaime-Figueroa S et al (2018) The advantages of targeted protein degradation over inhibition: an RTK case study. Cell Chem Biol 25(1):67–77

Cecchini C, Tardy S, Scapozza L (2022) Linkers as game-changers in PROTAC technology: emphasizing general trends in PROTAC pharmacokinetics for their rational design. Chimia 76:341–345

Chen Y, Tandon I, Heelan W, Wang Y, Tang W, Hua Q (2023) Proteolysis-targeting chimera (PROTAC) delivery system: advancing protein degraders towards clinical translation. Chem Soc Rev 51:5330

Chirnomas D, Hornberger KR, Crews CM (2023) Protein degraders enter the clinic—a new approach to cancer therapy. Nat Rev Clin Oncol 20(4):265–278

Cotton AD, Nguyen DP, Gramespacher JA, Seiple IB, Wells JA (2021) Development of antibody-based PROTACs for the degradation of the cell-surface immune checkpoint protein PD-L1. J Am Chem Soc 143(2):593–598

Douglass EF, Miller CJ, Sparer G, Shapiro H, Spiegel DA (2012) A comprehensive mathematical model for three-body binding equilibria. J Am Chem Soc 135:6092

Fitsiou E, Soto-Gamez A, Demaria M (2022) Biological functions of therapy-induced senescence in cancer. Semin Cancer Biol 81:5–13

Flanagan JJ, Qian Y, Gough SM, Andreoli M, Bookbinder M, Cadelina G, Bradley J, et al (2018) ARV-471, an oral estrogen receptor PROTAC™ protein degrader for breast cancer. In: San Antonio breast cancer symposium, San Antonio

Gao X, Burris HA, Vuky J, Dreicer R, Sartor O, Sternberg CN, Percent I, et al (2022) Phase 1/2 study of ARV-110, an androgen receptor PROTAC degrader, in metastatic castration-resistant prostate cancer. In: ASCO genitourinary cancers symposium, San Francisco, vol 40, p 17

Grover A, Benet LZ (2011) Intermittent drug dosing intervals guided by the operational multiple dosing half lives for predictable plasma accumulation and fluctuation. J Pharmakokinet Pharmacodyn 38(3):369–388

Hamilton E, Vahdat L, Han HS, Ranciato J, Gedrich R, Keung CF, Chirnomas D, Hurvitz S (2021) First-in-human safety and activity of ARV-471, a novel PROTAC® estrogen receptor degrader,

in ER+/HER2− locally advanced or metastatic breast cancer. In: San Antonio breast cancer symposium, San Antonio, TX
He S, Gao F, Ma J, Ma J, Dong G, Sheng C (2021) Aptamer-PROTAC conjugates (APCs) for tumor-specific targeting in breast cancer. Angew Chem 60:2–9
He M, Cao C, Ni Z, Liu Y, Song P, Hao S, He Y, Sun X, Rao Y (2022) PROTACs: great opportunities for academia and industry (an update from 2020 to 2021). Signal Transduct Target Ther 7(1):181
Ibrahim KH, Gunderson BW, Hermsen ED, Hovde LB, Rotschafer JC (2004) Pharmacodynamics of pulse dosing versus standard dosing: in vitro metronidazole activity against Bacteroides fragilis and Bacteroides thetaiotaomicron. Antimicrob Agents Chemother 48:4195–4199
Jia Y, Han L, Ramage CL, Wang Z, Weng CC, Yang L, Colla S et al (2023) Co-targeting BCL-XL and BCL-2 by PROTAC 753B eliminates leukemia cells and enhances efficacy of chemotherapy by targeting senescent cells. Haematologica 108:2626–2638
Khan S, Zhang X, Lv D, Zhang Q, He Y, Zhang P, Liu X et al (2019) A selective BCL-XL PROTAC degrader achieves safe and potent antitumor activity. Nat Med 25:1938
Khan S, Wiegand J, Zhang P, Hu W, Thummuri D, Budamagunta V, Hua N et al (2022) BCL-XL PROTAC degrader DT2216 synergizes with sotorasib in preclinical models of KRASG12C-mutated cancers. J Hematol Oncol 15(1):23
Koravović M, Marković B, Kovačević M, Rmandić M, Tasić G (2022) Protein degradation induced by PROTAC molecules as an emerging drug discovery strategy. J Serbian Chem Soc 87:785–811
Kregel S, Wang C, Han X, Xiao L, Fernandez-Salas E, Bawa P, McCollum BL et al (2020) Androgen receptor degraders over come common resistance mechanisms developed during prostate cancer treatment. Neoplasia 22(2):111–119
LeVasseur N, Chia SK (2017) Sequential versus concurrent chemotherapy for adjuvant breast cancer: does dose intensity matter? Br J Cancer 117:157–158
Lilja H, Ulmert D, Vickers AJ (2008) Prostate-specific antigen and prostate cancer: prediction, detection and monitoring. Nat Rev Cancer 8(4):268–278
Lim SL, Damnernsawad A, Shyamsunder P, Chng WJ, Han BC, Xu L, Pan J et al (2019) Proteolysis targeting chimeric molecules as therapy for multiple myeloma: efficacy, biomarker and drug combinations. Haematologica 104(6):1209–1220
Liu M, Liu H, Chen J (2018) Mechanism of the CDK4/6 inhibitor palbociclib (PD 0332991) and its future application in cancer treatment. Oncol Rep 39(3):901–911
Liu J, Ma J, Liu Y, Xia J, Yuyun L, Peter Wang Z, Wei W (2020) PROTACs: a novel strategy for cancer therapy. Semin Cancer Biol 67:171
Liu F, Wang X, Duan J, Wen S, Wang Z, Liu Q (2022) A temporal PROTAC cocktail-mediated sequential degradation of AURKA abrogates acute myeloid leukemia stem cells. Adv Sci 9(22):e2104823
Lu Y, Qiu Y, Qi J, Feng M, Ju D, Wu W (2015) Biomimetic reassembled chylomicrons as novel association model for the prediction of lymphatic transportation of highly lipophilic drugs via the oral route. Int J Pharm 483(1-2):69–76.
Mato A, Danilov A, Patel MR, Tees M, Flinn I, Ai W, Patel K, et al (2022) A first-in-human phase 1 trial of NX-2127, a first-in-class oral BTK degrader wih immunomodulatory activity, in patients with relapsed and refractory B-cell malignancies. In: ASCO annual meeting, Chicago, vol 6, p 547
Montaya S, Bourcier J, Thompson MC, Noviski M, Tan M, Wang E, Mi X et al (2022) Kinase dead BTK mutations confer resistance to covalent and noncovalent BTK inhibitors but are susceptible to clinical stage BTK degraders. Blood 140:1811
Moreau K, Coen M, Zhang AX, Fiona P, Paola Castaldi M, Dahl G, Boyd H, Scott C, Newham P (2020) Proteolysis-targeting chimeras in drug development: a safety perspective. Br Pharmacol Soc 177:1709–1718
Nagashima T, Inamura K, Nishizono Y, Suzuki A, Tanaka H, Yoshinari T, Yamanaka Y (2022) ASP3082, a first-in-class novel KRAS G12D degrader, exhibits remarkable anti-tumour activity in KRAS G12D mutated cancer models. Eur J Cancer 174:S30

Neklesa KT, James Winkler D, Craig Crews M (2017) Targeted protein degradation by PROTACs. Pharmacol Ther 174:138–144

Neklesa TK, Snyder LB, Willard RR, Vitale N, Pizzano J, Gordon DA, Bookbinder M, et al (2019) ARV-110: an oral androgen receptor PROTAC degrader for prostate cancer. In: ASCO genitourinary cancers symposium, San Francisco, vol 37, p 259

Ouellette MM, Zhou S, Yan Y (2022) Cell signaling pathways that promote radioresistance of cancer cells. Diagnostics 12:1–18

Padma VV (2015) An overview of targeted cancer therapy. Biomedicine 5(4):19–16

Petrylak D, Gao X, Vogelzang N, Garfield M, Taylor I, Moore M, Peck RA, Burris HA III (2020) First-in-human phase I study of ARV-110, an androgen receptor (AR) PROTAC degrader in patients (pts) with metastatic castrate-resistant prostate cancer (mCRPC) following enzalutamide (ENZ) and/or abiraterone (ABI). J Clin Oncol 38:3500

Rehman Y, Rosenberg JE (2012) Abiraterone acetate: oral androgen biosynthesis inhibitor for treatment of castration-resistant prostrate cancer. Drug Des Devel Ther 6:13–18

Saleh T, Bloukh S, Carpenter VJ, Alwohoush E, Bakeer J, Darwish S, Azab B, Gewirtz DA (2020) Therapy-induced senescence: an "old" friend becomes the enemy. Cancers 12(4):822

Sincere NI, Anand K, Ashique S, Yang J, You C (2023) PROTACs: emerging targeted protein degradation approaches for advanced druggable strategies. Molecules 28:4014

Snyder LB, Neklesa TK, Chen X, Dong H, Ferraro C, Gordon DA, Macaluso J, et al (2021) Discovery of ARV-110, a first in class androgen receptor degrading PROTAC® for the treatment of men with metastatic castration resistant prostate cancer. In: AACR annual meeting. Virtual, vol 81, p 43

Sowa ME, Kreger B, Baddour J, Liang Y, Simard JR, Pollock RM (2022) Preclinical evaluation of CFT1946 as a selective degrader of mutant BRAF for the treatment of BRAF driven cancers. In: AACR annual meeting, New Orleans, vol 2

Sun B, Fiskus W, Qian Y, Rajapakshe K, Raina K, Coleman KG, Crew AP et al (2018) BET protein proteolysis targeting chimera (PROTAC) exerts potent lethal activity against mantle cell lymphoma cells. Leukemia 32:343–352

Tamatam R, Shin D (2023) Emerging strategies in proteolysis—targeting chimeras (PROTACs): highlights from 2022. Int J Mol Sci 24:5190

Tirona RG, Kim RB (2017) Introduction to clinical pharmacology. In: Clinical and translational science (second edition). Wiley, New York, pp 365–388

Troup RI, Fallan C, Baud MGJ (2020) Current strategies for the design of PROTAC linkers: a critical review. Explor Target Antitumor Ther 1(5):273–312

Volak LP, Duevel HM, Humphreys S, Nettleton D, Phipps C, Pike A, Zientek M (2023) Industry perspective on the pharmacokinetic and ADME characterization of heterobifunctional protein degraders. Drug Metab Dispos 51:792

Watt GF, Scott-Stevens P, Gaohua L (2019) Targeted protein degradation in vivo with proteolysis targeting chimeras: current status and future considerations. Drug Discov Today Technol 31:69–80

Yang J, Li Y, Aguilar A, Liu Z, Yang C-Y, Wang S (2019) Simple structural modifications converting a bona fide MDM2 PROTAC degrader into a molecular glue molecule: a cautionary tale in the design of PROTAC degraders. J Med Chem 62:9471–9487

Zagidullin A, Milyukov V, Rizvanov A, Bulatov E (2020) Novel approaches for the rational design of PROTAC linkers. Explor Target Antitumor Ther 1(5):381–390

Zheng S, Tao W (2020) Targeting Cullin-RING E3 ligases for radiosensitization: from NEDDylation inhibition to PROTACs. Front Oncol 10:1–12

Toxicological Aspects of PROTACs 10

Rohini Pujari and Urvashi Soni

Abstract

The present chapter provides an in-depth exploration of PROTACs, a unique class of drug-like compounds that induce targeted protein degradation within cells. It discusses the novel mechanism of action of PROTACs compared to traditional small-molecule inhibitors, highlighting their potential to target a wide range of proteins, including previously "undruggable" ones. The chapter emphasizes the importance of thorough safety evaluations in the development of PROTAC-based therapies, addressing challenges such as off-target effects, immunogenicity, and cytotoxicity. By examining the toxicological studies conducted on PROTACs, the chapter underscores the critical need to assess and mitigate potential toxicity issues to ensure the safety and efficacy of these promising molecules in therapeutic applications.

Keywords

PROTACs · Toxicity · Safety · Efficacy · Off-target effects · Hook effect

Abbreviations

ADAs	Anti-drug antibodies
BRD	Bromodomain protein
BTK	Bruton tyrosine kinase
CDK6	Cyclin dependent kinase 6
CRBN	Cereblon

R. Pujari (✉) · U. Soni
School of Health Sciences and Technology, Dr. Vishwanath Karad MIT World Peace University, Kothrud, Pune, Maharashtra, India

M. Nandave, P. Jain (eds.), *PROTAC-Mediated Protein Degradation: A Paradigm Shift in Cancer Therapeutics*, https://doi.org/10.1007/978-981-97-5077-1_10

DDIs	Drug-drug interactions
HDAC6	Histone deacetylase 6
HIF-1α	Hypoxia-inducible factor 1-alpha
IKZF	IKAROS family zinc finger
IMiDs	Immunomodulatory drugs
mRNA	Messenger ribonucleic acid
PD	Pharmacodynamic
PK	Pharmacokinetic
POI	Protein of interest
PROTACs	Proteolysis-targeting chimeras
shRNA	Short hairpin ribonucleic acid
siRNA	Small interfering ribonucleic acid

10.1 Introduction

PROTACs, or proteolysis-targeting chimeras, are a unique family of drug-like compounds that work inside cells to cause the breakdown of certain target proteins. A target protein of interest (POI) is recruited by an E3 ubiquitin ligase by the action of PROTACs, which then ubiquitinates the POI and causes its proteasomal destruction (Békés et al. 2022). This targeted protein degradation approach offers a unique mechanism of action compared to traditional small-molecule inhibitors, allowing PROTACs to target a wide variety of proteins, inclusive of those considered previously as "undruggable". The ability of PROTACs to degrade specific proteins has generated significant interest in the pharmaceutical industry, with the potential to revolutionize drug development and therapy (Burslem and Crews 2020). Interest in PROTACs as a therapeutic modality has surged, particularly after Arvinas initiated Phase I clinical trials with two PROTAC candidate drugs in 2019. There is growing anticipation that PROTACs could become a ground-breaking therapy, with the potential to be a blockbuster treatment. This enthusiasm has led to significant investments by pharmaceutical companies in the development of PROTACs as therapeutic drugs. Nevertheless, like any efficacious new drug therapy, there are challenges to address, such as the need to recognize and assess clinical safety risks using appropriate preclinical studies (Churcher 2018). These challenges highlight the importance of thorough safety evaluations in the development and advancement of PROTAC-based therapies. As with any new class of therapeutic agents, the toxicological aspects of PROTACs are critical considerations in their development and clinical application. While PROTACs offer a unique and promising approach to targeted protein degradation, potential toxicity issues must be thoroughly evaluated to ensure the safety of these molecules (Gao et al. 2020). This chapter shall cover the toxicological aspects of the PROTACs based on the toxicological studies carried out in the past as per the documented literature (Fig. 10.1).

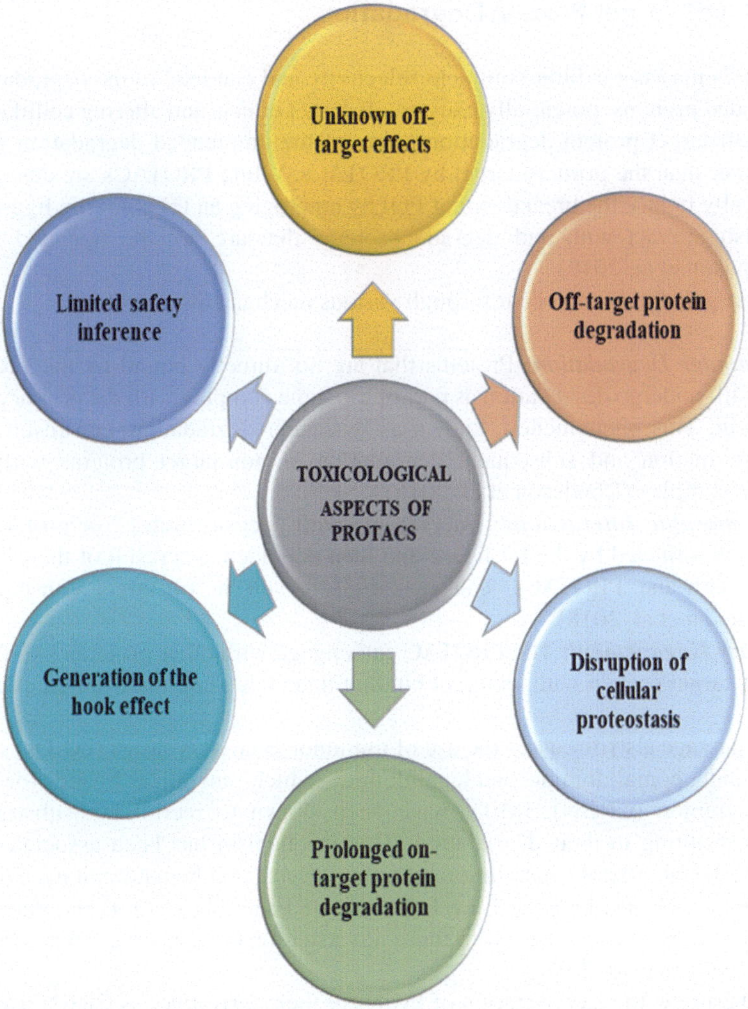

Fig. 10.1 Toxicological aspects of PROTACs

10.2 Toxicological Aspects of PROTACs

The safety challenges associated with PROTACs encompass various aspects, ranging from off-target effects to immunogenicity and cytotoxicity. Following is a detailed exploration of these safety challenges:

10.2.1 Off-Target Protein Degradation

PROTACs may not exhibit complete selectivity and can lead to the degradation of unintended proteins, potentially causing off-target effects and altering cellular functions. Off-target protein degradation refers to the unintended degradation of proteins other than the primary target by PROTACs. While PROTACs are designed to specifically induce the breakdown of POI by employing an E3 ubiquitin ligase, they may also interact with and degrade proteins that are not the intended targets (Edmondson et al. 2019).

Off-target effects can occur through various mechanisms as follows:

- ***Bystander Degradation***: Proteins that are not directly bound by the PROTAC can still undergo degradation as part of the same complex with the primary target protein. This phenomenon, known as bystander degradation, can result in the ubiquitination and subsequent degradation of non-target proteins within the same complex (Bondeson et al. 2015).
- ***Neo-morphic Interactions***: Interactions with "neosubstrates," or proteins that are ubiquitinated by the E3 ligase and then degraded as a result of their interaction with the PROTACs, can potentially result in off-target consequences (Donovan et al. 2018).
- ***Binary Engagement***: The PROTAC can engage with other proteins like the primary target protein with its target binding ligand, leading to off-target effects.

The passage also discusses the use of immunomodulatory drugs (IMiDs) such as thalidomide, pomalidomide, and lenalidomide, which function as ligands for the E3 ligase cereblon (CRBN). IMiDs have been shown to recruit neo-substrates to CRBN, resulting in their degradation. This mechanism has been associated with various adverse effects, including teratogenic effects and hematotoxicity. The degradation of proteins like Spalt-Like Transcription Factor 4 (SALL4) and Ikaros proteins by IMiDs has been linked to these adverse effects (Asatsuma-Okumura et al. 2019; Donovan et al. 2018).

Furthermore, the incorporation of IMiDs or their derivatives as CRBN warheads in PROTAC molecules raises concerns about potential safety liabilities. Studies have shown that PROTACs utilizing IMiDs as warheads can induce the degradation of IMiD substrates, highlighting the importance of assessing the safety risks associated with these interactions. It is crucial to evaluate the impact of PROTACs on hematopoiesis and other relevant endpoints through in vitro assays and preclinical toxicology studies to mitigate potential safety concerns (Asatsuma-Okumura et al. 2019).

Examples demonstrate the unintended degradation of IMiD substrates, including IKZF1, IKZF3, ZNF827, and ZFP91, by PROTACs targeting different proteins (BTK, CDK6, HDAC6) that incorporate pomalidomide or lenalidomide as E3 ligase warheads. This highlights the broad impact of specific E3 ligase warheads on protein degradation beyond the intended targets (Wu et al. 2019).

While the effects of these PROTACs on hematopoiesis were not explicitly reported, it is crucial to assess the safety risk associated with off-target effects, particularly on hematopoietic processes. Conducting in vitro assays and preclinical toxicology studies is essential to evaluate the safety implications of PROTACs incorporating IMiD compounds as CRBN warheads and to address potential safety concerns (Brand et al. 2019).

A class of small molecules known as anti-cancer sulfonamides, including compounds like indisulam, E7820, and CQS, induce targeted protein degradation by attaching E3 ligase DCAF15 to the neo-substrates for degradation. This mechanism is analogous to how IMiDs function. The mRNA splicing factor RBM39 is the most well-studied neo-substrate that sulfonamides break down. It plays a key role in the anti-proliferative effects observed with anti-cancer sulfonamide treatment (Han et al. 2017). Novartis has elucidated the complex structure of DCAF15-DDB1-DDA1-Indisulam-RBM39, revealing the specific mode of action through which indisulam facilitates the interaction between DCAF15 and RBM39. Indisulam forms a tripartite complex with DCAF15 and RBM39, primarily through non-polar interactions. This molecular glue mechanism demonstrates the challenge of targeting DCAF15 as a novel E3 ligase (Han et al. 2017; Uehara et al. 2017).

While indisulam has shown high selectivity and efficacy in degrading RBM39 as the primary target, further studies have indicated that other proteins are also downregulated by anti-cancer sulfonamides. Clinical trials with anti-cancer sulfonamides have faced limitations due to dose-limiting myelosuppression, including neutropenia and thrombocytopenia, along with observed side effects like injection site reactions, alopecia, and fatigue. Thus, anti-cancer sulfonamides, documented from preclinical studies to be active against DCAF15 should be mandatorily subjected to assessment and mitigation of potential toxicities, ensuring a safe profile for clinical development in specific patient populations (Bondeson et al. 2015).

Global proteomics techniques play a crucial role in understanding the effects of PROTAC treatment, including identifying previously unknown off-target proteins and monitoring downstream protein dynamics and pathway effects. By characterizing the binding and degradation selectivity of PROTACs using chemical and global proteomics, safety hazards based on the degradation profile associated with particular binding events can be assessed. Additionally, relative stoichiometry and complete quantitation of E3 ligase and POI levels and turnover rates of the POI, are important parameters for understanding and predicting ternary complex formation (Donovan et al. 2018).

To investigate safety risks associated with off-target effects of PROTACs, understanding the role of the broken-down proteins and the physiological effects of lowering their concentration in various tissues is crucial. Utilizing databases, in silico tools, and experimental approaches like shRNA/siRNA and genetic techniques can help in developing hypotheses of potential toxicities and testing them in cell and animal models. This information can guide the design of in vivo investigative studies to evaluate organ-specific toxicities, determine exposure levels relative to efficacy, and regulate the reversibility of toxicological effects of PROTAC withdrawal (Pickett and Reddel 2015).

In summary, a comprehensive proteomics strategy is crucial for investigating PROTAC safety, including assessing degradation selectivity, identifying off-target effects, and understanding the physiological consequences of protein degradation to ensure the development of safe and effective PROTAC-based therapies.

10.2.2 Unknown Off-Target Effects

These refer to potential unintended interactions and consequences that may arise from the degradation of proteins other than the primary target by PROTACs. These effects can occur when PROTACs interact with proteins that were not initially identified or predicted as targets, leading to off-target degradation and potentially causing unforeseen biological responses or toxicities (Xiao et al. 2022).

The complexity of cellular pathways and protein interactions makes it challenging to predict all possible off-target effects of PROTACs. These unknown off-target effects may result from interactions with proteins that have not been thoroughly characterized or whose functions are not well understood. Additionally, the dynamic nature of protein-protein interactions within cells can lead to unexpected consequences when using PROTACs to induce targeted protein degradation. To address the issue of unknown off-target effects, researchers and drug developers employ various strategies, including computational modeling, bioinformatics analyses, and experimental validation to predict and assess potential off-target interactions of PROTAC molecules. Additionally, preclinical studies using cell-based assays and animal models can help identify and evaluate any unforeseen off-target effects of PROTACs before advancing them to clinical trials (He et al. 2021).

Overall, understanding and mitigating unknown off-target effects of PROTACs are essential for ensuring the safety and efficacy of these novel therapeutic agents in drug development and clinical applications. Continued research and comprehensive safety assessments are crucial to uncover and address any potential off-target effects that may arise from PROTAC-mediated protein degradation.

10.2.3 Disruption of Cellular Proteostasis

The potential risks associated with PROTACs in terms of competing with natural substrates for binding to E3 ligases, lead to the accumulation of substrates and potential perturbation of cellular pathways. Apart from this, PROTACs may also enhance the cellular levels of ubiquitinated proteins, potentially saturating the proteasome and affecting cellular homeostasis. The proteasome plays a crucial role in regulating various cellular processes, and alterations in its activity due to PROTAC treatment could have deleterious effects, including implications for immune response, cell cycle regulation, and metabolic activity (Thibaudeau and Smith 2019).

While there is a theoretical concern about blocking natural E3 substrates and proteasomes, experimental observations suggest that at low concentrations, PROTACs do not significantly interfere with natural E3 ligase activity. This is because PROTACs act catalytically and engage with only a sub-population of E3 ligase molecules, allowing the majority of E3 ligases to continue their natural functions. Studies with VHL-based PROTACs have exhibited a large concentration between stabilization and degradation of HIF-1α, indicating minimal interference with natural substrate degradation (Rousseau and Bertolotti 2018).

However, achieving appropriate pharmacokinetic (PK) properties for PROTACs, given their distinct characteristics from traditional small-molecule drugs, can be challenging. Ensuring the desired exposure and exposure duration, especially for targets with high turnover rates, may require plasma concentration values significantly higher than the minimum effective concentration. This could increase the potential for off-target effects and narrow the therapeutic margin, particularly if the off-target is necessary for cellular homeostasis (Santambrogio et al. 2019).

Further, in vivo studies are necessary to understand the relationship between PK and pharmacodynamic (PD) effects on the target and key off-targets. Detailed investigations are needed to determine if PROTACs induce saturation of proteasome or affect natural levels of E3 ligase substrate and the potential toxicologic aspects of these events. Global proteomics approaches can be utilized to monitor changes in E3 ligase substrate concentrations and proteasome activity following PROTAC treatment, providing valuable insights into potential toxicity liabilities and guiding focused toxicological assessments (Swatek et al. 2019).

In summary, while anti-cancer sulfonamides show promise in inducing targeted protein degradation, their potential toxicities and side effects must be carefully evaluated to support their safe and effective use in drug development.

10.2.4 Prolonged On-Target Protein Degradation

This refers to the continuous and sustained degradation of the intended target protein by PROTACs, which can impact the duration and extent of target inhibition. To evaluate safety hazards, it is crucial to track the global proteome's degradation after PROTAC therapy in cell models or tissues from in vivo preclinical research. It is advised to use a multiplexed isobaric tandem mass tag labeling technique to simultaneously identify and quantify hundreds of proteins, which has become a cornerstone technology in PROTAC drug discovery. This technique allows for the measurement of degradation selectivity and identification of off-target effects across a wide range of proteins as a result of PROTAC treatment.

The ability of PROTACs to effectively degrade their primary target protein is remarkable. However, protracted degradation may occur depending on the compound's pharmacokinetic (PK) profile and the target protein's rate of resynthesis in the relevant disease tissue. PROTACs have an advantage over reversible small-molecule inhibitors in that they are more effective because they cause significant and prolonged target degradation, which results in a deeper and longer-lasting inhibition of the target protein. PROTACs operate through a non-equilibrium pharmacological mechanism that shares similarities with irreversible antagonists. This mechanism can result in a pharmacodynamic (PD) response that (Strelow 2017) surpasses what is predicted by the compound's PK profile. Irreversible inhibitors have been successfully utilized in various therapeutic areas, demonstrating that prolonged target inhibition can be tolerated, although its effectiveness depends on the specific target and context (Khan et al. 2019).

In the context of oncology, selecting PROTAC targets that drive synthetic lethality can enhance tolerability, as tumors may become more prone to target degradation in comparison to normal tissues. Additionally, choosing an E3 ligase that is especially expressed in diseased tissues over normal tissues can improve the therapeutic margin for on-target toxicity. Optimizing the dose and schedule of PROTAC administration is crucial for achieving effective target degradation in disease tissues, especially when on-target toxicity is a concern. This strategic approach aims to maximize the therapeutic benefits of PROTACs while minimizing potential adverse effects, highlighting the importance of target selection and dosing strategies in the development of PROTAC-based therapies (He et al. 2020; Li et al. 2023).

10.2.5 Generation of the Hook Effect

At high concentrations, PROTACs may form binary complexes instead of the desired ternary complex, resulting in reduced target degradation and potential implications for efficacy. The "hook" effect observed with PROTACs at high concentrations can lead to the development of binary complexes instead of the desired ternary complex, hindering target ubiquitination and degradation (Paiva and Crews 2019; Pettersson and Crews 2019).

This phenomenon raises two main concerns:

- ***Increase in Off-target Degradation***: The formation of binary complexes may result in the E3-PROTAC complex binding to lower affinity off-targets, potentially leading to off-target degradation activity (Pettersson and Crews 2019).
- ***Pharmacological Response***: There is a risk of a pharmacological effect produced by the interaction between the PROTAC-POI binary complex, which could differ from the intended target degradation. Androgen receptor antagonists may act as agonists for targets such as androgen receptors that have mutations in the ligand binding region, which raises major concerns (Paiva and Crews 2019).

While the hook effect is an intrinsic property of PROTACs, its implications for safety assessments are not yet fully understood. The reduced degradation of the target at higher concentrations of PROTACs could potentially lead to paradoxically greater toxicity at lower doses due to inadequate "tox the target." Therefore, careful exploration of a broad dose range is essential to understand the potential toxicity implications (Prekovic et al. 2016).

Developing pharmacokinetic-pharmacodynamic (PK-PD) models specific to PROTAC properties could help predict the onset of the hook effect and guide dose selection. However, the accuracy of such predictions will depend on the experimental determination of key parameters in relevant cell-based assays. Strategies to mitigate the hook effect, such as enhancing protein-protein interactions or cooperativity in the ternary complex, have shown promise in vitro but require further validation in vivo to determine their effectiveness in addressing this issue (Smith et al. 2019).

10.2.6 Immunogenicity

Immunogenicity is a critical consideration in the development of PROTACs due to their potential to induce immune responses against non-human protein components or synthetic ligands potentially leading to adverse immune reactions or allergic responses. In preclinical studies of PROTACs targeting oncogenic proteins, immune responses were observed in animal models, indicating potential immunogenicity concerns that need to be addressed for clinical translation (Bondeson et al. 2015). While there are limited direct examples of immunogenicity associated with PROTACs in the literature, the potential for immunogenicity is recognized and addressed during preclinical and clinical development. Following are the immunogenicity considerations:

- ***Non-Human Protein Components***:
  PROTACs may incorporate non-human protein domains, such as those derived from bacterial or viral proteins, for targeting or E3 ligase recruiting purposes. These non-human components can potentially trigger immune responses when administered to humans.

  In a study by Bondeson et al. (2015), a small-molecule PROTAC was designed to induce the degradation of FKBP12, a human protein. While the study demonstrated efficient degradation of FKBP12, it did not specifically address the immunogenicity of the PROTAC components. However, the use of non-human protein domains in PROTAC design raises concerns regarding potential immunogenicity (Bondeson et al. 2015).
- ***Synthetic Ligands***:
  Synthetic ligands utilized in PROTACs to target specific proteins or recruit E3 ubiquitin ligases may possess chemical structures that are novel to the human immune system, potentially eliciting immune responses.

  Although specific examples are limited, the incorporation of synthetic ligands in PROTACs has been a common strategy. For instance, heterobifunctional molecules incorporating small molecules like thalidomide derivatives or kinase inhibitors have been utilized in PROTAC design. These synthetic ligands may introduce immunogenicity concerns that need to be evaluated during development (Burslem and Crews 2020; Winter et al. 2015).
- ***Preclinical Immunogenicity Assessments***:
  Preclinical studies typically include assessments of immunogenicity to evaluate the potential for immune responses against PROTAC components. These assessments may involve measuring antibody responses, T-cell activation, or cytokine production in animal models.

  In preclinical studies of PROTACs targeting various disease-relevant proteins, researchers may conduct immunogenicity assessments alongside efficacy and safety evaluations. While specific examples may not be readily available in the literature, preclinical immunogenicity assessments are integral to the development of PROTAC-based therapies (Churcher 2018).

- ***Clinical Immunogenicity Studies***:
  Immunogenicity evaluations continue into clinical development, where PROTACs are tested in human subjects. Clinical trials include monitoring for immune responses, such as the development of anti-drug antibodies (ADAs), cytokine release syndrome, or infusion reactions.
  Clinical trials of PROTAC-based therapies include immunogenicity assessments as part of their safety monitoring protocols. While specific examples may vary depending on the clinical trial, these assessments contribute to understanding the immunological profile of PROTACs in humans (Burslem and Crews 2020).

10.2.7 Cytotoxicity

Cytotoxicity is a significant challenge in the development of PROTACs, as these molecules have the potential to induce cellular stress or disrupt essential cellular functions, leading to adverse effects on cell viability. Excessive activation of the ubiquitin-proteasome system by PROTAC-induced protein degradation may lead to cellular stress, cytotoxicity, or induction of apoptotic pathways, resulting in tissue damage or impaired cellular function. High concentrations of PROTACs targeting critical cellular proteins may induce cytotoxic effects in vitro, leading to reduced cell viability or cell death, highlighting the importance of optimizing dosing regimens to minimize cytotoxicity (Winter et al. 2015).

Following are the illustrations of the cytotoxicity challenge associated with PROTACs:

- ***Dose-dependent Cytotoxicity***:
  The cytotoxic effects of PROTACs can be dose-dependent, with higher concentrations leading to more pronounced cytotoxicity. Excessive activation of the ubiquitin-proteasome system by PROTAC-induced protein degradation may overwhelm cellular protein degradation pathways, resulting in cytotoxic effects.
  In a study by Bondeson et al. (2015), the authors demonstrated the dose-dependent cytotoxicity of a small-molecule PROTAC targeting FKBP12. At higher concentrations, the PROTAC-induced significant cytotoxic effects in cellular models, suggesting a dose-dependent relationship between PROTAC concentration and cytotoxicity (Bondeson et al. 2015).
- ***Off-target Effects and Cellular Dysfunction***:
  Description: Off-target effects of PROTACs may lead to the unintended degradation of non-specific proteins, disrupting vital cellular processes and inducing cytotoxicity or cellular dysfunction. Zengerle et al. (2015) reported off-target effects of a small-molecule PROTAC targeting the bromodomain protein BRD4. While the PROTAC efficiently degraded BRD4, it also induced cytotoxicity in cellular models, indicating potential off-target effects contributing to cellular dysfunction and cytotoxicity (Zengerle et al. 2015).
- ***Cell-type Specificity of Cytotoxicity***:
  The cytotoxic effects of PROTACs may vary depending on the cell type and cellular context. Differences in protein expression profiles, metabolic pathways, or

susceptibility to cellular stress can influence the cytotoxic response to PROTAC treatment. PROTACs may exhibit tissue-specific effects, leading to differential toxicity or responses in various tissues or organs, which can arise from factors such as off-target effects, pharmacokinetics, or tissue distribution. In preclinical studies of PROTACs targeting kinase proteins, differential toxicity was observed across different tissues due to variations in target expression levels, highlighting the need for comprehensive safety evaluation in relevant animal models (Zhang et al. 2020).

Winter et al. (2015) investigated the cytotoxicity of a PROTAC targeting the oncogenic protein BCR-ABL. While the PROTAC demonstrated efficacy in inducing target protein degradation, its cytotoxic effects varied across different cancer cell lines, highlighting cell-type-specific differences in cytotoxicity response to PROTAC treatment (Winter et al. 2015).

- ***Mitigation Strategies***:
  Strategies to mitigate cytotoxicity associated with PROTACs include optimizing PROTAC design to enhance specificity and reduce off-target effects.
  Burslem et al. (2018) described the optimization of a PROTAC targeting receptor tyrosine kinases (RTKs) to minimize off-target effects and cytotoxicity. Through rational design and structure-activity relationship studies, the researchers identified a lead PROTAC with improved specificity and reduced cytotoxicity compared to earlier iterations (Burslem and Crews 2020).

10.2.8 Pharmacokinetic Properties

Optimal pharmacokinetic properties are essential for the safety and efficacy of PROTACs, including metabolic stability, bioavailability, and tissue distribution. Poor pharmacokinetics may lead to suboptimal exposure or accumulation of PROTACs in specific tissues, increasing the risk of adverse events. In a pharmacokinetic study of a PROTAC targeting an oncogenic protein, rapid metabolism and clearance were observed, limiting its efficacy and necessitating modifications to improve pharmacokinetic properties (Gu et al. 2018).

10.2.9 Drug-Drug Interactions

PROTACs may interact with other drugs or compounds, leading to altered pharmacokinetics, efficacy, or toxicity, which can result from competition for metabolic pathways, protein binding sites, or transporters. Co-administration of PROTACs with other medications may lead to drug-drug interactions, affecting the metabolism or clearance of both the PROTAC and the co-administered drugs, necessitating careful consideration of potential interactions during drug development (He et al. 2022).

Drug-drug interactions (DDIs) are potential concerns with PROTACs due to their ability to interact with other drugs or compounds, potentially altering their pharmacokinetics, efficacy, or toxicity. Following are the illustrations of drug-drug interactions associated with PROTACs:

- ***Competition for Metabolic Pathways***:
  PROTACs and co-administered drugs may compete for the same metabolic pathways, leading to alterations in drug metabolism and clearance.
  Hughes et al. (2021) investigated the drug-drug interaction potential of a PROTAC targeting an oncogenic protein in combination with other anti-cancer agents. Co-administration of the PROTAC with drugs metabolized by the same cytochrome P450 enzymes resulted in altered pharmacokinetics and efficacy of both the PROTAC and the co-administered drugs (Hughes et al. 2021).
- ***Protein Binding Interactions***:
  PROTACs and other drugs may compete for binding sites on plasma proteins, affecting their distribution and pharmacokinetics.
  In a study by Sun et al. (2019), the researchers investigated the drug-drug interaction potential of a PROTAC targeting Bruton's tyrosine kinase (BTK) in combination with other BTK inhibitors. Co-administration of the PROTAC with BTK inhibitors resulted in displacement from plasma protein binding sites, leading to altered pharmacokinetics and potential toxicity (Sun et al. 2019).

10.2.9.1 Transporter-Mediated Interactions

PROTACs may interact with drug transporters, influencing their absorption, distribution, and elimination. Zhang et al. (2020) evaluated the drug-drug interaction potential of a PROTAC targeting an oncogenic kinase in combination with transporter substrates. Co-administration of the PROTAC with drugs that are substrates for efflux transporters resulted in altered pharmacokinetics and tissue distribution, affecting the efficacy and toxicity of both the PROTAC and the co-administered drugs (Zhang et al. 2020).

10.2.9.2 Pharmacodynamic Interactions

PROTAC and co-administered drugs may have synergistic or antagonistic pharmacodynamic effects, altering their therapeutic efficacy or toxicity.

In preclinical studies by Winter et al. (2015), the combination of a PROTAC targeting an oncogenic protein with chemotherapeutic agents resulted in synergistic antitumor effects. However, the combination also led to increased toxicity compared to either treatment alone, highlighting the potential for pharmacodynamic interactions (Winter et al. 2015).

10.2.10 Limited Safety Inference

It refers to the challenges associated with predicting and assessing the safety profile of these novel drug molecules based on existing safety data and traditional drug development approaches. Several factors contribute to the limited safety inference of PROTACs:

- ***Novel Mechanism of Action***: PROTACs operate through a unique mechanism of inducing targeted protein degradation, which differs from traditional small-molecule inhibitors. This novel mode of action presents challenges in extrapolating safety data from existing drug classes, as PROTACs may have distinct off-target effects and safety considerations (Gao et al. 2020).

- ***Lack of Historical Safety Data***: Since PROTACs represent a relatively new drug modality, there is a scarcity of historical safety data and clinical experience compared to established drug classes. This limited data hinders the ability to predict potential safety issues and long-term effects of PROTACs based on previous knowledge (Hemkens et al. 2023).
- ***Complex Protein Interactions***: The intricate interactions between PROTACs, target proteins, E3 ligases, and other cellular components can lead to unforeseen biological responses and off-target effects. Understanding the full spectrum of these interactions and their implications for safety requires comprehensive studies and analyses (Gao et al. 2020).
- ***Unique Chemical Properties***: PROTACs possess distinct chemical structures and properties that may influence their pharmacokinetics, biodistribution, and toxicity profiles. These unique characteristics make it challenging to draw direct comparisons with conventional drug molecules and infer safety outcomes based on traditional drug development paradigms (Sun et al. 2019).

To address the limited safety inference of PROTACs, researchers and drug developers employ a combination of in vitro studies, preclinical evaluations, computational modeling, and safety assessments to characterize the safety profile of PROTAC molecules comprehensively. By conducting rigorous safety analyses and leveraging emerging technologies, such as predictive toxicology tools and advanced screening methods, efforts are made to enhance the understanding of PROTAC safety and mitigate potential risks associated with their use in therapeutic applications.

10.3 Development of Safe and Efficacious PROTACs

The safety insights gained from the first generation of PROTACs are crucial for directing efforts toward developing safer PROTAC molecules, especially for non-oncology uses where higher drug safety margins are essentially required. This is particularly important for chronic therapy and non-life-threatening diseases, where maintaining a favorable balance between safety and efficacy is a significant challenge for clinical usage. While clinical safety data for PROTAC molecules are not yet available, there is a growing interest in developing "safer" PROTACs (Brand et al. 2019).

Optimizing the E3 ligase warhead to lower off-target degradation activity and raise the therapeutic index of these compounds is one way to improve the safety profile of PROTACs. Studies have shown that different CRBN ligands, such as pomalidomide and thalidomide, exhibit distinct degradation profiles, suggesting that optimizing the E3 ligase warhead could lead to decreased off-target effects (Wu et al. 2019). Additionally, modifying the linker and its orientation of attachment to the E3 ligase warhead has been demonstrated to enhance selectivity in several studies. By adjusting the linker region, researchers have successfully developed PROTACs with improved selectivity, reducing the degradation of unintended proteins. These modifications offer promise in minimizing off-target degradation and enhancing the safety profile of PROTAC molecules (Donovan et al. 2018).

Furthermore, utilizing E3 ligases with unique tissue expression can help restrict PROTAC activity to specific tissues, potentially avoiding degradation in tissues wherein the E3 ligase is not expressed. This targeted approach has been demonstrated in studies targeting specific diseases, such as cancer, to reduce on-target toxicity in non-diseased tissues. While many pharmaceutical and biotech companies are exploring the discovery of new E3 ligases for PROTAC applications, the identification of suitable ligands for these ligases remains a challenge due to the complexity and time-intensive nature of the process (Li et al. 2023).

In conclusion, advancements in optimizing E3 ligase warheads, modifying linkers, and targeting tissue-specific E3 ligases offer promising avenues for developing safer PROTAC molecules with enhanced selectivity and reduced off-target effects. These strategies are essential for improving the efficacy/safety balance of PROTACs, particularly in non-oncology indications where stringent safety requirements are paramount (Mansour 2018).

10.4 Conclusion

In conclusion, the chapter emphasizes the importance of understanding and addressing safety risks associated with PROTACs, a promising new drug modality that exploits the ubiquitin-proteasome system for targeted protein degradation. The review highlights key safety considerations, such as off-target degradation, intracellular accumulation of natural substrates, proteasome saturation, and potential liabilities associated with PROTACs. It underscores the need for thorough safety assessments during the discovery, optimization, and clinical development of PROTACs to ensure their efficacy and safety in therapeutic applications, especially in non-oncology indications where stringent safety requirements are essential. The review also discusses strategies to mitigate safety risks, such as optimizing E3 ligase warheads, modifying linkers, and targeting tissue-specific E3 ligases to enhance the selectivity and safety profile of PROTAC molecules.

References

Asatsuma-Okumura T, Ando H, De Simone M, Yamamoto J, Sato T, Shimizu N, Asakawa K, Yamaguchi Y, Ito T, Guerrini L, Handa H (2019) p63 is a cereblon substrate involved in thalidomide teratogenicity. Nat Chem Biol 15(11):1077–1084. https://doi.org/10.1038/s41589-019-0366-7

Békés M, Langley DR, Crews CM (2022) PROTAC targeted protein degraders: the past is prologue. Nat Rev Drug Discov 21(3):181–200. https://doi.org/10.1038/s41573-021-00371-6

Bondeson DP, Mares A, Smith IE, Ko E, Campos S, Miah AH, Mulholland KE, Routly N, Buckley DL, Gustafson JL, Zinn N, Grandi P, Shimamura S, Bergamini G, Faelth-Savitski M, Bantscheff

M, Cox C, Gordon DA, Willard RR et al (2015) Catalytic in vivo protein knockdown by small-molecule PROTACs. Nat Chem Biol 11(8):611–617. https://doi.org/10.1038/nchembio.1858

Brand M, Jiang B, Bauer S, Donovan KA, Liang Y, Wang ES, Nowak RP, Yuan JC, Zhang T, Kwiatkowski N, Müller AC, Fischer ES, Gray NS, Winter GE (2019) Homolog-selective degradation as a strategy to probe the function of CDK6 in AML. Cell Chem Biol 26(2):300–306.e309. https://doi.org/10.1016/j.chembiol.2018.11.006

Burslem GM, Smith BE, Lai AC, Jaime-Figueroa S, McQuaid DC, Bondeson DP, Toure M, Dong H, Qian Y, Wang J, Crew AP, Hines J, Crews CM (2018) The Advantages of Targeted Protein Degradation Over Inhibition: An RTK Case Study. Cell Chem Biol 25(1):67–77. e3. https://doi.org/10.1016/j.chembiol.2017.09.009.

Burslem GM, Crews CM (2020) Proteolysis-targeting chimeras as therapeutics and tools for biological discovery. Cell 181(1):102–114. https://doi.org/10.1016/j.cell.2019.11.031

Churcher I (2018) Protac-induced protein degradation in drug discovery: breaking the rules or just making new ones? J Med Chem 61(2):444–452. https://doi.org/10.1021/acs.jmedchem.7b01272

Donovan KA, An J, Nowak RP, Yuan JC, Fink EC, Berry BC, Ebert BL, Fischer ES (2018) Thalidomide promotes degradation of SALL4, a transcription factor implicated in Duane Radial Ray syndrome. elife 7:e38430. https://doi.org/10.7554/eLife.38430

Edmondson SD, Yang B, Fallan C (2019) Proteolysis targeting chimeras (PROTACs) in 'beyond rule-of-five' chemical space: recent progress and future challenges. Bioorg Med Chem Lett 29(13):1555–1564. https://doi.org/10.1016/j.bmcl.2019.04.030

Gao H, Sun X, Rao Y (2020) PROTAC technology: opportunities and challenges. ACS Med Chem Lett 11(3):237–240. https://doi.org/10.1021/acsmedchemlett.9b00597

Gu S, Cui D, Chen X, Xiong X, Zhao Y (2018) PROTACs: an emerging targeting technique for protein degradation in drug discovery. BioEssays 40(4):e1700247. https://doi.org/10.1002/bies.201700247

Han T, Goralski M, Gaskill N, Capota E, Kim J, Ting TC, Xie Y, Williams NS, Nijhawan D (2017) Anticancer sulfonamides target splicing by inducing RBM39 degradation via recruitment to DCAF15. Science 356(6336):eaal3755. https://doi.org/10.1126/science.aal3755

He Y, Koch R, Budamagunta V, Zhang P, Zhang X, Khan S, Thummuri D, Ortiz YT, Zhang X, Lv D, Wiegand JS, Li W, Palmer AC, Zheng G, Weinstock DM, Zhou D (2020) DT2216—a Bcl-xL-specific degrader is highly active against Bcl-xL-dependent T cell lymphomas. J Hematol Oncol 13(1):95. https://doi.org/10.1186/s13045-020-00928-9

He M, Lv W, Rao Y (2021) Opportunities and challenges of small molecule induced targeted protein degradation. Front Cell Dev Biol 9:685106. https://doi.org/10.3389/fcell.2021.685106

He M, Cao C, Ni Z, Liu Y, Song P, Hao S, He Y, Sun X, Rao Y (2022) PROTACs: great opportunities for academia and industry (an update from 2020 to 2021). Signal Transduct Target Ther 7(1):181. https://doi.org/10.1038/s41392-022-00999-9

Hemkens M, Stamp K, Loberg LI, Moreau K, Hart T (2023) Industry perspective on the non-clinical safety assessment of heterobifunctional degraders. Drug Discov Today 28(8):103643. https://doi.org/10.1016/j.drudis.2023.103643

Hughes SJ, Testa A, Thompson N, Churcher I (2021) The rise and rise of protein degradation: opportunities and challenges ahead. Drug Discov Today 26(12):2889–2897. https://doi.org/10.1016/j.drudis.2021.08.006

Khan S, Zhang X, Lv D, Zhang Q, He Y, Zhang P, Liu X, Thummuri D, Yuan Y, Wiegand JS, Pei J, Zhang W, Sharma A, McCurdy CR, Kuruvilla VM, Baran N, Ferrando AA, Kim YM, Rogojina A et al (2019) A selective BCL-X(L) PROTAC degrader achieves safe and potent antitumor activity. Nat Med 25(12):1938–1947. https://doi.org/10.1038/s41591-019-0668-z

Li X-M, Zhao Z-Y, Yu X, Xia Q-D, Zhou P, Wang S-G, Wu H-L, Hu J (2023) Exploiting E3 ubiquitin ligases to reeducate the tumor microenvironment for cancer therapy. Exp Hematol Oncol 12(1):34. https://doi.org/10.1186/s40164-023-00394-2

Mansour MA (2018) Ubiquitination: friend and foe in cancer. Int J Biochem Cell Biol 101:80–93. https://doi.org/10.1016/j.biocel.2018.06.001

Paiva SL, Crews CM (2019) Targeted protein degradation: elements of PROTAC design. Curr Opin Chem Biol 50:111–119. https://doi.org/10.1016/j.cbpa.2019.02.022

Pettersson M, Crews CM (2019) Proteolysis targeting chimeras (PROTACs)—past, present and future. Drug Discov Today Technol 31:15–27. https://doi.org/10.1016/j.ddtec.2019.01.002
Pickett HA, Reddel RR (2015) Molecular mechanisms of activity and derepression of alternative lengthening of telomeres. Nat Struct Mol Biol 22(11):875–880. https://doi.org/10.1038/nsmb.3106
Prekovic S, van Royen ME, Voet AR, Geverts B, Houtman R, Melchers D, Zhang KY, Van den Broeck T, Smeets E, Spans L, Houtsmuller AB, Joniau S, Claessens F, Helsen C (2016) The effect of F877L and T878A mutations on androgen receptor response to enzalutamide. Mol Cancer Ther 15(7):1702–1712. https://doi.org/10.1158/1535-7163.Mct-15-0892
Rousseau A, Bertolotti A (2018) Regulation of proteasome assembly and activity in health and disease. Nat Rev Mol Cell Biol 19(11):697–712. https://doi.org/10.1038/s41580-018-0040-z
Santambrogio L, Berendam SJ, Engelhard VH (2019) The antigen processing and presentation machinery in lymphatic endothelial cells [Mini Review]. Front Immunol 10:1033. https://doi.org/10.3389/fimmu.2019.01033
Smith BE, Wang SL, Jaime-Figueroa S, Harbin A, Wang J, Hamman BD, Crews CM (2019) Differential PROTAC substrate specificity dictated by orientation of recruited E3 ligase. Nat Commun 10(1):131. https://doi.org/10.1038/s41467-018-08027-7
Strelow JM (2017) A perspective on the kinetics of covalent and irreversible inhibition. SLAS Discov 22(1):3–20. https://doi.org/10.1177/1087057116671509
Sun X, Gao H, Yang Y, He M, Wu Y, Song Y, Tong Y, Rao Y (2019) PROTACs: great opportunities for academia and industry. Signal Transduct Target Ther 4(1):64. https://doi.org/10.1038/s41392-019-0101-6
Swatek KN, Usher JL, Kueck AF, Gladkova C, Mevissen TET, Pruneda JN, Skern T, Komander D (2019) Insights into ubiquitin chain architecture using Ub-clipping. Nature 572(7770):533–537. https://doi.org/10.1038/s41586-019-1482-y
Thibaudeau TA, Smith DM (2019) A practical review of proteasome pharmacology. Pharmacol Rev 71(2):170–197. https://doi.org/10.1124/pr.117.015370
Uehara T, Minoshima Y, Sagane K, Sugi NH, Mitsuhashi KO, Yamamoto N, Kamiyama H, Takahashi K, Kotake Y, Uesugi M, Yokoi A, Inoue A, Yoshida T, Mabuchi M, Tanaka A, Owa T (2017) Selective degradation of splicing factor CAPERα by anticancer sulfonamides. Nat Chem Biol 13(6):675–680. https://doi.org/10.1038/nchembio.2363
Winter GE, Buckley DL, Paulk J, Roberts JM, Souza A, Dhe-Paganon S, Bradner JE (2015) Drug development: phthalimide conjugation as a strategy for in vivo target protein degradation. Science 348(6241):1376–1381. https://doi.org/10.1126/science.aab1433
Wu H, Yang K, Zhang Z, Leisten ED, Li Z, Xie H, Liu J, Smith KA, Novakova Z, Barinka C, Tang W (2019) Development of multifunctional histone deacetylase 6 degraders with potent antimyeloma activity. J Med Chem 62(15):7042–7057. https://doi.org/10.1021/acs.jmedchem.9b00516
Xiao M, Zhao J, Wang Q, Liu J, Ma L (2022) Recent advances of degradation technologies based on PROTAC mechanism. Biomol Ther 12(9):1257. https://doi.org/10.3390/biom12091257
Zengerle M, Chan K-H, Ciulli A (2015) Selective small molecule induced degradation of the BET bromodomain protein BRD4. ACS Chem Biol 10(8):1770–1777. https://doi.org/10.1021/acschembio.5b00216
Zhang X, Xu F, Tong L, Zhang T, Xie H, Lu X, Ren X, Ding K (2020) Design and synthesis of selective degraders of EGFR(L858R/T790M) mutant. Eur J Med Chem 192:112199. https://doi.org/10.1016/j.ejmech.2020.112199

PROTACs in the Management of Prostate Cancer

11

Deepa S. Mandlik and Satish K. Mandlik

Abstract

Worldwide, prostate cancer (PC) is still a dangerous kind of cancer that affects males. Targeted therapy for cancer has attracted a lot of interest because of its excellent selectivity and minimal toxicity. Numerous illnesses, including cancer, metabolic problems, and neurological diseases, have been linked to ubiquitination abnormalities. By creating a chimeric molecule called a proteolysis-targeting chimera, it is possible to control the balance between the creation and breakdown of proteins in the treatment of cancer. PROteolysis TArgeting Chimeras (PROTACs), or proteolysis-targeting chimeras, are a tool for carrying out therapeutic intervention. The development of cancer treatments has given particular attention to PROTACs because of their distinct mechanism of action, capacity to target proteins that are difficult to manipulate, and targeted engagement of targets. Its mode of action is distinct from that of traditional small-molecule inhibitors, as it specifically degrades the target protein via the ubiquitin–proteasome scheme that causes illnesses. PC is the utmost common non-cutaneous cancer in men among all cancer types. Drug resistance to conventional treatments is a result of immune response suppression caused by genetic changes and the expression of numerous genes in PC. The overview of PROTACs, historical turning points, the biological mechanism, benefits, and latest developments, as well as their role in PC and advantage over traditional inhibitors, are all highlighted in

D. S. Mandlik (✉)
Department of Pharmacology, Bharati Vidyapeeth (Deemed to be University), Poona College of Pharmacy, Pune, Maharashtra, India
e-mail: deepa.mandlik@bharatividyapeeth.edu

S. K. Mandlik
Department of Pharmaceutics, Bharati Vidyapeeth (Deemed to be University), Poona College of Pharmacy, Pune, Maharashtra, India

M. Nandave, P. Jain (eds.), *PROTAC-Mediated Protein Degradation: A Paradigm Shift in Cancer Therapeutics*, https://doi.org/10.1007/978-981-97-5077-1_11

this review. We also trace the different PROTAC targets in PC, such as the androgen receptor and further important oncoproteins.

Keywords

PROTACs · Prostate cancer · Protein degradation · Androgen receptor

Abbreviations

AR	Androgen receptor
CDK	Cyclin-dependent kinase
CRL	Cullin-RING-Ligase
CRPC	Castration-resistant prostate cancer
DBD	DNA-binding domain
PC	Prostate cancer
PROTAC	PROteolysis TArgeting Chimera
RING	Really Interesting New Gene finger
RNAi	RNA interference
SMIs	Small molecule inhibitors
UPS	Ubiquitin-proteasome system

11.1 Introduction

In recent years, several cutting-edge cancer treatment approaches have been developed. These include using CRISPR technology to directly correct DNA mutations, RNA interference (RNAi) to degrade mRNA species, and antibodies or small molecule inhibitors (SMIs) to block the biological function of protein targets. According to data, 52 SMI for protein kinases have been approved by the FDA. These inhibitors are mostly created by wide-ranging screening and are optimised to achieve a superior interaction with each protein target's well-defined active regions to suppress their biological activity (Roskoski 2019). Some proteins, such as scaffold protein and transcriptional factors, have long been referred to as non-druggable targets because they lack specific active sites, unlike kinase (Ferguson and Gray 2018; Fischer 2017). Antibodies have been widely used in cancer treatment due to their biological specificity (Chau et al. 2019). However, their main target is proteins that are attached to the cell membrane, and their huge molecular weight is a drawback that restricts their use in clinical settings (Coats et al. 2019; Wolska-Washer and Robak 2019). Both intracellular and plasma membrane proteins can be targeted by small molecule inhibitors (SMIs); however, one of SMIs' main drawbacks is that SMIs might cause drug resistance over time (Table 11.1) (Ohashi et al. 2013).

Another novel tactic, PROteolysis TArgeting Chimera (PROTAC), was developed in 2001 (Sakamoto et al. 2001) to get around the lack of SMIs and antibodies for cancer therapy. It works by specifically degrading the protein of interest (POI) by using the endogenous ubiquitin-proteasome system (UPS) (Neklesa et al. 2017; Guo et al. 2019; Paiva and Crews 2019). A ternary chemical complex called

Table 11.1 A comparison among PROTACs, monoclonal antibodies (mAbs), and small molecule inhibitors (SMIs)

	PROTACs	mAbs	SMIs
Advantage	Can target scaffolding proteins	Low toxicity	Orally bioavailable
	Can degrade "undruggable" targets	Highly specificity	Extensive tissue penetration
	Extensive tissue penetration		
	Orally bioavailable		
Disadvantage	Mutation in UPS leads to drug resistance	Orally non-bioavailable	Cannot target scaffolding protein
	E3 ligase is essential for the stimulation	Can only target cell proteins	Cannot degrade undruggable target
		Mutation in target genes leads to drug resistance	Mutation in target genes leads to drug resistance
		Restricted tissue penetration	

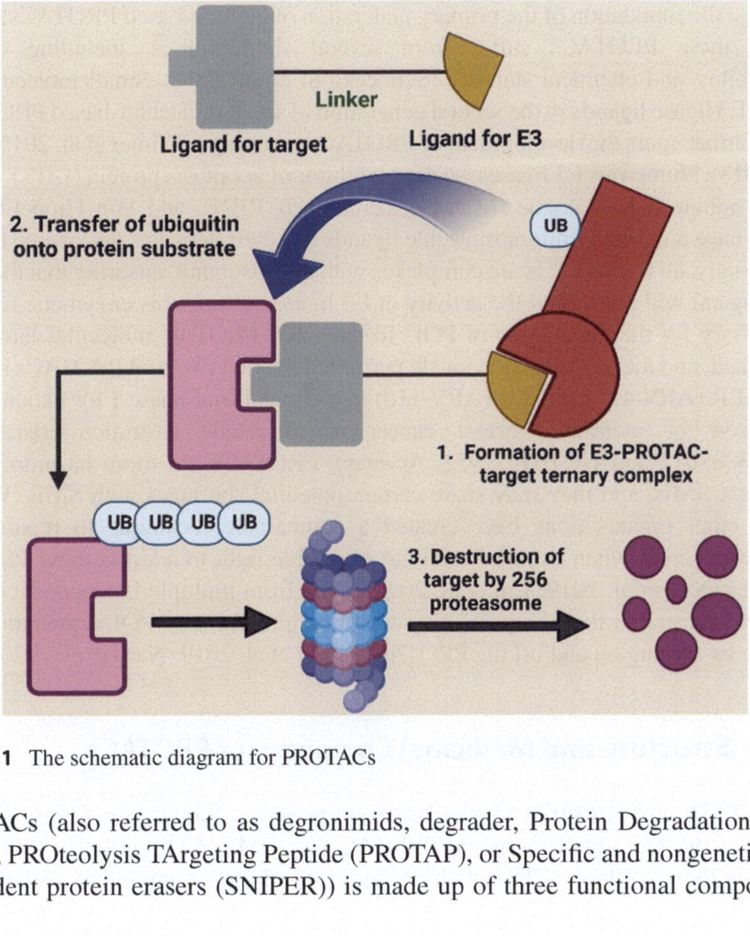

Fig. 11.1 The schematic diagram for PROTACs

PROTACs (also referred to as degronimids, degrader, Protein Degradation Probe (PDP), PROteolysis TArgeting Peptide (PROTAP), or Specific and nongenetic IAP-dependent protein erasers (SNIPER)) is made up of three functional components:

a POI-binding ligand, an E3 ligase-recruiting ligand, and a linker (Pettersson and Crews 2019) (Fig. 11.1). According to Wang et al. (2014), the UPS is made up of three stages of enzymatic processes that include ubiquitin-activating E1, ubiquitin-conjugating E2, and ubiquitin-ligase E3. According to Ciechanover (2015), in this mechanism, E3 ligases identify particular ubiquitination proteins and destruction via the 26S proteasome. By using the natural E3 ligase to ubiquitinate POI, PROTACs form a ternary complex with POI and E3 ligase. It then facilitates the transfer of ubiquitin to POI, which leads to the 26S proteasome's eventual destruction of POI.

Over 600 E3 ligases are found in the human genome and are categorised into 4 families: U-box, plant homeodomain (PHD) finger, Really Interesting New Gene (RING) finger, and homologous to the E6AP carboxyl terminus (HECT) (Bernassola et al. 2008; Deshaies and Joazeiro 2009). Cullin-RING-Ligase (CRL) and anaphase-promoting complex/cyclosome (APC/C) subfamilies make up the biggest RING finger E3 ligase family (Sackton et al. 2014). The discovery of degron (including hydroxyl-Hypoxia inducible factor (HIF) degron and phospho-degron) for those E3 ligases encourages the progress of PROTACs. The CRL E3 ligase complexes are made up of several subunits, and adaptor proteins like F-box protein influence the specificity of the substrate (Winston et al. 1999). The well-known substrates degron and structural evidence are the foundation of the primary generation of peptide-based PROTACs; nevertheless, these PROTACs suffer from several shortcomings, including cellular permeability and chemical stability (Schneekloth et al. 2004). Small molecules are used as E3 ligase ligands in the second generation of small molecular-based PROTACs, which further spurs the development of PROTAC technology (Winter et al. 2015; Testa et al. 2018). Numerous E3 ligases, such as inhibitor of apoptosis protein (IAPs), mouse double minute 2 homologue (MDM2), cereblon (CRBN), and Von Hippel-Lindau (VHL), have a wealth of minor molecule ligands and their derivates discovered to date. Interestingly, all of those E3s are complexes with multi-subunit, ensuring that the binding of ligand will not inhibit the activity of E3 ligase because this enzymatic function is necessary for the destruction of POI. To date, 100 PROTAC molecules have been established, and the FDA has previously permitted the first two oral PROTACs in 2019 against ER (ARV-471) and AR (ARV-110) in a clinical trial phase I for patients with progressive or metastatic breast cancer or metastatic castration-resistant PC (NCT03888612 and NCT04072952, Arvinas). PROTACs are more hazardous than equivalent SMIs, and they may share certain potential shortages with SMIs. We and several other organisations have created a comparable technique to regulate the PROTAC's action when exposed to UV-A or visible light to address these PROTAC shortages (Xue et al. 2019; Liu et al. 2020). Data from multiple independent experiments demonstrated that the precise spatiotemporal control of POI degradation was possible by turning on and off the PROTACs (Pfaff et al. 2019; Naro et al. 2020).

11.2 Structure and Medicinal Chemistry of PROTACs

Before 2008, most PROTACs were based on peptides; though, the Crews group's release of the first small molecule PROTAC using MDM2 signified a substantial change in their synthesis. The sensible strategy of all three components—the linker, the target-binding warhead, and the E3 ligase warhead (E3 ligand)—is essential for the development of PROTACs (Schneekloth et al. 2008).

11.2.1 Design of E3 Ligands

Over 600 E3 ubiquitin ligases have been found in the human genome to date, but the lack of small compounds that have a high affinity and specificity for these ligases limits their potential (Li et al. 2008). A few crucial elements need to be taken into account when creating an E3 ligase ligand for targeted degradation of protein, including the target protein's E3 ligase selectivity and the expression levels of various E3 ligases in the tissue where the protein is concerted. However, because of their broad range of activity, VHL and CRBN ligands are an excellent place to start for any particular target protein (Guenette et al. 2022). Utilising contemporary methods like structure-based drug design, fragment-based drug design, DNA-encoded libraries, high-throughput screening, and phage display expertise, numerous analogues that mimic the activity of the currently available VHL and CRBN ligands were reported and have been used in the development of novel PROTACs (Ishida and Ciulli 2021).

11.2.2 Design of Linker

In PROTAC design, the linker is essential for the creation of ternary complexes, with different linker components participating in distinct interactions. Therefore, one of the main objectives in the creation of PROTACs is to optimise the linker, taking into account elements like length, composition, stiffness, and attachment locations (Troup et al. 2020). When creating linkers, two crucial factors to take into account are the linker's length composition and its capacity to target the attachment site of the protein (the ligand's solvent-exposed site). Inappropriate linker length can alter the distance between target proteins and E3 ligase, which in turn can alter activity. When a tertiary structure forms between the E3 ligase, PROTAC, and target protein, the linker's composition affects PROTAC's pharmacokinetic characteristics (Cao et al. 2022). For instance, PROTACs' kinetic and dynamic characteristics are greatly enhanced by the linker's stiffness (Xiang et al. 2021).

11.2.3 Target Protein-Binding Ligand Design

The majority of the ligands utilised in the design of this PROTAC constituent are centred on knowledge. Typically, they employ ligands that have been shown to have a strong binding affinity for the target protein or those that have previously been licenced as POI inhibitors. Unlike typical inhibitors, the protein's active binding site need not be taken into account while creating a ligand. Unless it is not causing degradation, it can bind to any place on the protein (Cao et al. 2022).

11.3 PROTACs Targeting Prostate Cancer

The pathogenesis of PC has been thoroughly researched and comprehended throughout the last few decades. Even though it tends to build resistance over time, blocking the AR is still regarded as one of the most successful treatments for PC. Other strategies have been put out to deal with this problem, such as the deployment of PROTACs, which attack and destroy the AR. Since the discovery of ARV-110, the elimination of the protein has garnered a lot of attention in the treatment of PC since it lowers the likelihood of acquiring resistance. About 170 PROTACs have been identified to target different proteins linked to PC. With a focus on medicinal chemistry and drug optimisation, we discuss the progress of different non-AR and AR target-degrading PROTACs and their capacity to treat PC in this paper.

11.3.1 AR Targeting PROTACs

One of the foremost reasons for cancer-related death is PC (Campos et al. 2018). For patients with a great threat of PC, the first-line treatments are surgery, radiation therapy, and androgen deprivation therapies. In addition to being essential for healthy prostate development, AR signalling promotes the progress and existence of PC cells. Inhibiting AR function with AR antagonists like enzalutamide and apalutamide and blocking androgen production with medications like abiraterone have been the main strategies employed in the past to successfully target AR signalling. However, due to AR gene mutations, amplification, and alternate splicing, such small molecule inhibitors are useless against progressive prostate tumour cases (Kahn et al. 2014; Abbasi et al. 2021).

11.3.1.1 CRBN-Based PROTACs

Today, a substantial number of novel AR degraders that operate based on target protein degradation (Proteolysis-targeting chimera) have been identified. Lukas et al. created three series of degraders dependent on enzalutamide, an AR inhibitor approved by the FDA. They changed the linker composition and the benzamide, phenoxy, and piperidine scaffolds in the enzalutamide exit vector section. At a dose of 1 μM, the subseries with biphenyl/phenoxy type exit vector showed efficient degradation of AR. It was found that a PROTAC comprising polyethylene glycol (PEG) linker was more energetic than a straightforward alkyl chain. The best molecules were produced by phenoxy exit vector PROTACs and additional stiff linkers found in the last subseries. Moreover, because the ternary complexes' geometrical orientation was probably changed, the stereochemistry of the molecules affected the degradation of the target. Since PROTAC 35 (compound 1) considerably deteriorated even at 0.1 mM, yielding DC_{50} values of 77 and 75 nM in cell lines like LNCaP and A549, correspondingly, it was selected for further in-depth analysis. Furthermore, according to the studies on cell specificity, LNCaP PC cell lines were more selective for compound 1 (Gockel et al. 2022). Takwale et al. (2020) created and assessed several CRBN-mediated AR PROTACs that mixed an AR antagonist based on the tetramethylcyclobutane with the TD-106 CRBN binder for the treatment of metastatic castration-resistant PC.

Compound 2, TD-802, demonstrated the highest efficacy in breaking down AR proteins among the synthesised compounds, with a DC_{50} of 12.5 nM in the LNCaP cell lines with 93% maximum degradation. The placement of the linker within the TD-106 (CRBN binder) has a significant impact on how well AR degradation works. Furthermore, after 30 min of incubation, more than 70% of the degraders, counting TD-802, showed better microsomal steadiness, with more than 4 h half-life and significant exposure to plasma after intraperitoneal and intravenous injections, according to pharmacokinetic experiments conducted in mice. Lastly, in a xenograft mouse model, TD-802 triggered AR degradation and showed substantial antitumor activity, suggesting that CRBN-based AR degraders might find application in the creation of novel AR therapeutics (Takwale et al. 2020). Several novel PROTACs dependent on derivatives of RU59063 and cereblon ligands were created and evaluated by Liang et al. (2021) to see if they could break down the AR protein in LNCaP (AR+) cells. The majority of PROTACs demonstrated strong binding affinities and efficacies against AR. Compound 3 (A16) was the most promising, showing an 85% efficacy as an AR antagonist compared to 89% for enzalutamide. Furthermore, A16 only slightly reduced the amount of AR protein in LNCaP cells, according to the Western blot study. A16's binding characteristics are comparable to those of other non-steroidal AR ligands, according to molecular docking studies conducted with the DDB1-CRBN E3 ubiquitin ligase complex (PDB: 2AXA) and AR (Liang et al. 2021). Kim et al. (2021) created, synthesised, and assessed a set of PROTACs that included thalidomides (E3 ligase recruiters) as new AR degraders and PEG chain-linked analogues of bicalutamide.

Compound 4 (13b) was the most effective of the synthesised compounds in reducing the AR-dependent transcriptional genes mRNA levels, such as prostate-specific antigens and TMPRSS2. Additionally, compound 5 (13c), exhibited marginally higher AR inhibitory action in comparison to bicalutamide. According to the author's results, bicalutamide's AR antagonist activity might be enhanced by substituting ether for the trifluoromethyl group and iodine for the sulphide group. The ether linkage might be utilised as the linker moiety for attaching the AR ligand with thalidomide, as evidenced by the fact that the AR antagonist action remained unaffected when the hydroxyl group was substituted with fluorine. Compounds 4 and 5, also known as compounds 13b and 13c, were considerably broken down by the AR in LNCaP cells in a way that was dependent on both dose and time. It is interesting to note that compound 13c could affect the AR variation 7 (AR-V) form more than the wild type. According to these findings, compound 13c does not alter the expression of the mRNA; instead, it directly binds to and destroys AR (Kim et al. 2021). An extremely effective and strong AR degrader, Compound 14 (ARD-61), for the management of PC, was reported by researchers under the direction of Shaomeng Wang in a different investigation. When given orally to mice, it was discovered to be ineffective. Because cereblon has a smaller molecular weight and better drug-like qualities than the VHL ligand, the authors chose to use it as the E3 ligase ligand to increase the oral bioavailability of ARD-61.

With both semi-rigid and stretchy linkers, the AR antagonists of thalidomide and ARD-61 were used to create the first series of compounds. It was shown that restricted linkers had a similar efficacy as flexible and linear linkers. To lower the activity, the phenyl ring position of the cereblon ligand's connecting location was changed from the meta to the ortho in the subsequent series of molecules. The greatest hopeful molecule from this series was then examined with altered AR antagonists that had a similar CRBN ligand and linker, demonstrating that AR antagonist ARD-61 is necessary for its function. Using the VCaP cell line, the DC_{50} and DC_{90} values of the favourable compounds were determined to be 0.5 and 5 nM for activity. Testing was done on the mechanisms of action and potencies of these substances. Additional pharmacokinetic characteristics demonstrated the high oral bioavailability of the chosen drugs. Further testing revealed that the greatest encouraging molecule, Compound 6 (ARD-2128), effectively reduced the AR, PSA, TMPRSS2, and FKBP5 levels. Furthermore, ARD-2128 was found to be far more effective than enzalutamide at decreasing tumour sizes, even at lower dosages (Han et al. 2021). The creation of a strong active chemical for degradation of AR was reported by Xiang et al. (2021). The CRBN ligand, AR antagonists from Pfizer and Arvinas, and linkers of different sizes were used to create the first series. Pfizer's molecule was attached to the isoindoline-1,3-dione moiety of thalidomide at position 5 by hooking its phenyl group para-positionally. This resulted in the first series of compounds, which had linear alkyl groups of dissimilar sizes and amine groups on both ends. All the compounds with a DC_{50} of 1.6 nM and a D_{max} of 98% in the VCaP cell line, demonstrated encouraging outcomes comparable to ARV-110; however, their oral bioavailability was not optimal. Semi-rigid and rigid linkers were utilised to synthesise additional chemicals. Substantial progress in the DC_{50}, kinetic, and D_{max} characteristics were obtained with rigid linkers. By changing the cyclohexane ring and the linker atom that connects the phenyl and cyclohexane rings, the lead compound's AR antagonist molecule was further optimised. The outcomes showed how crucial and extremely effective the tertiary nitrogen (as the linker atom) and the cyclohexane ring are for the activity. The changed compounds with varying stiff linkers were shown to be very effective against AR; their DC_{50} values ranged from 0.01 to 1.4 nM, and their D_{max} values were found to be between 93% and 100%. Compound 7 (ARD-2585) (Fig. 11.2) was shown to be the most promising chemical among the chosen AR degraders in LNCaP cell lines after undergoing additional investigation, such as AR degradation kinetics, tissue distribution, and pharmacodynamics studies (Xiang et al. 2021).

11.3.1.2 VHL-Based PROTACs

The relationship between the effectiveness of the resulting PROTAC degraders and the VHL ligand component binding affinity of the VHL protein was examined by Han et al. The study found that powerful and very effective degraders could be created by consuming VHL ligands that have a low binding affinity (Ki = 2–3 μM) to

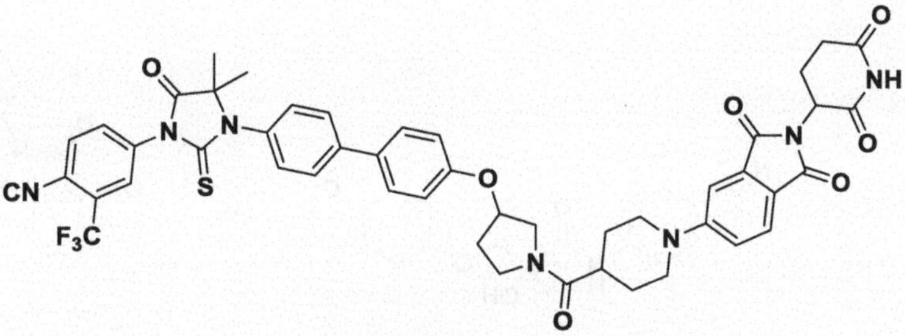

Compound 1

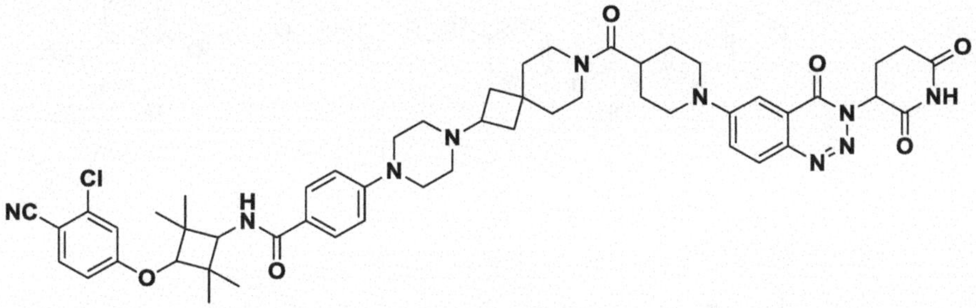

Compound 2

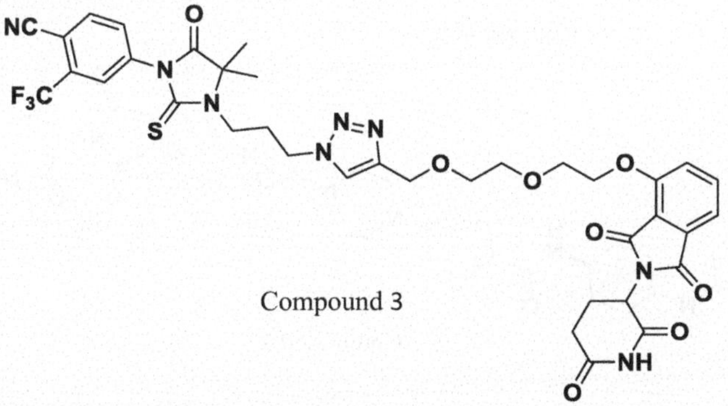

Compound 3

Fig. 11.2 The structures of CRBN-based PROTACs that target the androgen receptor classically show the POI ligand, the E3 ligase ligand, and the linker region

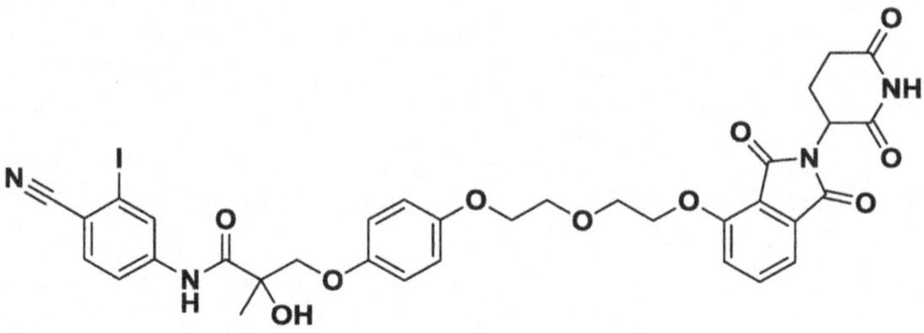

Compound 4

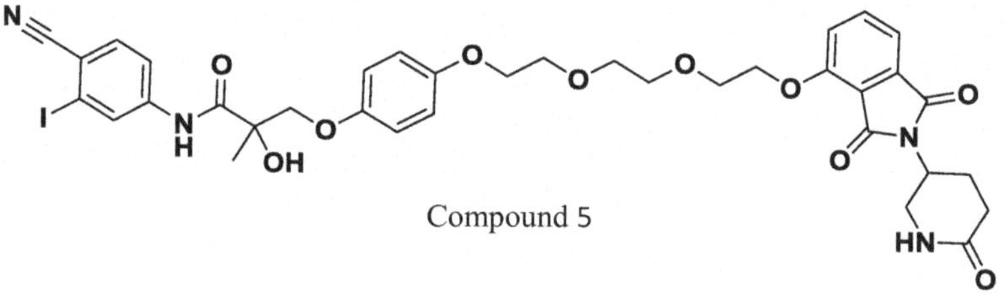

Compound 5

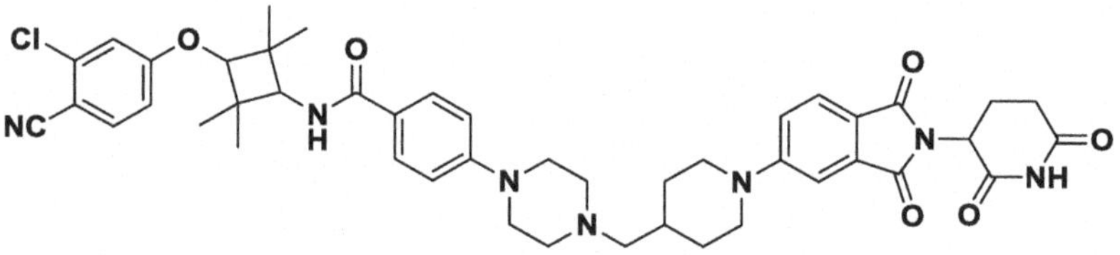

Compound 6

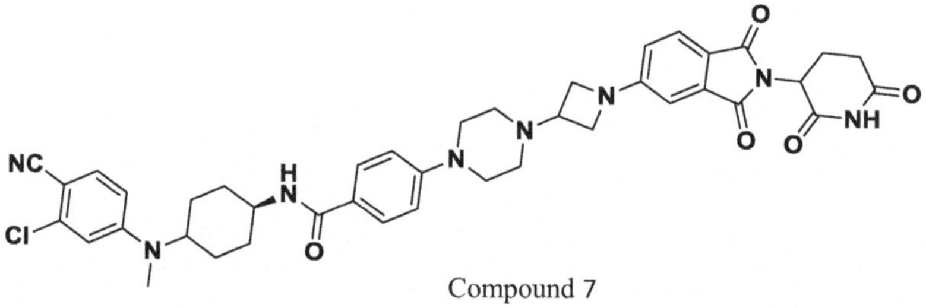

Compound 7

Fig. 11.2 (continued)

the VHL protein. Previous studies have reported that VHL ligands and VHL protein co-crystal structures show that the ligand's 4-methylthiazole group fixes the VHL protein's hydrophobic region. Therefore, several groups, including bromine, nitrile, hydrogen, chlorine, and fluorine were added to the area, and the subsequent compounds' propensity for binding to the VHL protein was examined. Several novel PROTAC AR degraders were created employing VHL ligands with a variety of binding affinities to VHL, utilising Compound 14 (ARD-61) as a template. Nearly every synthetic molecule was shown to be efficient at dropping AR at every confirmed concentration; three compounds, despite their modest binding affinity to VHL, demonstrated excellent action, with more than 80% degradation at 10 nM. Molecule 8 (ARD-266) has a weak VHL binding affinity. To further optimise the molecule, the linker was changed by removing some of the atoms inside the rings and rings themselves. This resulted in a further effective compound. According to the cell culture evaluation, ARD-266 had DC_{50} values of 0.5, 1, and 0.2 nM in the LNCaP, VCaP, and 22Rv1 cell lines, respectively. ARD-266 was able to decrease the amount of AR protein by over 95% in the VCaP and LNCaP cell lines after treatment of 6 h at doses of 10 and 30 nM, respectively, in the 22Rv1 cell line.

Han et al. (2019a, b) discovered that ARD-266 had greater potency when compared to traditional AR inhibitors. Its IC_{50} value was 6 nM, while the AR antagonist's was 182 nM. A031, among other AR degraders, was described by Linrong Chen et al. based on four E3 ligands and two different AR antagonists (azabicyclic and fused bicyclic compounds). They created PROTACs with a range of linking choices, including different warheads, and they altered the linker by adding an alkyl chain and a variety of heterocyclic and phenyl ring combinations, which led to a moderate increase in activity. There was no increase in activity as a result of the CRBN ligand's attachment site changes. However, alterations to the component of AR antagonist resulted in a decrease in action. Interestingly, better compounds like A031 (compound 9) were produced by decreasing the linker length and replacing the E3 ligase ligand with VHL-2. As little as 0.125 μM, the chemical A031 showed encouraging action even at lower concentrations. Furthermore, A031 has five times reduced toxicity and equivalent efficacy, according to additional biological tests when matched to EZLA (the positive control) (Chen et al. 2021). VHL-032 is an E3 ligase-binding warhead used in a sequence of niclosamide-based PROTACs that Munoz et al. (2022) reported for the management of AR + ve PC. When tested against the LNCaP, PC-3, and 22Rv1 cell lines, the compounds showed modest IC_{50} values. Even at concentrations of more than 1 μM, Western blotting experiments demonstrated that their mechanism did not include AR degradation. In contrast, AR degradation was shown by the clinical candidate ARV-110. Nonetheless, in the LNCaP cell line, Niclo-Click Compound 10 (PROTAC 5) demonstrated a respectable IC_{50} of 1.04 ± 0.18 μM (Munoz et al. 2022).

Using 4-(4-phenylthiazol-2-yl) morpholine, Lee et al. (2021) created a novel PROTAC known as Compound 11 (MTX-23) that targets and destroys both the AR splice variant 7 (AR-V7) and full-length AR (AR-FL). MTX-23 demonstrated a DC_{50} of 2 μM for AR-FL and 0.37 μM for AR-V7 in immunoblotting experiments. It was discovered that the substance specifically increased apoptosis and inhibited proliferation in PC cells that are androgen-responsive. Tests with MTX-23 showed encouraging outcomes in cells resistant to the FDA-approved SAT drugs (abiraterone, apalutamide, enzalutamide, and darolutamide). By destroying both AR-FL and AR-V7, it slowed down tumour growth and cellular proliferation in both in vitro and in vivo settings, which may make it useful against prostate cancer that is resistant to castration (Lee et al. 2021). Using VHL as an E3 ligase-binding warhead, Salami et al. (2018a, b) created a sequence of PROTACs built on enzalutamide to target the AR. With a D_{max} of 98% and a DC_{50} of 5 nM, Compound 12 (ARCC-4) was the most potent of the compounds that were synthesised. Compared to its parent drug, enzalutamide, ARCC-4 was further successful in prompting apoptosis and suppressing the proliferation of AR-amplified PC cells in cellular models of castration-resistant prostate cancer (CRPC). Furthermore, in an environment rich in androgen, ARCC-4 was still able to damage AR and prevent the propagation of cells. The Tandem Ubiquitin-Binding Element (TUBE1) pull-down test was used to confirm that polyubiquitination of AR leads to proteasomal degradation, which is the mode of action for ARCC-4. Additionally, clinically significant AR mutants, including AR-T877A, AR-F876L, AR-H874Y, AR-L702H, and AR-M896V, were successfully degraded by ARCC-4. Even at increasing quantities of the R1881 (synthetic androgen) up to 1 nM, ARCC-4 was successful in breaking down the AR protein in VCaP cells (Salami et al. 2018a, b). A thorough optimisation approach was carried out by Han et al. (2019a, b) to create an extremely effective and potent AR-degrading PROTAC. They created a set of PROTACs by using an amide linker with different chain lengths and enzalutamide as the AR-binding ligand and VHL as the E3 ligand. The most efficient AR degradation was found with an ideal linker length of 11–12 atoms, according to biological investigations conducted on the synthesised compounds. The compound's activity and solubility were greatly enhanced by the addition of a pyridine group that was straight bonded to the ethynyl group. The optimisation data showed that the PROTAC's activity is increased when the linker's stiffness is increased. Subsequently, the scientists altered the VHL stereochemistry to investigate its function in binding to the VHL protein. Compound 13 (ARD-69) was ultimately chosen for more study. In LNCaP and VCaP cell lines, ARD-69 produced over 95% AR degradation at 10 nM and had DC_{50} values of 0.76 and 0.86 nM, respectively (Han et al. 2019a, b).

The creation of the AR degrader Compound 14 (ARD-61), as described by Kregel et al. (2020) showed promise in tackling resistance that frequently surfaces during the management of PC. The findings showed that, irrespective of AR levels, expressions of AR splice variant or AR mutational status, ARD-61 induced PARP cleavage in all AR-dependent cell lines. In model systems resistant to enzalutamide and AR splice variants, such as ARD-61, AR-V7 was similarly successful. ARD-61 can suppress tumour cell proliferation in models overexpressing AR-V7, despite its inability to fix and degrade AR-V7. Furthermore, each of the four CRISPR-induced clones that overexpressed ARD-61 and AR-V7 was degraded to full-length AR while having no impact on AR-V7 production. In the CWR-R1 EnzR xenograft model that states great levels of AR-V7 and is metastatic, aggressive, castration-, and enzalutamide-resistant, ARD-61 was found to be highly efficient in vivo. Such results imply that full-length AR is necessary for the survival and development of castration- and enzalutamide-resistant PC, even if AR-V7 may improve AR signalling (Fig. 11.3) (Kregel et al. 2020).

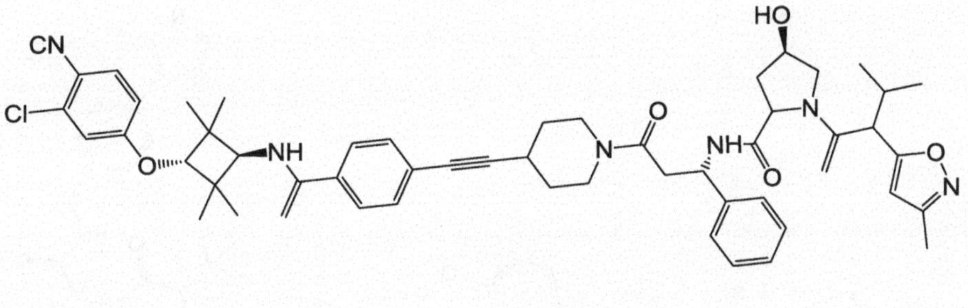

Compound 8

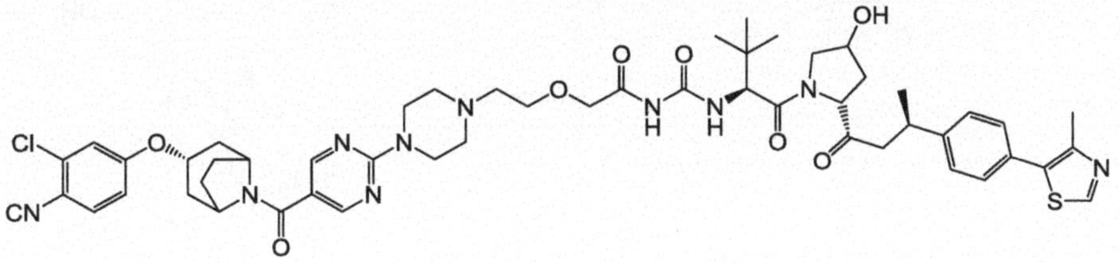

Compound 9

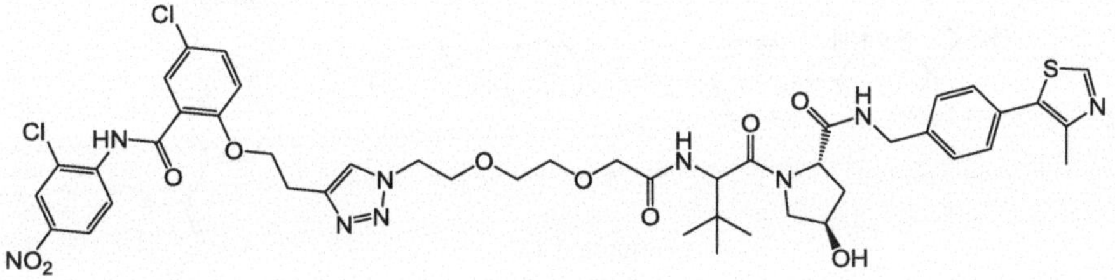

Compound 10

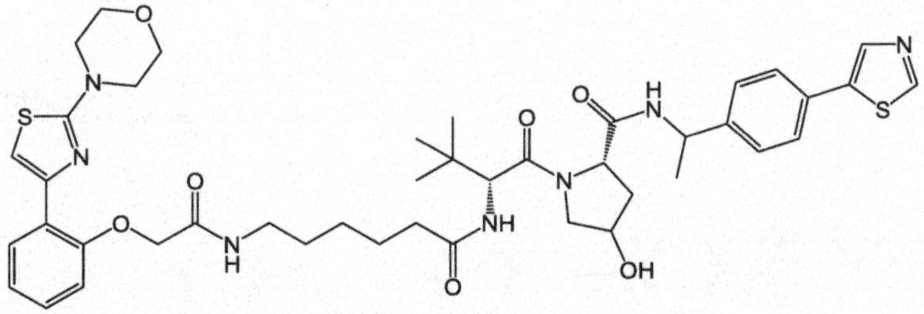

Compound 11

Fig. 11.3 The chemical structures of VHL-based PROTACs that target the androgen receptor characteristically exhibited the POI ligand, the E3 ligase ligand, and the linker region

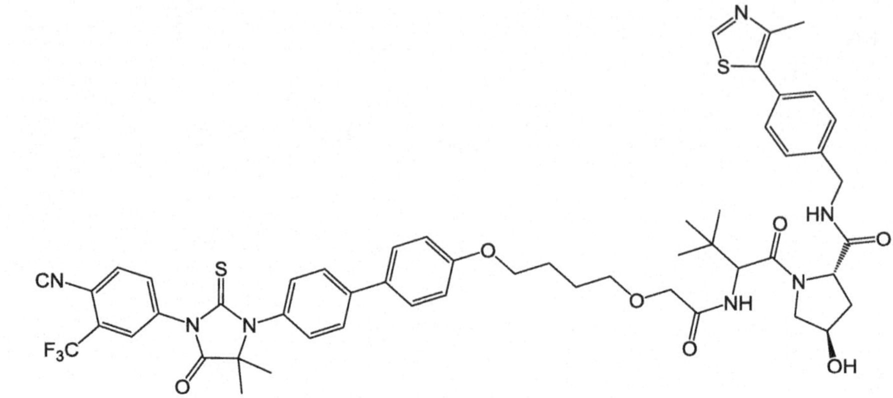

Compound 12

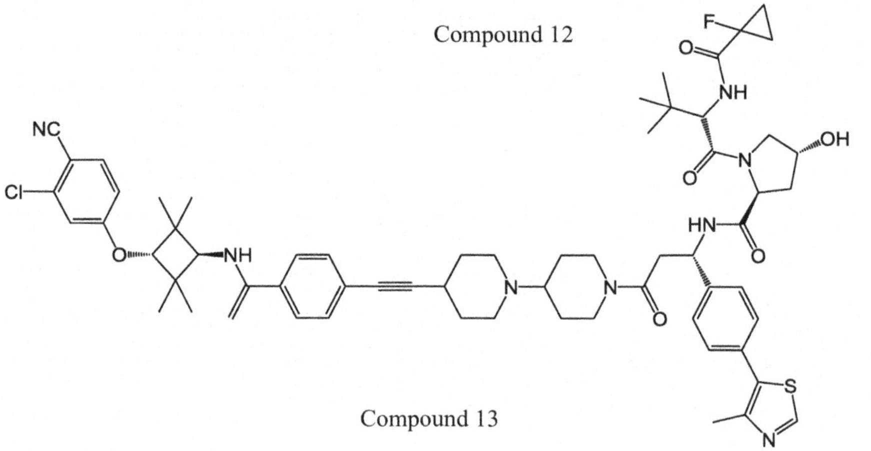

Compound 13

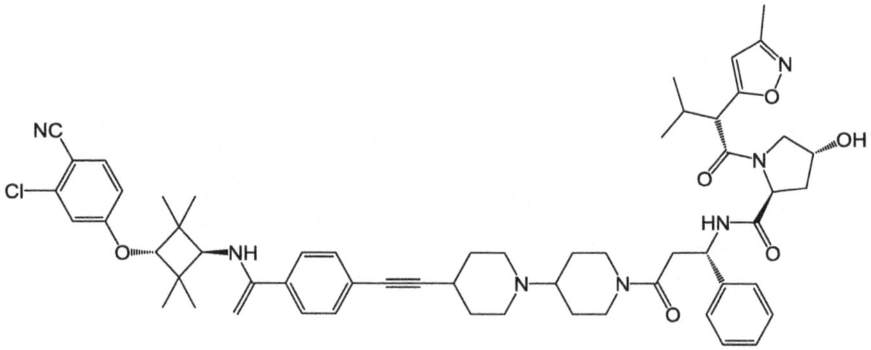

Compound 14

Fig. 11.3 (continued)

11.3.1.3 IAP-Based PROTACs

In 2018, Shibata et al. created a range of new IAP-dependent PROTACs that target AR and are derived from IAP ligands. PROTAC 16 demonstrated the most efficient protein knockdown efficacy against AR among them. In line with the breakdown of the AR protein, PROTAC 16 suppressed the production of AR-mediated genes and the growth of androgen-dependent PC cells. Furthermore, in PC cells treated with AR antagonists, there was no evidence of caspase activation or apoptosis triggered by PROTAC 16. According to these findings, PROTAC 16 may be a promising candidate for an anti-cancer medication targeting prostate tumours that demonstrated AR-dependent propagation.

11.3.1.4 Miscellaneous

A peptide-dependent PROTAC with ultra-small gold nanoparticles as a delivery mechanism was created by Ma et al. (2022). The artificial degraders cause the degradation of both AR and AR-V7 by specifically targeting the DNA-binding domain (DBD) of AR. Solid-phase peptide synthesis was used to create the appropriate PROTAC sequence, and amid the AR-DBD- and MDM2-targeting sequences, a flexible linker was added. The binding affinities of AR-pep-PROTAC and AR-V7 were found to be 12.2 and 49.6 nM, correspondingly, using isothermal titration calorimetry. With IC_{50} values of 230.8, 248.1, and 126.9 nM, respectively, it was revealed that the Au-AR pep-PROTAC stimulated the AR and AR-V7 degradation and inhibited the proliferation of AR- and AR-V7-positive PC cell lines, counting C4-2, LNCaP, and CWR22Rv1. In CWR22rv1, LNCaP, and C4-2 cells, the half-maximal degradation DC_{50} values were determined to be 48.8 nM (AR)/79.2 nM (AR-V7), 175.2, and 193.0 nM, respectively. Moreover, it was found that the Au-AR pep-PROTAC combination causes AR to be broken down in the nuclear and cytoplasmic compartments (Ma et al. 2022). Two series of selective AR degraders based on the AR agonist RU59063 with hydrophobic degrons were synthesised, as described by Xie et al. (2020). Comparing compounds with and without phenyl groups, the structure–activity relationship research showed that compounds with alkyl chain linkers exhibited greater anti-proliferative action. The activities of the compounds were likewise greatly affected by the presence of the triazole group. It was discovered that compound 15, or A9, shared an AR-binding affinity with enzalutamide and exhibited a potent inhibitory impact on LNCaP cells. According to the molecular docking investigation, the rate of degradation may be influenced by the interaction between the synthesised compounds' long hydrophobic tagging chains and the AR (Fig. 11.4) (Xie et al. 2020).

11.3.2 Non-AR-Targeting PROTACs

Many different proteins have been suggested as possible targets for PC treatment. In addition to the AR, PROTACs have been designed to specifically target few of the proteins. For instance, protein kinases and c-Myc are commonly overexpressed in PC and have been connected to the advancement of cancer. There is a chance that

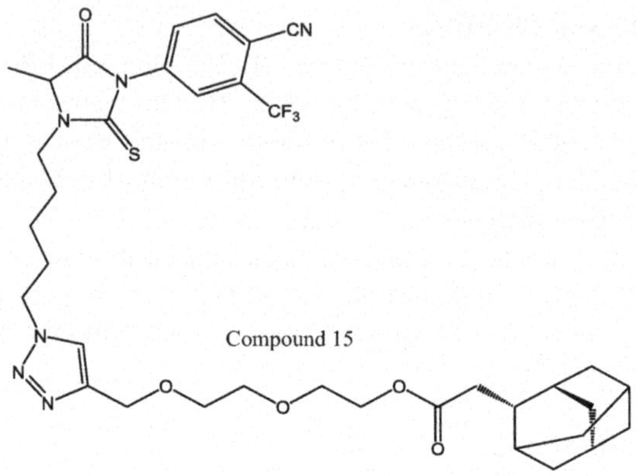

Compound 15

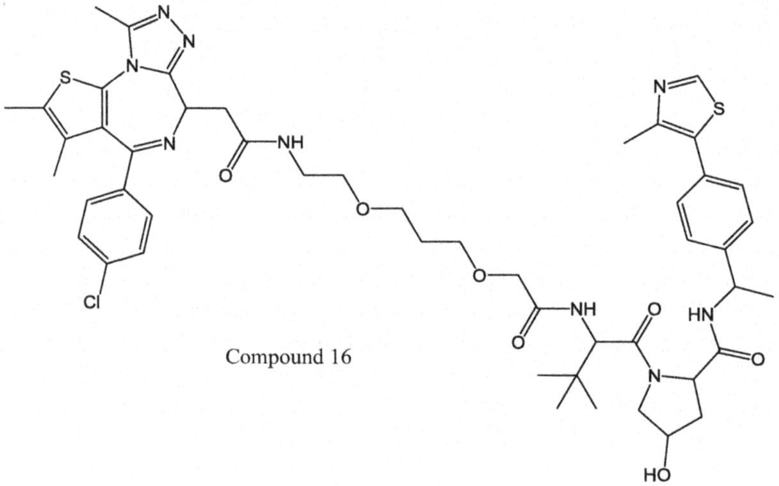

Compound 16

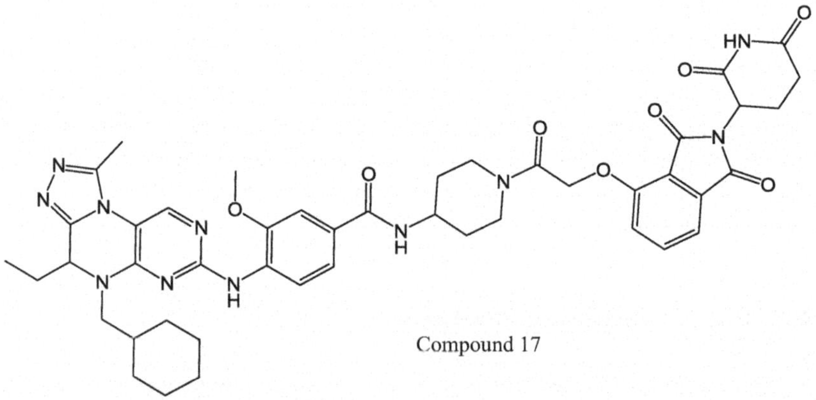

Compound 17

Fig. 11.4 Chemical structures of miscellaneous E3 ligase ligand-based PROTAC (Compound 15)-targeting androgen receptor and PROTACs targeting non-AR targets (Compound 16–20) with POI ligand, the E3 ligase ligand, and the linker

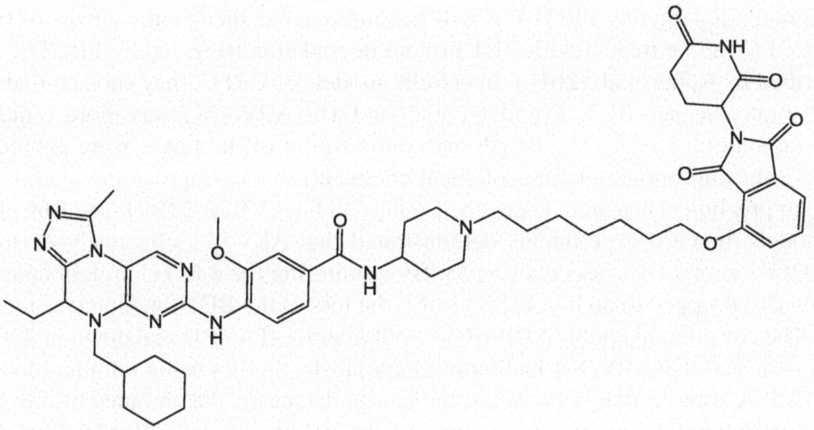

Compound 18

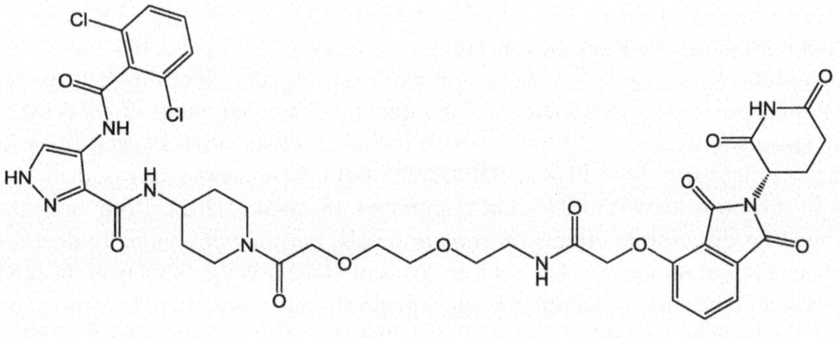

Compound 19

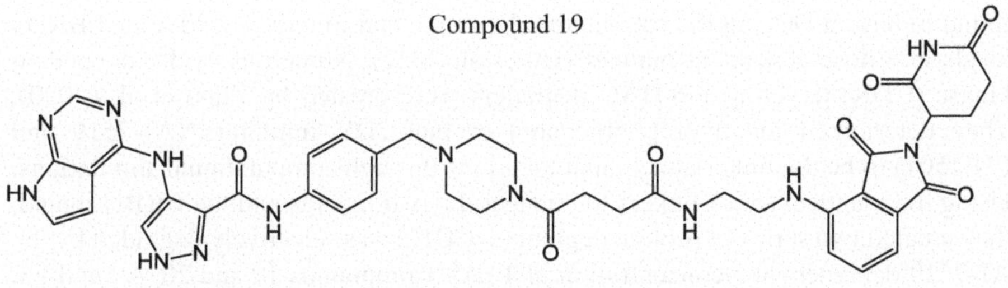

Fig. 11.4 (continued)

these non-AR-targeting PROTACs will be employed as therapeutic agents to treat PC. CRPC can be treated with BET protein degradation triggered by PROTAC, as described by Raina et al. (2016). In cellular models of CRPC, they showed that the small-molecule pan-BET degrader compound 16 (ARV-771) was more efficient than inhibition of BET. The length and composition of the linker were altered to achieve the appropriate pharmacological characteristics to improve the lead molecule for preclinical research. Using the CRPC cell lines VCaP, 22Rv1, and LnCaP95 as models, in vitro experiments demonstrated that ARV-771 effectively destroys BRD2/3/4 with a DC_{50} less than 5 nM. By monitoring the c-MYC protein concentration that dropped to an IC_{50} below 1 nM, the loss of the BET function was identified. The considerable poly (ADP-ribose) polymerase (PARP) breakdown in 22Rv1 cells indicated that ARV-771 had induced apoptosis. Studies using immunoblotting and ELISA showed that ARV-771 significantly decreased the amounts of AR protein in VCaP cells. Following treatment of the VCaP cells with 10 nM ARV-771, there was a reduction in the amounts of FL-AR and AR-V7 mRNA. AR-V7 levels in the tumours were significantly reduced in the preclinical trials of ARV-771, which were given at a 10 mg/kg dose to mice with AR-V7+ 22Rv1 tumour xenografts (Raina et al. 2016). Using WNY0824, a dual BET/PLK1 inhibitor, Hu et al. (2022) developed and synthesised the selective BRD4-PROTAC (WWL0245). Based on computational investigations, it was possible to insert an E3 ligand, like thalidomide or lenalidomide, into the solvent region-exposed 1-methylpiperidin-4-yl motif of WNY0824. Selective cytotoxicity was demonstrated by compound 17 (WWL0245) in BETi-sensitive cancer cell lines, which included AR-positive PC cell lines. The researchers assessed how PLK1, BRD2/3/4, and c-Myc were expressed in VCaP cells in response to WWL0245 and compound 18 (WWL0202). Both substances exhibited no discernible effects on protein levels, but they dramatically decreased BRD4 levels and inhibited c-Myc at low doses of 10 nM. WWL0202 was chosen for more research because of its remarkable anti-proliferative potential, low-molecular-weight, and selectivity. When evaluated on 22Rv1 and VCaP cells, WWL0202 was found to have a DC_{50} in the sub-nanomolar range and to reduce c-Myc and BRD4 levels in a dose-dependent manner (Hu et al. 2022). Numerous cyclin-dependent kinase (CDK)-targeting PROTAC degraders were created by Zhou et al. (2020). They determined the hydrophobic area of pan-CDK inhibitors (AT-7519 and FN-1501) to be the linker attachment locations through computational simulations. Using the identical set of linkers to connect the two ligands and the CRBN ligand, they created two series of protein degraders. CDK2 was selectively degraded by the AT-7519 degraders at a concentration of 1 μM. Compounds 19 and 20, A9 and F3, demonstrated potency with IC_{50} values of 0.84 and 0.12 μM, correspondingly, in a CCK8 experiment against PC-3 cell lines. With DC_{50} values of 62 and 33 nM, correspondingly, mechanistic investigations demonstrated that it might efficiently and selectively lower the CDK2 and CDK9 levels in a dose-dependent way. Additional synthesis of F3-based compounds did not lead to an increase in activity. At 250 nM, compound F3 was shown to considerably prolong the phase of DNA synthesis by 9.8%; at greater concentrations, cell cycle arrest occurred at the G2/M phase. These results offer a solid foundation for PC treatment targeting non-AR targets with the targeted protein degradation approach (Fig. 11.5) (Zhou et al. 2020).

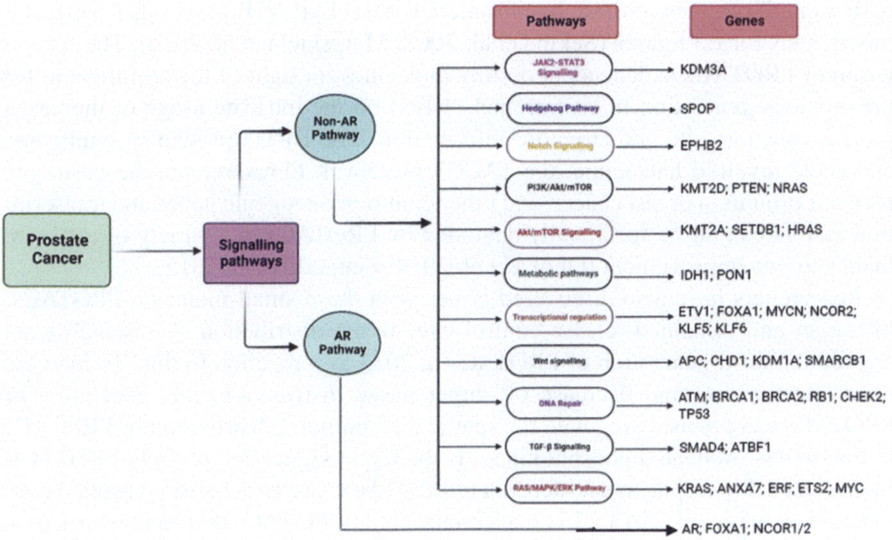

Fig. 11.5 The signalling pathways comprising both androgen receptor (AR) and non-AR pathways and their corresponding genes have been connected in the progression of PC

11.4 Research Progress of Proteolytic Targeted Chimera Technology

The chemical breakdown of proteins was initially discovered by Professors Deshaies and Crews' research group in 2001 (Sakamoto et al. 2001). Subsequently, the intracellular ubiquitination and degradation of particular proteins were greatly increased and employed. PROTAC-1 is any of the original PROTAC molecules, belonging to the PROTACs first generation. This represents the primary iteration of the peptide-dependent PROTACs expertise that uses structural data and in-depth study of substrate-binding sites to identify proteins. Ovalicin is used by one end of the PROTAC-1 molecule to collaborate with the histidine at the 231st act amino peptidase-2 (MetAP-2) peptide. In the E3 ubiquitin ligase protein complex, β-TRCP can attach to the opposite end of the phospho-peptide of the nuclear factor κB inhibitor α (NF-κB inhibitor α) structure. The MetAP-2 protein is degraded by stimulating the ubiquitin–proteasome degradation path, which is triggered by PROTAC-1's strong binding of the β-TRCP complex and MetAP-2 protein. A few other peptide-based PROTACs have been developed, namely PROTAC-2, which links the AR ligand dihydrotestosterone to the IκBα phospho-peptide to break down the AR. Estradiol, an agonist of the oestrogen receptor, and IκBα phospho-peptide are used by PROTAC-3 to break down the oestrogen receptor (Sakamoto et al. 2003). Furthermore, the E3 ubiquitin ligand's higher molecular weight causes an unstable structure and reduced cell permeability. Researchers are still finding and creating small-molecule ligands for E3 ligases and the derivative structures that go along with them. Examples of these ligands are CRBN, cIAP, VHL, and other particular tiny ligands for E3 ligases (Sekine et al. 2008; Matyskiela et al. 2016). The second group of PROTACs is dependent on tiny molecules, in light of the requirement for the previous generation to be extremely based on the intricate usage of the target protein substrate site and structure information. The BRD4 protein in lymphoma and acute myeloid leukaemia, the TACC3 protein in fibrosarcoma, the oestrogen receptor protein in breast cancer, and other common oncogenic genes and transcription factors can all be specifically degraded by PROTACs in a variety of tumours; thanks to this improvement (Ohoka et al. 2014; Zengerle et al. 2015).

Researchers have also discovered issues with these small-molecule PROTACs' off-target and sustained-release control over tissue distribution. A research team created the third generation of PROTACs in 2013 as a reaction to this. To increase their effectiveness and decrease off-target tissue distribution, this generation of PROTACs can precisely regulate the spatial and temporal distribution of PROTACs in the tissues. Activated phosphokinase is the target signal that phosphoPROTACs, often referred to as phosphate-dependent PROTACs, use to achieve controlled degradation (Hines et al. 2013). To create controllable PROTACs that can be turned on and off by light, there are likewise PROTACs that employ UV light sources in photodynamic treatment (Xue et al. 2019). Researchers are now focusing on third-generation controlled PROTACs for medication development. In terms of molecular structure, PROTACs are usually chimeras, with the linker joining the two ends of the molecule. The length, type, and binding site of the linker—which affects the molecules at both ends—as well as other factors also have a significant influence on PROTAC activity (Lu et al. 2015). Using oestrogen receptor-targeted PROTACs as a case, PROTACs with a linker having 16 atom chains have the maximum degradation properties when the linker creates a specific breach amid the target protein binding site and the E3 ubiquitin ligase (Cyrus et al. 2010). It is anticipated that in the ensuing decades, controllable medications for PROTACs will become a reality. Drugs that use PROTACs as the general strategy concept are becoming more and more prevalent (Buckley et al. 2015; Scudellari 2019).

The benefits and drawbacks of PROTACs since its debut, PROTAC has benefited greatly, especially in terms of its distinct targeting and chemical degradation. According to Xia et al. (2019), the site-directed targeting of PROTACs does not depend on blocking the kinase's active site or the interaction site to prevent protein function. Targeted site design does not require a detailed analysis of the protein's active and interaction sites, nor does it take the site's overall spatial folding structure into account. In theory, PROTACs can be designed using this method as long as there is no involvement of a functional site. The ubiquitin–proteasome protein degradation mechanism within the cell is responsible for the chemical breakdown of PROTACs. Instead of focusing on tiny molecules that block the active sites, this degradation leads to the protein "disappearing" within the cell following ubiquitination and hydrolysis. Because of this chemical breakdown, the short half-life of small-molecule compounds can be less problematic, and substantial degradation effects can be achieved with lower doses of PROTACs. Since the targeted breakdown of the protein requires resynthesis in order to reinstate its distribution and function, the resultant chemical degradation ought to be more robust and effective than small-molecule drugs that obstruct its active site.

The beneficial effects of BRD4 inhibitors and BRD4 degraders alone were compared by Winter et al. (2015). BRD4 degraders can further selectively and effectively lead to the degradation of protein in vivo and in vitro, in addition to slowing the course of leukaemia in mice, when BRD4 inhibitors are used alone (Winter et al. 2015). The possibilities for creating PROTACs have increased exponentially with the ongoing identification of tiny ligands for E3 ubiquitin ligase (Schapira et al. 2019). Nevertheless, PROTACs have flaws, and numerous possible issues have been found in clinical settings. Because of their high molecular weight, several PROTACs have poor pharmacokinetics and are difficult to absorb orally. Oral bioavailability was enhanced and this deficiency was rectified by the latter small-molecule PROTACs. The kinds of small-molecule ligands that could be appropriately complemented and constructed are still restricted, despite the discovery of multiple E3 ubiquitin ligases. E3 ubiquitin ligase activity and the ubiquitin–proteasome protein degradation pathway are essential for PROTAC function. Drug adaptation or acquired resistance may happen if the degradation process malfunctions or if the E3 ubiquitin ligase mutates. Predictive biomarkers must thus be included in PROTACs' clinical treatment regimen to solve these problems (Ottis et al. 2018). Following improvement, the current small-molecule PROTACs ought to possess a low molecular weight. Additionally, they possess the ability to relatively freely enter and exit a vast amount of tissue cells within the human body, leading to a broad distribution in tissue and potentially harmful side effects. Although third-generation PROTACs are a more sensible approach, their use is restricted since UVA and other ultraviolet photons have a low penetration rate and the potential to damage DNA. Thus, in order to increase the range of applications and decrease side effects, it may be feasible to think about using further light sources with more penetration to exactly regulate the release of PROTACs (Chen et al. 2018).

11.5 Application of Proteolytic Targeting Chimera Technology in PC

The AR is a crucial transcription factor and protein molecule in PC that is connected to the progress and carcinogenesis of the disease. It is also a target for several drugs, including androgen deprivation therapy (Pagliarulo et al. 2012; Culig and Santer 2014). Regretfully, after a year or 2 of starting endocrine therapy, the mainstream of patients develop CRPC, move on to a challenging malignant stage, and ultimately pass away from the illness (Grasso et al. 2012; Kirby et al. 2011). The introduction of AR antagonist medications like enzalutamide has significantly improved the clinical outcomes of CRPC patients (Scott 2018; Hussain et al. 2018). This implies that a resolution to the still-active AR signalling pathway is desperately needed. Professor Crews created a selective androgen receptor degrader (SARD; no. SARD279) as early as 2015. According to Gustafson et al. (2015), this tiny molecule has a hydrophobic group connected to an AR ligand that may cause AR proteins to degrade, decrease the production of AR target genes, and stop the growth of androgen-dependent PCa cells. Certain resistance mechanisms common to old-style medications can be circumvented by this novel degradation common. To create a more effective AR degrading agent, known as ARCC-4, Professor Crews' research group combined this method with PROTAC technology in 2018 (Salami et al. 2018a, b). A low-molecular-weight AR degradation agent called ARCC-4 is capable of breaking down over 95% of the AR protein in PCa cells. Prostate tumour cells (VCaP, LNCaP, PC3, etc.) can be considerably inhibited from proliferating by it. Additionally, certain clinically related AR point mutants (AR-T877A, AR-H874Y, AR-F876L, AR-L702H, AR-M896V, and other point mutants) can be degraded. In contrast to enzalutamide, ARCC-4 exhibits a noteworthy anti-proliferative potential when exposed to high testosterone concentrations. This implies that utilising this degrading expertise to address resistance to enzalutamide could potentially be a game-changer for the treatment of these patients. Simultaneously, an AR-targeted PROTAC degradation agent (ARD-69) was created by a research team under the direction of Professor Wang from the University of Michigan (Han et al. 2019a, b). In vitro studies have demonstrated that ARD-69 causes dose- and time-dependent degradation of AR protein in AR-positive PCa cell lines (VCaP, LNCaP, and 22Rv1). ARD-69 concurrently suppresses the expression of downstream regulatory genes in the AR signalling path and lowers the endogenous AR protein in PCa cell lines to less than 95%. Compared to the AR antagonist enzalutamide, ARD-69 has an effect that is more than 100 times greater and can effectively suppress cell proliferation. A sole dose of ARD-69 can efficiently lower the AR protein levels in mouse xenograft tumour tissue, according to in vivo investigations. These findings imply that AR-targeted degradation therapy dependent on PROTACs might become a further practical choice for individuals with CRPC who test positive for AR. Additionally, a new AR degrader (AR PROTAC degrader, no. ARD-61) base PROTACs has been produced by Professor Chinnaiyan's research group, and its mechanism of action has been thoroughly explained. By straight way decreasing the AR protein in breast cancer and PCa cell lines with AR-positive characteristics, ARD-61 circumvents the current anti-hormone therapy resistance mechanism (Kregel et al. 2020; Zhao et al. 2020). Research conducted both in vivo and in vitro has revealed that ARD-61, in contrast to enzalutamide, exhibits more potent anti-proliferation and pro-apoptotic effects. Additionally, it has been observed to significantly suppress the control of AR signalling in cancer cells, all while maintaining its inhibitory potential in the enzalutamide resistance model. Subsequent investigation revealed that ARD-61 can suppress the progression of tumour cells expressing AR-V7 overexpression even if it is unable to bind to or target AR-V7 (AR splice variant 7). According to studies, AR is still the primary cause of PCa, and individuals who are resistant to AR antagonists may benefit more clinically from new AR degraders. The US Food and Drug Administration granted permission to Arvinas in 2019 to start a phase I clinical trial (NCT03888612) using oral PROTACs (ARV-110) that target the AR protein in patients with metastatic colorectal cancer (CRPC). Patients with ER-positive/HER2-negative locally advanced or metastatic breast cancer are eligible to participate in phase I clinical trial (NCT04072952) using another oral PROTAC, ARV-471, which targets the oestrogen receptor protein.

Table 11.2 The basic chemical structures, biological activities, and physiochemical properties of the compounds

Name	Target	E3 ligase	Molecular weight (g/mol)	Molecular formula	Degradation activity
ARCC-4	AR	VHL	1024.2	C53H56F3N7O7S2	AR protein degradation in VCaP cells
CST620	CDK4/6	VHL	1064.5	C55H73N11O9S	Degradation of CDK4/6 in various cancer cell lines such as myeloma, leukaemia, and breast
ARD61	AR	VHL	1095.8	C61H71ClN8O7S	AR protein degradation in LNCaP/VCaP cells
ARD69	AR	VHL	1129.8	C62H74ClFN8O7S	AR protein degradation in LNCaP/VCaP/22Rv1 cells

AR androgen receptor, *VHL* Von Hippel-Lindau, *CDK4/6* cyclin-dependent kinase 4/6

A sub-family of the mammalian cyclin-dependent kinases (CDKs) known as CDK1, CDK2, CDK4, and CDK6 is associated with the cell cycle (Asghar et al. 2015). Clinicians are finding that CDK4/6 inhibitors are an effective class of drugs for treating oestrogen receptor-positive breast cancer (Kase et al. 2020). Considering the impact of AR signalling on the cell cycle in PCa, targeting the cell cycle amounts to a core attack on a characteristic that defines cancer. This is a crucial part of cancer treatment. Considering the significance of CDK4 and CDK6, we suggest that CDK4/6 inhibitors and innovative strategic combinatorial therapy may enhance patients' quality of life and overall survival (Laroche-Clary et al. 2017). A VHL-based PROTAC with double action against CDK4 and CDK6 was created by Steinebach et al. (2020). It demonstrated strong and sustained degrading potential in both mouse and human cells and suppressed the growth of many myeloma, leukaemia, and breast cancer cell lines. This appealing strategy for CDK4/6 targeted degradation could be investigated in PCa in more detail. In conclusion, the time during which transcription factors are "undruggable" comes to an end with the start of oral PROTAC clinical trials that target transcription factors. Future clinical applications have a lot of potential; thanks to the creation of these innovative and fascinating clinical studies. Table 11.2 lists the fundamental characteristics of the compounds discussed above.

11.6 Conclusion and Future Prospective

Genetic changes in the AR gene have been found in a sizeable fraction of prostate cancer patients, and AR plays a critical role in the onset and spread of PC. Different versions of the AR protein may arise from these genetic alterations, which may impact the protein's functionality and response to therapy. Because there is a wide range of AR variants, it can be challenging to create medications that specifically target the receptor because no single treatment may be able to suppress every variation. Targeting all AR abnormalities with small-molecule therapies for PC presents a significant hurdle. However, targeted protein breakdown has consistently shown this theory to be false. Twenty two years have passed since the targeted protein degradation strategy—which uses PROTACs to induce protein breakdown—was introduced (Hughes et al. 2021). Genetic changes in the AR gene have been found in a sizeable fraction of prostate cancer patients, and AR plays a critical part in the onset and spread of PC. Different versions of the AR protein may arise from these genetic alterations, which may impact the protein's functionality and response to therapy. Because there is a wide range of AR variants, it can be challenging to create medications that specifically target the receptor because no sole treatment might be able to suppress every variation. Targeting all AR abnormalities with small-molecule therapies for PC presents a significant hurdle. However, targeted protein breakdown has consistently shown this theory to be false. Numerous clinical trials have been initiated, demonstrating the clinical translation of PROTACs and their potential for treatment. In addition, a wide range of PROTACs that specifically target PC has been reported in the literature, targeting both AR and non-AR proteins. The efficacious degradative capacity of specific PROTACs towards these AR variants underscores their promise as a viable approach to surmount management resistance in PC and other ailments. Notably, non-AR proteins also have a large role in the pathogenesis of PC, accounting for 60% of the disease, even though most PROTACs produced for PC have focused on AR. Non-AR proteins are important, but no PROTACs have been created to target them particularly. Small-molecule inhibitors have a hard time targeting these non-AR proteins since they frequently display a wide range of genetic abnormalities.

We outlined the genetic changes in non-AR proteins in the review article that is now available, and we talked about how using PROTACs to target these changes could be a useful approach to treating PC because they have the potential to circumvent resistance mechanisms linked to more established targeted therapies. Furthermore, an interesting field of study is the application of PROTACs as a precision medicine strategy for PC management. With the ability to specifically destroy particular disease-causing proteins, PROTACs may provide a more individualised and focused approach to treating prostate cancer (PC) in each patient. Better patient outcomes could result from this therapeutic approach by increasing treatment efficacy and specificity. The utilisation of PROTACs may also result in the discovery of previously unidentified novel therapeutic targets. Preclinical and clinical research on PROTACs' potential as a precision medicine strategy for treating PC is necessary. Additional research into the growth of ternary complex co-crystal structures using silico methods, the identification of supplementary E3 ligase ligands, the optimisation of synthetic processes, and the improvement of PROTAC pharmacokinetic properties present challenges for the arena of PROTAC synthesis and design. Enhancing the ligands' binding affinity and specificity as well as learning more about the PROTAC mechanism of action are also necessary.

Moreover, in silico methods like PRosettaC, Rosetta, PROTAC-Model, and others can help improve the efficiency of PROTAC generation and rationalise PROTAC-mediated ternary complex synthesis. In the arena of drug discovery, using these skills for the design, development, and optimisation of PROTACs may show to be a hopeful and expedient approach. Furthermore, new approaches that enhance degradation based on proximity, including molecular glue, ATTEC, AUTAC, LYTAC, and RIBOTAC are expanding the possibilities for drug progress and might have beneficial benefits (Dey and Jaffrey 2019). Research on the application of artificial intelligence to target protein degraders (PROTACs) is a field that is expanding quickly and has a lot of promise. The development of highly targeted and effective PROTACs can be sped up by utilising AI approaches like molecular dynamics simulations, computer-aided drug design, and machine learning. Deep learning algorithms were used in a recent work to find new E3 ubiquitin ligase-based PROTACs. The results showed enhanced binding affinity and specificity over old-fashioned designs (Li et al. 2022). A machine learning technique was used to optimise the linker component of PROTACs, improving specificity and stability (Bemis et al. 2021). These results show that AI can greatly improve PROTAC design and development efficiency and success rate. To fully benefit from AI and subsidise the development of beneficial solutions for a variety of ailments, more research in this area is essential. AI is anticipated to become more and more significant in the design and progress of PROTACs as its application in medicinal chemistry grows. The application of AI to PROTAC research has the potential to completely change how we go about finding and developing drugs, resulting in quicker and more efficient treatments for a range of illnesses.

Acknowledgement The authors would like to acknowledge Dr. A. P. Pawar, Principal, Poona College of Pharmacy, Bharati Vidyapeeth (Deemed to be) University, Pune for providing helpful article writing suggestions.

Author's Contribution All authors contributed to the manuscript. DM has collected the relevant articles. DM and SM were responsible for the literature search, drafting and critically revising of the manuscript. All the authors have read and approved the manuscript carefully.

Funding No funding was received for this research work.

Conflict of Interest Statement The authors declare that they have no conflict of interest.

References

Abbasi A, Movahedpour A, Amiri A, Najaf MS, Mostafavi-Pour Z (2021) Darolutamide as a second-generation androgen receptor inhibitor in the treatment of prostate cancer. Curr Mol Med 21(4):332–346

Asghar U, Witkiewicz AK, Turner NC, Knudsen ES (2015) The history and future of targeting cyclin-dependent kinases in cancer therapy. Nat Rev Drug Discov 14:130–146

Bemis TA, La Clair JJ, Burkart MD (2021) Unraveling the role of linker design in proteolysis targeting chimeras: miniperspective. J Med Chem 64(12):8042–8052

Bernassola F, Karin M, Ciechanover A, Melino G (2008) The HECT family of E3 ubiquitin ligases: multiple players in cancer development. Cancer Cell 14(1):10–21

Buckley DL, Raina K, Darricarrere N, Hines J, Gustafson JL, Smith IE, Miah AH, Harling JD, Crews CM (2015) HaloPROTACS: use of small molecule PROTACs to induce degradation of HaloTag fusion proteins. ACS Chem Biol 10:1831–1837

Campos C, Sotomayor P, Jerez D, González J, Schmidt CB, Schmidt K, Banzer W, Godoy AS (2018) Exercise and prostate cancer: from basic science to clinical applications. Prostate 78(9):639–645

Cao C, He M, Wang L, He Y, Rao Y (2022) Chemistries of bifunctional PROTAC degraders. Chem Soc Rev 51(16):7066–7114

Chau CH, Steeg PS, Figg WD (2019) Antibody–drug conjugates for cancer. Lancet 394(10200):793–804

Chen S, Weitemier AZ, Zeng X, He L, Wang X, Tao Y, Huang AJ, Hashimotodani Y, Kano M, Iwasaki H, Parajuli LK (2018) Near-infrared deep brain stimulation via upconversion nanoparticle-mediated optogenetics. Science (New York, NY) 359:679–684

Chen L, Han L, Mao S, Xu P, Xu X, Zhao R, Wu Z, Zhong K, Yu G, Wang X (2021) Discovery of A031 as effective proteolysis targeting chimera (PROTAC) androgen receptor (AR) degrader for the treatment of prostate cancer. Eur J Med Chem 216:113307

Ciechanover A (2015) The unravelling of the ubiquitin system. Nat Rev Mol Cell Biol 16(5):322–324

Coats S, Williams M, Kebble B, Dixit R, Tseng L, Yao NS, Tice DA, Soria JC (2019) Antibody–drug conjugates: future directions in clinical and translational strategies to improve the therapeutic index. Clin Cancer Res 25(18):5441–5448

Culig Z, Santer FR (2014) Androgen receptor signaling in prostate cancer. Cancer Metastasis Rev 33:413–427

Cyrus K, Wehenkel M, Choi EY, Lee H, Swanson H, Kim KB (2010) Jostling for position: optimizing linker location in the design of estrogen receptor-targeting PROTACs. ChemMedChem 5:979–985

Deshaies RJ, Joazeiro CA (2009) RING domain E3 ubiquitin ligases. Annu Rev Biochem 78:399–434

Dey SK, Jaffrey SR (2019) RIBOTACs: small molecules target RNA for degradation. Cell Chem Biol 26:1047–1049

Ferguson FM, Gray NS (2018) Kinase inhibitors: the road ahead. Nat Rev Drug Discov 17(5):353–377

Fischer PM (2017) Approved and experimental small-molecule oncology kinase inhibitor drugs: a mid-2016 overview. Med Res Rev 37(2):314–367

Gockel LM, Pfeifer V, Baltes F, Bachmaier RD, Wagner KG, Bendas G, Gütschow M, Sosič I, Steinebach C (2022) Design, synthesis, and characterization of PROTACs targeting the androgen receptor in prostate and lung cancer models. Arch Pharm 355(5):2100467

Grasso CS, Wu YM, Robinson DR, Cao X, Dhanasekaran SM, Khan AP, Quist MJ, Jing X, Lonigro RJ, Brenner JC, Asangani IA (2012) The mutational landscape of lethal castration-resistant prostate cancer. Nature 487:239–243

Guenette RG, Yang SW, Min J, Pei B, Potts PR (2022) Target and tissue selectivity of PROTAC degraders. Chem Soc Rev 51(14):5740–5756

Guo J, Liu J, Wei W (2019) Degrading proteins in animals: "PROTAC"tion goes in vivo. Cell Res 29(3):179–180

Gustafson JL, Neklesa TK, Cox CS, Roth AG, Buckley DL, Tae HS, Sundberg TB, Stagg DB, Hines J, McDonnell DP, Norris JD (2015) Small-molecule-mediated degradation of the androgen receptor through hydrophobic tagging. Angew Chem 54:9659–9662

Han X, Wang C, Qin C, Xiang W, Fernandez-Salas E, Yang CY, Wang M, Zhao L, Xu T, Chinnaswamy K, Delproposto J (2019a) Discovery of ARD-69 as a highly potent proteolysis targeting chimera (PROTAC) degrader of androgen receptor (AR) for the treatment of prostate cancer. J Med Chem 62(2):941–964

Han X, Zhao L, Xiang W, Qin C, Miao B, Xu T, Wang M, Yang CY, Chinnaswamy K, Stuckey J, Wang S (2019b) Discovery of highly potent and efficient PROTAC degraders of androgen receptor (AR) by employing weak binding affinity VHL E3 ligase ligands. J Med Chem 62(24):11218–11231

Han X, Zhao L, Xiang W, Qin C, Miao B, McEachern D, Wang Y, Metwally H, Wang L, Matvekas A, Wen B (2021) Strategies toward discovery of potent and orally bioavailable proteolysis

targeting chimera degraders of androgen receptor for the treatment of prostate cancer. J Med Chem 64(17):12831–12854
Hines J, Gough JD, Corson TW, Crews CM (2013) Posttranslational protein knockdown coupled to receptor tyrosine kinase activation with phosphoPROTACs. Proc Natl Acad Sci U S A 110:8942–8947
Hu R, Wang WL, Yang YY, Hu XT, Wang QW, Zuo WQ, Xu Y, Feng Q, Wang NY (2022) Identification of a selective BRD4 PROTAC with potent antiproliferative effects in AR-positive prostate cancer based on a dual BET/PLK1 inhibitor. Eur J Med Chem 227:113922
Hughes SJ, Testa A, Thompson N, Churcher I (2021) The rise and rise of protein degradation: opportunities and challenges ahead. Drug Discov Today 26:2889–2897
Hussain M, Fizazi K, Saad F, Rathenborg P, Shore N, Ferreira U, Ivashchenko P, Demirhan E, Modelska K, Phung D, Krivoshik A (2018) Enzalutamide in men with nonmetastatic, castration-resistant prostate cancer. N Engl J Med 378:2465–2674
Ishida T, Ciulli A (2021) E3 ligase ligands for PROTACs: how they were found and how to discover new ones. SLAS Discov 26(4):484–502
Kahn B, Collazo J, Kyprianou N (2014) Androgen receptor as a driver of therapeutic resistance in advanced prostate cancer. Int J Biol Sci 10:588–595
Kase AM, Copland Iii JA, Tan W (2020) Novel therapeutic strategies for CDK4/6 inhibitors in metastatic castrate-resistant prostate cancer. Onco Targets Ther 13:10499–10513
Kim GY, Song CW, Yang YS, Lee NR, Yoo HS, Son SH, Lee SJ, Park JS, Lee JK, Inn KS, Kim NJ (2021) Chemical degradation of androgen receptor (AR) using bicalutamide analog–thalidomide PROTACs. Molecules 26(9):2525
Kirby M, Hirst C, Crawford ED (2011) Characterizing the castration-resistant prostate cancer population: a systematic review. Int J Clin Pract 65:1180–1192
Kregel S, Wang C, Han X, Xiao L, Fernandez-Salas E, Bawa P, McCollum BL, Wilder-Romans K, Apel IJ, Cao X, Speers C (2020) Androgen receptor degraders overcome common resistance mechanisms developed during prostate cancer treatment. Neoplasia 22(2):111–119
Laroche-Clary A, Chaire V, Algeo MP, Derieppe MA, Loarer FL, Italiano A (2017) Combined targeting of MDM2 and CDK4 is synergistic in dedifferentiated liposarcomas. J Hematol Oncol 10(1):123
Lee GT, Nagaya N, Desantis J, Madura K, Sabaawy HE, Kim WJ, Vaz RJ, Cruciani G, Kim IY (2021) Effects of MTX-23, a novel PROTAC of androgen receptor splice variant-7 and androgen receptor, on CRPC resistant to second-line antiandrogen therapy. Mol Cancer Ther 20(3):490–499
Li W, Bengtson MH, Ulbrich A, Matsuda A, Reddy VA, Orth A, Chanda SK, Batalov S, Joazeiro CA (2008) Genome-wide and functional annotation of human E3 ubiquitin ligases identifies MULAN, a mitochondrial E3 that regulates the organelle's dynamics and signaling. PLoS One 3(1):e1487
Li F, Hu Q, Zhang X, Sun R, Liu Z, Wu S, Tian S, Ma X, Dai Z, Yang X, Gao S (2022) DeepPROTACs is a deep learning-based targeted degradation predictor for PROTACs. Nat Commun 13:7133
Liang JJ, Xie H, Yang RH, Wang N, Zheng ZJ, Zhou C, Wang YL, Wang ZJ, Liu HM, Shan LH, Ke Y (2021) Designed, synthesized and biological evaluation of proteolysis targeting chimeras (PROTACs) as AR degraders for prostate cancer treatment. Bioorg Med Chem 45:116331
Liu J, Chen H, Ma L, He Z, Wang D, Liu Y, Lin Q, Zhang T, Gray N, Kaniskan HÜ, Jin J (2020) Light-induced control of protein destruction by opto-PROTAC. Sci Adv 6(8):eaay5154
Lu J, Qian Y, Altieri M, Dong H, Wang J, Raina K, Hines J, Winkler JD, Crew AP, Coleman K, Crews CM (2015) Hijacking the E3 ubiquitin ligase cereblon to efficiently target BRD4. Chem Biol 22:755–763
Ma B, Fan Y, Zhang D, Wei Y, Jian Y, Liu D, Wang Z, Gao Y, Ma J, Chen Y, Xu S (2022) De novo design of an androgen receptor DNA binding domain-targeted peptide PROTAC for prostate cancer therapy. Adv Sci 9(28):2201859
Matyskiela ME, Lu G, Ito T, Pagarigan B, Lu CC, Miller K, Fang W, Wang NY, Nguyen D, Houston J, Carmel G (2016) A novel cereblon modulator recruits GSPT1 to the CRL4 (CRBN) ubiquitin ligase. Nature 535:252–257

Munoz E, Chen G, Hossain A, Wu S, Nava EO, Hang J, Lee T, Zhang Q, Wang G, Chen QH (2022) Synthesis and biological evaluation of niclosamide PROTACs. Bioorg Med Chem Lett 72:128870

Naro Y, Darrah K, Deiters A (2020) Optical control of small molecule-induced protein degradation. J Am Chem Soc 142(5):2193–2197

Neklesa TK, Winkler JD, Crews CM (2017) Targeted protein degradation by PROTACs. Pharmacol Ther 174:138–144

Ohashi K, Maruvka YE, Michor F, Pao W (2013) Epidermal growth factor receptor tyrosine kinase inhibitor-resistant disease. J Clin Oncol 31(8):1070–1080

Ohoka N, Nagai K, Hattori T, Okuhira K, Shibata N, Cho N, Naito M (2014) Cancer cell death induced by novel small molecules degrading the TACC3 protein via the ubiquitin-proteasome pathway. Cell Death Dis 5:e1513

Ottis P, Palladino C, Thienger P, Britschgi A, Heichinger C, Berrera M, Julien-Laferriere A, Roudnicky F, Kam-Thong T, Bischoff JR, Martoglio B (2018) Cellullar resistance mechanisms to targeted protein degradation converge toward impairment of the engaged ubiquitin transfer pathway. Science 359:679–684

Pagliarulo V, Bracarda S, Eisenberger MA, Mottet N, Schröder FH, Sternberg CN, Studer UE (2012) Contemporary role of androgen deprivation therapy for prostate cancer. Eur Urol 61:11–25

Paiva SL, Crews CM (2019) Targeted protein degradation: elements of PROTAC design. Curr Opin Chem Biol 50:111–119

Pettersson M, Crews CM (2019) PROteolysis TArgeting chimeras (PROTACs)—past, present and future. Drug Discov Today Technol 31:15–27

Pfaff P, Samarasinghe KT, Crews CM, Carreira EM (2019) Reversible spatiotemporal control of induced protein degradation by bistable PhotoPROTACs. ACS Cent Sci 5(10):1682–1690

Raina K, Lu J, Qian Y, Altieri M, Gordon D, Rossi AM, Wang J, Chen X, Dong H, Siu K, Winkler JD (2016) PROTAC-induced BET protein degradation as a therapy for castration-resistant prostate cancer. Proc Natl Acad Sci 113(26):7124–7149

Roskoski R Jr (2019) Properties of FDA-approved small molecule protein kinase inhibitors. Pharmacol Res 144:19–50

Sackton KL, Dimova N, Zeng X, Tian W, Zhang M, Sackton TB, Meaders J, Pfaff KL, Sigoillot F, Yu H, Luo X (2014) Synergistic blockade of mitotic exit by two chemical inhibitors of the APC/C. Nature 514(7524):646–649

Sakamoto KM, Kim KB, Kumagai A, Mercurio F, Crews CM, Deshaies RJ (2001) Protacs: chimeric molecules that target proteins to the Skp1–Cullin–F box complex for ubiquitination and degradation. Proc Natl Acad Sci 98(15):8554–8559

Sakamoto KM, Kim KB, Verma R, Ransick A, Stein B, Crews CM, Deshaies RJ (2003) Development of Protacs to target cancer-promoting proteins for ubiquitination and degradation. Mol Cell Proteomics 2:1350–1308

Salami J, Alabi S, Willard RR, Vitale NJ, Wang J, Dong H, Jin M, McDonnell DP, Crew AP, Neklesa TK, Crews CM (2018a) Androgen receptor degradation by the proteolysis-targeting chimera ARCC-4 outperforms enzalutamide in cellular models of prostate cancer drug resistance. Commun Biol 1(1):1–9

Salami J, Alabi S, Willard RR, Vitale NJ, Wang J, Dong H, Jin M, McDonnell DP, Crew AP, Neklesa TK, Crews CM (2018b) Androgen receptor degradation by the proteolysis-targeting chimera ARCC-4 outperforms enzalutamide in cellular models of prostate cancer drug resistance. Commun Biol 1:100

Schapira M, Calabrese MF, Bullock AN, Crews CM (2019) Targeted protein degradation: expanding the toolbox. Nat Rev Drug Discov 18:949–963

Schneekloth JS Jr, Fonseca FN, Koldobskiy M, Mandal A, Deshaies R, Sakamoto K, Crews CM (2004) Chemical genetic control of protein levels: selective in vivo targeted degradation. J Am Chem Soc 126(12):3748–3754

Schneekloth AR, Pucheault M, Tae HS, Crews CM (2008) Targeted intracellular protein degradation induced by a small molecule: en route to chemical proteomics. Bioorg Med Chem Lett 18(22):5904–5908

Scott LJ (2018) Enzalutamide: a review in castration-resistant prostate cancer. Drugs 78:1913–1924
Scudellari M (2019) Protein-slaying drugs could be the next blockbuster therapies. Nature 567:298–300
Sekine K, Takubo K, Kikuchi R, Nishimoto M, Kitagawa M, Abe F, Nishikawa K, Tsuruo T, Naito M (2008) Small molecules destabilize cIAP1 by activating auto-ubiquitylation. J Biol Chem 283:8961–8969
Steinebach C, Ng YLD, Sosič I, Lee CS, Chen S, Lindner S, Vu LP, Bricelj A, Haschemi R, Monschke M, Steinwarz E (2020) Systematic exploration of different E3 ubiquitin ligases: an approach towards potent and selective CDK6 degraders. Chem Sci 11:3474–3478
Takwale AD, Jo SH, Jeon YU, Kim HS, Shin CH, Lee HK, Ahn S, Lee CO, Du Ha J, Kim JH, Hwang JY (2020) Design and characterization of cereblon-mediated androgen receptor proteolysis-targeting chimeras. Eur J Med Chem 208:112769
Testa A, Lucas X, Castro GV, Chan KH, Wright JE, Runcie AC, Gadd MS, Harrison WT, Ko EJ, Fletcher D, Ciulli A (2018) 3-Fluoro-4-hydroxyprolines: synthesis, conformational analysis, and stereoselective recognition by the VHL E3 ubiquitin ligase for targeted protein degradation. J Am Chem Soc 140(29):9299–9313
Troup RI, Fallan C, Baud MG (2020) Current strategies for the design of PROTAC linkers: a critical review. Explor Target Antitumor Ther 1(5):273–312
Wang Z, Liu P, Inuzuka H, Wei W (2014) Roles of F-box proteins in cancer. Nat Rev Cancer 14(4):233–247
Winston JT, Strack P, Beer-Romero P, Chu CY, Elledge SJ, Harper JW (1999) The SCFβ-TRCP–ubiquitin ligase complex associates specifically with phosphorylated destruction motifs in IκBα and β-catenin and stimulates IκBα ubiquitination in vitro. Genes Dev 13(3):270–283
Winter GE, Buckley DL, Paulk J, Roberts JM, Souza A, Dhe-Paganon S, Bradner JE (2015) DRUG DEVELOPMENT. Phthalimide conjugation as a strategy for in vivo target protein degradation. Science 348(6241):1376–1381
Wolska-Washer A, Robak T (2019) Safety and tolerability of antibody-drug conjugates in cancer. Drug Saf 42(2):295–314
Xia L, Liu W, Song Y, Zhu H, Duan Y (2019) The present and future of novel protein degradation technology. Curr Top Med Chem 19:1784–1788
Xiang W, Zhao L, Han X, Qin C, Miao B, McEachern D, Wang Y, Metwally H, Kirchhoff PD, Wang L, Matvekas A (2021) Discovery of ARD-2585 as an exceptionally potent and orally active PROTAC degrader of androgen receptor for the treatment of advanced prostate cancer. J Med Chem 64(18):13487–13509
Xie H, Liang JJ, Wang YL, Hu TX, Wang JY, Yang RH, Yan JK, Zhang QR, Xu X, Liu HM, Ke Y (2020) The design, synthesis and anti-tumor mechanism study of new androgen receptor degrader. Eur J Med Chem 204:112512
Xue G, Wang K, Zhou D, Zhong H, Pan Z (2019) Light-induced protein degradation with photocaged PROTACs. J Am Chem Soc 141(46):18370–18374
Zengerle M, Chan KH, Ciulli A (2015) Selective small molecule induced degradation of the BET bromodomain protein BRD4. ACS Chem Biol 10:1770–1777
Zhao L, Han X, Lu J, McEachern D, Wang S (2020) A highly potent PROTAC androgen receptor (AR) degrader ARD-61 effectively inhibits AR-positive breast cancer cell growth in vitro and tumor growth in vivo. Neoplasia 22:522–532
Zhou F, Chen L, Cao C, Yu J, Luo X, Zhou P, Zhao L, Du W, Cheng J, Xie Y, Chen Y (2020) Development of selective mono or dual PROTAC degrader probe of CDK isoforms. Eur J Med Chem 187:111952

Exploring the Role of PROTACs for the Treatment of Breast Cancer

12

Arzoo Pannu, Ramesh K. Goyal, and Shikha Goswami

Abstract

The emergence of PROTACs as a transformative class of therapeutic agents has prompted an intense investigation into their potential role in breast cancer. This chapter consolidates current knowledge on the application of PROTACs in breast cancer treatment. The primary objective is to elucidate the mechanisms by which PROTACs operate, emphasizing their unique ability to induce targeted protein degradation. Through a systematic review of the literature, this article identifies key proteins and pathways implicated in breast cancer that serve as potential targets for PROTACs. Clinical trials investigating the efficacy of PROTACs in breast cancer are summarized, providing insights into the current status of their development. The chapter also discusses challenges associated with PROTAC therapy in breast cancer and proposes future research directions. In conclusion, this comprehensive review underscores the promising role of PROTACs in reshaping breast cancer treatment paradigms and highlights the critical areas for further investigation in this rapidly evolving field.

Keywords

Breast cancer · PROTACs · Clinical trials · Breast cancer gene · Mechanism · Possible targets

A. Pannu · R. K. Goyal · S. Goswami (✉)
Department of Pharmacology, Delhi Pharmaceutical Sciences and Research University, New Delhi, Delhi, India

M. Nandave, P. Jain (eds.), *PROTAC-Mediated Protein Degradation: A Paradigm Shift in Cancer Therapeutics*, https://doi.org/10.1007/978-981-97-5077-1_12

Abbreviation

AKT	Protein kinase B
BET	Bromodomain
BRCA	Breast cancer gene
BUB1	Budding uninhibited by benzimidazole 1
CDKs	Cyclin-dependent kinases
CHK	Checkpoint kinases
EGFR	Epidermal growth factor receptor
ER	Estrogen receptor
FAK	Focal adhesion kinase
HDACs	Histone deacetylases
HER2/ErbB2	Human epidermal growth factor receptor 2
HIF-1α	Hypoxia-inducible factor 1-alpha
HIPK2	Homeodomain-interacting protein kinase 2
IDH	Isocitrate dehydrogenases
IKBα	Inhibitor of nuclear factor kappa B
JAK-STAT	Janus kinase-signal transducer and activator of transcription
MAPK	Mitogen-activated protein kinase
MCL-1	Myeloid cell leukemia 1
MET	Mesenchymal-epithelial transition factor receptor
MetAP-2	Methionine aminopeptidase 2
mTOR	Mammalian target of rapamycin
NAE	NEDD8-activating enzyme
NRF2	Nuclear factor erythroid 2-related factor 2
PAKs	p21-activated kinases
PIM	Proviral integration sites for Moloney murine leukemia virus
PK	Protein kinase
POI	Protein of interest
PPARγ	Peroxisome proliferator-activated receptor gamma
PPPI3K	Phosphoinositide 3-kinase
PR	Progesterone receptors
PROTACs	Proteolysis targeting chimeras
PTP1B	Protein tyrosine phosphatase 1B
RAS	Ras/Raf/MAPK pathway
SCF	Skp1-cullin-F-box
β-TRCP	Beta-transducin repeats-containing proteins

12.1 Introduction

Breast cancer, a prevalent and heterogeneous disease, continues to pose substantial challenges in the realm of oncology. Despite significant advances in diagnostic techniques and therapeutic strategies, the complexity of breast cancer demands constant innovation. Current treatment modalities, while effective in many cases, face limitations such as drug resistance, adverse effects, and the inability to comprehensively address the intricate molecular landscape of the disease. The burden of breast cancer is not only clinical but also economic and societal, underscoring the critical need for novel therapeutic approaches (Harbeck et al. 2019). As we navigate the evolving landscape of breast cancer research and treatment, the quest for innovative strategies becomes imperative. This introduction serves to illuminate the persisting challenges within the existing treatment paradigm and sets the stage for exploring a groundbreaking avenue: the role of PROTACs in breast cancer therapy.

PROTACs represent a cutting-edge frontier in drug development, offering a revolutionary approach to modulating cellular protein levels. The idea of PROTAC was initially introduced in 2001, featuring the inaugural degrader as a peptide. Subsequently, in 2008, the first collection of small-molecule degraders was reported (Sakamoto et al. 2003; Schneekloth et al. 2004). Over the following decade (2008–2018), numerous non-orally available PROTAC degraders were identified, demonstrating robust biological activity both in vitro and in vivo (Nabet et al. 2020). Unlike traditional small-molecule inhibitors that merely block protein activity, PROTACs leverage the ubiquitin-proteasome system to induce the targeted degradation of specific proteins. This unique mechanism of action holds the promise of addressing challenges associated with drug resistance, enhancing selectivity, and expanding the druggable target space. PROTACs operate by recruiting the cellular machinery responsible for protein degradation, resulting in the ubiquitin-mediated destruction of the target protein (Schneekloth et al. 2004). This approach not only addresses challenges associated with drug resistance but also holds the potential to impact undruggable targets, broadening the scope of therapeutic possibilities. The adaptability of PROTACs to target a diverse array of proteins positions them as a versatile tool in the oncologist's armamentarium.

The purpose of this chapter is to comprehensively examine the evolving role of PROTACs in the context of breast cancer treatment. By synthesizing current literature, we aim to elucidate the molecular mechanisms underpinning PROTAC function and explore their potential advantages in the intricate landscape of breast cancer. Through a critical analysis of preclinical and clinical studies, we seek to provide insights into the current status of PROTAC research, identify challenges, and propose future directions for harnessing the full therapeutic potential of PROTACs in breast cancer management. This review serves as a guide for researchers, clinicians, and pharmaceutical developers navigating the promising intersection of PROTACs and breast cancer therapeutics.

12.2 Structure and Mechanisms of PROTACs

PROTAC, functioning as a heterobifunctional molecule, exhibits a tripartite structure composed of three integral components, as illustrated in Fig. 12.1, a ligand designed for the target protein, a ligand tailored for the E3 ubiquitin ligase, and a connecting linker. This linker facilitates the formation of a "three-body" polymer known as the target protein ligand-linker-E3 ligand. The mechanism of action

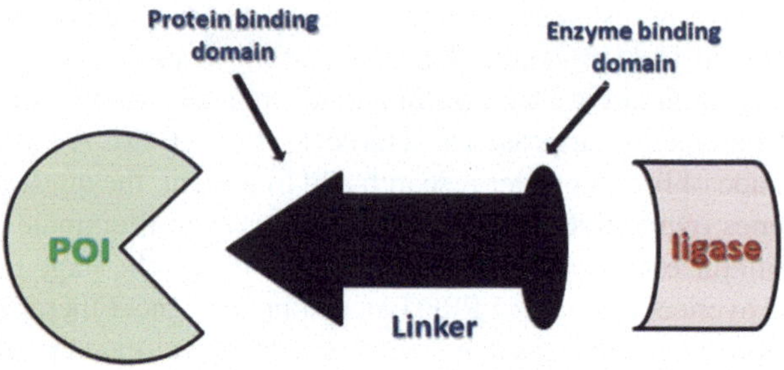

Fig. 12.1 Basic structure of PROTAC

involves harnessing the ubiquitin-proteasome pathway to achieve targeted protein degradation. There are four major steps in PROTAC-induced degradation. Initially, PROTAC binds to both the POI and the E3 ligase, forming a ternary complex, i.e., POI-PROTAC-E3 ligase. Subsequently, the E3 ligase recruits the E2 ligase, facilitating the transfer of ubiquitin from the E2 to lysine residues on the POI. Thirdly, multiple rounds of ubiquitination occur, resulting in the formation of polyubiquitin chains on the POI's surface. Finally, the ternary complex dissociates, and the polyubiquitinated POI undergoes degradation through the 26S proteasome. Following dissociation, PROTAC becomes available for a new cycle of target protein degradation (Lai et al. 2016).

In 2001, Sakamoto et al. pioneered the concept of PROTACs, introducing the first successful design and synthesis of a bifunctional PROTAC molecule for the targeted degradation of MetAP-2 (Sakamoto et al. 2003). In its early stages, PROTAC molecules primarily comprised peptides. The inaugural PROTAC-1 molecule, designed to target MetAP-2, featured a modular composition that included the angiogenesis inhibitor ovalicin and a segment binding to the E3 ubiquitin ligase complex SCF. This module also incorporated a phosphopeptide derived from the IKBα protein, specifically recognized by β-TRCP within the SCF complex. Experimental results demonstrated the ability of PROTAC-1 to recruit both MetAP-2 and SCFβ-TRCP (He et al. 2022). Since this seminal work, the field of PROTAC technology has progressed significantly. Various PROTACs have been designed for diverse proteins, including members of the BET protein family, EGFR, and aperiodic CDKs (Sakamoto et al. 2003; Sun et al. 2018; Zhang et al. 2020). Many of these proteins play crucial roles in the growth and metastasis of malignant tumors, representing existing or potential therapeutic targets. Despite the development of alternative methods, such as clustered regularly interspaced short palindromic repeats, siRNA, small-molecule inhibitors, and monoclonal antibodies for regulating the expression or function of tumor-related proteins, PROTAC stands out due to its advantages, including a broad spectrum of targets, rapid action, and high specificity for selectively degrading pathogenic variant proteins while preserving the functions of normal proteins (Prozzillo et al. 2020; Zhao and Burgess 2019). The mechanism of PROTACs is characterized by a precise and orchestrated process, offering a unique approach to modulating protein levels within cells.

12.2.1 Recognition and Binding (Li and Song 2020; Li et al. 2022; Qi et al. 2021; Zhao et al. 2022)

12.2.1.1 Ligand Design for Target Protein Recognition

The initial step in the mechanism of PROTACs involves the recognition and binding of the PROTAC molecule to the target protein. This process is driven by the specific design of the ligand that constitutes one moiety of the bifunctional PROTAC. The ligand is meticulously crafted to recognize and bind selectively to a particular epitope on the surface of the target protein.

12.2.1.2 High Affinity and Specificity

The ligand's design is guided by principles of high affinity and specificity, ensuring a strong and selective interaction with the target protein. Computational methods, structural biology techniques such as X-ray crystallography or nuclear magnetic resonance, and iterative medicinal chemistry optimization play crucial roles in tailoring ligands for optimal binding. The goal is to achieve a tight association that withstands cellular conditions and fluctuations.

12.2.1.3 Epitope Accessibility

The success of PROTAC recognition and binding relies on the accessibility of the chosen epitope on the target protein. The ligand must be designed to interact with a region that is sufficiently exposed and accessible in the three-dimensional structure of the protein. This consideration ensures that the ligand can effectively engage the target, facilitating the formation of a stable complex.

12.2.1.4 Rational Selection of Target Proteins

The selection of target proteins is a critical aspect of the recognition and binding step. Target proteins are often chosen based on their relevance to disease biology, their involvement in key signaling pathways, and their potential as therapeutic targets. Rational target selection enhances the clinical significance of PROTAC-mediated protein degradation.

12.2.1.5 Cellular Permeability

In addition to specificity and affinity, ligand design must account for cellular permeability. The ligand needs to traverse cell membranes to reach the target protein within the cellular context. Medicinal chemistry approaches aim to optimize the physicochemical properties of the ligand to ensure effective cellular penetration.

12.2.1.6 Dynamic Nature of Binding

The recognition and binding process is dynamic, and the interaction between the ligand and the target protein is not static. Conformational changes and adaptability are inherent in the binding dynamics, allowing the ligand to adjust to the target protein's structural nuances and conformations.

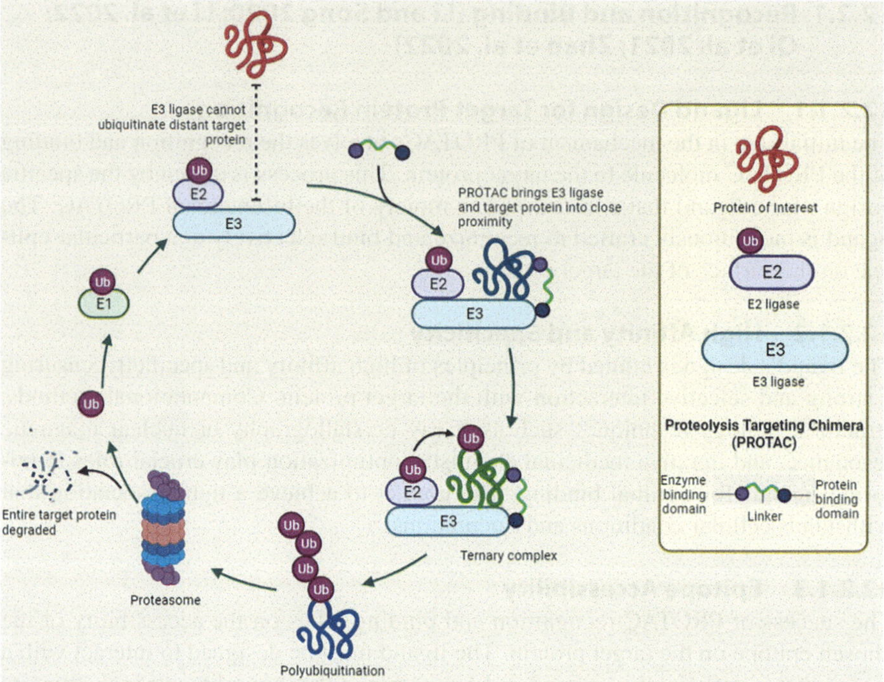

Fig. 12.2 Demonstrating the PROTAC mechanism: It involves the use of a ligand designed to bind target. This ligand is connected to another ligand, which binds to the ubiquitin ligase enzyme, through a short chemical linker. This arrangement results in the formation of a ternary complex consisting of the POI, PROTAC, and E3 ligase. Subsequently, this complex facilitates the tagging of the POI with ubiquitin groups. The polyubiquitination signal generated is then recognized by the proteasome, initiating the degradation of the targeted protein. Notably, the PROTAC molecule can be recycled for subsequent degradation cycles

12.2.2 Binding to Target Protein (Li and Song 2020; Li et al. 2022; Qi et al. 2021; Zhao et al. 2022)

The efficacy of PROTACs hinges on their ability to form specific interactions with the target POI. The molecular interaction begins with the binding of one moiety of the bifunctional PROTAC to the target protein. This ligand is carefully designed to recognize and bind to a unique epitope on the surface of the target protein, ensuring high affinity and specificity (Fig. 12.2).

12.2.2.1 Engaging the E3 Ubiquitin Ligase

Simultaneously, the other moiety of the PROTAC is designed to engage an E3 ubiquitin ligase. The choice of E3 ligase is critical, as it determines the specificity of the PROTAC-induced ubiquitination process. Different E3 ligases have distinct substrate preferences, and the rational selection of an E3 ligase enhances the precision of PROTAC-mediated protein degradation.

12.2.2.2 Ternary Complex Formation

The power of PROTACs lies in their ability to bring the target protein and the E3 ligase into proximity, initiating the formation of a ternary complex. This complex is a dynamic and transient association, facilitated by the simultaneous binding of the PROTAC to both the target protein and the E3 ligase. The ternary complex formation is a pivotal step that sets the stage for the subsequent ubiquitination and degradation process.

12.2.2.3 Optimization of Ligand Design

The success of molecular interactions in PROTACs relies heavily on the rational design of ligands. Ligands are carefully optimized to achieve a delicate balance of affinity, specificity, and cellular permeability. Computational approaches, structural biology, and medicinal chemistry play crucial roles in tailoring ligands to ensure optimal binding to both the target protein and the E3 ligase.

12.2.2.4 Importance of Proximity

The concept of induced proximity is central to the molecular interactions orchestrated by PROTACs. By physically bringing the target protein and the E3 ligase close together, PROTACs exploit the cellular machinery responsible for ubiquitination and subsequent degradation. This induced proximity enhances the efficiency and specificity of the degradation process, providing a unique and targeted strategy for modulating protein levels within cells.

12.2.3 Ubiquitin-Mediated Degradation (Li and Song 2020; Li et al. 2022; Qi et al. 2021; Zhao et al. 2022)

12.2.3.1 Ubiquitination Process Initiation

The central feature of PROTACs is their ability to induce targeted protein degradation through the ubiquitin-proteasome system. After the successful recognition and binding of the PROTAC to the target protein, the next crucial step is the initiation of the ubiquitination process.

12.2.3.2 E3 Ubiquitin Ligase Catalysis

The ubiquitination process is catalyzed by the E3 ubiquitin ligase, the other moiety of the bifunctional PROTAC. This ligase serves as the orchestrator of ubiquitin transfer from the ubiquitin-conjugating enzyme (E2) to the lysine residues of the target protein. The choice of E3 ligase is pivotal, as different E3 ligases have distinct substrate preferences, contributing to the specificity of the degradation process.

12.2.3.3 Polyubiquitin Chain Formation

The transfer of ubiquitin molecules onto the lysine residues of the target protein results in the formation of a polyubiquitin chain. This chain serves as a recognition signal for the cellular proteasome, a large proteolytic complex responsible for degrading tagged proteins. The polyubiquitin chain functions as a molecular tag, marking the target protein for proteasomal degradation.

12.2.3.4 Recognition by the Proteasome

The polyubiquitinated target protein is then recognized by the 26S proteasome, a large multisubunit complex. The proteasome unfolds the ubiquitinated protein and translocates it into its catalytic core, where the protein is degraded into smaller peptides.

12.2.3.5 Generation of Peptides

Within the proteasome, the ubiquitinated target protein is subjected to proteolysis, leading to the generation of small peptides. These peptides are then released into the cytoplasm and can be further processed for antigen presentation or recycled for the synthesis of new proteins.

12.2.3.6 Recycling of PROTAC Components

Crucially, PROTACs exhibit a unique property of recyclability. After the ubiquitin-mediated degradation of the target protein, the PROTAC molecule is released. Both ligands of the PROTAC become available for subsequent rounds of target engagement and degradation, contributing to the sustainability and efficiency of the overall process.

12.2.3.7 Importance of Timing and Dynamics

The timing and dynamics of the ubiquitin-mediated degradation process are tightly regulated within the cell. The controlled and reversible nature of this process ensures that cellular homeostasis is maintained, and unwanted degradation events are minimized.

12.2.3.8 Precision and Selectivity

One of the distinctive features of PROTACs is their ability to induce targeted protein degradation with precision and selectivity. Unlike traditional small-molecule inhibitors that may lack specificity, PROTACs leverage the cellular machinery to precisely target proteins for degradation. This minimizes off-target effects, reducing the risk of adverse events associated with broader protein inhibition.

12.2.4 Recycling and Reusability (Li and Song 2020; Li et al. 2022; Qi et al. 2021; Zhao et al. 2022)

12.2.4.1 Releasing PROTAC After Degradation

One of the distinctive features of PROTACs is their inherent recyclability. After successfully inducing the ubiquitin-mediated degradation of the target protein, the PROTAC molecule is released from the ternary complex. This release marks the completion of the degradation process and allows for the reutilization of the PROTAC components in subsequent rounds of target engagement.

12.2.4.2 Liberation of Ligands

Upon release, both ligands of the bifunctional PROTAC become available for recycling. The ligand initially binding to the target protein is freed, and the ligand engaging the E3 ubiquitin ligase is also released. This liberation of ligands is a critical aspect of the recyclability of PROTACs and contributes to their sustained and efficient functionality.

12.2.4.3 Ligand Reusability in New Ternary Complex Formation

The liberated ligands retain their original specificity and affinity for their respective targets—target protein and E3 ligase. This enables the reutilization of these ligands in the formation of new ternary complexes with additional molecules of the target protein. The ligands can engage in multiple rounds of target binding, ternary complex formation, and subsequent ubiquitin-mediated degradation.

12.2.4.4 Sustained Therapeutic Impact

The recyclability of PROTACs contributes to their sustained therapeutic impact. As opposed to irreversible inhibitors that may require continuous administration to maintain their effects, PROTACs can achieve prolonged and efficient target degradation through multiple cycles of binding and release. This characteristic aligns with the principles of sustainable and adaptable therapeutic interventions.

12.2.4.5 Enhanced Efficiency and Cost-Effectiveness

The ability of PROTACs to recycle their components enhances their overall efficiency and cost-effectiveness. By reusing the ligands, the need for a continuous supply of new PROTAC molecules is reduced. This property has implications for the feasibility and practicality of PROTAC-based therapies, potentially offering advantages in terms of manufacturing and therapeutic dosing.

12.2.4.6 Biological Implications

The recycling and reusability of PROTACs reflect a dynamic and adaptable nature that aligns with cellular processes. This feature also mirrors the ability of cells to regulate protein levels continuously. The recycling process ensures that PROTACs can respond to the dynamic changes within the cellular environment and maintain their efficacy over time.

12.2.4.7 Considerations for Therapeutic Development

While the recyclability of PROTACs presents a promising feature, considerations for therapeutic development include potential challenges such as off-target effects and unintended degradation events. Careful optimization of ligands, dosing regimens, and monitoring of treatment response are crucial to harnessing the full potential of PROTACs in a therapeutic context.

12.3 Versatility in Targeting

The versatility of PROTACs lies in their modular design, allowing for the targeted degradation of a diverse array of proteins. This modular structure consists of two distinct ligands—one designed to selectively bind to the target protein, and the other tailored to engage an E3 ubiquitin ligase. The modularity of PROTACs provides flexibility in customizing ligands to address a wide spectrum of target proteins (Li et al. 2022).

12.3.1 Overcoming Limitations of Traditional Inhibitors (Burslem and Crews 2020)

Traditional small-molecule inhibitors often face challenges in targeting proteins deemed "undruggable" due to factors such as lack of binding pockets or undesirable pharmacokinetic properties. PROTACs, with their ability to induce degradation, overcome these limitations. By leveraging the cellular ubiquitin-proteasome system, PROTACs can target proteins that are traditionally difficult to modulate with conventional inhibitors.

12.3.2 Addressing Challenging Targets (Bondeson et al. 2015)

The versatility of PROTACs extends to the targeting of proteins considered challenging in the context of cancer and other diseases. This includes proteins involved in transcriptional regulation, epigenetic control, and protein-protein interactions, which are often critical drivers of disease progression. The ability of PROTACs to induce targeted degradation broadens the druggable target space, offering new opportunities for therapeutic intervention.

12.3.3 Adaptable to Diverse Cellular Pathways (Ottis and Crews 2017)

PROTACs can be designed to target proteins involved in various cellular pathways, including signaling cascades, cell cycle regulation, and DNA repair mechanisms. This adaptability allows researchers and drug developers to tailor PROTACs to address specific disease contexts, making them valuable tools in the pursuit of precision medicine.

12.3.4 Expanding Therapeutic Potential (Zorba et al. 2018)

The versatility in targeting provided by PROTACs has the potential to revolutionize therapeutic strategies. By expanding the range of druggable targets, PROTACs offer opportunities for more precise and personalized treatment approaches. This is particularly significant in the context of diseases with complex and heterogeneous molecular profiles, such as cancer.

12.3.5 Rational Design for Specificity (Raina and Crews 2017)

While PROTACs are versatile, their success relies on rational design for specificity. The careful selection of ligands for the target protein and the E3 ligase ensures that the induced degradation is focused on the intended target. This specificity minimizes off-target effects and enhances the therapeutic precision of PROTACs.

12.3.6 Implications for Drug Discovery (Buhimschi et al. 2018)

The versatility of PROTACs in targeting a wide range of proteins has profound implications for drug discovery. Researchers can explore novel targets and pathways, paving the way for the development of innovative therapies. The modular design of PROTACs also offers a platform for exploring combination therapies, where multiple targets can be simultaneously addressed for synergistic effects.

In summary, the versatility of PROTACs in targeting diverse proteins positions them as powerful tools in drug discovery and therapeutic development. This flexibility opens new avenues for addressing challenging targets and advancing precision medicine approaches.

12.4 Role in Breast Cancer

Breast cancer's heterogeneity, driven by a diverse array of molecular pathways and genetic alterations, necessitates innovative therapeutic strategies. The distinctive mechanism of PROTACs holds significant relevance to breast cancer due to its capacity to address specific molecular features inherent to the disease. In many breast cancers, the overexpression of hormone receptors, such as ER and PR, plays a pivotal role in driving tumor growth (Zhao et al. 2022). PROTACs offer a targeted approach to degrade these hormone receptors, disrupting signaling pathways crucial for cancer progression (Zeng et al. 2020; Zhao et al. 2022) by circumventing traditional inhibitory strategies, PROTACs may mitigate the emergence of drug-resistant phenotypes commonly observed in endocrine therapies. PROTACs exhibit promise in overcoming resistance mechanisms, a persistent challenge in breast cancer treatment (Burslem and Crews 2020). Through the degradation of specific proteins implicated in resistance, such as members of the HER2 or PI3K pathways, PROTACs may re-sensitize tumors to existing therapies. This adaptability is particularly relevant in addressing the evolving landscape of resistance encountered in advanced breast cancer stages (Burslem and Crews 2020; Zhao and Jensen 2009).

A hallmark of breast cancer is the presence of proteins deemed "undruggable" using traditional pharmacological approaches. PROTACs, with their ability to induce the degradation of proteins rather than inhibiting their activity, present a novel strategy to target these challenging proteins (Qi et al. 2021). This expands the repertoire of potential therapeutic targets, offering avenues for intervention in pathways previously considered inaccessible. Additionally, breast cancer cells often exploit specific oncogenic dependencies for survival and proliferation. By precisely targeting these dependencies, PROTACs have the potential to induce a more profound and sustained impact on cancer cells compared to traditional therapies (Maniaci et al. 2017). This approach aligns with the evolving understanding of cancer biology and the importance of personalized and targeted interventions. Moreover, the specificity of PROTACs in targeting specific proteins for degradation reduces the likelihood of off-target effects, a concern associated with traditional chemotherapeutic agents. This selectivity is particularly advantageous in the context of breast cancer, where minimizing collateral damage to healthy tissues is crucial for mitigating side effects and enhancing the overall safety profile of therapeutic interventions (Churcher 2018).

In short, the unique mechanism of PROTACs aligns with the intricate molecular landscape of breast cancer. Their ability to target hormone receptors, overcome resistance mechanisms, address undruggable proteins, minimize off-target effects, and exploit oncogenic dependencies positions PROTACs at the forefront of precision medicine in breast cancer therapeutics. As we delve into the specific molecular pathways involved, the potential of PROTACs to revolutionize treatment strategies becomes increasingly evident, marking a paradigm shift in the pursuit of more effective and tailored breast cancer therapies.

12.5 Possible Targets in Breast Cancer: Identification and Exploration for PROTAC Targeting

Breast cancer, a complex and heterogeneous disease, encompasses various subtypes characterized by distinct molecular profiles. The molecular diversity of these subtypes underscores the need for precise, tailored therapeutic interventions. Identifying specific proteins and signaling pathways within breast cancer that contribute to tumorigenesis becomes paramount for the development of targeted therapies, with PROTACs emerging as a promising modality. In the pursuit of suitable targets for PROTAC-mediated degradation, a thorough examination of the breast cancer proteome is imperative. Numerous proteins and signaling cascades have been implicated in the initiation, progression, and maintenance of breast cancer (Raina and Crews 2017; Sakamoto et al. 2003) (Figs. 12.3 and 12.4). Among these, receptors such as the ER, HER2, and members of the PI3K-AKT-mTOR pathway stand out as pivotal players.

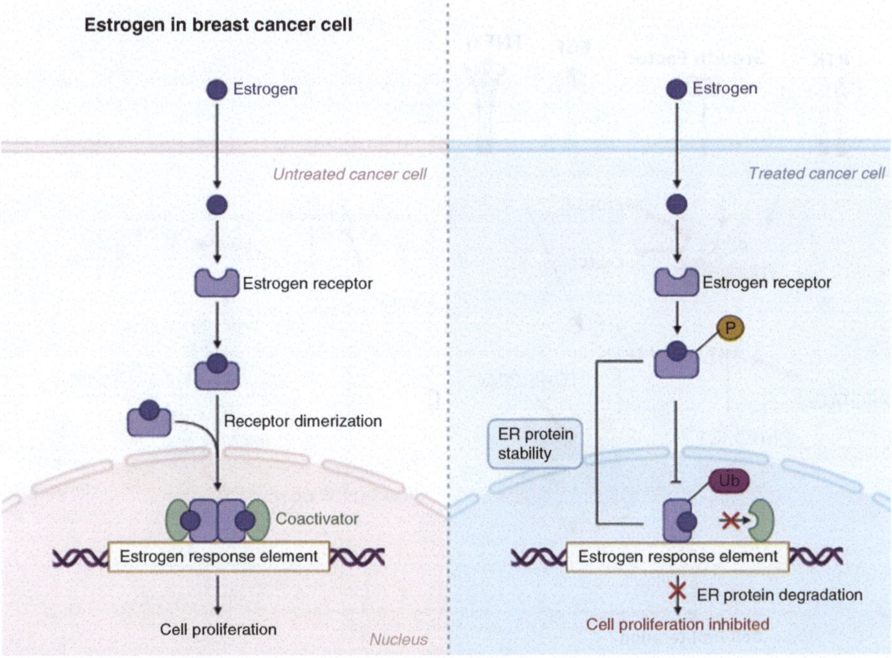

Fig. 12.3 PROTAC targets various pathways in cancer: in breast tumor growth inhibition and apoptosis promotion, HER2, PPPI3K, and PROTAC-AKT operate through the classic PI3K/AKT signaling pathway. Heterobifunctional peptide-PROTAC facilitates ER degradation, suppressing the transcription of downstream target genes and exerting a lethal effect on ER-positive breast cancer. Suppression of E-cadherin expression, release of β-catenin, and its translocation to the nucleus occur, actions antagonized by PROTACX-protein

12.5.1 ER: A Classical Target (Finn et al. 2009)

The estrogen receptor, a transcription factor responsive to estrogenic stimuli, plays a central role in hormone receptor-positive breast cancers. PROTAC-mediated degradation of ER presents a compelling strategy to abrogate estrogen signaling, a cornerstone in the pathogenesis of hormone-sensitive breast cancers. By employing PROTACs designed to engage ER and recruit E3 ligases, targeted ER degradation becomes a potential avenue for disrupting the growth and survival of hormone receptor-positive breast cancer cells.

12.5.2 HER2: Targeting Amplification in Aggressive Subtypes (Cromm and Crews 2017)

Amplification of the HER2 represents a defining feature in a subset of breast cancers, leading to heightened proliferative signaling. PROTACs designed to induce the degradation of HER2 could offer a precise means to attenuate this oncogenic

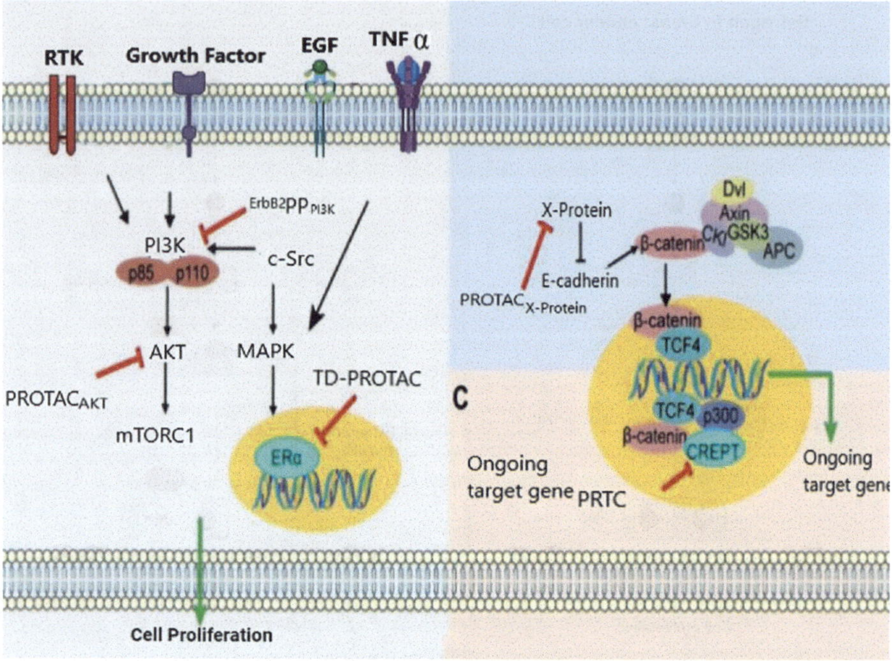

Fig. 12.4 PROTAC targets various pathways in cancer: in breast tumor growth inhibition and apoptosis promotion, HER2, PPPI3K, and PROTAC-AKT operate through the classic PI3K/AKT signaling pathway. Suppression of E-cadherin expression, release of β-catenin, and its translocation to the nucleus occur, actions antagonized by PROTACX-protein. PRTC effectively inhibits cell proliferation and motility by blocking transcription initiation in the MAPK signaling pathway, along with suppressing downstream target gene cyclin D1, leading to cycle arrest and inducing DNA damage

signaling axis. Through the targeted removal of HER2, PROTACs may impede the aberrant cellular processes driven by HER2 amplification, providing a tailored approach for treating HER2-positive breast cancers.

12.5.3 PI3K-AKT-mTOR Pathway: Intersecting Nodes of Intricacy (Cromm and Crews 2017)

The PI3K-AKT-mTOR pathway, a signaling nexus frequently dysregulated in breast cancer, orchestrates crucial cellular functions related to survival, proliferation, and metabolism. Identifying specific components within this pathway as PROTAC targets holds promise for disrupting the intricate network of signaling events contributing to breast cancer progression. PROTACs designed to degrade key proteins in the PI3K-AKT-mTOR pathway could intervene at multiple nodes, exerting a comprehensive effect on the aberrant signaling cascades associated with malignant transformation.

12.5.4 Beyond Conventional Targets: Exploring Novel Candidates (Burslem and Crews 2020)

In addition to well-established targets, ongoing exploration aims to uncover novel proteins and pathways specific to certain breast cancer subtypes. These may include targets associated with therapy resistance, metastasis, or immune evasion. PROTACs, with their adaptable design, offer a flexible platform to address emerging targets, providing a dynamic and evolving strategy in the pursuit of precision breast cancer therapy.

12.5.5 BRCA1 and BRCA2: Precision Targeting in Hereditary Breast Cancers (Lord and Ashworth 2017)

In the context of hereditary breast cancers, mutations in the BRCA1 and BRCA2 genes represent critical contributors. These mutations impair DNA repair mechanisms, leading to genomic instability and an increased predisposition to breast cancer. PROTACs designed to selectively degrade mutant BRCA proteins could potentially capitalize on synthetic lethality, disrupting the survival advantage conferred by these mutations. By targeting BRCA1 and BRCA2 for degradation, PROTACs may offer a unique avenue for therapeutic intervention in hereditary breast cancers.

12.5.6 CDKs: Orchestrating Cell Cycle Regulation (Finn et al. 2009; Malumbres and Barbacid 2009)

Dysregulated cell cycle progression is a hallmark of cancer, and targeting key regulators of this process, such as CDKs, holds promise in breast cancer therapy. PROTACs designed to degrade specific CDKs, especially those associated with the G1-S transition, could arrest cell cycle progression, inducing cell cycle exit or apoptosis in breast cancer cells. This approach may be particularly relevant in hormone receptor-positive breast cancers, where CDK4/6 inhibitors have shown efficacy.

12.5.7 Immune Checkpoint Proteins: Enhancing Immunotherapy Strategies (Pardoll 2012; Ribas and Wolchok 2018)

The intersection between PROTACs and immunotherapy in breast cancer extends to immune checkpoint proteins. In the tumor microenvironment, proteins like programmed cell death protein 1 and programmed death ligand 1 play crucial roles in evading immune surveillance. PROTACs targeting these immune checkpoint proteins may enhance the immune response against breast cancer cells, potentially synergizing with existing immunotherapeutic strategies and overcoming resistance mechanisms.

12.5.8 Epigenetic Regulators: Unraveling Chromatin Dynamics (Dawson and Kouzarides 2012; West and Johnstone 2014)

Epigenetic dysregulation is a pervasive feature in breast cancer, influencing gene expression patterns and cellular phenotypes. PROTACs targeting epigenetic regulators, such as histone deacetylases or DNA methyltransferases, could induce selective degradation, leading to alterations in chromatin structure and gene expression. This approach may offer a nuanced strategy to modulate the epigenetic landscape of breast cancer cells, influencing their behavior and response to therapy.

12.5.9 Heat Shock Proteins: Modulating Cellular Stress Responses (Jego et al. 2013)

Heat shock proteins play crucial roles in maintaining protein homeostasis and aiding cellular stress responses. In breast cancer, elevated levels of heat shock proteins are often associated with increased aggressiveness and resistance to therapy. PROTACs designed to degrade specific heat shock proteins, such as HSP90, could disrupt the chaperoning of client proteins involved in cancer cell survival and proliferation. Targeting HSPs with PROTACs may offer a unique strategy to induce proteotoxic stress and sensitize breast cancer cells to therapeutic interventions.

12.5.10 MET Receptor: Navigating Invasive Phenotypes (Gherardi et al. 2012; Organ and Tsao 2011)

The MET receptor, implicated in promoting invasive and metastatic phenotypes, represents a potential target for PROTAC-mediated degradation in breast cancer. By selectively degrading MET, PROTACs may interfere with the signaling cascades that drive cellular migration and invasion. This approach could be particularly relevant in triple-negative breast cancers, where metastasis is a major determinant of poor prognosis.

12.5.11 Cancer Stem Cell Markers: Addressing Tumor Heterogeneity (Clevers 2011; O'Brien et al. 2007)

Cancer stem cells contribute to tumor heterogeneity and are associated with therapy resistance and recurrence. Identifying and targeting cancer stem cells markers with PROTACs may present a strategy to eliminate the subpopulation of cells driving tumor initiation and relapse. By degrading proteins associated with the stem cell phenotype, PROTACs could potentially reduce the regenerative capacity of breast cancer cells, improving the overall effectiveness of therapeutic interventions.

12.5.12 DNA Damage Response Proteins: Enhancing Radiotherapy Sensitivity (Jackson and Bartek 2009; Lord and Ashworth 2017)

Components of the DNA damage response pathway, including proteins involved in DNA repair, are critical for the survival of cancer cells exposed to genotoxic stress. PROTACs targeting DNA damage response pathway proteins could sensitize breast cancer cells to radiotherapy by impairing their ability to repair DNA damage. This synergistic approach may enhance the therapeutic efficacy of radiotherapy and overcome resistance mechanisms in certain breast cancer subtypes.

12.5.13 Neovascularization Factors: Targeting Angiogenesis (Bergers and Hanahan 2008; Ferrara et al. 2003)

Angiogenesis, the formation of new blood vessels, is essential for tumor growth and metastasis. Proteins involved in angiogenesis, such as vascular endothelial growth factor, represent potential targets for PROTAC-mediated degradation. By disrupting angiogenic signaling pathways, PROTACs could hinder the establishment of a vascular network, restricting the nutrient supply to breast tumors and impeding their growth.

12.5.14 Microenvironment Modulators: Influencing Tumor-Stromal Interactions (Kalluri 2016; Pickup et al. 2014)

The tumor microenvironment plays a pivotal role in cancer progression, influencing tumor-stromal interactions, immune responses, and therapeutic resistance. PROTACs targeting proteins involved in modulating the tumor microenvironment, such as stromal cell-derived factor 1 or transforming growth factor-beta, may alter the dynamics of the tumor niche. This approach could impact various aspects of breast cancer biology, including metastatic potential and responsiveness to immunotherapy.

12.5.15 Hormone Receptor Variants: Addressing Therapy-Resistant Phenotypes (Musgrove and Sutherland 2009)

In hormone receptor-positive breast cancers, the emergence of hormone receptor variants contributes to therapy resistance. PROTACs designed to selectively degrade these variant receptors could offer a strategy to overcome resistance mechanisms and restore sensitivity to endocrine therapies. Targeting specific hormone receptor variants with PROTACs may represent a precision medicine approach in the management of hormone receptor-positive breast cancer.

12.5.16 Autophagy Regulators: Balancing Cellular Recycling Mechanisms (Galluzzi et al. 2015; Yang et al. 2011)

Autophagy, a cellular recycling process, plays a dual role in cancer by promoting survival or inducing cell death, depending on context. Targeting autophagy regulators with PROTACs could tip the balance toward detrimental autophagy, leading to the degradation of critical pro-survival factors. This approach may selectively induce autophagic cell death in breast cancer cells, providing a novel avenue for therapeutic intervention.

12.5.17 Glycolytic Enzymes: Disrupting Metabolic Reprogramming (Liberti and Locasale 2016; Vander Heiden et al. 2009)

Metabolic reprogramming is a hallmark of cancer, and many breast cancers exhibit increased reliance on glycolysis for energy production. PROTACs targeting key glycolytic enzymes, such as hexokinase or lactate dehydrogenase, could disrupt this metabolic shift, inducing energy crises and compromising the survival of breast cancer cells. Combining metabolic targeting with PROTAC-mediated degradation presents an innovative strategy to undermine the adaptive metabolic strategies of cancer cells.

12.5.18 RNA Binding Proteins: Modulating Transcriptomic Landscapes (Gerstberger et al. 2014; Pereira et al. 2017)

RNA binding proteins play crucial roles in posttranscriptional gene regulation, influencing the expression of key genes involved in cancer progression. PROTACs designed to degrade specific RNA binding proteins could modulate the transcriptomic landscapes of breast cancer cells, influencing cell fate decisions and response to therapy. Targeting RNA binding proteins may represent a nuanced approach to intervene in the intricate gene regulatory networks governing breast cancer biology.

12.5.19 Notch Signaling Components: Governing Cell Fate Decisions (Bray 2016; Kopan and Ilagan 2009)

The Notch signaling pathway governs cell fate decisions, impacting differentiation, proliferation, and survival. Dysregulation of Notch signaling is implicated in breast cancer pathogenesis. PROTACs targeting Notch receptors or downstream signaling components could perturb this pathway, influencing cell fate decisions and disrupting the balance between self-renewal and differentiation. Targeting Notch signaling with PROTACs may hold therapeutic potential, especially in breast cancer subtypes where Notch activity is aberrantly elevated.

12.5.20 Niche-Specific Targets: Addressing Metastatic Niches (Cox and Erler 2011; Pickup et al. 2013)

Metastasis involves the colonization of distant organs by cancer cells, a process facilitated by the establishment of niche-specific interactions. PROTACs targeting proteins involved in metastatic niches, such as integrins or extracellular matrix components, could disrupt these interactions, impeding the colonization and growth of breast cancer cells in secondary sites. This approach may offer a unique strategy to prevent or treat metastatic disease, a major cause of mortality in breast cancer.

12.5.21 Proteins Involved in DNA Replication Stress: Overcoming Genomic Instability (Macheret and Halazonetis 2015; Zeman and Cimprich 2014)

DNA replication stress is a common feature in cancer, leading to genomic instability and therapeutic resistance. PROTACs targeting proteins involved in maintaining DNA replication fidelity, such as components of the replication fork, could induce replication stress and compromise the survival of breast cancer cells. This strategy may selectively sensitize cancer cells to DNA-damaging therapies and overcome resistance mechanisms associated with genomic instability.

12.5.22 Apoptotic Regulators: Inducing Programmed Cell Death (Hanahan and Weinberg 2011; Oltersdorf et al. 2005)

Dysregulation of apoptotic pathways contributes to cancer cell survival and therapy resistance. PROTACs targeting antiapoptotic proteins, such as Bcl-2 family members, could induce programmed cell death in breast cancer cells. By selectively degrading inhibitors of apoptosis, PROTACs may shift the balance toward apoptotic cell death, providing a mechanism to enhance the efficacy of existing therapeutic strategies.

12.5.23 DNA Methyltransferases: Epigenetic Modulators in Breast Cancer (Baylin and Jones 2016; Yoo and Jones 2006)

Epigenetic modifications, including DNA methylation, play a pivotal role in breast cancer development and progression. DNA methyltransferases are key enzymes responsible for adding methyl groups to DNA, leading to altered gene expression patterns. PROTACs targeting DNA methyltransferases could induce selective degradation, influencing the epigenetic landscape of breast cancer cells. This approach may offer a unique strategy to reverse aberrant DNA methylation patterns, potentially reactivating tumor-suppressive genes and restraining cancer progression.

12.5.24 RAS Signaling: Tackling an Elusive Oncogenic Pathway (Hobbs et al. 2016; Stephen et al. 2014)

The RAS signaling pathway is frequently dysregulated in various cancers, contributing to uncontrolled cell proliferation and survival. In breast cancer, RAS mutations or aberrant RAS activation are associated with aggressive phenotypes. PROTACs targeting components of the RAS pathway, such as mutant RAS proteins, may provide a means to disrupt this elusive oncogenic signaling cascade. Precision degradation of RAS proteins with PROTACs could represent a novel strategy to intervene in RAS-driven breast cancers.

12.5.25 Integrins: Disrupting Tumor-Stromal Interactions (Hamidi et al. 2016; Seguin et al. 2015)

Integrins play a crucial role in mediating interactions between cancer cells and the surrounding stromal microenvironment. Aberrant integrin signaling contributes to tumor progression, invasion, and metastasis. PROTACs targeting specific integrins could disrupt these interactions, potentially hindering the invasive potential of breast cancer cells. Precision degradation of integrins may provide a targeted approach to modulate tumor-stromal crosstalk and impede metastatic spread.

12.5.26 GTPase Proteins: Regulating Cell Migration and Invasion (Jaffe and Hall 2005; Seguin et al. 2015)

Rho family GTPases, including RhoA, Rac1, and Cdc42, play key roles in regulating cytoskeletal dynamics, cell migration, and invasion. Dysregulation of these GTPases is implicated in cancer metastasis. PROTACs designed to degrade specific GTPases could interfere with the signaling pathways governing cell motility and invasion. Targeting GTPases with PROTACs may represent a precise strategy to curb the metastatic potential of breast cancer cells.

12.5.27 Wnt Signaling Components: Balancing Cell Fate and Tumor Progression (Clevers and Nusse 2012; MacDonald et al. 2009)

The Wnt signaling pathway is integral to cell fate determination and tissue homeostasis. Dysregulated Wnt signaling is associated with breast cancer initiation and progression. PROTACs targeting components of the Wnt pathway, such as β-catenin, may disrupt aberrant signaling and modulate cell fate decisions in breast cancer cells. Precision degradation of Wnt signaling components with PROTACs could offer a strategic approach to intervene in the intricate balance between self-renewal and differentiation.

12.5.28 RNA Helicases: Modulating RNA Processing in Breast Cancer (Abdelhaleem 2005; Fuller-Pace 2006)

RNA helicases play critical roles in RNA processing and gene expression regulation. Dysregulated RNA helicase activity is associated with cancer progression, including breast cancer. PROTACs targeting specific RNA helicases could induce their degradation, influencing RNA processing and altering the expression of genes involved in breast cancer development. Precision degradation of RNA helicases with PROTACs may provide a unique avenue to modulate the transcriptomic landscape and disrupt cancer-associated gene expression patterns.

12.5.29 Hedgehog Signaling: Navigating Developmental Pathways in Breast Cancer (Jiang and Hui 2008; Katoh and Katoh 2006)

The Hedgehog signaling pathway, critical for embryonic development, is often reactivated in cancer, including breast cancer. Aberrant Hedgehog signaling is associated with tumor initiation and maintenance of cancer stem cells. PROTACs targeting components of the Hedgehog pathway could selectively degrade key proteins, potentially disrupting cancer stem cell populations and impeding breast cancer progression. Precision degradation of Hedgehog signaling components with PROTACs may represent a targeted approach to navigate developmental pathways in breast cancer cells.

12.5.30 MAPK/ERK Pathway: Navigating Proliferation and Survival Signaling (Roberts and Der 2007; Roskoski 2012)

The mitogen-activated protein kinase/extracellular signal-regulated kinase (MAPK/ERK) pathway is a crucial signaling cascade involved in cell proliferation and survival. Dysregulated MAPK/ERK signaling is a common feature in breast cancer, contributing to uncontrolled cell growth. PROTACs targeting key components of the MAPK/ERK pathway, such as RAF or MEK proteins, could selectively induce degradation, disrupting the aberrant signaling cascade and potentially inhibiting the proliferation of breast cancer cells. Precision degradation of MAPK/ERK pathway components with PROTACs may offer a targeted strategy to navigate proliferative signaling in breast cancer.

12.5.31 Aurora Kinases: Regulating Mitotic Progression (Marumoto et al. 2005; Vader and Lens 2008)

Aurora kinases are essential regulators of mitotic progression, ensuring proper chromosome segregation during cell division. Aberrant expression of Aurora kinases is associated with genomic instability in cancer, including breast cancer. PROTACs targeting specific Aurora kinase isoforms could induce their degradation, disrupting mitotic progression and potentially inducing cell death in breast cancer cells. Precision degradation of Aurora kinases with PROTACs may present a targeted approach to modulate mitotic fidelity in breast cancer.

12.5.32 c-Myc: Master Regulator of Cellular Processes (Conacci-Sorrell et al. 2014; Dang 2012)

c-Myc, a transcription factor, serves as a master regulator of various cellular processes, including proliferation, apoptosis, and metabolism. Dysregulated c-Myc expression is implicated in breast cancer pathogenesis. PROTACs designed to degrade c-Myc could influence the expression of downstream target genes, disrupting the oncogenic functions of c-Myc in breast cancer cells. Targeting c-Myc with PROTACs may provide a strategic approach to modulate the multifaceted roles of this transcription factor in cancer progression.

12.5.33 NRF2: Balancing Oxidative Stress Responses (Hayes and McMahon 2009; Rojo de la Vega et al. 2018)

NRF2 is a transcription factor that plays a crucial role in regulating oxidative stress responses and cellular detoxification. Aberrant NRF2 activation is associated with cancer cell survival and therapy resistance, including in breast cancer. PROTACs targeting NRF2 could induce its degradation, potentially disrupting the antioxidant defense mechanisms and sensitizing breast cancer cells to oxidative stress and therapeutic interventions. Precision degradation of NRF2 with PROTACs may represent a targeted strategy to modulate oxidative stress responses in breast cancer.

12.5.34 DNA-PK: Influencing DNA Damage Repair Pathways (Blackford and Jackson 2017; Davis and Chen 2013)

DNA-dependent protein kinase is a key component of the DNA double-strand break repair machinery. Dysregulated DNA repair processes contribute to genomic instability and therapy resistance in breast cancer. PROTACs targeting DNA-PK could induce its degradation, potentially impairing DNA repair mechanisms and sensitizing breast cancer cells to genotoxic stress. Precision degradation of DNA-PK with PROTACs may offer a targeted approach to modulate DNA damage repair pathways in breast cancer.

12.5.35 PIM Kinases: Orchestrating Cell Survival Signaling (Warfel et al. 2016; Zhao et al. 2022)

PIM kinases are implicated in cell survival and antiapoptotic signaling. Overexpression of PIM kinases is observed in breast cancer and is associated with aggressive phenotypes. PROTACs targeting specific PIM kinase isoforms could induce their degradation, potentially disrupting cell survival signaling and sensitizing breast cancer cells to apoptotic stimuli. Precision degradation of PIM kinases with PROTACs may present a targeted strategy to modulate survival pathways in breast cancer.

12.5.36 CHK1 and CHK2: Influencing Cell Cycle Checkpoints (Ma et al. 2011; Zachos et al. 2007)

CHK1 and CHK2 play crucial roles in regulating cell cycle checkpoints in response to DNA damage. Dysregulated CHK1 and CHK2 activity is associated with genomic instability and therapy resistance in breast cancer. PROTACs targeting CHK1 or CHK2 could induce their degradation, potentially disrupting cell cycle checkpoints and sensitizing breast cancer cells to DNA-damaging agents. Precision degradation of CHK1 and CHK2 with PROTACs may offer a targeted approach to modulate cell cycle regulation in breast cancer.

12.5.37 HIPK2: Modulating Cell Fate and DNA Damage Responses (Calzado et al. 2007; Puca et al. 2010)

HIPK2 is a kinase involved in regulating cell fate decisions, apoptosis, and DNA damage responses. Dysregulation of HIPK2 has been observed in breast cancer and is associated with altered DNA repair mechanisms. PROTACs targeting HIPK2 could induce its degradation, potentially influencing both cell fate decisions and DNA damage responses in breast cancer cells. Precision degradation of HIPK2 with PROTACs may offer a targeted strategy to modulate these interconnected processes in breast cancer.

12.5.38 FAK: Orchestrating Focal Adhesion Signaling (Canel et al. 2013; Sulzmaier et al. 2014)

FAK is a key player in focal adhesion signaling, influencing cellular processes such as migration, invasion, and survival. Overexpression and dysregulation of FAK are implicated in breast cancer progression. PROTACs targeting FAK could induce its degradation, potentially disrupting focal adhesion signaling and impairing the migratory and invasive capacities of breast cancer cells. Precision degradation of FAK with PROTACs may represent a targeted approach to modulate the dynamic interplay between cancer cells and their microenvironment.

12.5.39 BUB1: Regulating Mitotic Checkpoints (Jemaà et al. 2016)

BUB1, a mitotic checkpoint kinase, is critical for proper chromosome segregation during cell division. Aberrant BUB1 expression is associated with aneuploidy and genomic instability in breast cancer. PROTACs targeting BUB1 could induce its degradation, potentially disrupting mitotic checkpoints and inducing cell death in breast cancer cells. Precision degradation of BUB1 with PROTACs may offer a targeted strategy to modulate mitotic fidelity and genomic stability in breast cancer.

12.5.40 IDH1 and IDH2: Influencing Metabolic Reprogramming (Rohle et al. 2013; Wang et al. 2013)

IDH1 and IDH2 play key roles in cellular metabolism and are frequently mutated in various cancers, including breast cancer. Mutant IDH enzymes contribute to metabolic reprogramming and epigenetic alterations. PROTACs targeting mutant IDH1 or IDH2 could induce their degradation, potentially disrupting metabolic pathways and mitigating the oncogenic effects of these mutations in breast cancer cells.

Precision degradation of mutant IDH enzymes with PROTACs may represent a targeted approach to modulate metabolic reprogramming in breast cancer.

12.5.41 MCL-1: Regulating Apoptotic Pathways (Kotschy et al. 2016; Ramsey et al. 2018)

MCL-1, an antiapoptotic protein, is frequently overexpressed in cancer and contributes to therapy resistance. PROTACs targeting MCL-1 could induce its degradation, potentially promoting apoptotic cell death in breast cancer cells. Precision degradation of MCL-1 with PROTACs may offer a targeted strategy to overcome the antiapoptotic defenses employed by breast cancer cells, enhancing the efficacy of apoptotic-inducing therapies.

12.5.42 JAK-STAT Pathway Components: Modulating Immune Responses (O'Shea et al. 2015)

The JAK-STAT pathway plays a crucial role in regulating immune responses and inflammation. Dysregulated JAK-STAT signaling is implicated in breast cancer progression and immune evasion. PROTACs targeting components of the JAK-STAT pathway could induce their degradation, potentially modulating the immune microenvironment and sensitizing breast cancer cells to immune-mediated responses. Precision degradation of JAK-STAT pathway components with PROTACs may represent a targeted approach to enhance the effectiveness of immunotherapeutic strategies in breast cancer.

12.5.43 PPARγ: Influencing Lipid Metabolism and Tumorigenesis (Li et al. 2023; Mal et al. 2021)

PPARγ is a nuclear receptor that plays a role in lipid metabolism and cellular differentiation. Dysregulation of PPARγ is implicated in breast cancer tumorigenesis. PROTACs targeting PPARγ could induce its degradation, potentially disrupting lipid metabolism and inhibiting the growth of breast cancer cells. Precision degradation of PPARγ with PROTACs may offer a targeted strategy to modulate metabolic pathways and impede the oncogenic effects of PPARγ dysregulation in breast cancer.

12.5.44 GLI Transcription Factors: Targeting Hedgehog Signaling Effectors (Rimkus et al. 2016; Stecca and Ruizi Altaba 2010)

GLI transcription factors are downstream effectors of the Hedgehog signaling pathway and play pivotal roles in cancer progression, including breast cancer. PROTACs targeting GLI proteins could induce their degradation, potentially disrupting Hedgehog signaling and impeding breast cancer growth. Precision degradation of GLI transcription factors with PROTACs may represent a targeted strategy to modulate the downstream effects of aberrant Hedgehog pathway activation in breast cancer.

12.5.45 HDACs: Regulating Chromatin Dynamics and Gene Expression (Falkenberg and Johnstone 2014; West and Johnstone 2014)

HDACs are enzymes that modulate chromatin structure and gene expression. Dysregulated HDAC activity is implicated in breast cancer, influencing cellular processes such as proliferation and apoptosis. PROTACs targeting specific HDAC isoforms could induce their degradation, potentially altering chromatin dynamics and gene expression profiles in breast cancer cells. Precision degradation of HDACs with PROTACs may offer a targeted approach to modulate epigenetic regulation and impede the oncogenic effects of dysregulated HDAC activity.

12.5.46 PAK Proteins: Orchestrating Cytoskeletal Dynamics (Molli et al. 2009)

PAKs are involved in regulating cytoskeletal dynamics, cell motility, and invasion. Dysregulated PAK activity is associated with metastatic potential in breast cancer. PROTACs targeting specific PAK isoforms could induce their degradation, potentially disrupting cytoskeletal rearrangements and impairing the invasive capabilities of breast cancer cells. Precision degradation of PAK proteins with PROTACs may represent a targeted strategy to modulate the migratory and invasive features of breast cancer.

12.5.47 NAE: Targeting Cullin-RING Ligases (Soucy et al. 2009; Swords et al. 2010)

NAE is a key regulator of Cullin-RING ligases (CRLs), involved in ubiquitin-mediated protein degradation. Dysregulated CRL activity is implicated in cancer, including breast cancer. PROTACs targeting NAE could induce the degradation of CRL components, potentially disrupting the ubiquitin-proteasome system and influencing the stability of key proteins in breast cancer cells. Precision degradation of

NAE with PROTACs may offer a targeted approach to modulate protein turnover and impact the proteostasis of breast cancer cells.

12.5.48 PTP1B: Regulating Signaling Pathways and Cellular Responses (Julien et al. 2011; Tonks 2006)

PTP1B is involved in regulating signaling pathways by dephosphorylating key tyrosine residues. Dysregulated PTP1B activity is associated with tumorigenesis and therapy resistance in breast cancer. PROTACs targeting PTP1B could induce its degradation, potentially impacting signaling cascades and sensitizing breast cancer cells to therapeutic interventions. Precision degradation of PTP1B with PROTACs may represent a targeted strategy to modulate cellular responses and overcome resistance mechanisms in breast cancer.

12.5.49 HIF-1α: Influencing Hypoxia-Induced Responses (Keith and Simon 2007; Semenza 2012)

HIF-1α plays a central role in orchestrating cellular responses to low oxygen conditions. Dysregulated HIF-1α is implicated in breast cancer progression and metastasis. PROTACs targeting HIF-1α could induce its degradation, potentially disrupting hypoxia-induced responses and inhibiting the adaptation of breast cancer cells to low-oxygen environments. Precision degradation of HIF-1α with PROTACs may offer a targeted approach to modulate the hypoxic microenvironment and impede tumor aggressiveness in breast cancer.

Resistance to standard therapies, a recurring challenge in breast cancer treatment, often involves the upregulation or activation of specific proteins. PROTACs designed to degrade these resistance-associated proteins may represent a valuable strategy to overcome therapeutic obstacles. This includes proteins involved in resistance to endocrine therapy, chemotherapy, or targeted therapies, providing a tailored approach to sensitize resistant breast cancer cells to treatment. Moreover, advancements in functional genomics have unraveled novel vulnerabilities and dependencies in breast cancer cells. The integration of PROTAC technology with insights from functional genomics enables the exploration of uncharted territories, identifying new targets that could be selectively degraded for therapeutic benefit (Mittendorf et al. 2020; Mosele et al. 2020). This dynamic and exploratory aspect of PROTAC research aligns with the evolving understanding of breast cancer biology, presenting opportunities for innovative and personalized treatment strategies. Hence, the identification of targets for PROTAC-mediated degradation in breast cancer extends beyond conventional markers. From hereditary factors to immune evasion, cell cycle regulation, epigenetic dynamics, and therapy resistance, the diverse landscape of breast cancer offers a spectrum of proteins and pathways that could be strategically targeted by PROTACs. The exploration of additional targets for

PROTAC-mediated degradation in breast cancer unveils new dimensions of precision therapeutics (Mittendorf et al. 2020; Mosele et al. 2020). As the understanding of molecular vulnerabilities in breast cancer deepens, PROTACs stand poised at the forefront of innovation, offering a dynamic and versatile platform to develop targeted interventions that transcend the boundaries of conventional treatment modalities.

12.6 Clinical Status of PROTACs in Breast Cancer Research

The translation of promising preclinical findings into clinical applications marks a critical juncture in the evolution of PROTACs as a therapeutic modality for breast cancer. While many of these agents have undergone preclinical design and evaluation, only ARV-471 has transitioned into the clinical setting. Serving as an ER degrader, ARV-471 entered clinical trials as a monotherapy for individuals with ER^+/HER2—locally advanced or metastatic breast cancer. Phase 1 data for ARV-471 indicated well-tolerated high levels of ER degradation (89%) across various dose levels (30 mg to 700 mg). Interim data further unveiled a manageable tolerability profile, robust signals of clinical efficacy, a 42% clinical benefit rate in a heavily pretreated population, and superior ER degradation compared to fulvestrant and other clinical-stage selective ER degraders at the same developmental stage (Snyder et al. 2021). ARV-471 has progressed to phase II (VERITAC; NCT04072952) in metastatic breast cancer as a single agent. Additionally, a phase Ib study evaluating the combination of ARV-471 with the cyclin-dependent kinase 4/cyclin-dependent kinase 6 (CDK4/CDK6) inhibitor palbociclib has commenced. The collaborative development of ARV-471 is currently underway, involving Arvinas and Pfizer, aiming to address ER+/HER2− metastatic breast cancer (Snyder et al. 2021).

12.7 Challenges and Considerations

12.7.1 Current Limitations: Navigating the Complex Landscape of PROTACs in Breast Cancer Treatment

While the promise of PROTACs in reshaping breast cancer therapy is evident, the journey from bench to bedside is fraught with challenges and limitations. This section critically examines the current hurdles associated with the utilization of PROTACs in the complex terrain of breast cancer treatment. One significant challenge lies in ensuring the efficient intracellular delivery of PROTACs to their intended targets. Cellular membranes and internalization processes may impede the access of PROTAC molecules to specific proteins, affecting the overall efficacy of the therapeutic intervention (Burslem and Crews 2020). Balancing specificity and off-target effects emerges as another significant obstacle, as the precise design of PROTACs is juxtaposed with concerns about the unintended degradation of nontarget proteins and their potential impact on normal cellular functions (Olson et al.

2018). Furthermore, the dynamic nature of cancer cells poses a persistent challenge, with PROTACs encountering resistance mechanisms that necessitate ongoing research efforts for understanding and counteracting these adaptive responses (Maniaci et al. 2017). The pharmacokinetics and stability of PROTAC molecules add complexity, with issues such as short half-lives and susceptibility to degradation posing potential limitations in clinical efficacy (Bondeson et al. 2015). Biomarker identification for predicting patient response introduces an additional layer of complexity, given the heterogeneous nature of breast cancer (Xie et al. 2014). Lastly, the strategic integration of PROTACs with existing treatment modalities requires meticulous consideration, demanding a nuanced understanding of synergistic interactions to optimize efficacy and minimize potential side effects (Raina et al. 2016). As the field of PROTACs in breast cancer treatment advances, ongoing research endeavors aim to surmount these challenges, holding the promise of refining and tailoring therapeutic approaches for more effective combat against breast cancer using this innovative modality.

12.7.2 Toxicity and Side Effects: Navigating the Balance of Therapeutic Impact

The integration of PROTACs into breast cancer treatment regimens brings forth considerations related to toxicity and potential side effects. As with any therapeutic intervention, understanding and managing the impact of PROTACs on normal cellular functions is pivotal for ensuring patient safety and the overall success of the treatment strategy.

12.7.2.1 Off-Target Effects: Unintended Consequences and Cellular Consequences (Bondeson et al. 2018)

A paramount concern in the context of toxicity involves potential off-target effects leading to unintended protein degradation. While PROTACs are designed for precision, the inherent complexity of cellular signaling pathways may give rise to unanticipated consequences, affecting nontarget proteins and perturbing normal cellular functions.

12.7.2.2 Organ-Specific Toxicity: Assessing Impact on Vital Organs (Mullard 2019)

Considerations extend to the potential impact of PROTACs on vital organs. The intricate network of proteins within different tissues may respond divergently to PROTAC treatment, raising concerns about organ-specific toxicities that could compromise patient well-being.

12.7.2.3 Immunogenicity: Evaluating Immune Responses to PROTACs (Zorba et al. 2018)

The immunogenicity of PROTAC molecules adds another layer of complexity. Immune responses triggered by PROTACs may vary among individuals, potentially leading to allergic reactions or autoimmune responses. Understanding and mitigating these immunological considerations is crucial for ensuring the safety of patients undergoing PROTAC-based therapies.

12.7.2.4 Cumulative Toxicity: Examining Long-Term Effects and Treatment Durations (Cromm et al. 2018)

As PROTACs are considered for longer-term treatment regimens, assessing cumulative toxicity becomes imperative. Prolonged exposure may reveal latent side effects or cumulative toxicities that were not apparent in shorter-duration studies, emphasizing the importance of vigilant monitoring and long-term safety assessments.

12.7.2.5 Dose-Dependent Effects: Balancing Efficacy and Tolerability (Ottis et al. 2017)

The relationship between dosage and toxicity is a delicate balance that requires meticulous consideration. Determining the optimal therapeutic dose that achieves maximal efficacy while maintaining an acceptable safety profile is critical for the successful clinical implementation of PROTACs in breast cancer treatment.

12.7.2.6 Mitigation Strategies: Proactive Approaches to Minimize Adverse Events (Lai et al. 2016)

Proactive mitigation strategies form an integral component of addressing toxicity and side effects associated with PROTACs. These may include refining drug formulations, optimizing dosing schedules, or developing co-administration strategies to enhance therapeutic efficacy while minimizing adverse events.

Navigating the complexities of toxicity and side effects represents an ongoing challenge in the development of PROTACs for breast cancer therapy. Rigorous preclinical and clinical assessments, coupled with the implementation of strategic mitigation approaches, aim to strike a balance between achieving therapeutic impact and safeguarding patient well-being in the pursuit of innovative cancer treatments.

12.8 Future Perspectives

12.8.1 Emerging Strategies: Pioneering the Next Frontier in PROTAC Development for Breast Cancer

The landscape of PROTACs in breast cancer treatment is dynamic, continuously evolving with advancements in scientific understanding and innovative technological approaches. This section delves into emerging strategies that hold promise for reshaping the future of PROTAC development for breast cancer.

12.8.1.1 Enhanced Targeting Specificity: Precision Beyond Conventional Boundaries (Gadd et al. 2017)

The refinement of PROTAC design to achieve enhanced targeting specificity represents a pivotal avenue for future exploration. Innovations in molecular engineering, such as the incorporation of advanced recognition motifs or sophisticated delivery systems, aim to further minimize off-target effects and elevate the precision of PROTAC-mediated protein degradation.

12.8.1.2 Dual Targeting PROTACs: Addressing Molecular Redundancy (Zeng et al. 2020)

Emerging strategies embrace the concept of dual targeting PROTACs, designed to simultaneously engage multiple proteins within interconnected signaling pathways. This approach is particularly relevant in addressing molecular redundancy, a phenomenon often encountered in breast cancer, where alternative pathways can compensate for the loss of a single target.

12.8.1.3 PROTAC Combinations: Synergistic Integration with Other Therapies (Raina et al. 2016)

Combining PROTACs with existing therapeutic modalities holds considerable potential. Synergistic interactions may be harnessed through strategic combinations with chemotherapy, immunotherapy, or other targeted agents, offering a multifaceted approach to combatting breast cancer and potentially overcoming resistance mechanisms.

12.8.1.4 Advanced Delivery Systems: Navigating Intracellular Barriers (Burslem et al. 2018)

Innovative delivery systems are emerging as critical components of PROTAC development. Nanoparticle-based carriers, cell-penetrating peptides, or other advanced delivery vehicles aim to enhance the intracellular delivery of PROTACs, overcoming barriers and ensuring effective engagement with target proteins.

12.8.1.5 PROTACs for Undruggable Targets: Expanding Therapeutic Horizons (Bondeson et al. 2015)

Efforts are intensifying to extend the application of PROTACs to traditionally "undruggable" targets in breast cancer. By leveraging the unique mechanism of PROTAC-induced degradation, researchers seek to unlock new possibilities for intervention, addressing previously uncharted facets of the disease.

12.8.1.6 Patient-Specific PROTACs: Toward Personalized Treatment Approaches (Churcher 2018)

Advancements in molecular profiling and patient stratification may pave the way for patient-specific PROTAC development. Tailoring PROTAC interventions based on the unique molecular landscape of individual breast cancer cases holds promise for a more personalized and effective therapeutic paradigm.

12.8.2 Translational Considerations: Bridging Bench to Bedside (Manasanch and Orlowski 2017)

As these emerging strategies unfold, careful consideration of translational aspects becomes paramount. Rigorous preclinical validation, predictive biomarker identification, and innovative clinical trial designs are integral to ensuring the seamless translation of these strategies from bench to bedside, ultimately benefiting breast cancer patients.

In the ever-evolving realm of PROTAC development for breast cancer, these emerging strategies exemplify the spirit of innovation and the relentless pursuit of more effective, targeted, and personalized therapeutic interventions. The future holds exciting possibilities as researchers and clinicians collaborate to unlock the full potential of PROTACs in the fight against breast cancer.

12.8.2.1 Potential Combinations: Synergizing PROTACs with Existing Therapeutic Modalities

The integration of PROTACs into breast cancer treatment strategies opens avenues for synergistic combinations with existing modalities. This section explores the potential synergies and strategic integration of PROTACs with other therapeutic approaches, presenting a holistic vision for enhanced breast cancer management.

12.8.2.2 Chemotherapy Combinations: Fortifying Cytotoxic Impact (Zeng et al. 2020)

Combining PROTACs with traditional chemotherapeutic agents represents a compelling strategy. The synergistic interaction between targeted protein degradation and the cytotoxic effects of chemotherapy may lead to enhanced anti-tumor efficacy, potentially overcoming resistance mechanisms and broadening the therapeutic window.

12.8.2.3 Immunotherapy Synergies: Harnessing Immune Modulation (Bondeson et al. 2015)

The immunomodulatory potential of PROTACs aligns with the principles of immunotherapy. Strategic combinations could potentiate immune responses, fostering a more robust anti-cancer milieu. PROTACs may be integrated with immune checkpoint inhibitors or other immunotherapeutic agents to amplify the immune system's role in eliminating breast cancer cells.

12.8.2.4 Hormone Therapy Integration: Tailoring PROTAC Approaches to Receptor Status (Bardia et al. 2019)

In the context of hormone receptor-positive breast cancer, combining PROTACs with hormone therapies offers a tailored approach. Dual targeting of hormone receptors alongside specific protein degradation may address both the proliferative and survival pathways, presenting a comprehensive strategy for improved therapeutic outcomes.

12.8.2.5 Targeted Therapy Collaborations: Addressing Molecular Subtypes (Mullard 2019)

Strategic collaborations with existing targeted therapies provide an opportunity to address specific molecular subtypes of breast cancer. PROTACs may be combined with agents targeting specific pathways or oncogenic mutations, offering a nuanced approach to combat the diverse molecular landscapes observed in breast cancer.

12.8.2.6 Radiotherapy Enhancement: Augmenting Localized Treatment Impact (Prozzillo et al. 2020)

Exploring combinations with radiotherapy aims to augment localized treatment impact. The precision of PROTACs in degrading specific proteins may complement the spatial accuracy of radiotherapy, potentially enhancing the efficacy of localized breast cancer treatments.

12.8.2.7 Personalized Approaches: Tailoring Combinations to Patient Profiles (Coleman and Crews 2018)

The era of personalized medicine invites the consideration of patient-specific combinations. Integration of PROTACs into tailored therapeutic regimens, guided by individual molecular profiles and treatment histories, holds promise for optimizing treatment outcomes while minimizing adverse effects.

12.8.2.8 Overcoming Resistance Mechanisms: Dynamic Adaptations for Prolonged Efficacy (Raina et al. 2016)

Combining PROTACs with various modalities addresses the dynamic nature of resistance mechanisms. The multifaceted impact of these combinations may hinder the emergence of resistance, prolonging the efficacy of treatment strategies over time.

As we navigate the potential combinations of PROTACs with existing therapeutic modalities, a nuanced understanding of the molecular intricacies of breast cancer is paramount. Collaborative efforts between researchers, clinicians, and industry partners are essential to translate these theoretical synergies into clinically impactful strategies, bringing us closer to a more comprehensive and effective approach to breast cancer treatment.

12.9 Conclusion

In conclusion, the exploration of PROTACs in the context of breast cancer reveals a transformative potential in the landscape of cancer therapy. From their precision in targeted protein degradation demonstrated across diverse breast cancer subtypes in preclinical models to the ongoing clinical trials offering glimpses into their real-world efficacy, PROTACs emerge as a promising and innovative intervention. The ability of PROTACs to redefine "undruggable" targets, expand therapeutic horizons, and synergize with existing modalities positions them as a multifaceted tool in the oncologist's arsenal. Moreover, the prospect of patient-centric strategies, tailoring interventions based on individual molecular profiles, offers a vision of more effective and personalized breast cancer therapy. As we stand at the cusp of this promising frontier, the implications are profound, heralding a future where the precision and versatility of PROTACs could significantly impact the way we understand, approach, and treat breast cancer.

References

Abdelhaleem M (2005) RNA helicases: regulators of differentiation. Clin Biochem 38(6):499–503. https://doi.org/10.1016/j.clinbiochem.2005.01.010

Bardia A, Mayer IA, Vahdat LT, Tolaney SM, Isakoff SJ, Diamond JR, Kalinsky K (2019) Sacituzumab govitecan-hziy in refractory metastatic triple-negative breast cancer. N Engl J Med 380(8):741–751. https://doi.org/10.1056/NEJMoa1814213

Baylin SB, Jones PA (2016) Epigenetic determinants of cancer. Cold Spring Harb Perspect Biol 8(9):19505. https://doi.org/10.1101/cshperspect.a019505

Bergers G, Hanahan D (2008) Modes of resistance to anti-angiogenic therapy. Nat Rev Cancer 8(8):592–603. https://doi.org/10.1038/nrc2442

Blackford AN, Jackson SP (2017) ATM, ATR, and DNA-PK: the trinity at the heart of the DNA damage response. Mol Cell 66(6):801–817. https://doi.org/10.1016/j.molcel.2017.05.015

Bondeson DP, Mares A, Smith IE, Ko E, Campos S, Miah AH, Crews CM (2015) Catalytic in vivo protein knockdown by small-molecule PROTACs. Nat Chem Biol 11(8):611–617. https://doi.org/10.1038/nchembio.1858

Bondeson DP, Smith BE, Burslem GM, Buhimschi AD, Hines J, Jaime-Figueroa S, Crews CM (2018) Lessons in PROTAC design from selective degradation with a promiscuous warhead. Cell Chem Biol 25(1):78–87. https://doi.org/10.1016/j.chembiol.2017.09.010

Bray SJ (2016) Notch signaling in context. Nat Rev Mol Cell Biol 17(11):722–735. https://doi.org/10.1038/nrm.2016.94

Buhimschi AD, Armstrong HA, Toure M, Jaime-Figueroa S, Chen TL, Lehman AM, Crews CM (2018) Targeting the C481S ibrutinib-resistance mutation in Bruton's tyrosine kinase using PROTAC-mediated degradation. Biochemistry 57(26):3564–3575

Burslem GM, Crews CM (2020) Proteolysis-targeting chimeras as therapeutics and tools for biological discovery. Cell 181(1):102–114

Burslem GM, Smith BE, Lai AC, Jaime-Figueroa S, McQuaid DC, Bondeson DP, Crews CM (2018) The advantages of targeted protein degradation over inhibition: an RTK case study. Cell Chem Biol 25(1):67–77. https://doi.org/10.1016/j.chembiol.2017.09.009

Calzado MA, Renner F, Roscic A, Schmitz ML (2007) HIPK2: a versatile switchboard regulating the transcription machinery and cell death. Cell Cycle 6(2):139–143. https://doi.org/10.4161/cc.6.2.3788

Canel M, Serrels A, Frame MC, Brunton VG (2013) E-cadherin-integrin crosstalk in cancer invasion and metastasis. J Cell Sci 126(Pt 2):393–401. https://doi.org/10.1242/jcs.100115

Churcher I (2018) Protac-induced protein degradation in drug discovery: breaking the rules or just making new ones? J Med Chem 61(2):444–452

Clevers H (2011) The cancer stem cell: premises, promises and challenges. Nat Med 17(3):313–319. https://doi.org/10.1038/nm.2304

Clevers H, Nusse R (2012) Wnt/β-catenin signaling and disease. Cell 149(6):1192–1205. https://doi.org/10.1016/j.cell.2012.05.012

Coleman KG, Crews CM (2018) Proteolysis-targeting chimeras: harnessing the ubiquitin-proteasome system to induce degradation of specific target proteins. Ann Rev Cancer Biol 2:41–58

Conacci-Sorrell M, McFerrin L, Eisenman RN (2014) An overview of MYC and its interactome. Cold Spring Harb Perspect Med 4(1):a014357. https://doi.org/10.1101/cshperspect.a014357

Cox TR, Erler JT (2011) Remodeling and homeostasis of the extracellular matrix: implications for fibrotic diseases and cancer. Dis Model Mech 4(2):165–178. https://doi.org/10.1242/dmm.004077

Cromm PM, Crews CM (2017) Targeted protein degradation: from chemical biology to drug discovery. Cell Chem Biol 24(9):1181–1190. https://doi.org/10.1016/j.chembiol.2017.05.024

Cromm PM, Samarasinghe KTG, Hines J, Crews CM (2018) Addressing kinase-independent functions of Fak via PROTAC-mediated degradation. J Am Chem Soc 140(49):17019–17026. https://doi.org/10.1021/jacs.8b08008

Dang CV (2012) MYC on the path to cancer. Cell 149(1):22–35. https://doi.org/10.1016/j.cell.2012.03.003

Davis AJ, Chen DJ (2013) DNA double strand break repair via non-homologous end-joining. Transl Cancer Res 2(3):130–143. https://doi.org/10.3978/j.issn.2218-676X.2013.04.02

Dawson MA, Kouzarides T (2012) Cancer epigenetics: from mechanism to therapy. Cell 150(1):12–27

Falkenberg KJ, Johnstone RW (2014) Histone deacetylases and their inhibitors in cancer, neurological diseases and immune disorders. Nat Rev Drug Discov 13(9):673–691. https://doi.org/10.1038/nrd4360

Ferrara N, Gerber HP, LeCouter J (2003) The biology of VEGF and its receptors. Nat Med 9(6):669–676. https://doi.org/10.1038/nm0603-669

Finn RS, Dering J, Conklin D, Kalous O, Cohen DJ, Desai AJ, Fowst C (2009) PD 0332991, a selective cyclin D kinase 4/6 inhibitor, preferentially inhibits proliferation of luminal estrogen receptor-positive human breast cancer cell lines in vitro. Breast Cancer Res 11(5):1–13

Fuller-Pace FV (2006) DExD/H box RNA helicases: multifunctional proteins with important roles in transcriptional regulation. Nucleic Acids Res 34(15):4206–4215. https://doi.org/10.1093/nar/gkl460

Gadd MS, Testa A, Lucas X, Chan KH, Chen W, Lamont DJ, Ciulli A (2017) Structural basis of PROTAC cooperative recognition for selective protein degradation. Nat Chem Biol 13(5):514–521. https://doi.org/10.1038/nchembio.2329

Galluzzi L, Pietrocola F, Bravo-San Pedro JM, Amaravadi RK, Baehrecke EH, Cecconi F, Kroemer G (2015) Autophagy in malignant transformation and cancer progression. EMBO J 34(7):856–880. https://doi.org/10.15252/embj.201490784

Gerstberger S, Hafner M, Tuschl T (2014) A census of human RNA-binding proteins. Nat Rev Genet 15(12):829–845. https://doi.org/10.1038/nrg3813

Gherardi E, Birchmeier W, Birchmeier C, Vande Woude G (2012) Targeting MET in cancer: rationale and progress. Nat Rev Cancer 12(2):89–103. https://doi.org/10.1038/nrc3205

Hamidi H, Pietilä M, Ivaska J (2016) The complexity of integrins in cancer and new scopes for therapeutic targeting. Br J Cancer 115(9):1017–1023. https://doi.org/10.1038/bjc.2016.312

Hanahan D, Weinberg RA (2011) Hallmarks of cancer: the next generation. Cell 144(5):646–674. https://doi.org/10.1016/j.cell.2011.02.013

Harbeck N, Penault-Llorca F, Cortes J, Gnant M, Houssami N, Poortmans P, Cardoso F (2019) Breast Cancer Nat Rev Dis Primers 5:66

Hayes JD, McMahon M (2009) NRF2 and KEAP1 mutations: permanent activation of an adaptive response in cancer. Trends Biochem Sci 34(4):176–188. https://doi.org/10.1016/j.tibs.2008.12.008

He M, Cao C, Ni Z, Liu Y, Song P, Hao S, Rao Y (2022) PROTACs: great opportunities for academia and industry (an update from 2020 to 2021). Signal Transduct Target Ther 7(1):181

Hobbs GA, Der CJ, Rossman KL (2016) RAS isoforms and mutations in cancer at a glance. J Cell Sci 129(7):1287–1292. https://doi.org/10.1242/jcs.182873

Jackson SP, Bartek J (2009) The DNA-damage response in human biology and disease. Nature 461(7267):1071–1078. https://doi.org/10.1038/nature08467

Jaffe AB, Hall A (2005) Rho GTPases: biochemistry and biology. Annu Rev Cell Dev Biol 21:247–269. https://doi.org/10.1146/annurev.cellbio.21.020604.150721

Jego G, Hazoumé A, Seigneuric R, Garrido C (2013) Targeting heat shock proteins in cancer. Cancer Lett 332(2):275–285. https://doi.org/10.1016/j.canlet.2010.10.014

Jemaà M, Manic G, Lledo G, Lissa D, Reynes C, Morin N, Abrieu A (2016) Whole-genome duplication increases tumor cell sensitivity to MPS1 inhibition. Oncotarget 7(1):885–901. https://doi.org/10.18632/oncotarget.6432

Jiang J, Hui CC (2008) Hedgehog signaling in development and cancer. Dev Cell 15(6):801–812. https://doi.org/10.1016/j.devcel.2008.11.010

Julien SG, Dubé N, Hardy S, Tremblay ML (2011) Inside the human cancer tyrosine phosphatome. Nat Rev Cancer 11(1):35–49. https://doi.org/10.1038/nrc2980

Kalluri R (2016) The biology and function of fibroblasts in cancer. Nat Rev Cancer 16(9):582–598. https://doi.org/10.1038/nrc.2016.73

Katoh Y, Katoh M (2006) Hedgehog signaling pathway and gastrointestinal stem cell signaling network (review). Int J Mol Med 18(6):1019–1023

Keith B, Simon MC (2007) Hypoxia-inducible factors, stem cells, and cancer. Cell 129(3):465–472. https://doi.org/10.1016/j.cell.2007.04.019

Kopan R, Ilagan MX (2009) The canonical Notch signaling pathway: unfolding the activation mechanism. Cell 137(2):216–233. https://doi.org/10.1016/j.cell.2009.03.045

Kotschy A, Szlavik Z, Murray J, Davidson J, Maragno AL, Le Toumelin-Braizat G, Geneste O (2016) The MCL1 inhibitor S63845 is tolerable and effective in diverse cancer models. Nature 538(7626):477–482. https://doi.org/10.1038/nature19830

Lai AC, Toure M, Hellerschmied D, Salami J, Jaime-Figueroa S, Ko E, Crews CM (2016) Modular PROTAC design for the degradation of oncogenic BCR-ABL. Angew Chem Int Ed Engl 55(2):807–810. https://doi.org/10.1002/anie.201507634

Li X, Song Y (2020) Proteolysis-targeting chimera (PROTAC) for targeted protein degradation and cancer therapy. J Hematol Oncol 13(1):1–14

Li X, Pu W, Zheng Q, Ai M, Chen S, Peng Y (2022) Proteolysis-targeting chimeras (PROTACs) in cancer therapy. Mol Cancer 21(1):1–30

Li DH, Liu XK, Tian XT, Liu F, Yao XJ, Dong JF (2023) PPARG: a promising therapeutic target in breast cancer and regulation by natural drugs. PPAR Res 2023:4481354. https://doi.org/10.1155/2023/4481354

Liberti MV, Locasale JW (2016) Correction to: 'The Warburg effect: how does it benefit cancer cells?': [Trends in Biochemical Sciences, 41 (2016) 211]. Trends Biochem Sci 41(3):287. https://doi.org/10.1016/j.tibs.2016.01.004

Lord CJ, Ashworth A (2017) PARP inhibitors: synthetic lethality in the clinic. Science 355(6330):1152–1158

Ma CX, Janetka JW, Piwnica-Worms H (2011) Death by releasing the breaks: CHK1 inhibitors as cancer therapeutics. Trends Mol Med 17(2):88–96. https://doi.org/10.1016/j.molmed.2010.10.009

MacDonald BT, Tamai K, He X (2009) Wnt/beta-catenin signaling: components, mechanisms, and diseases. Dev Cell 17(1):9–26. https://doi.org/10.1016/j.devcel.2009.06.016

Macheret M, Halazonetis TD (2015) DNA replication stress as a hallmark of cancer. Annu Rev Pathol 10:425–448. https://doi.org/10.1146/annurev-pathol-012414-040424

Mal S, Dwivedi AR, Kumar V, Kumar N, Kumar B, Kumar V (2021) Role of peroxisome proliferator-activated receptor gamma (PPARγ) in different disease states: recent updates. Curr Med Chem 28(16):3193–3215. https://doi.org/10.2174/0929867327666200716113136

Malumbres M, Barbacid M (2009) Cell cycle, CDKs and cancer: a changing paradigm. Nat Rev Cancer 9(3):153–166

Manasanch EE, Orlowski RZ (2017) Proteasome inhibitors in cancer therapy. Nat Rev Clin Oncol 14(7):417–433. https://doi.org/10.1038/nrclinonc.2016.206

Maniaci C, Hughes SJ, Testa A, Chen W, Lamont DJ, Rocha S, Ciulli A (2017) Homo-PROTACs: bivalent small-molecule dimerizers of the VHL E3 ubiquitin ligase to induce self-degradation. Nat Commun 8(1):830. https://doi.org/10.1038/s41467-017-00954-1

Marumoto T, Zhang D, Saya H (2005) Aurora-A - a guardian of poles. Nat Rev Cancer 5(1):42–50. https://doi.org/10.1038/nrc1526

Mittendorf EA, Zhang H, Barrios CH, Saji S, Jung KH, Hegg R, Harbeck N (2020) Neoadjuvant atezolizumab in combination with sequential nab-paclitaxel and anthracycline-based chemotherapy versus placebo and chemotherapy in patients with early-stage triple-negative breast cancer (IMpassion031): a randomised, double-blind, phase 3 trial. Lancet 396(10257):1090–1100. https://doi.org/10.1016/s0140-6736(20)31953-x

Molli PR, Li DQ, Murray BW, Rayala SK, Kumar R (2009) PAK signaling in oncogenesis. Oncogene 28(28):2545–2555. https://doi.org/10.1038/onc.2009.119

Mosele F, Remon J, Mateo J, Westphalen CB, Barlesi F, Lolkema MP, André F (2020) Recommendations for the use of next-generation sequencing (NGS) for patients with metastatic cancers: a report from the ESMO Precision Medicine Working Group. Ann Oncol 31(11):1491–1505. https://doi.org/10.1016/j.annonc.2020.07.014

Mullard A (2019) First targeted protein degrader hits the clinic. Nat Rev Drug Discov. https://doi.org/10.1038/d41573-019-00043-6

Musgrove EA, Sutherland RL (2009) Biological determinants of endocrine resistance in breast cancer. Nat Rev Cancer 9(9):631–643. https://doi.org/10.1038/nrc2713

Nabet B, Ferguson FM, Seong BKA, Kuljanin M, Leggett AL, Mohardt ML, Mancias JD (2020) Rapid and direct control of target protein levels with VHL-recruiting dTAG molecules. Nat Commun 11(1):4687

O'Brien CA, Pollett A, Gallinger S, Dick JE (2007) A human colon cancer cell capable of initiating tumor growth in immunodeficient mice. Nature 445(7123):106–110. https://doi.org/10.1038/nature05372

O'Shea JJ, Schwartz DM, Villarino AV, Gadina M, McInnes IB, Laurence A (2015) The JAK-STAT pathway: impact on human disease and therapeutic intervention. Annu Rev Med 66:311–328. https://doi.org/10.1146/annurev-med-051113-024537

Olson CM, Jiang B, Erb MA, Liang Y, Doctor ZM, Zhang Z, Gray NS (2018) Pharmacological perturbation of CDK9 using selective CDK9 inhibition or degradation. Nat Chem Biol 14(2):163–170. https://doi.org/10.1038/nchembio.2538

Oltersdorf T, Elmore SW, Shoemaker AR, Armstrong RC, Augeri DJ, Belli BA, Rosenberg SH (2005) An inhibitor of Bcl-2 family proteins induces regression of solid tumors. Nature 435(7042):677–681. https://doi.org/10.1038/nature03579

Organ SL, Tsao MS (2011) An overview of the c-MET signaling pathway. Ther Adv Med Oncol 3(1 Suppl):S7–S19. https://doi.org/10.1177/1758834011422556

Ottis P, Crews CM (2017) Proteolysis-targeting chimeras: induced protein degradation as a therapeutic strategy. ACS Chem Biol 12(4):892–898

Ottis P, Toure M, Cromm PM, Ko E, Gustafson JL, Crews CM (2017) Assessing different E3 ligases for small molecule induced protein ubiquitination and degradation. ACS Chem Biol 12(10):2570–2578. https://doi.org/10.1021/acschembio.7b00485

Pardoll DM (2012) The blockade of immune checkpoints in cancer immunotherapy. Nat Rev Cancer 12(4):252–264

Pereira B, Billaud M, Almeida R (2017) RNA-binding proteins in cancer: old players and new actors. Trends Cancer 3(7):506–528. https://doi.org/10.1016/j.trecan.2017.05.003

Pickup MW, Laklai H, Acerbi I, Owens P, Gorska AE, Chytil A, Moses HL (2013) Stromally derived lysyl oxidase promotes metastasis of transforming growth factor-β-deficient mouse mammary carcinomas. Cancer Res 73(17):5336–5346. https://doi.org/10.1158/0008-5472.can-13-0012

Pickup MW, Mouw JK, Weaver VM (2014) The extracellular matrix modulates the hallmarks of cancer. EMBO Rep 15(12):1243–1253. https://doi.org/10.15252/embr.201439246

Prozzillo Y, Fattorini G, Santopietro MV, Suglia L, Ruggiero A, Ferreri D, Messina G (2020) Targeted protein degradation tools: overview and future perspectives. Biology 9(12):421

Puca R, Nardinocchi L, Givol D, D'Orazi G (2010) Regulation of p53 activity by HIPK2: molecular mechanisms and therapeutical implications in human cancer cells. Oncogene 29(31):4378–4387. https://doi.org/10.1038/onc.2010.183

Qi S-M, Dong J, Xu Z-Y, Cheng X-D, Zhang W-D, Qin J-J (2021) PROTAC: an effective targeted protein degradation strategy for cancer therapy. Front Pharmacol 12:692574

Raina K, Crews CM (2017) Targeted protein knockdown using small molecule degraders. Curr Opin Chem Biol 39:46–53

Raina K, Lu J, Qian Y, Altieri M, Gordon D, Rossi AM, Coleman KG (2016) PROTAC-induced BET protein degradation as a therapy for castration-resistant prostate cancer. Proc Natl Acad Sci USA 113(26):7124–7129. https://doi.org/10.1073/pnas.1521738113

Ramsey HE, Fischer MA, Lee T, Gorska AE, Arrate MP, Fuller L, Savona MR (2018) A novel MCL1 inhibitor combined with venetoclax rescues venetoclax-resistant acute myelogenous leukemia. Cancer Discov 8(12):1566–1581. https://doi.org/10.1158/2159-8290.cd-18-0140

Ribas A, Wolchok JD (2018) Cancer immunotherapy using checkpoint blockade. Science 359(6382):1350–1355

Rimkus TK, Carpenter RL, Qasem S, Chan M, Lo HW (2016) Targeting the sonic hedgehog signaling pathway: review of smoothened and GLI inhibitors. Cancer 8(2):22. https://doi.org/10.3390/cancers8020022

Roberts PJ, Der CJ (2007) Targeting the Raf-MEK-ERK mitogen-activated protein kinase cascade for the treatment of cancer. Oncogene 26(22):3291–3310. https://doi.org/10.1038/sj.onc.1210422

Rohle D, Popovici-Muller J, Palaskas N, Turcan S, Grommes C, Campos C, Mellinghoff IK (2013) An inhibitor of mutant IDH1 delays growth and promotes differentiation of glioma cells. Science 340(6132):626–630. https://doi.org/10.1126/science.1236062

Rojo de la Vega M, Chapman E, Zhang DD (2018) NRF2 and the Hallmarks of Cancer. Cancer Cell 34(1):21–43. https://doi.org/10.1016/j.ccell.2018.03.022

Roskoski R (2012) ERK1/2 MAP kinases: structure, function, and regulation. Pharmacol Res 66(2):105–143. https://doi.org/10.1016/j.phrs.2012.04.005

Sakamoto KM, Kim KB, Verma R, Ransick A, Stein B, Crews CM, Deshaies RJ (2003) Development of Protacs to target cancer-promoting proteins for ubiquitination and degradation. Mol Cell Proteomics 2(12):1350–1358

Schneekloth, John S, Fonseca FN, Koldobskiy M, Mandal A, Deshaies R, Sakamoto K, Crews CM (2004) Chemical genetic control of protein levels: selective in vivo targeted degradation. J Am Chem Soc 126(12):3748–3754

Seguin L, Desgrosellier JS, Weis SM, Cheresh DA (2015) Integrins and cancer: regulators of cancer stemness, metastasis, and drug resistance. Trends Cell Biol 25(4):234–240. https://doi.org/10.1016/j.tcb.2014.12.006

Semenza GL (2012) Hypoxia-inducible factors: mediators of cancer progression and targets for cancer therapy. Trends Pharmacol Sci 33(4):207–214. https://doi.org/10.1016/j.tips.2012.01.005

Snyder LB, Flanagan JJ, Qian Y, Gough SM, Andreoli M, Bookbinder M, Taylor I (2021) Abstract 44: The discovery of ARV-471, an orally bioavailable estrogen receptor degrading PROTAC for the treatment of patients with breast cancer. Cancer Res 81(13):44. https://doi.org/10.1158/1538-7445.AM2021-44

Soucy TA, Smith PG, Milhollen MA, Berger AJ, Gavin JM, Adhikari S, Langston SP (2009) An inhibitor of NEDD8-activating enzyme as a new approach to treat cancer. Nature 458(7239):732–736. https://doi.org/10.1038/nature07884

Stecca B, Ruizi Altaba A (2010) Context-dependent regulation of the GLI code in cancer by HEDGEHOG and non-HEDGEHOG signals. J Mol Cell Biol 2(2):84–95. https://doi.org/10.1093/jmcb/mjp052

Stephen AG, Esposito D, Bagni RK, McCormick F (2014) Dragging ras back in the ring. Cancer Cell 25(3):272–281. https://doi.org/10.1016/j.ccr.2014.02.017

Sulzmaier FJ, Jean C, Schlaepfer DD (2014) FAK in cancer: mechanistic findings and clinical applications. Nat Rev Cancer 14(9):598–610. https://doi.org/10.1038/nrc3792

Sun B, Fiskus W, Qian Y, Rajapakshe K, Raina K, Coleman KG, Mill CP (2018) BET protein proteolysis targeting chimera (PROTAC) exerts potent lethal activity against mantle cell lymphoma cells. Leukemia 32(2):343–352

Swords RT, Kelly KR, Smith PG, Garnsey JJ, Mahalingam D, Medina E, Carew JS (2010) Inhibition of NEDD8-activating enzyme: a novel approach for the treatment of acute myeloid leukemia. Blood 115(18):3796–3800. https://doi.org/10.1182/blood-2009-11-254862

Tonks NK (2006) Protein tyrosine phosphatases: from genes, to function, to disease. Nat Rev Mol Cell Biol 7(11):833–846. https://doi.org/10.1038/nrm2039

Vader G, Lens SM (2008) The Aurora kinase family in cell division and cancer. Biochim Biophys Acta 1786(1):60–72. https://doi.org/10.1016/j.bbcan.2008.07.003

Vander Heiden MG, Cantley LC, Thompson CB (2009) Understanding the Warburg effect: the metabolic requirements of cell proliferation. Science 324(5930):1029–1033. https://doi.org/10.1126/science.1160809

Wang F, Travins J, DeLaBarre B, Penard-Lacronique V, Schalm S, Hansen E, Yen KE (2013) Targeted inhibition of mutant IDH2 in leukemia cells induces cellular differentiation. Science 340(6132):622–626. https://doi.org/10.1126/science.1234769

Warfel NA, Sainz AG, Song JH, Kraft AS (2016) PIM kinase inhibitors kill hypoxic tumor cells by reducing Nrf2 signaling and increasing reactive oxygen species. Mol Cancer Ther 15(7):1637–1647. https://doi.org/10.1158/1535-7163.mct-15-1018

West AC, Johnstone RW (2014) New and emerging HDAC inhibitors for cancer treatment. J Clin Invest 124(1):30–39. https://doi.org/10.1172/jci69738

Xie T, Lim SM, Westover KD, Dodge ME, Ercan D, Ficarro SB, Gray NS (2014) Pharmacological targeting of the pseudokinase Her3. Nat Chem Biol 10(12):1006–1012. https://doi.org/10.1038/nchembio.1658

Yang ZJ, Chee CE, Huang S, Sinicrope FA (2011) The role of autophagy in cancer: therapeutic implications. Mol Cancer Ther 10(9):1533–1541. https://doi.org/10.1158/1535-7163.mct-11-0047

Yoo CB, Jones PA (2006) Epigenetic therapy of cancer: past, present and future. Nat Rev Drug Discov 5(1):37–50. https://doi.org/10.1038/nrd1930

Zachos G, Black EJ, Walker M, Scott MT, Vagnarelli P, Earnshaw WC, Gillespie DA (2007) Chk1 is required for spindle checkpoint function. Dev Cell 12(2):247–260. https://doi.org/10.1016/j.devcel.2007.01.003

Zeman MK, Cimprich KA (2014) Causes and consequences of replication stress. Nat Cell Biol 16(1):2–9. https://doi.org/10.1038/ncb2897

Zeng M, Xiong Y, Safaee N, Nowak RP, Donovan KA, Yuan CJ, Gray NS (2020) Exploring targeted degradation strategy for oncogenic KRAS(G12C). Cell Chem Biol 27(1):19–31. https://doi.org/10.1016/j.chembiol.2019.12.006

Zhang H, Zhao H-Y, Xi X-X, Liu Y-J, Xin M, Mao S, Zhang S-Q (2020) Discovery of potent epidermal growth factor receptor (EGFR) degraders by proteolysis targeting chimera (PROTAC). Eur J Med Chem 189:112061

Zhao B, Burgess K (2019) PROTACs suppression of CDK4/6, crucial kinases for cell cycle regulation in cancer. Chem Commun 55(18):2704–2707

Zhao Y, Jensen ON (2009) Modification-specific proteomics: strategies for characterization of post-translational modifications using enrichment techniques. Proteomics 9(20):4632–4641. https://doi.org/10.1002/pmic.200900398

Zhao L, Zhao J, Zhong K, Tong A, Jia D (2022) Targeted protein degradation: mechanisms, strategies and application. Signal Transduct Target Ther 7(1):113

Zorba A, Nguyen C, Xu Y, Starr J, Borzilleri K, Smith J, Calabrese MF (2018) Delineating the role of cooperativity in the design of potent PROTACs for BTK. Proc Natl Acad Sci USA 115(31):E7285–E7292. https://doi.org/10.1073/pnas.1803662115

Role of PROTACs in Hematological Malignancies

13

Ashif Iqubal, Kashif Iqubal, Sanjeev Yadav, V. A. Arun, and Kundan Mishra

Abstract

Hematological malignancies (HM) are among the most difficult types of cancer as far as management and treatment are concerned. There have been advancements in the treatment of HM, but despite the availability of targeted therapy or small molecules, clinical outcomes are not up to the mark. Relapse or refractory types of HM are even more difficult to treat. At the same time, the pathogenic role of various proteins has also been identified, and the available therapeutic approaches are only able to inhibit its function. Proteolysis targeting chimeras (PROTACs) are a group of hetero-bifunctional small molecules that cause pathogenic protein degradation instead of their inhibition by using a ubiquitin-proteasome system (UPS). PROTACs are considered superior to small molecules in terms of unique mechanisms of action, selectivity, and efficiency. Considering these facts, PROTACs emerged as a novel approach for targeted drug development for hematological malignancies. Currently, a few PROTACs are in the initial phase of a clinical trial, and many are under preclinical development.

A. Iqubal
Department of Pharmacology, Jamia Hamdard University, New Delhi, Delhi, India

K. Iqubal
Department of Pharmaceutical Sciences, Irma Lerma Rangel College of Pharmacy, Texas A&M Health Science Center, Texas A&M University, College Station, TX, USA

S. Yadav
Department of Hematology, Sanjay Gandhi Postgraduate Institute of Medical Sciences, Lucknow, India

V. A. Arun
Department of Medicine, Naval Hospital Asvini Colaba, Mumbai, Maharashtra, India

K. Mishra (✉)
Department of Clinical Hematology and BMT, Command Hospital, Lucknow, India

M. Nandave, P. Jain (eds.), *PROTAC-Mediated Protein Degradation: A Paradigm Shift in Cancer Therapeutics*, https://doi.org/10.1007/978-981-97-5077-1_13

Therefore, in this section, we have precisely discussed the various protein targets or proteins of interest that have pathogenic in hematological malignancies and enlightened various PROTACs in terms of their efficacy, safety, developmental status, limitations, and future prospects.

Keywords

Leukemia · Lymphoma · PROTACs · Drug development · Clinical trial and cancer

Abbreviations

ALL	Acute lymphoblastic leukemia
AML	Acute myeloid leukemia
CALR	Calreticulin gene
CRBN	Cereblon
HECT	Homologous to E6-AP carboxy terminus
HM	Hematological malignancies
IAPs	Inhibitor of apoptosis proteins (IAPs)
NPM1	Nucleophosmin
PROTACs	Proteolysis targeting chimeras
RBR	Ring between ring
RING	Really interesting new gene
SMIs	Small molecule inhibitors
UPS	Ubiquitin-proteasome system
VHL	Von Hippel-Lindau

13.1 Introduction

Hematological malignancies (HM) are among the most common types of diagnosed cancer caused by the abnormality in the differentiation and function of hematopoietic stem cells (Khan et al. 2024). No doubt, a lot of work is going on, and many advancements have occurred in the diagnosis and treatment of HM. However, the use of chemotherapy is still considered the primary treatment regimen (Khan et al. 2024). It is evident that various chemotherapeutic drugs have improved clinical outcomes. Nonetheless, their use is related to multiple side effects that compromise clinical outcomes, negatively affect the quality of life of patients, and organ toxicities. Looking into the side effects of these chemotherapeutic drugs, targeted therapies have been developed and are currently in use (Huang et al. 2020). These targeted therapies are primarily monoclonal antibodies or antibody-drug conjugates. It is worth mentioning that these targeted therapies possess significant selectivity, potent binding affinity, increased release pattern or sustained release characteristic, improved clinical outcome, and above all, minimal side effects and favorable safety profile (Huang et al. 2020). However, these monoclonal antibodies also possess certain disadvantages, such as being bulkier molecules, unable to cross the cell membrane, and mainly acting on the receptors or targets present on the cell's surface and not within the cell. Additionally, the bioavailability of these monoclonal antibodies is significantly lower when given orally, and they also have limited permeability in crossing the epithelium of the intestinal membrane (Huang et al. 2020). It is also pertinent to note that these molecules are proteinous and prone to proteolytic degradation by enzymes in the stomach or intestine. Apart from monoclonal antibodies, small molecule inhibitors (SMI) can be given by oral route, and being small in size, they cross cell membranes through the porin channels and have intrinsic permeability (Zhong et al. 2021). These SMI act on the pockets or sites within the target protein, and these targets or sites are lacking in the human proteins and the transcription factor, etc. (An and Fu 2018). Interestingly, few published reports have shown the efficacy of these SMI in various types of hematological malignancies, but the use of Proteolysis Targeting Chimeras (PROTACs) to degrade the target protein is emerging as a promising modality in the management and treatment of various types of hematological malignancies (He et al. 2020a). The PROTACs approach comprises a ubiquitin-proteasome system (UPS) for targeting the proteins of interest and degrading them. It is worth mentioning that several PROTACs are under different phases of clinical trials against various types of hematological malignancies (He et al. 2020a). NX-2127 is a BTK degrader, and it is currently under phase 1a/b clinical trial for treating relapsed or refractory B cell malignancies (NCT04830137).

13.2 Ubiquitin-Proteasome System: An Overview

In eukaryotes, UPS is the primary proteolytic system that, on the one hand, regulates the degradation of various proteins, but on the other hand, also controls the physiological processes such as cellular proliferation, repair of DNA, stress response, apoptosis, etc. Henceforth, UPS emerged as a potent and essential machinery in regulating various proteinous systems and overall homeostasis (Park et al. 2020). Considering the mechanism of protein degradation via UPS, first of all, modification with ubiquitin tag takes place, followed by the signal recognition and degradation by the proteasome 26S subunit (Park et al. 2020). In the ubiquitination process, ubiquitin gets attached to the lysine residue on the surface of the substrate protein via three sequential steps with the help of E1-E3 enzymes. First of all, ubiquitin gets activated via E1 ubiquitin-activating enzyme (E1). In the next step, the carboxyl-terminal of ubiquitin and cysteine residue of E1 undergoes covalent linking and forms a thioester bond, represented as E1-ubq. Now, the ubiquitin gets transferred to E2-conjugated enzymes, defined as E2-ubq, and finally, E3 ligase causes the shifting of E2 to the substrate (Park et al. 2020). Considering E3 ligase, there exist three structurally and functionally different E3 ligases: (1) The really interesting new gene (RING), (2) The homologous to E6-AP carboxy terminus (HECT), and (3) The ring between ring (RBR) family. RING E3 ligase is the most prominent family, and Culluns is the primary molecular scaffold that functions as a recruiter of the targeted substrate and also helps in bringing the substrate in close proximity to E2-ubq. Hence, ubiquitin gets directly transferred from E-2 ubq to the substrate, and thioester bond formation is necessary (Mani and Gelmann 2005). HECT E3 ligase undergoes trans-thiolation, which is a catalytically cysteine-dependent process (BouMalhab et al. 2023). Considering the RBR E3 ligase, there exist two canonical) domains, (a) RING1 and (b) RING2, that are linked in between the RING domain. Indeed, ubiquitination is a crucial and tightly controlled process, but, on the contrary, it is also very much a dynamic and reversible mechanism. This process sometimes gets compromised by the deubiquitinating enzymes (DUBs) that basically cause the removal of the ubiquitin chain from the protein substrate and, hence, prevent its degradation (Komander et al. 2009).

13.3 An Overview of PROTAC Technology

PROTACs is one of the novel approaches that causes targeted protein degradation by using the ubiquitin-proteasome approach, as shown in Fig. 13.1.

Structurally, PROTACs are hetero-bifunctional, and small molecules containing three different elements such as a ligand that binds with the targeted protein, another ligand that causes recreation of E3 ligase, and the third, a linker that causes conjugation of these aforementioned structures (Gao et al. 2020). Compared to the small molecules that, in general, inhibit the target protein's function, PROTACs mechanistically cause the degradation of the protein of interest (Gao et al. 2020). Mechanistically, upon the formation of protein/substrate-PROTACs-E3 ligase complex, the E3 ligase transfer causes ubiquitin transfer to the substrate, leading to its degradation (Paiva and Crews 2019). As per the published literature, more than 600 E3 ligases are present in the human cell, but only limited have been used via this technology for protein degradation (Fisher and Phillips 2018). In 2002, Sakamoto et al. (2001), for the first time, used the PROTAC approach using phosphopeptide moiety to degrade methionine aminopeptidase-2. In 2003, Sakamoto et al. also developed another PROTAC to target and degrade estrogen receptor-α and androgen receptor (Sakamoto et al. 2003). These initially developed PROTACs are of higher molecular weight, peptide-based, and possess limited permeability toward cell membranes (Sakamoto et al. 2003). Post 2003, there has been substantial development, and numerous E3 ligases for receptors of the substrate have been discovered, including the inhibitor of apoptosis proteins (IAPs), Von Hippel-Lindau (VHL), Cereblon (CRBN), and many more. Among these, Cereblon and Von Hippel-Lindau have been used as E3 ligases in hematological malignancies (BouMalhab et al. 2023).

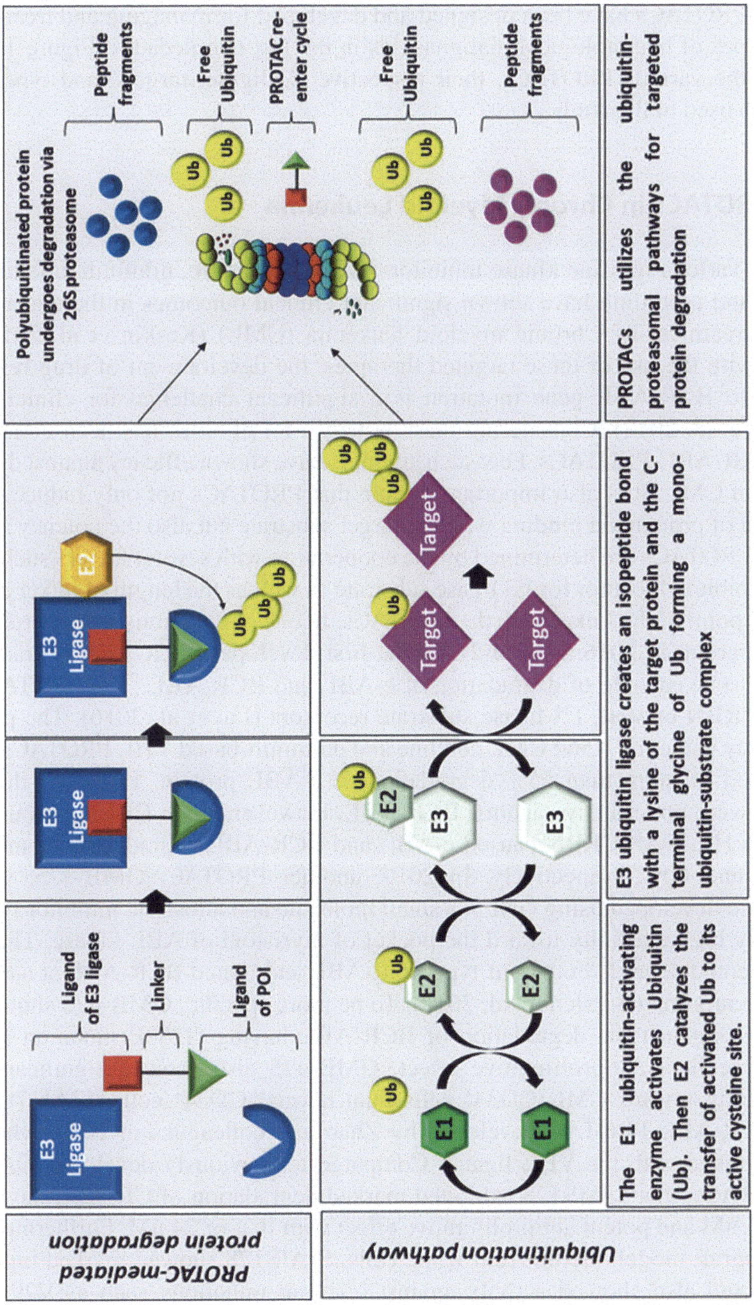

Fig. 13.1 Showing mechanism of PROTACs that causes protein degradation via Linkage of E3 ligase to target protein, resulting in proteasomal degradation (BouMalhab et al. 2023)

13.4 PROTACs in Hematological Malignancies

Numerous PROTACs have been designed and developed for managing and treating multiple types of hematological malignancies in the last two decades. Figure 13.2 represents the various PROTACs, their respective E3 ligase, targets, and type of cancer cells used in the study.

13.4.1 PROTACs in Chronic Myeloid Leukemia

As of now, various tyrosine kinase inhibitors such as dasatinib, nilotinib, imatinib, bosutinib, and asciminib have shown significant clinical outcomes in the management and treatment of Chronic myeloid leukemia (CML) (Keskin et al. 2016). However, with the use of these targeted therapies, the development of drug resistance due to BCR-ABL gene mutation is a significant challenge for clinicians (Hantschel et al. 2012). Considering these issues, a lot of work has been done to develop BCR-ABL-PROTACs. Few such attempts have shown efficacy against drug resistance in CML. It is also important to note that PROTACs not only induce the degradation of protein via binding with the target substrate but also the potency and activity of PROTACs are determined by the cooperation with several factors such as warhead inhibitor, receptor for E3 ligase substrate as well as the length of linker and attachment point of the linker with the PROTACs. In one of the published reports by the Crew's group in 2016, DAS-6-2-2-6, the first developed BCR-ABL degrader, was found to be capable of degradation of c-ABL and BCR-ABL. The PROTACs acted via CRBN or VHL E3 ligase substrate receptors (Lai et al. 2016). The preclinical study using the K562 CML cell line and dasatinib-based VHL PROTACs (1 μM) resulted in more than 65% degradation of c-ABL protein. However, these PROTACs were not effective against BCR-ABL. However, when CRBN was used instead of VHL, DAS-CRBN caused c-ABL and BCR-ABL degradation by more than 85% and 60%, respectively. In 2019, another PROTAC, GMB-475, was designed and developed using GNF5, a small molecule and allosteric inhibitor with significantly higher affinity toward the pocket of myristoyl of ABL kinase. These PROTACs could degrade both wild-type BCR-ABL and muted BCR-ABL at nanomolar concentration (Burslem et al. 2019). To be more specific, GMB-475 showed effectiveness against the degradation of BCR-ABL having G250E mutation and exhibited excellent antiproliferative effect. GMB-475 also showed significantly higher toxicity toward CML $CD34^+$ cells than normal $CD34^+$ cells. SIAI178 is another BCR-ABL PROTAC developed by Zhao and colleagues in 2019, where DAS was linked with the VHL ligand. Compared to previously developed DAS-VHL PROTACs, this SIAIS178 exhibited marked degradation of CR-ABL having DC_{50} of 8.5 nM and potent antiproliferative effect with IC_{50} of 24 nM. Furthermore, in the xenograft model, driven from K562 cells, SIAIS178 showed marked tumor regression and also showed activity against resistant mutations such as V299L, F317V, and G250E but failed to exhibit activity against T3151. Recently, Yang et al. (2020), showed that P19P degrades T3151 and asciminib-resistant V468F mutation.

PROTACs	E3 ligases	Targets	Types of cancer (cells used)
DAS-6-2-2-6-VHL	VHL	c-Abl	CML (K562)
DAS-6-2-2-6-CRBN	CRBN	c-Abl Bcr-Abl	
GMB-475	VHL	Bcr-Abl Bcr-Abl G250E	CML (Primary CML $CD34^+$, Ba/F3-BCR-ABL1)
SAIS178	VHL	Bcr-Abl Bcr-Abl G250E, V299L, F317L, and F317V mutants	CML (K562)
^{PMI}Bcr/Abl-R6	MDM2	Bcr-Abl and various Bcr-Abl mutants	CML; ALL (KU-812, SUP-B15)
P19P	CRBN	Bcr-Abl mutants: V468F, T315I	CML (K562)
FLT-3 PROTAC	VHL	FLT-3	AML (MV4-11, MOLM-14)
DT2216	VHL	BCL-xL	ALL (MOLT-4, RS4)
PZ15227	CRBN		
753b	VHL	BCL-xL/Bcl-2	AML (Kasumi-1)
GT19630 GT19715	CRBN	MYC	AML (HL-60)
PROTAC2	VHL	EED, EZH2 and SUZ12 (PCR2 subunits)	DLBCL (Karpas422)
UNC6852	VHL	EED, EZH2, SUZ12 EZH2-Y641 mutant	DLBCL
ARV-825	CRBN	BRD2/3/4	T-ALL, BL (6T-CEM; MOLT-4; Jurkat)
MZ1	VHL	BRD2/3/4	AML (Kasumi-1, MV4-11, NB4)
dBET1	CRBN	BRD2/3/4	AML (Kasumi-1, THP-1, MV4-11, NB4)
NX-2127	CRBN	BTK wt BTK C481S mutant IKZF3	MCL, DLBCL
NX-5948	CRBN	BTK wild type BTK-C481S mutant	Lymphoma (TMD8)
P131	CRBN	BTK wt BTK C481S mutant	NHL (HBL1, RAMOS, Mino cells)
L18I	CRBN	BTK C481S/T/A/G/W	DLBCL (HBL1), MCL (Mino; Z138)
MT-802	CRBN	BTK wt BTK mutants: C481S, E41K, C481R, C481Y, C481T, and C481F.	CLL, BL, DLBCL.
DD-03-171	CRBN	BTK, IKFZ1, and IKFZ3	MCL and other B-cell lymphomas
MS4077	CRBN	NPM-ALK	Lymphoma (SU-DHL-1)
MS4078		EML4-ALK	NCCLC (NCI-H2228)
TD-004	VHL	NPM-ALK	Lymphoma (SU-DHL-1)
		EML4-ALK	NSCLC (H3122).

Abbreviations: CML, chronic myeloid leukemias; AML, acute myeloid leukemia; ALL, acute lymphoblastic leukemia; CLL, chronic lymphocytic leuk diffuse large B-cell lymphoma; NHL, non-Hodgkin's lymphoma; BL, Burkitt's lymphoma; NSCLC, non-small cell lung cancer.

Fig. 13.2 Showing various PROTACs, E3 ligase, and targets in various hematological malignancies (BouMalhab et al. 2023)

Another newly developed PROTAC, PMIBcr/Abl-R6, by Ma et al. (2022) causes degradation of BCR_ABL, irrespective of the type of mutation. PMIBcr/Abl-R6 is dual-acting PROTAC and also causes activation of p53, leading to induction of apoptosis and significant outcomes in CML and Acute lymphoblastic leukemia (Ma et al. 2022).

13.4.2 PROTACs in Acute Myeloid Leukemia (AML) and Acute Lymphoblastic Leukemia (ALL)

13.4.2.1 FLT-3 is a Potential Target for PROTACs

FLT-3 stands for FMS-like tyrosine kinase-3. Expression of FLT-3 is mainly seen on $CD34^+$ hematopoietic stem cells and plays a pivotal role in normal physiological processes (Ma et al. 2022). These membrane-bound receptors have two mutated forms, and their expression is also seen in various types of leukemia. In 30% of CML patients, within the juxta-membrane domain of FLT-3 oncoprotein, the Internal Tandem Duplication (ITD) gene is found, and in the case of AML, a mutation in ITD accounts for 23% and positively correlated with the significantly higher release or refractory rate as well as reduced overall survival (Levis et al. 2005). Moreover, studies have shown the activation of FLT-3 during leukemogenesis (Levis et al. 2005). Activation of FLT-3 modulates the signaling pathways, such as MAPK/ERK and STAT5 pathways, resulting in increased proliferation, reduced apoptosis, and differentiation. When considering FLT-3, particularly AML, its wild-type confers reduced overall survival and compromised clinical outcomes (Bienz et al. 2005). Therefore, FLT protein is regarded as an emerging target for managing and treating leukemia (Levis and Small 2003; Bienz et al. 2005). Currently, gilteritinib and midostaurin are two drugs that FDA approves as FLT-3 inhibitors for patients who are diagnosed with FLT-3 mutation in AML (Antar et al. 2020). In one of the studies by Burslem et al. (2019), VHL-FLT-3 PROTACs employed the FLT-3 inhibitor quizartinib and degraded FLT-3ITD in MOLM-14 cells are a human acute myeloid leukemia (AML) cell line (Burslem et al. 2019). These two drugs were designed and developed using quizartinib with CRBN ligand and PEG as a linker. Surprisingly, the outcome of the study increased the expression of FLT-3 in MV-4-11 and MOLAM-14 cells and in the in vivo model (Burslem et al. 2019). Another study, on the contrary, where FLT-3 was combined with FF-10101 using CRBN ligand, showed minimal IC_{50} value.

13.4.2.2 BCL-2 and BCL-xL as a Potential Target for PROTACs in Leukemia

Apoptosis is one of the extensively studied and tightly controlled mechanisms that play an essential role in maintaining normal homeostasis. Apoptosis is regulated via mitochondrial and non-mitochondrial pathways, and any stimulus that reduces the apoptosis results in the initiation and progression of tumorigenesis (Hafezi and Rahmani 2021). The mitochondrial pathway of apoptosis is regulated by the movement of the BCL-2 protein, which modulates the permeabilization of the outer mitochondrial membrane. BCL-2 proteins are classified as antiapoptotic (BCL-xL, Mcl-1, BFL-1) and proapoptotic proteins (BAD, PUMA, BAX, HRK, BID, BIK, and BIM) (Hafezi and Rahmani 2021). Considering this, BCL-xL/BCL-2 inhibitors, such as navitoclax, venetoclax, have been developed. However, drugs targeting apoptotic proteins exhibit platelet toxicities and therefore a more refined/optimized form of dBETi-1 is needed. Considering this, PROTACs such as DT2216 and PZ15227 were designed and developed by Khan and colleagues. It showed promising efficacy against BCL-xL and reduced platelet toxicities compared to ABT-263 (Khan et al. 2019). In AML, DT2216 showed a superior antitumor effect than the ABT-263 in the cell and xenograft models (Khan et al. 2019). Moreover, no significant concern for thrombocytopenia was seen in the in vivo model (Khan et al. 2019). Khan and colleagues modified this PROTAC by using other linkers and attachment points on E3 ligase and ABT-263. This approach results in the dual-acting PROTAC against BCL-xL and BCL-2, respectively. Another PROTAC, PZ15227, is also ABT-263 based and CBRN E ligase and selectively targets BCL-xL degradation (He et al. 2020b). When the cell line and in vivo study were conducted for AML using PZ15227, increased selectivity with reduced platelet toxicity and moderate thrombocytopenia were found (He et al. 2020b).

13.4.2.3 Myc PROTACs as a Potential Target for PROTACs in Leukemia

Similar to apoptotic proteins, Myc is also one of the extensively deregulated genes in various types of solid and hematological malignancies. Its overexpression via multiple mechanisms manifests in the rearrangement of chromosomes and tumorigenesis. No doubt, the mechanism and role of Myc in tumorigenesis are well understood, but its targeting has been challenging. Additionally, PROTAC technology has not been promising for targeting Myc protein and its degradation. Mechanistically, the most appropriate and successful approach to target Myc was its inhibition via targeting BET bromodomain protein, a crucial transcription regulator. Notably, some PROTACs, such as GT19630 and GT19715, which are CRBN-based, showed promising activity in the preclinical study of the AML model (Nishida et al. 2022). In HL-60 cells, IC50 was found to be 1.4 nM and 1.8 nM, whereas, in the in vivo model, GTN1963085 inhibited the tumor growth at the very low dose of 0.3 mg/kg twice a day (Nishida et al. 2022).

13.4.2.4 BRD PROTACs as a Potential Target for PROTACs in Leukemia

BRD4 stands for the bromodomain and extra terminal domain (BET) family protein and encompasses BRD1-BRD-4. All these four types of BRD consist of bromodomains that act as epigenetic readers. BET inhibitor/BRD-4 inhibitors, such as JQ1, OTX015, and I-BET-151, are available and showed promising antitumor activity in the in vitro and in vivo model (Andrieu et al. 2016). Despite being a potent small molecule, these drugs exhibit acquired resistance, resulting in a compromised antitumor effect (Antonarakis et al. 2014). Additionally, when high doses of BETi were used because of developed resistance, dose-dependent toxicity was found that further negatively affected the quality of life and worsened the clinical outcome (Antonarakis et al. 2014). Another challenge with using a higher dose of BETi was an accumulation of BRD4 protein, leading to compromised Myc suppression (Lu et al. 2015). Considering these issues, novel PROTACs were developed to target BET selectively. ZXH-3-26 was first developed PROTAC by the Ciulli group,' but this PROTAC failed to demonstrate any significant antitumor effect (He et al. 2020a). Bradner's group, 2015 reported the antitumor effect of bBETi, which was a small molecule-based BRD-4 degrader (Winter et al. 2015). dBET-6n was a more refined/optimized form of dBETi-1 and exhibited a marked antitumor effect against T-ALL when compared with the JQ-1 (Winter et al. 2015). Crews' group also developed ARV-825, which was CRBN-based PROTAC, and interestingly, OTX015 was used as a warhead and exhibited an antitumor effect against T cell ALL (Lu et al. 2015). ARV771 is another PROTAC based on VHL, but the warhead was similar, OTX015, as in the case of ARV-825 (Raina et al. 2016). BETd-260 and QCA570 are PROTAC developed by Wang et al. in 2018, which was CRBN-based and demonstrated a potent effect in the degradation of BRD2-4 and also causes the degradation of RS4-11 B cell ALL (Qin et al. 2018; Zhou et al. 2018). In 2019, Crews'group also developed MDM2-based PROTAC known as A1874 that, on the one hand, causes degradation of BRD4, while on the other hand, also causes stabilization of proapoptotic protein, p53 an audience, exhibited potent antiproliferative effect in myeloid leukemia (Hines et al. 2019).

TD-428 is another recently developed BRD4-PROTAC that consists of TD-106, a CRBN binding moiety, and JQ1, a BET binding moiety (Kim et al. 2019). This compound exhibited promising antitumor effects compared to ARV-825 and interestingly showed no hook effect, which is an immunogenic problem (Kim et al. 2019). Compound 15, which was CRBN-based BRD PROTAC developed by Jiang et al., causes degradation of BRD2 and BRD 4 but not BRD3 in the in vitro model of AML, whereas I-BET726 which was VHL-based PROTAC, causes more selective degradation of BRD 3 and BRD 4 but no effect on BRD 2 degradation was found in the in vitro model of AML (Jiang et al. 2020).

13.4.2.5 CDK PROTACs as a Potential Target for PROTACs in Leukemia

CDK stands for cyclin-dependent kinase and plays a vital role in the progression of the cell cycle. Various isoforms of CDK exist, but CDK 4/6, particularly, is more clinically crucial as CDK6 causes leukemia stem cell activation that further manifests in AML and CML (Scheicher et al. 2015). CDK 4/6 is found to be a more promising target for the management and treatment of breast cancer. Hence, the selective CDK 4/6 inhibitors, palbociclib, ribociclib, and abimabaciclib, are approved by the FDA for the treatment of metastatic breast cancer. In mixed lineage leukemia, CDK6 plays pivotal pathogenic roles (Scheicher et al. 2015). Therefore, considering the pathogenic role of CDKs in hematological malignancies, various SMI for ATP-binding domains have been designed and developed (He et al. 2020a). However, the SMI is nonselective toward CDK4 and CDK6 because of similar/identical ATP domains and, hence, are found to be clinically irrelevant (He et al.

2020a). Considering this fact, CDK PROTAC, i.e., OYX-2-107, has been developed using Palbociclib and CRBN as ligands for managing and treating CML and Ph^+ ALL (Dominici et al. 2020). This PROTC specifically caused the degradation of CDK6 instead of CDK4 and showed promising results in the in vivo study (Dominici et al. 2020).

13.4.2.6 STAT3 PROTACs as a Potential Target for PROTACs in Leukemia

STAT stands for signal transducer and activator of transcription 3 (STAT3) and belongs to the STAT family. Abnormal activation of STAT3 positively correlates with the various types of solid and hematological malignancies since STAT3 is actively involved in regulating and functioning different pro and anti-oncogenes, leading to initiation, progression, and invasion. Several SMIs are available as STAT3 inhibitors that mechanistically inhibit the dimerization of STAT3. These SMIs are currently in the initial phase of clinical development (Johnson et al. 2018). Considering the importance of STAT3 in hematological malignancies, STA3 PROTC, SD-6 has been developed (Bai et al. 2019). SD-6 has been reported to cause degradation of STAT3 in the AML in vitro model of MOLM-16 cells and ALCL cell lines. Moreover, SD-6 demonstrates promising pharmacodynamic and pharmacokinetic properties. Moreover, This STA3 PROTAC exhibits a potent effect and better safety profile in the mice model of AML (Johnson et al. 2018).

13.4.3 PROTAC in Lymphoma

13.4.3.1 BTK PROTAC as a Potential Target in Lymphoma

Lymphoma is cancer of the lymphatic system, lymph nodes, spleen, thymus, and bone marrow. BTK stands for Bruton's Tyrosine Kinase, a soluble protein that plays a pivotal role in the initiation and maturation progression of B cell malignancies and lymphoma (Zhao et al. 2021). Ibrutinib is one of the BTK inhibitors used for treating mantle cell lymphoma alone or combined with another anticancer drug (Zhao et al. 2021). A significant challenge with the use of this drug is the development of drug resistance because of the missense mutation of C481S (Woyach et al. 2014). Hence, few BTK PROTAC have been developed to overcome such clinical limitations. NX-2127 and NVX-5948 are recently developed CRBN-based BTK PROTAC that have been evaluated in phase 1a/b clinical trial for the patient of B cell malignancies of the patient in whom the disease has progressed on two lines of therapy NCT04830137 and NCT05131022 respectively). P131 and L18I are other BTK PROTAC developed for treating ibrutinib-resistant patients. This drug showed a promoting antitumor effect among patients of ibrutinib-resistant and wild-type muted C481S protein at very low concentrations (Woyach et al. 2014). It is also important to mention that L181, a second-generation BTK PROTAC, showed efficacy against C481S mutation and other clinically meaningful C481 mutations in B cell cancer (George et al. 2020). MT-802 is another potent BTK PROTAC developed by the Crew's group and showed an antitumor effect against cell lines obtained from CLL patients who were confirmed with C481S mutation at DC50 of 14.9 nM (Buhimschi et al. 2018). DD-04-015 is another recently developed BTK PROTAC that exhibits the potent and rapid onset of action, i.e., causes degradation of BTK within four hours of treatment (Huang et al. 2018). DD-04-015 was further modified as DD-03-171 and caused degradation of C418S mutant degradation in vitro model of MCL at 5.1 nM and significant efficacy toward patient-derived xenografts in-vivo (Dobrovolsky et al. 2019).

13.4.3.2 ALK PROTAC as a Potential Target in Lymphoma

ALK stands for Anaplastic lymphoma kinase, a type of tyrosine receptor kinase. ALK is found to be activated in many tumor types of solid and hematological malignancies, such as NSCLC, ALCL, or DLBCL (Zia et al. 2023). Activation of ALK is manifested by gene alteration, such as chromosomal substitution, rearrangement, or amplification. In the case of NSCLC, EML4-ALK rearrangement is commonly seen, whereas, in the case of lymphoma, NPM-ALK fusion is frequently seen (Zia et al. 2023). Currently, ALK inhibitors such as crizotinib, ceritinib, oclacitinib, and lorlatinib are FDA-approved and used to treat specific mutation types in NSCLS. Some of the aforementioned ALK inhibitors are also under various phases of clinical trial for treating ALCL or other types of lymphoma [NCT02465060; NCT00939770; NCT03719898]. It has been observed that the various resistance mechanisms compromise the potency or efficacy of ALK inhibitors. Therefore, the PROTAC approach was used to overcome such limitations and improve the clinical outcome. Gray's group reported the first ALK PROTAC that was effective at DC50 of 10 nM in the H3122 NSCLC cell and effectively degraded the ALCL cells (Powell et al. 2018). A significant limitation of this ALK PROTAC was that it was not ALK specific and showed ineffectiveness in the degradation of EML4-ALK fusion protein. MS4077 and MS4078 are two other ALK PROTAC that effectively degraded NOM-ALK and EML4-ALK in ALCL and NCI-H2228 NSCLC cells (Zhang et al. 2018; Zia et al. 2023). TD-004 is another ALK PROTAC that causes degradation of ALK and additionally showed an antiproliferative effect against SUDHL-1 and H3122 in the in vitro and in vivo model (Kang et al. 2018). SIAIS117 is another ALK PROTAC that is primarily VHL-recruiting and based on brigatinib. This ALK PROTAC was more potent than brigatibib against the G1202R mutation, which is also known as a solvent front mutation and is very difficult to treat (Sun et al. 2020). More detailed information about the ALK PROTAC has been discussed by Kong et al. (2019).

13.4.3.3 PRC2 PROTAC as a Potential Target in Lymphoma

PRC2 stands for Polycomb repressive complex two and is an extensively studied protein in various hematological malignancies (Iwama 2017). Structurally, PRC2 proteins comprise various subunits, such as EHZ2, SUZ12, EED, AEBP12, and various other subunits (Herviou et al. 2016). The catalytic subunit of PRC2 is EZH2, which causes methylation of H3 histone, and EED, as well as SUZ12, are needed for this process (Cao et al. 2002). Interestingly, EZH2 and SUZ12 are responsible for the coding of the PCR2 component, but the loss of function mutation in it is accountable for T cell ALL (Zhang et al. 2012). Currently, EZH2 inhibitors exist, and the preclinical studies exhibited resistance due to secondary mutation in mutant EZH2 alleles (Gibaja et al. 2016). Considering these facts, Bloecher's group developed PROTAC2 that causes EZH2 SUZ12 and ED degradation and exhibited an antiproliferative effect (Hsu et al. 2020). UNC6852 is another PCR2 PROTAC that causes selective degradation of EZH2 and shows a potent antitumor effect (Potjewyd et al. 2020).

13.5 Emerging Oncoprotein for PROTAC Targeting in Hematological Malignancies

Apart from the oncoproteins above for PROTAC targeting in hematological malignancies, various other oncoproteins exist as potential targets and can be taken from bench to bedside (BouMalhab et al. 2023). In the case of Ph-negative myeloproliferative neoplasm, polycythemia, and myelofibrosis, Calreticulin gene (CALR) mutations are seen. However, this mutation is exclusive to the Jak 2 mutation. This mutation is preset on exon 9. CALR mutation has been reported to be pathogenic via activation of JAK2-STAT pathways (Marty et al. 2016). Therefore, the CALR mutant protein appears to be a potential target for PROTAC. Nucleophosmin (NPM1) is another common mutant gene in the AML and is found in 20% of children and 50% of adults (Falini et al. 2005; Schlenk et al. 2008). NPM1 mutation is present on exon-12 (Alcalay et al. 2005). An unmet need exists for a drug that can target NPM1 and cause proteasomal degradation (Nabbouh et al. 2017). It is well-known that chromosome translocations play a major pathogenic role in the process of solids and hematological malignancies. One of the most common mechanisms of malignancies in chromosomal translocation is via the formation of oncogenic fusion protein or some tumor promotor (Mitelman et al. 2007). Some of the deactivated and fusion proteins are Bcl-6, Bcl-4, and BCR-ABL, which have been targeted by PROTAC technology in various hematological malignancies. Moreover, other fusion proteins exist, such as PML-RAR-alpha, AML-1-ETO, and MLL fusions in AML, while TCF3-PBX1 fusion protein potential targets PROTAC in B cell ALL (Rowley 2008).

13.6 Conclusion and Prospect of PROTACs in Hematological Malignancies

Despite advancements in the management and treatment of hematological malignancies, there still exists a gap in the significant clinical outcome and improvement in the quality of life. The situation becomes worse when there is a relapse and only a few options are left, and the issue of drug resistance also occurs. It is well-known that the degradation of various pathogenic proteins by using PROTAC technology is a novel approach in the field of drug discovery and finding an appropriate modality of hematological malignancies (He et al. 2020a). The first PROTACs were reported almost two decades back, and since then, a lot of advancement has occurred because of consistent advancement in molecular-cellular and analytical techniques. Since the first PROTAC was developed, we have found significant improvement and somewhat superior response than SMIs. Despite being an important therapeutic approach, the low solubility and high molecular weight of PROTACs confer low efficacy in the preclinical models (An and Fu 2018). It is also important to highlight that balanced cellular distribution, tissue permeability, stability, and scalability must be maintained for taking these PROTACs from bench to bedside. It is also worth mentioning that PROTACs that are developed using VHL or CRBN are found to be associated with inefficient tumor suppressor activities (He et al. 2020a). This can be understood from the fact that in kidneys, VHL is muted, and hence, PROTACs based on VHL cannot used among patients with renal carcinoma having defects in the VHL gene. Studies have also reported the development of resistance to CRBN as well as VHL-based PROTACs when used for longer periods (Zhang et al. 2019). Proper dose optimization is also equally important for the PROTACs in order to achieve optimum antitumor effect. Considering the clinical advancement of PROTACs, as of now, two PROTACs are under phase I clinical trials that specifically target androgen receptors and estrogen receptors in prostate and breast cancer. Moreover, as discussed above, many PROTACs have demonstrated marked antitumor response in the initial phase of preclinical studies and can be further translated into clinical trials (He et al. 2020a). Above all, the safety of PROTACs must be of prime importance so that tumor-specific and very selective PROTACs can be developed and can be translated into clinical applications. Recently, approaches have been taken to delete the specific RNA or oncogenic RNA using small molecules, and they are known as RIBOTACS (Zhang et al. 2019; He et al. 2020a). Hence, it can be concluded that PROTACs are a novel approach for selectively targeting the proteins responsible for hematological malignancies. These PROTACs are indeed very specific and cause protein degradation via ubiquitination.

Conflict of Interest Authors declare no potential conflict of interest.

References

Alcalay M, Tiacci E, Bergomas R, Bigerna B, Venturini E, Minardi SP, Meani N et al (2005) Acute myeloid leukemia bearing cytoplasmic nucleophosmin (NPMc+ AML) shows a distinct gene expression profile characterized by up-regulation of genes involved in stem-cell maintenance. Blood 106(3):899–902. https://doi.org/10.1182/BLOOD-2005-02-0560

An S, Fu L (2018) Small-molecule PROTACs: an emerging and promising approach for the development of targeted therapy drugs. EBioMedicine 36:553. https://doi.org/10.1016/J.EBIOM.2018.09.005

Andrieu G, Belkina AC, Denis GV (2016) Clinical trials for BET inhibitors run ahead of the science. Drug Discov Today 19:45–50. https://doi.org/10.1016/J.DDTEC.2016.06.004

Antar AI, Otrock ZK, Jabbour E, Mohty M, Bazarbachi A (2020) FLT3 inhibitors in acute myeloid leukemia: ten frequently asked questions. Leukemia 34(3):682–696. https://doi.org/10.1038/S41375-019-0694-3

Antonarakis ES, Changxue L, Wang H, Luber B, Nakazawa M, Roeser JC, Chen Y et al (2014) AR-V7 and resistance to enzalutamide and abiraterone in prostate cancer. N Engl J Med 371(11):1028–1038. https://doi.org/10.1056/NEJMOA1315815

Bai L, Zhou H, Renqi X, Zhao Y, Chinnaswamy K, McEachern D, Chen J et al (2019) A potent and selective small-molecule degrader of STAT3 achieves complete tumor regression in vivo. Cancer Cell 36(5):498–511. https://doi.org/10.1016/J.CCELL.2019.10.002

Bienz M, Ludwig M, Mueller BU, OppligerLeibundgut E, Ratschiller D, Solenthaler M, Fey MF, Pabst T (2005) Risk assessment in patients with acute myeloid leukemia and a normal karyotype. Clin Cancer Res 11(4):1416–1424. https://doi.org/10.1158/1078-0432.CCR-04-1552

BouMalhab LJ, Alsafar H, Ibrahim S, Rahmani M (2023) PROTACs: walking through hematological malignancies. Front Pharmacol 14:1086946. https://doi.org/10.3389/FPHAR.2023.1086946/BIBTEX

Buhimschi AD, Armstrong HA, Toure M, Jaime-Figueroa S, Chen TL, Lehman AM, Woyach JA, Johnson AJ, Byrd JC, Crews CM (2018) Targeting the C481S ibrutinib-resistance mutation in Bruton's tyrosine kinase using PROTAC-mediated degradation. Biochemistry 57(26):3564–3575. https://doi.org/10.1021/ACS.BIOCHEM.8B00391

Burslem GM, Schultz AR, Bondeson DP, Eide CA, Samantha L, Stevens S, Druker BJ, Crews CM (2019) Targeting BCR-ABL1 in chronic myeloid leukemia by PROTAC-mediated targeted protein degradation. Cancer Res 79(18):4744–4753. https://doi.org/10.1158/0008-5472.CAN-19-1236

Cao R, Wang L, Wang H, Xia L, Erdjument-Bromage H, Tempst P, Jones RS, Zhang Y (2002) Role of histone H3 lysine 27 methylation in polycomb-group silencing. Science 298(5595):1039–1043. https://doi.org/10.1126/SCIENCE.1076997

Dobrovolsky D, Wang ES, Morrow S, Leahy C, Faust T, Nowak RP, Donovan KA et al (2019) Bruton tyrosine kinase degradation as a therapeutic strategy for cancer. Blood 133(9):952–961. https://doi.org/10.1182/BLOOD-2018-07-862953

Dominici M, Porazzi P, Xiao Y, Chao A, Tang HY, Kumar G, Fortina P et al (2020) Selective inhibition of Ph-positive ALL cell growth through kinase-dependent and -independent effects by CDK6-specific PROTACs. Blood 135(18):1560. https://doi.org/10.1182/BLOOD.2019003604

Falini B, Mecucci C, Tiacci E, Alcalay M, Rosati R, Pasqualucci L, La Starza R et al (2005) Cytoplasmic nucleophosmin in acute myelogenous leukemia with a normal karyotype. N Engl J Med 352(3):254–266. https://doi.org/10.1056/NEJMOA041974

Fisher SL, Phillips AJ (2018) Targeted protein degradation and the enzymology of degraders. Curr Opin Chem Biol 44:47–55. https://doi.org/10.1016/J.CBPA.2018.05.004

Gao H, Sun X, Rao Y (2020) PROTAC technology: opportunities and challenges. ACS Med Chem Lett 11(3):237–240. https://doi.org/10.1021/acsmedchemlett.9b00597

George B, Chowdhury SM, Hart A, Sircar A, Singh SK, Nath UK, Mamgain M, Singhal NK, Sehgal L, Jain N (2020) Ibrutinib resistance mechanisms and treatment strategies for B-cell lymphomas. Cancer 12(5):1328. https://doi.org/10.3390/CANCERS12051328

Gibaja V, Shen F, Harari J, Korn J, Ruddy D, Saenz-Vash V, Zhai H et al (2016) Development of secondary mutations in wild-type and mutant EZH2 alleles cooperates to confer resistance to EZH2 inhibitors. Oncogene 35(5):558–566. https://doi.org/10.1038/ONC.2015.114

Hafezi S, Rahmani M (2021) Targeting BCL-2 in cancer: advances, challenges, and perspectives. Cancer 13(6):1–13. https://doi.org/10.3390/CANCERS13061292

Hantschel O, Warsch W, Eckelhart E, Kaupe I, Grebien F, Wagner KU, Superti-Furga G, Sexl V (2012) BCR-ABL uncouples canonical JAK2-STAT5 signaling in chronic myeloid leukemia. Nat Chem Biol 8(3):285–293. https://doi.org/10.1038/NCHEMBIO.775

He Y, Khan S, Huo Z, Lv D, Zhang X, Liu X, Yuan Y et al (2020a) Proteolysis targeting chimeras (PROTACs) are emerging therapeutics for hematologic malignancies. J Hematol Oncol 13(1):103. https://doi.org/10.1186/S13045-020-00924-Z

He Y, Zhang X, Chang J, Kim HN, Zhang P, Wang Y, Khan S et al (2020b) Using proteolysis-targeting chimera technology to reduce navitoclax platelet toxicity and improve its senolytic activity. Nat Commun 11:1–14. https://doi.org/10.1038/s41467-020-15838-0

Herviou L, Cavalli G, Cartron G, Klein B, Moreaux J (2016) EZH2 in normal hematopoiesis and hematological malignancies. Oncotarget 7(3):2284–2296. https://doi.org/10.18632/ONCOTARGET.6198

Hines J, Lartigue S, Dong H, Qian Y, Crews CM (2019) MDM2-recruiting PROTAC offers superior, synergistic antiproliferative activity via simultaneous degradation of BRD4 and stabilization of p53. Cancer Res 79(1):251–262. https://doi.org/10.1158/0008-5472.CAN-18-2918

Hsu JHR, Rasmusson T, Robinson J, Pachl F, Read J, Kawatkar S, O'Donovan DH et al (2020) EED-targeted PROTACs degrade EED, EZH2, and SUZ12 in the PRC2 complex. Cell Chem Biol 27(1):41–46. https://doi.org/10.1016/J.CHEMBIOL.2019.11.004

Huang HT, Dobrovolsky D, Paulk J, Yang G, Weisberg EL, Doctor ZM, Buckley DL et al (2018) A chemoproteomic approach to query the degradable kinome using a multi-kinase degrader. Cell Chem Biol 25(1):88–99. https://doi.org/10.1016/J.CHEMBIOL.2017.10.005

Huang L, Huang J, Huang J, Xue H, Liang Z, Jun W, Chen C (2020) Nanomedicine – a promising therapy for hematological malignancies. Biomater Sci 8(9):2376–2393. https://doi.org/10.1039/D0BM00129E

Iwama A (2017) Polycomb repressive complexes in hematological malignancies. Blood 130(1):23–29. https://doi.org/10.1182/BLOOD-2017-02-739490

Jiang F, Wei Q, Li H, Li H, Cui Y, Ma Y, Chen H, Cao P, Tao L, Chen Y (2020) Discovery of novel small molecule induced selective degradation of the bromodomain and extra-terminal (BET) bromodomain protein BRD4 and BRD2 with cellular potencies. Bioorg Med Chem 28(1):115181. https://doi.org/10.1016/J.BMC.2019.115181

Johnson DE, O'Keefe RA, Grandis JR (2018) Targeting the IL-6/JAK/STAT3 signalling axis in cancer. Nat Rev Clin Oncol 15(4):234–248. https://doi.org/10.1038/NRCLINONC.2018.8

Kang CH, Lee DH, Lee CO, Du Ha J, Park CH, Hwang JY (2018) Induced protein degradation of anaplastic lymphoma kinase (ALK) by proteolysis targeting chimera (PROTAC). Biochem Biophys Res Commun 505(2):542–547. https://doi.org/10.1016/J.BBRC.2018.09.169

Keskin D, Sadri S, Eskazan AE (2016) Dasatinib for the treatment of chronic myeloid leukemia: patient selection and special considerations. Drug Des Devel Ther 10:3355–3361. https://doi.org/10.2147/DDDT.S85050

Khan S, Zhang X, Lv D, Zhang Q, He Y, Zhang P, Liu X et al (2019) A selective BCL-XL PROTAC degrader achieves safe and potent antitumor activity. Nat Med 25(12):1938–1947. https://doi.org/10.1038/S41591-019-0668-Z

Khan AN, Asija S, Pendhari J, Purwar R (2024) CAR-T cell therapy in hematological malignancies: where are we now and where are we heading for? Eur J Haematol 112(1):6–18. https://doi.org/10.1111/EJH.14076

Kim SA, Go A, Jo SH, Park SJ, Jeon YU, Kim JE, Lee HK et al (2019) A novel cereblon modulator for targeted protein degradation. Eur J Med Chem 166:65–74. https://doi.org/10.1016/J.EJMECH.2019.01.023

Komander D, Clague MJ, Urbé S (2009) Breaking the chains: structure and function of the deubiquitinases. Nat Rev Mol Cell Biol 10(8):550–563. https://doi.org/10.1038/NRM2731

Kong X, Pan P, Sun H, Xia H, Wang X, Li Y, Hou T (2019) Drug discovery targeting anaplastic lymphoma kinase (ALK). J Med Chem 62(24):10927–10954. https://doi.org/10.1021/ACS.JMEDCHEM.9B00446

Lai AC, Toure M, Hellerschmied D, Salami J, Jaime-Figueroa S, Ko E, Hines J, Crews CM (2016) Modular PROTAC design for the degradation of oncogenic BCR-ABL. Angew Chem 55(2):807–810. https://doi.org/10.1002/ANIE.201507634

Levis M, Small D (2003) Novel FLT3 tyrosine kinase inhibitors. Expert Opin Investig Drugs 12(12):1951–1962. https://doi.org/10.1517/13543784.12.12.1951

Levis M, Murphy KM, Pham R, Kim KT, Stine A, Li IMN, Douglas Smith B, Small D (2005) Internal tandem duplications of the FLT3 gene are present in leukemia stem cells. Blood 106(2):673. https://doi.org/10.1182/BLOOD-2004-05-1902

Lu J, Qian Y, Altieri M, Dong H, Wang J, Raina K, Hines J et al (2015) Hijacking the E3 ubiquitin ligase cereblon to efficiently target BRD4. Chem Biol 22(6):755–763. https://doi.org/10.1016/J.CHEMBIOL.2015.05.009

Ma B, Feng H, Feng C, Liu Y, Zhang H, Wang J, Wang W, He P, Niu F (2022) Kill two birds with one stone: a multifunctional dual-targeting protein drug to overcome imatinib resistance in Philadelphia Chromosome-positive leukemia. Adv Sci 9(13):2104850. https://doi.org/10.1002/ADVS.202104850

Mani A, Gelmann EP (2005) The ubiquitin-proteasome pathway and its role in cancer. J Clin Oncol 23(21):4776–4789. https://doi.org/10.1200/JCO.2005.05.081

Marty C, Pecquet C, Nivarthi H, El-Khoury M, Chachoua I, Tulliez M, Villeval JL et al (2016) Calreticulin mutants in mice induce an MPL-dependent thrombocytosis with frequent progression to myelofibrosis. Blood 127(10):1317–1324. https://doi.org/10.1182/BLOOD-2015-11-679571

Mitelman F, Johansson B, Mertens F (2007) The impact of translocations and gene fusions on cancer causation. Nat Rev Cancer 7(4):233–245. https://doi.org/10.1038/NRC2091

Nabbouh AI, Hleihel RS, Saliba JL, Karam MM, Hamie MH, Hsin Chieh JMW, Berthier CP et al (2017) Imidazoquinoxaline derivative EAPB0503: a promising drug targeting mutant nucleophosmin 1 in acute myeloid leukemia. Cancer 123(9):1662–1673. https://doi.org/10.1002/CNCR.30515

Nishida Y, Scruggs DA, Ayoub E, Patsilevas T, Ruvolo VR, Mak PY, Carter BZ et al (2022) AML-147 C-MYC targeting by degradation: novel dual c-Myc/GSPT1 degrader GT19715 exerts profound cell kill in vitro and in vivo in acute myeloid leukemia and lymphomas. Clin Lymphoma Myeloma Leuk 22:S218. https://doi.org/10.1016/S2152-2650(22)01230-7

Paiva SL, Crews CM (2019) Targeted protein degradation: elements of PROTAC design. Curr Opin Chem Biol 50:111. https://doi.org/10.1016/J.CBPA.2019.02.022

Park J, Cho J, Song EJ (2020) Ubiquitin-proteasome system (UPS) as a target for anticancer treatment. Arch Pharm Res 43(11):1144–1161. https://doi.org/10.1007/S12272-020-01281-8

Potjewyd F, Turner AMW, Beri J, Rectenwald JM, Norris-Drouin JL, Cholensky SH, Margolis DM, Pearce KH, Herring LE, James LI (2020) Degradation of polycomb repressive complex 2 with an EED-targeted bivalent chemical degrader. Cell Chem Biol 27(1):47–56. https://doi.org/10.1016/J.CHEMBIOL.2019.11.006

Powell CE, Gao Y, Tan L, Donovan KA, Nowak RP, Loehr A, et al. (2018) Chemically induced degradation of anaplastic lymphoma kinase (ALK). J Med Chem. 61:4249–4255.

Qin C, Yang H, Zhou B, Fernandez-Salas E, Yang CY, Liu, McEachern D et al (2018) Discovery of QCA570 as an exceptionally potent and efficacious proteolysis targeting chimera (PROTAC) degrader of the bromodomain and extra-terminal (BET) proteins capable of inducing complete and durable tumor regression. J Med Chem 61(15):6685–6704. https://doi.org/10.1021/ACS.JMEDCHEM.8B00506

Raina K, Jing L, Qian Y, Altieri M, Gordon D, Ann MK, Rossi JW et al (2016) PROTAC-induced BET protein degradation as a therapy for castration-resistant prostate cancer. Proc Natl Acad Sci USA 113(26):7124–7129. https://doi.org/10.1073/PNAS.1521738113

Rowley JD (2008) Chromosomal translocations: revisited yet again. Blood 112(6):2183–2189. https://doi.org/10.1182/BLOOD-2008-04-097931

Sakamoto KM, Kim KB, Kumagai A, Mercurio F, Crews CM, Deshaies RJ (2001) Protacs: chimeric molecules that target proteins to the Skp1-Cullin-F box complex for ubiquitination and degradation. Proc Natl Acad Sci USA 98(15):8554–8559. https://doi.org/10.1073/PNAS.141230798

Sakamoto KM, Kim KB, Verma R, Ransick A, Stein B, Crews CM, Deshaies RJ (2003) Development of Protacs to target cancer-promoting proteins for ubiquitination and degradation. Mol Cell Proteomics 2(12):1350–1358. https://doi.org/10.1074/MCP.T300009-MCP200

Scheicher R, Hoelbl-Kovacic A, Bellutti F, Sarmiza Tigan A, Prchal-Murphy M, Heller G, Schneckenleithner C et al (2015) CDK6 as a key regulator of hematopoietic and leukemic stem cell activation. Blood 125(1):90. https://doi.org/10.1182/BLOOD-2014-06-584417

Schlenk RF, Döhner K, Krauter J, Fröhling S, Corbacioglu A, Bullinger L, Habdank M et al (2008) Mutations and treatment outcome in cytogenetically normal acute myeloid leukemia. N Engl J Med 358(18):1909–1918. https://doi.org/10.1056/NEJMOA074306

Sun N, Ren C, Kong Y, Zhong H, Chen J, Li Y, Zhang J et al (2020) Development of a Brigatinib degrader (SIAIS117) as a potential treatment for ALK positive cancer resistance. Eur J Med Chem 193:112190. https://doi.org/10.1016/J.EJMECH.2020.112190

Winter GE, Buckley DL, Paulk J, Roberts JM, Souza A, Dhe-Paganon S, Bradner JE (2015) Drug development. Phthalimide conjugation as a strategy for in vivo target protein degradation. Science 348(6241):1376–1381. https://doi.org/10.1126/SCIENCE.AAB1433

Woyach JA, Furman RR, Liu T-M, Ozer HG, Zapatka M, Ruppert AS, Xue L et al (2014) Resistance mechanisms for the Bruton's tyrosine kinase inhibitor ibrutinib. N Engl J Med 370(24):2286–2294. https://doi.org/10.1056/NEJMOA1400029

Yang Y, Gao H, Sun X, Sun Y, Qiu Y, Weng Q, Rao Y (2020) Global PROTAC toolbox for degrading BCR-ABL overcomes drug-resistant mutants and adverse effects. J Med Chem 63(15):8567–8583. https://doi.org/10.1021/ACS.JMEDCHEM.0C00967

Zhang J, Ding L, Holmfeldt L, Gang W, Heatley SL, Payne-Turner D, Easton J et al (2012) The genetic basis of early T-cell precursor acute lymphoblastic leukaemia. Nature 481(7380):157–163. https://doi.org/10.1038/NATURE10725

Zhang C, Han XR, Yang X, Jiang B, Liu J, Xiong Y, Jin J (2018) Proteolysis targeting chimeras (PROTACs) of anaplastic lymphoma kinase (ALK). Eur J Med Chem 151:304–314. https://doi.org/10.1016/J.EJMECH.2018.03.071

Zhang L, Riley-Gillis B, Vijay P, Shen Y (2019) Acquired resistance to BET-PROTACs (proteolysis-targeting chimeras) caused by genomic alterations in core components of E3 ligase complexes. Mol Cancer Ther 18(7):1302–1311. https://doi.org/10.1158/1535-7163.MCT-18-1129

Zhao Y, Shu Y, Lin J, Chen Z, Xie Q, Bao Y, Lixue L, Sun N, Wang Y (2021) Discovery of novel BTK PROTACs for B-Cell lymphomas. Eur J Med Chem 225:113820. https://doi.org/10.1016/J.EJMECH.2021.113820

Zhong L, Li Y, Xiong L, Wang W, Ming W, Yuan T, Yang W et al (2021) Small molecules in targeted cancer therapy: advances, challenges, and future perspectives. Signal Transduct Target Ther 6(1):1–48. https://doi.org/10.1038/s41392-021-00572-w

Zhou B, Jiantao H, Fuming X, Chen Z, Bai L, Fernandez-Salas E, Lin M et al (2018) Discovery of a small-molecule degrader of bromodomain and extra-terminal (BET) proteins with picomolar cellular potencies and capable of achieving tumor regression. J Med Chem 61(2):462–481. https://doi.org/10.1021/ACS.JMEDCHEM.6B01816

Zia V, Lengyel CG, Tajima CC, de Mello RA (2023) Advancements of ALK inhibition of non-small cell lung cancer: a literature review. Transl Lung Cancer Res 12(7):1563–1574. https://doi.org/10.21037/TLCR-22-619/COIF

Artificial Intelligence and Machine Learning for Exploring PROTAC in Underutilized Cells

14

Ruchi Tandon and Parveen Kumar

Abstract

PROTACs consist of a protein-targeting ligand chemically linked to an E3 ligase ligand. PROTACs facilitate the ubiquitination and degradation of target proteins by utilizing E3 ligases. The capacity of PROTACs to selectively induce protein breakdown, as opposed to inhibition, presents novel opportunities for tackling diseases linked to proteins that are difficult to address using conventional small-molecule inhibitors. Designing PROTAC molecules involves a careful selection of the recognition element, linker, and ubiquitin ligase binding element to ensure specificity, efficacy, and minimal off-target effects. Computational methods, structure-activity relationship studies, and iterative optimization are often employed in the design process. This opens newer avenues for utilizing the rapidly evolving Artificial Intelligence in PROTAC development.

Recent exploration methods have mostly been centered around: (a) Using molecular docking to precisely model the protein-PROTAC-E3 ternary complex for PROTACs. This approach enables quick study of structure-activity relationships and enhances ligand affinity and selectivity. Virtual screening enables researchers to uncover potential ligands for the E3 ligase and target protein. (b) Generative methods explore chemical space to generate novel PROTAC structures. Deep learning models assist in predicting ligand-protein interactions and optimizing drug properties, and (c) The stability of the ternary complex generated by PROTACs, target proteins, and E3 ligases is assessed using AI-driven Molecular Dynamics (MD) simulations. In summary, AI and ML accelerate

R. Tandon (✉)
Translational Health Science and Technology Institute, Faridabad, India
e-mail: ruchi.tandon@thsti.res.in

P. Kumar
Oracle India Private Limited, Gurugram, India
e-mail: parveen.kumar@oracle.com

M. Nandave, P. Jain (eds.), *PROTAC-Mediated Protein Degradation: A Paradigm Shift in Cancer Therapeutics*, https://doi.org/10.1007/978-981-97-5077-1_14

PROTAC drug discovery by enhancing ligand selection, predicting interactions, and optimizing drug properties. This chapter explores the transformative potential of integration of PROTAC technology and the rapidly advancing field of AI-ML in drug discovery.

Keywords

PROTAC · Neural networks · Deep learning · Ubiquitin ligase binding element · Linker · Recognition element · LSTM · CNN · RNN · GAN

Abbreviations

AI	Artificial intelligence
ANN	Artificial neural networks
BCM	Binary classification model
CNN	Convolutional neural networks
DGM	Deep generative model
DL	Deep learning
GAN	Generative adversarial networks
GCN	Graph convolutional neural networks
GPU	Graphical processing units
LSTM	Long short-term memory
MAPD	Model free analysis of protein degradability
ML	Machine learning
MLP	Multi-layer perceptron
POI	Protein of interest
PROTAC	Proteolysis targeting chimeras
RNN	Recurrent neural networks
UPS	Ubiquitin-proteasome system

14.1 Introduction

Traditionally, small molecules have been utilized to modulate targets in the field of drug development. In Past, these methods have achieved success in a wide range of biological targets, mainly because pharmacological intervention primarily focuses on specific sites that have well- defined active regions capable of accepting small molecules. However, there are several proteins that do not fully meet this feature and are classified as "undruggable," making them difficult targets. And herein comes PROTACs with potential to bridge this gap.

PROTACs employ a distinct approach by promoting the breakdown of the target proteins. This is accomplished by the mechanism called ubiquitin-proteasome system. PROTACs have the capability to selectively degrade proteins instead of targeting to inhibit them. This offers new opportunities for treating disorders that include proteins which are difficult to address with conventional small-molecule inhibitors.

A PROTAC molecule typically consists of three essential components (Danishuddin et al. 2023):

1. Recognition element is responsible for specifically binding to the target protein. Its design is intended to selectively identify and attach to the specific protein of interest.
2. Linker joins the recognition element with the third component, which recruits the cellular machinery responsible for protein breakdown. The efficiency and specificity of PROTAC-induced degradation have been found to be significantly influenced by the length and composition of the linker.
3. Ubiquitin ligase binding element invokes an enzyme, E3 ubiquitin ligase, which binds ubiquitin molecules to the target protein. Ubiquitin functions as a molecular tag that signals the protein for breakdown by the proteasome.

Proteolysis Targeting Chimeras (PROTACs) are degraders that selectively remove certain proteins by using the ubiquitin-proteasome system (UPS). This strategy has evolved as a distinct method for inhibiting classic targets and disease-related proteins that were previously considered "undruggable" using small-molecule inhibitors (Guedeney et al. 2023). In contrast to protein inhibition, this approach utilizes the cell's intrinsic protein breakdown mechanism, known as the Ubiquitin-Proteasome System (UPS), to selectively eliminate labeled proteins. Ubiquitin-activating enzymes (E1), ubiquitin-conjugating enzymes (E2), and ubiquitin ligases (E3) together carry out a series of complex sequential steps adding Polyubiquitin chains to a target protein. This then acts as a recognition marker for its proteolytic degradation via the proteasome. By utilizing the UPS mechanism, PROTACs can induce the reduction of a wide range of proteins, such as skeletal proteins, enzymes, nuclear receptors, regulatory proteins, and transcription factors known to be anomalous in cancer and other disorders.

Designing PROTAC molecules involves a careful selection of the recognition element, linker, and ubiquitin ligase binding element to ensure specificity, efficacy, and minimal off-target effects. Computational methods, structure-activity relationship studies, and iterative optimization are often employed in the design process. Artificial intelligence and machine learning has been increasing utilized to predict the most effective PROTAC designs by leveraging existing information on protein structures and interactions. Some of the recent studies for cross utilizing developments in Artificial Intelligence and Machine learning and PROTACs have been around:

1. Ligand design: AI-powered computational chemistry techniques aid in the creation of small compounds that effectively bind to both the target protein and the

E3 ligase, promoting their interaction and eventual degradation. These tools can optimize properties such as potency, selectivity, and pharmacokinetics.
2. Predicting protein-protein interactions: The effectiveness of PROTAC depends on the creation of a ternary complex involving the target protein, PROTAC molecule, and E3 ligase. AI models can predict protein-protein interactions, helping in the selection of compatible ligands and optimizing the design of PROTAC molecules.
3. Off-target effects prediction: AI algorithms can predict potential off-target interactions of PROTAC molecules, reducing the risk of unintended side effects. This predictive capability enhances the safety profile of PROTACs during the drug development process.
4. Drug repurposing: AI-driven drug repurposing approaches can identify existing compounds with the potential to be repurposed as PROTACs, accelerating the drug discovery process by leveraging existing safety and pharmacokinetic data.
5. Structure-based drug design: AI-powered molecular modeling techniques, such as deep learning-based protein structure prediction and molecular docking, aid in understanding the structural basis of PROTAC-target interactions. This knowledge guides the rational design of more potent and selective PROTAC molecules.
6. Optimizing PROTAC linker chemistry: An important factor in PROTAC's efficacy is the linker that joins the target-binding and E3 ligase-binding portions. By considering properties like cell permeability, flexibility, and stability, AI algorithms can predict the best linker properties.
7. Clinical trial optimization: AI-powered predictive modeling can optimize clinical trial designs by identifying the patient groups who are most likely to experience positive outcomes with PROTAC therapy by predicting the response rates and optimize the dosage schedules.

14.2 PROTAC Design

A PROTAC functions by bringing the target protein and the E3 ubiquitin ligase in proximity. Once the PROTAC binds to the target protein, it invokes the E3 ligase to help transfer ubiquitin molecules to the target protein. Ubiquitination tags the protein for degradation by the proteasome, resulting in its degradation into smaller peptides. The design of PROTACs is an interdisciplinary process that involves a deep understanding of the target protein, expertise in medicinal chemistry, computational modeling, and iterative refinement based on experimental data. The incorporation of computational tools, artificial intelligence, and big data analytics is becoming more and more crucial in expediting and enhancing the PROTAC design process.

PROTAC design comprises three essential elements: The warhead, the E3 ligand, and the Linker. While the process of discovering warheads and E3 ligands shares similarities with that of small molecules, the design of linkers presents a particular challenge due to the inability of the target protein of interest (POI) and E3 ligases to connect without an effective PROTAC. The degradation efficiency is significantly

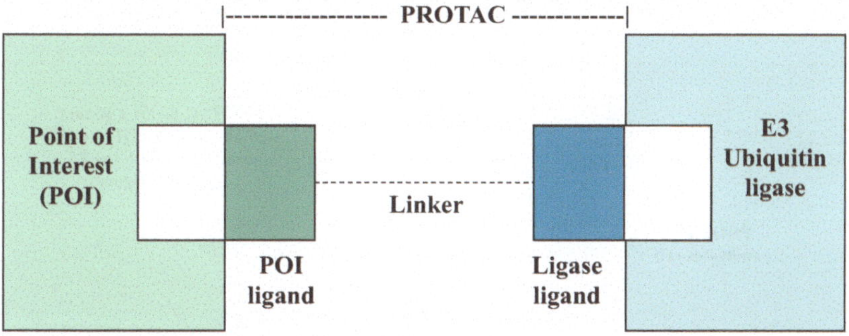

Fig. 14.1 PROTAC structure

influenced by the length, content, flexibility, and attachment sites of the linker. With a growing understanding of the significance of linkers in the physicochemical properties and degrading activity of PROTACs, many researchers have now started working on the development of de novo PROTAC designs.

Further, the clinical translation of PROTACs is impeded by other design challenges, including inadequate solubility, permeability, bioavailability, and unpredictable hook effects, which are observed in PROTACs. Hence, it is imperative to explore novel approaches to enhance the efficacy of PROTACs' discovery.

Structurally, as depicted in Fig. 14.1, the design of PROTACs is based on the combination of three components:

1. An E3 ligase binder
2. A target of interest (TOI) ligand
3. A linker that joins the above two

The two ligands, namely the protein of interest (POI) and the E3 ubiquitin ligase, are connected by a linker that can vary in both length and composition. The function of the ligase ligand is to identify an E3 ubiquitin ligase, while the POI ligand will mobilize the intended biological target. Schematically, PROTAC working can be depicted in Fig. 14.2.

To determine the optimal combination of these three factors, a thorough analysis of the structural properties of the E3 ligase and the TOI, along with molecular modeling and dynamics, is necessary. The spatial orientation and alignment of both parts play a critical role in the development of a ternary complex and are essential for effectively reducing the levels of ubiquitin-protein. Further, the stability of the latter, the speed at which the labeled target is processed by the proteasome, and the rates of newly synthesized proteins impact the effectiveness of PROTAC. PROTACs design steps are illustrated in Fig. 14.3.

Researchers/students can identify the computational steps and utilize machine learning to optimize some of these steps.

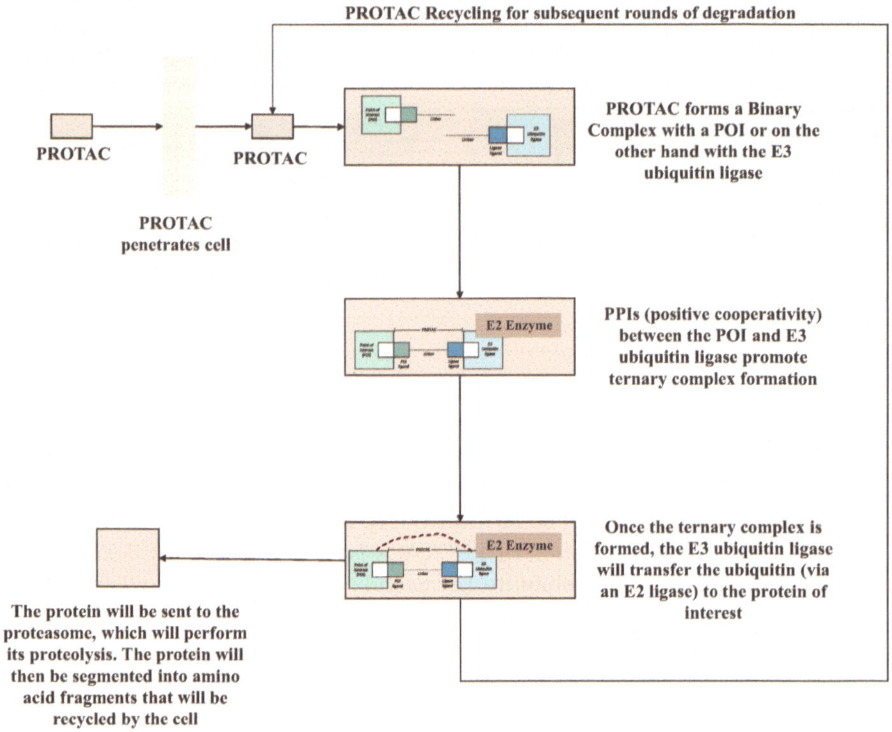

Fig. 14.2 How PROTAC works—a high level view

14.3 Artificial Intelligence and Machine Learning

The goal of machine learning is to equip computers with the ability to learn and improve upon existing data to accomplish activities that were previously reserved for humans, such as image classification, data analysis, and price prediction. These models tend to solve the standard problem depicted in Fig. 14.4.

In a typical Supervised machine learning model, the Input and the Output are available while the relation is learned by the machine from data. In unsupervised model, the input is available while model learns the hidden relations between the data elements. Currently, there are four distinct categories of machine learning that are commonly in use:

14.3.1 Supervised Machine Learning

Supervised machine learning trains algorithms using labeled data sets with tags describing each item of data. Thus, algorithms receive input with an "answer key" indicating how to interpret it. An algorithm may be, for example, fed photographs of diseased tissue with tags for each disease indicator type to better identify them

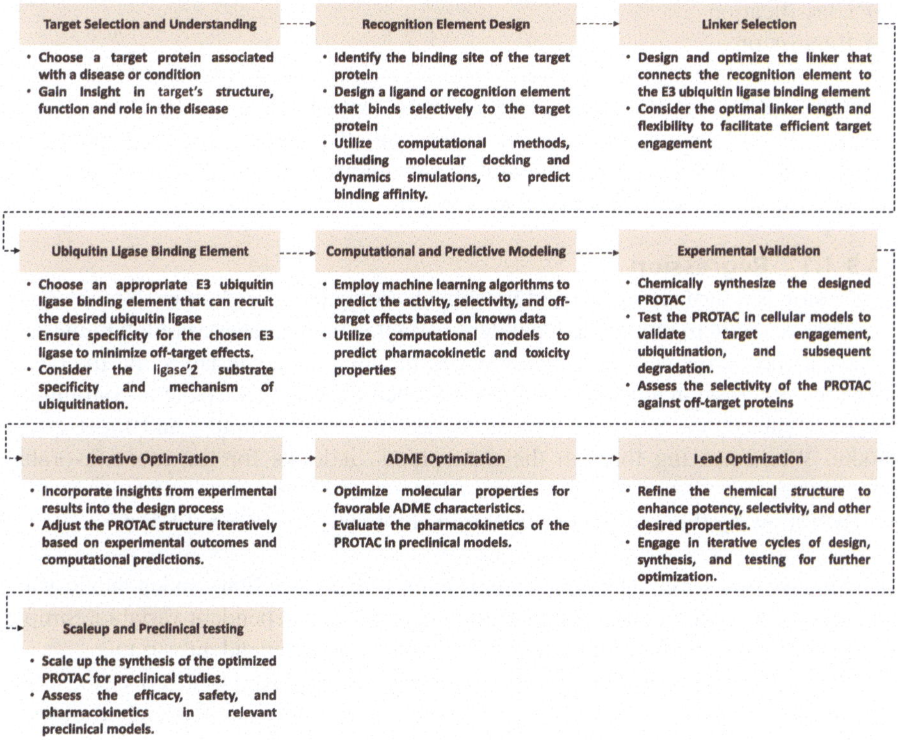

Fig. 14.3 Representative PROTAC design steps

Fig. 14.4. The equation

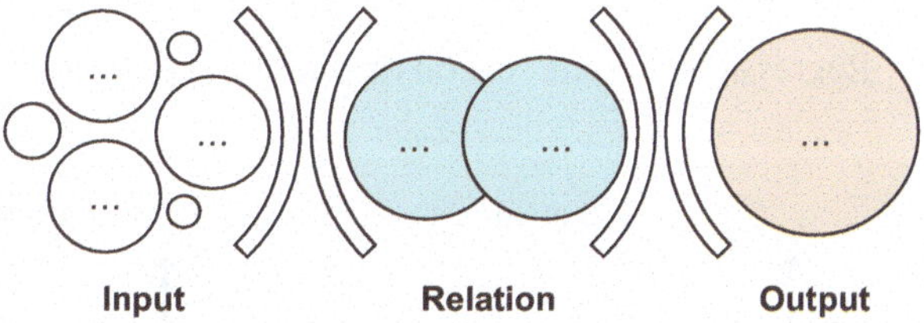

when fed a fresh shot. Prediction and classification models are commonly created using supervised machine learning.

Mathematically, the input variables (x) and an output variable (Y) are available while an algorithm is used to learn the relation function from the input to the output:

$$Y = f(X).$$

The goal is to approximate the relation function "f" so that when new input data (x) is passed through the function, it can predict the output variables (Y) with reasonable accuracy.

Following few paragraphs describe supervised learning in more detail. There are two main types of supervised learning:

(a) Classification
(b) Regression

These are distinguished by the target variable type. In classification cases, it is of categorical type, while in cases of regression, the target variable is numeric. Categorical Variables are finite in number and mainly in form of "strings" or "categories". Numerical Variables usually answer the quantity questions.

14.3.1.1 Regression

Regression is a form of supervised machine learning that uses known data to create predictions about an unknown dependent variable. To estimate or predict numerical values, a correlation is established between the input characteristics and the target variables. Regression analysis is employed when the output variable is continuous, such as "salary" or "weight." Linear regression is the simplest and most direct model. While moving through the data points, it looks for the most favorable hyperplane.

The following three forms of regression are widely used (Fig. 14.5):

(a) Simple regression: The simple regression utilizes a solitary independent variable to produce predictions regarding a continuous dependent variable. Simple linear regression is useful when there is only one independent variable.
(b) Multiple regression: The multiple regression is employed to discover the association between a dependent variable and two or more independent variables.
(c) Non-linear regression: Non-linear regression is employed when there is a lack of a linear relationship between the dependent variable and one or more independent variables.

Regression exhibits several distinctive characteristics:

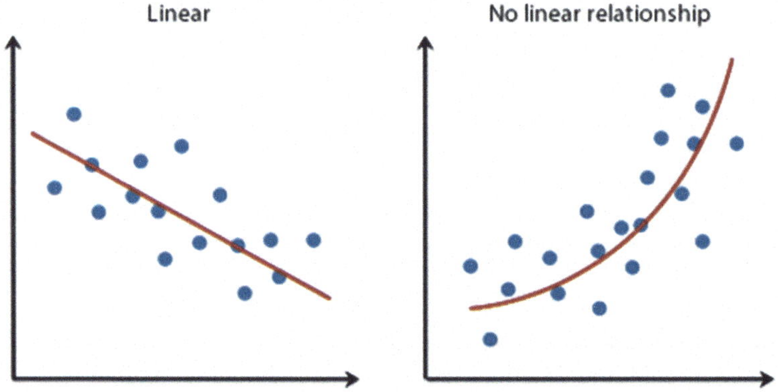

Fig. 14.5 Regression: linear and non-linear relations

(a) Continuous target variable: Regression involves the estimation of continuous target variables, which are numerical values. Some examples include forecasting residential property values, projecting sales numbers, or estimating patient recovery durations.
(b) Error measurement: The evaluation of regression models depends on their capacity to reduce the discrepancy between the anticipated and observed values of the target variable. Widely used error metrics include the mean absolute error (MAE), the mean squared error (MSE), and the root mean squared error (RMSE).
(c) Model complexity: Regression models vary in complexity, ranging from basic linear models to more involved non-linear models. The selection of model complexity is contingent upon the intricacy of the association between the input features and the target variable.
(d) Interpretability: This defines the extent to which humans can understand the reason why a machine learning model created a specific result. The interpretability of regression models is contingent upon the selection of the algorithm. Linear models possess a high degree of interpretability, however, more intricate models may present difficulties in interpretation.

14.3.1.2 Classification

Classification is the process of categorizing anything into smaller and more specific groups. Classification is a crucial component of supervised machine learning, where we utilize labeled or preclassified data to train the model. Machine learning encompasses four primary classification types:

(a) Binary classification: Binary classification is the procedure of classifying the input into two clearly defined groups or categories. For Instance, given the specified health conditions, classification results will identify if an individual subject is afflicted with a certain sickness or not.
(b) Multiclass classification: Multiclass classification involves the categorization of input data into multiple distinct classes or categories. An application of mul-

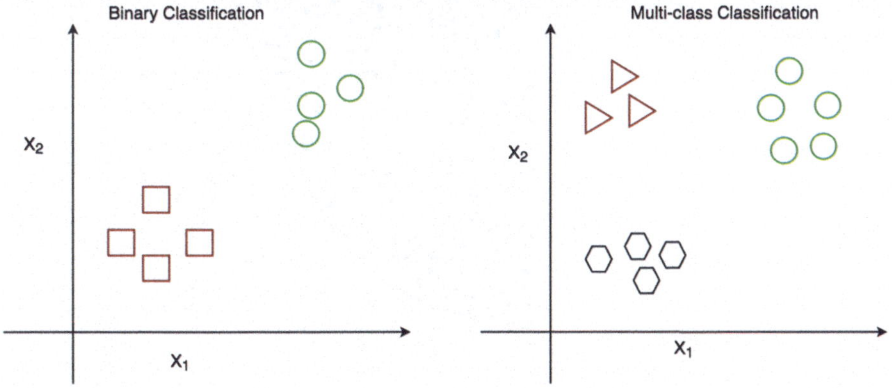

Fig. 14.6 Classification: binary and multiclass classification

ticlass classification can be when we want to separate out multiple flowers from various species according to their respective species. (Fig. 14.6).

(c) Multi-label classification: Multi-label classification involves selecting the most appropriate labels for a new data item from a predefined collection of multiple labels. Unlike multiclass classification, which assigns each data point to only one class, this situation is distinct in that there can be multiple labels for each data point. For instance, a multi-label classification method can be employed to categorize movies into one or more of the specified groups, for example, you could be tagged "movies" by genre: comedy, drama, romance, horror, etc. It may so happen that same movie could have comedy and horror components (Fig. 14.6).

(d) Imbalanced classification: Imbalanced classification entails the forecasting of whether a data point pertains to a minority class, despite the presence of a significantly larger number of samples from the majority class. An instance of this occurs when a medical diagnostic algorithm is employed to forecast if a patient is afflicted with a rare ailment, despite the presence of a larger number of patients suffering from more prevalent ailments.

Mathematically, there are various types of classifier algorithms, for example:

14.3.1.2.1 Linear Classifiers

Linear models employ a boundary that is linear to differentiate between different groups. They are simple and have efficient computation. Here are a few examples of linear classification models:

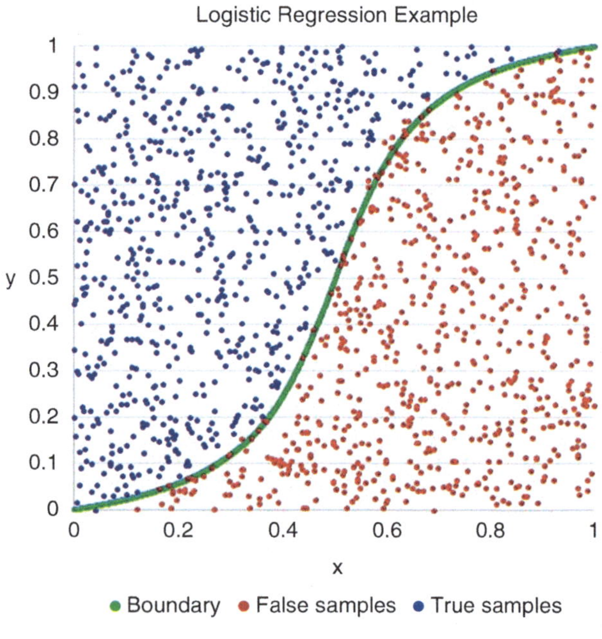

Fig. 14.7 Logistic regression

(a) Logistic regression: Logistic regression is used to predict the likelihood of an event happening, such as voting or not voting, by considering a group of independent factors. Given that the result is a probability, the variable that is influenced by other factors is limited to a spectrum of values between 0 and 1. Logistic regression involves applying a logit (sigmoid) transformation to the odds, which is the ratio between the probability of success and the probability of failure (Fig. 14.7). The term frequently used to refer to this is the log odds, which is the natural logarithm of the odds.

(b) Support vector machines: The goal of the support vector machine algorithm is to determine the most effective line, known as the decision boundary, that separates the data points belonging to distinct classes. Within feature spaces with many dimensions, this boundary is commonly known as a hyperplane. The goal is to maximize the margin, which is the distance between the hyperplane and the closest data points from each category. This enables the distinct categorization of data classes, as illustrated in Fig. 14.8.

(c) Single-layer perceptron: A single-layer perceptron (SLP) is a type of neural network that operates in a feed-forward fashion and employs a threshold transfer function. The SLP is a basic form of artificial neural network that is limited to classifying situations that can be separated by a straight line and have a binary outcome, either 1 or 0.

(d) Stochastic gradient descent (SGD) classifier: Stochastic Gradient Descent (SGD) is an adapted form of the Gradient Descent algorithm to address the computational inefficiencies inherent in traditional Gradient Descent methods when dealing with large datasets. Stochastic Gradient Descent (SGD) takes an alternative technique by randomly selecting only one training example (or a small batch) for each iteration to calculate the gradient and update the model parameters, instead of using the entire dataset. The incorporation of this arbitrary assortment introduces an element of uncertainty into the process of optimization, which is why the term "stochastic".

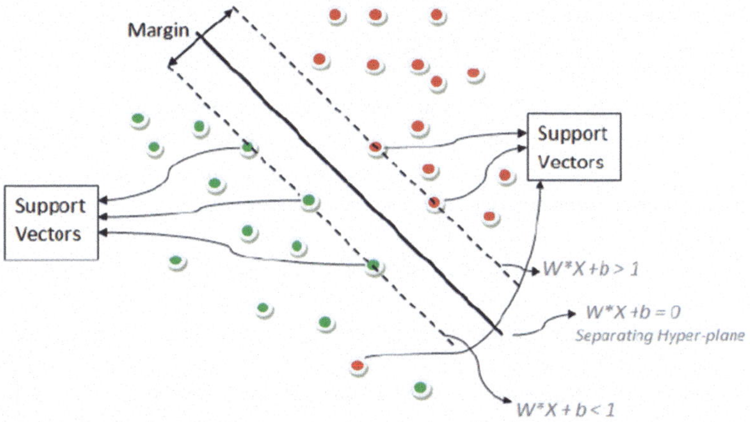

Fig. 14.8 Support vector machines

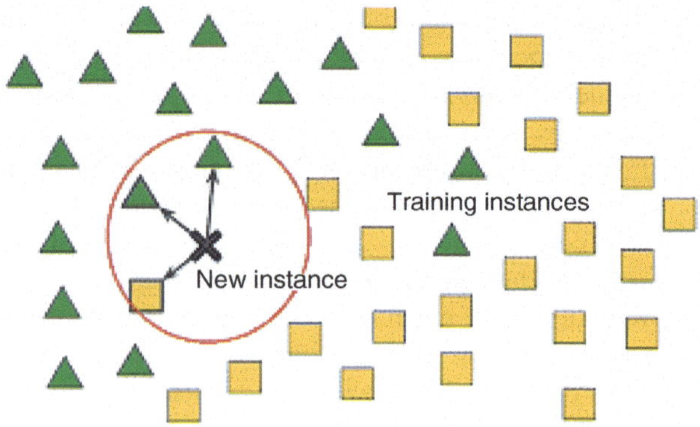

Fig. 14.9 K-Nearest neighbors

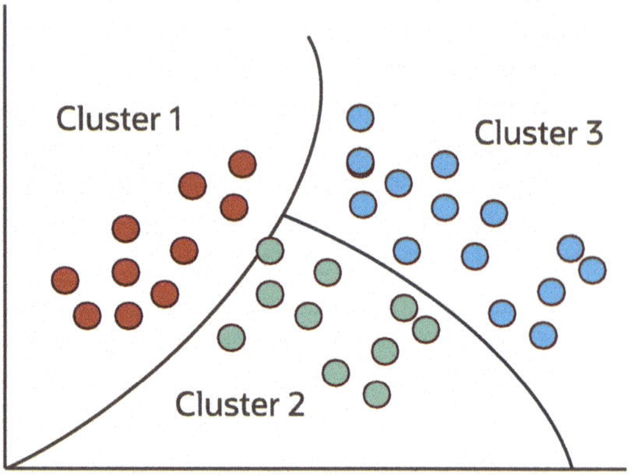

Fig. 14.10. Naïve Bayes

14.3.1.2.2 Non-linear Classifiers

In Non-Linear Models, the decision boundary between classes is modeled using non-linear functions. They can pick up on more complex and nuanced relations between the input characteristics and the output variable. Here are a few examples of non-linear classification models:

(a) K-Nearest neighbors

The key premise of the K-Nearest Neighbors model is that values and labels assigned to data points that are statistically comparable tend to be consistent (Fig. 14.9). In other words, a data point closer to a cluster of points classified as 'A' has a higher probability of getting classified as 'A'.

(b) Kernel SVM

The main principle of kernel approaches is transforming the input data into a feature space with higher dimensionality, facilitating the differentiation between classes or the generation of predictions. Kernel methods utilize a function, called the kernel function, to automatically transform the data into the feature space. The Support Vector Machine (SVM) is one of the most often used type of kernel technique. It is a binary classifier that identifies the optimal hyperplane for efficiently separating the two groups. SVMs employ a kernel function to effectively transfer the input data into a higher-dimensional space to determine the optimal hyperplane.

(c) Naïve Bayes

The Naïve Bayes algorithm is a type of supervised learning algorithm that utilizes Bayes' theorem to solve classification issues. It is mostly utilized in text classification tasks including a training dataset with many dimensions. The Naïve Bayes Classifier is a simple and highly efficient classification method that aids in constructing rapid machine learning models capable of making swift predictions (Fig. 14.10). It functions as a probabilistic classifier, making predictions based on the probability of an object. Naïve Bayes Algorithm is commonly employed in several applications such as spam filtration, sentimental analysis, and article classification.

A family of simple "probabilistic classifiers" known as Naïve Bayes classifiers employ Bayes theorem with naive assumptions about features being mutually independent. Here is the equation expressed as

$$P(h/d) = (P(d/h) * P(h)) / P(d)$$

where $P(h/d)$ = probability of hypothesis h happening given the data d (posterior), $P(d/h)$ = probability of data d given the hypothesis h (Likelihood), $P(h)$ = probability of hypothesis h (prior), $P(d)$ = probability of data d (evidence)

Using Bayesian Probability terminology, the above equation can be written as

$$\text{Posterior} = \text{Prior}^{*}\text{Likelihood} / \text{Evidence}$$

Figure 14.10 shows an example clustering using Naïve Bayes.

(d) Decision tree classifier

Although Decision Tree is a type of supervised learning method that can be applied to both classification and regression problems, it is primarily utilized to solve classification issues. The classifier is designed in the form of a tree, with internal nodes denoting the features of the dataset, branches signifying the decision criteria, and leaf nodes signifying the corresponding outcome. A decision tree is composed of two distinct nodes, namely the decision node and the leaf node. Leaf nodes are the results of those decisions and do not contain any additional branches, whereas decision nodes are utilized to make all decisions and contain multiple branches. Decisions or tests are executed in accordance with the characteristics of the provided dataset. It is a graphical representation that, given the conditions, returns every possible solution to a problem or decision.

Additional techniques such as Random Forests, AdaBoost, Bagging Classifier, Multi-layer Artificial Neural Networks, and many more are available. For a more comprehensive examination of the topic, please consult a machine learning textbook.

Assessing a classification model is a crucial step in determining its performance and capacity to generalize on new, unknown data. Various metrics and methodologies exist for assessing a classification model, which rely on the specific need at hand. Below are some of the frequently used evaluation metrics:

(a) Classification accuracy: Classification accuracy is defined as the ratio of correctly classified instances to the total number of instances in the test set. The metric is clear and comprehensible, but it might be misleading in datasets with imbalanced classes, where the accuracy score is strongly influenced by the dominant class.
(b) Confusion matrix: Confusion matrix is a table that presents the number of correctly identified positive instances, correctly identified negative instances, incorrectly identified positive instances, and incorrectly identified negative instances for each category. This tool is valuable for computing many evaluation metrics.
(c) Precision and recall: Precision measures the proportion of correctly identified positive cases out of all instances labeled as positive, whereas recall measures the proportion of correctly identified positive instances out of all actual positive occurrences. These measurements are useful in scenarios when one class is more important than the other, or where there is a requirement to strike a balance between false positive results and false negative results.
(d) The F1-Score is a statistic that quantifies the balance between precision and recall by calculating their harmonic mean. The derivation of this value is accomplished by applying the formula:

$$2 \times (\text{precision} \times \text{recall}) / (\text{precision} + \text{recall}).$$

It is a useful metric for datasets with imbalanced distribution, where both accuracy and recall are important.
(e) The terms ROC curve and AUC stand for Receiver Operating Characteristic curve and Area Under the Curve, respectively. The ROC curve is a visual depiction of how the true positive rate (recall) and the false positive rate (1-specificity) are related to different threshold values of the classifier's decision function. The Area Under the Curve (AUC) measures the overall performance of the classifier, ranging from 0.5 (random guess) to 1 (perfect prediction).
(f) Cross-validation: A technique that divides the data into multiple subsets or folds, trains the model on each subset/fold while evaluating its performance on the remaining subsets/folds. This approach is used to obtain a more accurate evaluation of the model's performance.

It is crucial to select the suitable assessment metric(s) according to the problem and needs and to prevent overfitting by assessing the model on independent test data.

14.3.2 Unsupervised Machine Learning

To train algorithms, unsupervised machine learning makes use of data sets that have not been labeled. By presenting it with untagged data, this method trains the algorithm to find patterns autonomously, without human intervention. To illustrate, a social media platform may collect massive amounts of unlabeled user data to train an algorithm to detect patterns in user behavior. Researchers and data scientists frequently employ unsupervised machine learning to discover patterns efficiently and rapidly in massive, unlabeled datasets.

Unsupervised Learning, commonly referred to as knowledge discovery, utilizes training data that is unlabeled, unclassified, and categorized. The primary objective of unsupervised learning is to uncover concealed and intriguing patterns in data that lacks labeling. Unsupervised learning methods cannot be directly used for regression or classification problems since the values for the output are unknown. Clustering is a prevalent unsupervised learning approach utilized for data analysis to uncover concealed patterns or groups within the data. Cluster analysis is utilized in several fields such as DNA sequence analysis, market research, and object identification. Unsupervised learning commonly employs techniques like clustering, anomaly detection, neural networks, and methods for learning latent variable models.

Unsupervised learning often entails watching several instances of a random vector x and striving to acquire knowledge about the probability distribution $p(x)$, or some notable characteristics of that distribution.

Clustering is the most popular unsupervised learning approach for discovering hidden patterns or groups in data through data analysis. Cluster analysis has several practical uses, such as in the fields of gene sequence analysis, market research, and object recognition. Unsupervised learning often makes use of neural networks, clustering, anomaly detection, and other methods for learning latent variable models.

Some examples of unsupervised learning techniques are:

14.3.2.1 Cluster Analysis

Clustering is the process of organizing data into sets where similar objects are grouped together. Data exploration and data mining make frequent use of clustering. Mixture models, hierarchical clustering, and k-means clustering are some of the most popular clustering algorithms.

14.3.2.2 Principal Component Analysis (PCA)

Principal Component Analysis (PCA) is a method for reducing the dimensionality of a dataset by retaining most of the data's variability while decreasing the number of features. It accomplishes this by identifying the orthogonal axes (principal components) that capture the most amount of variation in the data.

14.3.2.3 Anomaly Detection

Anomaly detection involves the identification of uncommon instances that deviate considerably from most of a dataset. These occurrences are referred to as anomalies or outliers. Popular anomaly detection techniques include k-nearest neighbor and isolation forests. Univariate anomaly detection refers to the process of identifying anomalies or outliers in a single variable or feature. Univariate anomaly identification involves providing an algorithm with a sequence of observations for a single variable, x. The program detects any observation that deviates considerably from the preceding observations. Anomaly detection can also be performed inside a multivariate context. Regression analysis can be employed to determine the relationship between two variables, as an illustration. An anomaly refers to a data point that deviates significantly from the regression line. It is important to understand that in both univariate and multivariate anomaly detection, the model does not receive labels indicating which training examples are anomalies. Instead, it is provided with a set of rules that define the characteristics of similar observations. The model then identifies the observations that deviate the most from the majority on its own.

14.3.2.4 Autoencoders

Autoencoders are neural networks employed for the purpose of unsupervised learning. The system comprises an encoder network that compresses the input data into a latent-space representation and a decoder network that reconstructs the original data from the latent representation. By optimizing the autoencoder to minimize the discrepancy between the original data and its reconstructed version, the model acquires the ability to accurately represent the crucial characteristics of the data.

14.3.2.5 Generative Adversarial Networks (GANs)

GANs are a type of generative model in which two neural networks, namely the generator and the discriminator, are trained concurrently. The generator acquires the ability to produce data that is indiscernible from authentic data, while the discriminator acquires the ability to discriminate between genuine and counterfeit data. GANs have been employed for purposes such as generating images, synthesizing data, and transferring styles.

14.3.3 Semi-supervised Machine Learning

Semi-supervised machine learning makes use of both unlabeled and labeled data sets to train learning algorithms. In most cases, when it comes to semi-supervised machine learning, algorithms are initially provided with a lesser quantity of labeled data to assist in the direction of their development. Subsequently, they are provided with significantly higher amounts of unlabeled data to finish the model. For example, to construct a machine learning model that is capable of voice recognition, an algorithm might be given a smaller quantity of labeled speech data and then trained on a much larger collection of unlabeled speech data. This would allow the algorithm to learn with both labeled and unlabeled data. When significant amounts of labeled data are not available, semi-supervised machine learning is frequently used to train algorithms for classification and prediction purposes.

14.3.4 Reinforcement Learning

For training the algorithms and developing models, reinforcement learning makes use of a process that involves trial and error. Inside the course of the training process, algorithms operate inside predetermined contexts, and they are subsequently provided with feedback following each outcome. In a manner that is analogous to the way in which a young child acquires knowledge, the algorithm gradually acquires an understanding of its environment and begins to optimize its actions to achieve the outcomes that are wanted. For instance, an algorithm can be improved by playing a series of chess games in a row. This gives the algorithm the opportunity to learn from its prior successes and failures in each game. Developing algorithms that need to efficiently perform a succession of decisions or actions to fulfill their objectives, such as playing a game or summarizing an entire text, is a frequent use of reinforcement learning.

14.3.5 Neural Networks and Deep Learning

14.3.5.1 Neural Networks

Neural Networks are currently the most widely employed machine learning techniques. Previously, the significant need for computing power posed a limitation. However, the advent of readily accessible computing capabilities, such as GPUs, has made them highly sought-after and appealing. A neural network is an artificial intelligence technique that enables computers to process data using algorithms inspired by the functioning of the human brain. Deep learning is a machine learning technique that uses interconnected nodes or neurons arranged in a layered framework, like the organization of the human brain. It establishes an adaptive framework that enables computers to learn from errors and enhance their performance consistently.

14.3.5.1.1 Basic Neural Network Architecture

A basic neural network has interconnected artificial neurons in three layers illustrated in Fig. 14.11 and classified as:

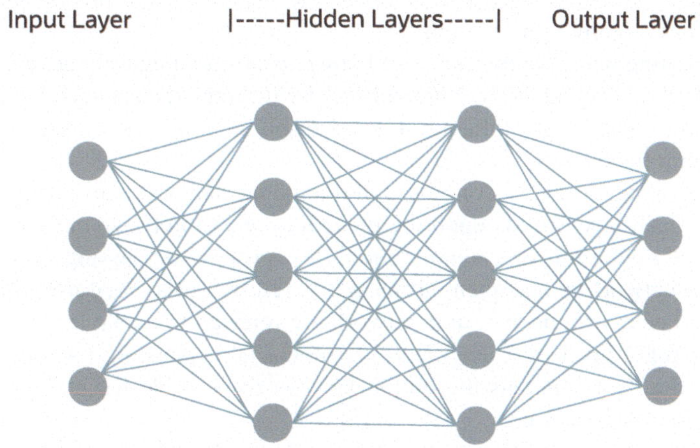

Fig. 14.11 A sample neural network

1. The input layer of the artificial neural network receives information from the external environment. Input nodes perform the tasks of processing, analyzing, or categorizing the data, and then transmit it to the subsequent layer.
2. The hidden layer receives input from the input layer or other hidden levels. Artificial neural networks can possess numerous hidden layers. Every hidden layer scrutinizes the output from the preceding layer, further processes it, and transmits it to the subsequent layer.
3. The output layer of an artificial neural network provides the outcome of the data processing. The layer can consist of either a single node or many nodes. In the case of a binary (yes/no) classification problem, the output layer will consist of a single output node that will provide the result as either 1 or 0. In the case of a multiclass classification problem, the output layer may have many output nodes.

14.3.5.2 Deep Learning

Deep learning is a specific kind of machine learning that involves the use of neural networks consisting of three or more layers. These neural networks aim to replicate the functioning of the human brain; even though they fall short of matching its capabilities. They have the capacity to "learn" from extensive datasets. Although a neural network with only one layer can still provide rough predictions, the inclusion of extra hidden layers can enhance optimization and improve accuracy. Deep learning obviates certain aspects of data preprocessing that are commonly associated with machine learning. These algorithms can take in and analyze unstructured data, such as text and images. They automate the process of extracting important characteristics, reducing the need for human expertise.

For example, deep learning algorithms can identify the most significant characteristics, such as ears, that differentiate one animal from another. To begin, the arrangement of features into a hierarchical structure is determined through manual intervention by a knowledgeable human expert. Subsequently, via the techniques of gradient descent and backpropagation, the deep learning system fine-tunes and adapts itself to enhance accuracy, enabling it to generate more precise predictions regarding a new animal photograph.

Deep learning neural networks, also known as artificial neural networks, aim to replicate the functioning of the human brain by utilizing data inputs, weights, and biases. These components collaborate to precisely identify, categorize, and depict objects within the dataset.

Deep neural networks are composed of interconnected nodes arranged in numerous layers. Each layer builds upon the preceding one to improve the accuracy and efficiency of prediction or categorization. The process of passing computations through the network in a sequential manner is referred to as forward propagation. The visible layers of a deep neural network are referred to as the input and output layers. The input layer of a deep learning model is responsible for receiving and processing the data, whereas the output layer is responsible for making the final prediction or classification.

Backpropagation is a technique that employs algorithms, such as gradient descent, to compute prediction errors. It then modifies the weights and biases of the function by traveling in reverse through the layers to train the model. Forward propagation and backpropagation function in tandem to enable a neural network to make predictions and rectify any faults in a systematic manner. Over time, the algorithm progressively improves its accuracy.

The above passages provide a concise explanation of the most basic form of deep neural networks. In practice, deep learning algorithms are quite complex, and there exist various types of neural networks designed to tackle certain issues or datasets. Convolutional neural networks (CNNs) are largely utilized in computer vision and image classification applications. They can identify characteristics and patterns within an image, allowing for tasks such as object detection or recognition. The concealed layers inside convolutional neural networks execute distinct mathematical operations, such as summarization or filtering, known as convolutions. Convolutional neural networks (CNNs) are highly effective for image classification due to their ability to extract pertinent characteristics from images, which are crucial for image recognition and classification. The new form is more streamlined for efficient processing while retaining essential features necessary for accurate prediction. Every concealed layer in the neural network identifies and analyses distinct visual attributes of the image, such as edges, color, and depth. Recurrent neural networks (RNNs) are commonly employed in applications involving natural language and speech recognition due to their ability to effectively utilize sequential or time series data.

14.4 PROTAC Development Using Machine Learning

In recent years, there has been a significant increase in the generation of large, high-quality datasets in biological systems. This has led to a faster adoption of artificial intelligence (AI) techniques in the field of drug discovery and development. Additionally, the widespread accessibility of expansive storage systems and graphical processing units (GPUs) enables the execution of computationally expensive tasks on a large scale, hence enhancing the speed of numerical calculations. Unlike the process of directly creating physical models, machine learning (ML) techniques employ mathematical algorithms to uncover intricate connections within existing information and generate precise predictions for fresh samples.

Artificial neural networks (ANNs) are widely regarded as one of the most potent machine learning frameworks. Due to the implementation of non-linear activation functions in each neuron unit and the presence of multi-layer structures, artificial neural networks (ANNs) can learn complex correlations between inputs and outputs in datasets. Deep learning (DL) utilizes complex and intricate architectures to extract meaningful features from vast quantities of training data. In itself, deep learning (DL) can be categorized into many subclasses, such as multilayer perceptron (MLP), convolutional neural networks (CNNs), recurrent neural networks (RNNs), graph convolutional neural networks (GCNs), and others, based on different network architectures. RNNs have been extensively employed in analyzing sequential data, such as natural language processing. On the other hand, CNNs have achieved significant accomplishments in handling regular Euclidean data, including images in the field of computer vision. Graph convolutional networks (GCNs) have demonstrated significant efficacy in extracting features from non-Euclidean geometric data, such as graphs and manifolds.

An important benefit of using AI in medicine delivery systems is its capacity to efficiently handle and analyze large volumes of data. Through the utilization of extensive datasets, AI algorithms have the capability to detect correlations and patterns in medication formulations that may not be readily observable using conventional methods. By utilizing a data-driven strategy, researchers can obtain valuable insights on the correlations among different formulation components, their concentrations, and their effects on the effectiveness of drug delivery. Artificial intelligence algorithms have the capability to analyze a wide range of data sources, such as chemical structures, physicochemical qualities, and biological activity data, to reveal concealed correlations. AI can improve drug delivery systems by analyzing patterns and trends to enhance effectiveness, safety, and stability.

For a considerable amount of time, researchers have been utilizing intricate AI models to decipher the mechanisms of diseases. For example, models like AlphaFold2, ESMFold, and MoLeR employ deep learning techniques to accurately forecast the configurations of almost all documented proteins, revolutionizing our comprehension of the associated ailments.

Obtaining efficient molecules with proper linkers for PROTACs can only be achieved through time-consuming trial-and-error techniques. Therefore, it is highly justifiable to implement DL technologies. When using machine learning (ML) to design PROTACs (Proteolysis Targeting Chimeras), several key parameters and features are considered to build predictive models. These parameters play a crucial role in training ML algorithms to predict the efficacy, selectivity, and other characteristics of PROTAC molecules. The below table (Table 14.1) gives some of the key parameters identified and used by various studies for PROTAC design.

The selection and combination of these parameters depend on the specific objectives of PROTAC design, and the iterative process involves refining the ML models based on experimental outcomes and additional data.

Several recent research have concentrated on optimizing the length of the linker to achieve a high level of selectivity and effectiveness in the degradation of proteins mediated by PROTAC. The following Table 14.2, outlines the main characteristics of linkers used in PROTAC development:

Table 14.3 gives a summary of models that have been recently used/proposed for predicting PROTACs.

Three-dimensional (3D) molecular modeling techniques can be used while optimally selecting the characteristics that make a target protein suitable for degradation by PROTACs. Some researchers have been approaching the concept as:

1. Molecular modeling: Visualize and manipulate molecular structures in 3D using software for dynamics simulations, docking, and structure-based drug design.
2. Target protein structure: Obtaining and analyzing the structure of the target protein in 3D, either experimentally or computationally. Understanding the pro-

Table 14.1 Some key parameters commonly used in ML approaches for PROTAC design

#	Feature	Description	Importance
1	Structural features	Representation of the molecular structure of the PROTAC, including information on atoms, bonds, and functional groups	Structural features help capture the three-dimensional arrangement of atoms, influencing binding affinity and selectivity
2	Binding site characteristics	Information about the binding site of the target protein, including amino acid residues, binding pocket properties, and electrostatic potential	Understanding the target binding site aids in predicting the affinity of the PROTAC for the intended target
3	Linker characteristics	Features related to the linker region, such as length, flexibility, and chemical composition	The linker is crucial in determining the proximity between the recognition element and the E3 ligase binding element, hence impacting target engagement and ubiquitination
4	Ubiquitin ligase binding characteristics	Characteristics pertaining to the interaction between the PROTAC and the recruited E3 ubiquitin ligase, such as the affinity of binding and the specificity	Effective application of the ligase is essential for the ubiquitination process and subsequent degradation of the target protein
5	Off-target binding prediction	Features related to potential interactions with unintended off-target proteins	Predicting off-target effects helps enhance the selectivity of PROTAC molecules, reducing the risk of adverse effects
6	Physicochemical properties	Parameters such as molecular weight, lipophilicity, solubility, and hydrogen bond donors/acceptors	The drug-like qualities of PROTACs are influenced by their physicochemical properties, which in turn affect their absorption, distribution, metabolism, and excretion (ADME) profiles
7	ADME-Tox properties	Predictions of toxicity based on machine learning models	Assessing ADME-Tox properties helps in predicting the pharmacokinetic behavior and potential safety concerns of PROTACs
8	Activity labels	Experimental data on the biological activity and efficacy of PROTACs, serve as labels for training ML models	Learning from experimental outcomes helps the ML model predict the activity of new PROTAC candidates
9	Iterative learning Metrics	Metrics assessing model performance during iterative learning, such as Confusion Matrix, ROC, F1 score	Continuous evaluation of model predictions guides the refinement of ML models over successive iterations

Table 14.2 Key features of linkers for PROTAC development

#	Feature	Description
1	Length	• The length of the linker is a crucial factor as it controls the spatial separation between the ligands that target the protein of interest (POI) and the E3 ligase
2	Flexibility	• Target proteins and their relative orientations determine the required flexibility of the linker to a significant level
3	Composition	• PROTAC is made up of different chemical components like polyethylene glycol, Alkyne, Triazole, etc. that impact their stability, solubility, and pharmacokinetics
4	Cleavability	• Cleavable linkers release ligands and degrade POIs in response to cellular conditions whereas non-cleavable linkers stay intact, enabling continued ternary complex synthesis
5	Cell Permeability	• Appropriate Cell permeability is very important for PROTAC to work on target cells
6	Hydrophilicity/ hydrophobicity	• Linker's hydrophilic or hydrophobic properties can significantly influence the PROTACs' solubility and cell membrane penetration capability
7	Specificity	• Specificity in terms of desired POI and E3 ligase is an indication of potential off-target effects
8	In vivo stability	• An indication of how successful a PROTAC could be in vivo is the stability of the linker in the physiological environment
9	Structural diversity	• More the Structural diversity better the potential customization for different targets

tein's structure is crucial for designing PROTAC molecules that can bind to it effectively.

3. Ligand design: Design ligands using molecular modeling methods. This involves predicting the binding features and ligand(s) orientation at the target protein and E3 ligase binding sites.
4. Linker design: Designing the linker component that links protein-targeting ligands to E3 ligases. For ternary complex formation and protein breakdown, the linker should have the right length, flexibility, and chemical characteristics.
5. Optimization: Iteratively refining the design of the PROTAC molecule through computational simulations and structure-activity relationship (SAR) analysis to enhance its potency, selectivity, and pharmacokinetic properties.
6. Validation: Experimentally validating the designed PROTAC molecules to assess their ability to induce target protein degradation in cell-based assays and animal models.

By integrating 3D molecular modeling techniques into PROTAC design, researchers can rationally engineer PROTAC molecules with improved efficacy and selectivity, accelerating the drug discovery process for targeted protein degradation therapies.

Generative artificial intelligence (AI) techniques like generative adversarial networks (GANs) and variational autoencoders (VAEs) have huge potential in the process of PROTAC drug discovery:

Table 14.3 Some recently used/proposed models for predicting PROTACs

#	Study/model	Underlying algorithms	Description
1	PROTAC-RL—deep learning and molecular simulations	Deep reinforcement learning, classifiers, and MD simulations (Nori et al. 2022)	• A deep generative model for designing PROTACs in a low-resource scenario
2	DeepPROTACs	Graph convolutional network (Li et al. 2022)	• A PROTACs deterioration predictor using deep learning • Modeling was done using PROTAC-DB dataset • Can anticipate if PROTACs can degrade a certain POI by recruiting an E3 ligase
3	Graph-based deep generative model (DGM)	DGM and policy gradient reinforcement learning (Nori et al. 2022)	• A graph- based deep generative model capable of designing PROTAC-like molecules atom by atom • This can help design PROTACs which will have minimized off-target effects
4	Model free analysis of protein degradability, MAPD	Naïve Bayes, KNN, LR LiblineaR, SVMLinear, SVMRadial, and random forest (Zhang et al. 2022)	• Predicts protein degradability based on base protein properties • Predict kinases that are degradable
5	Binary classification model, BCM	Random forest, decision tree, support vector machine, and kappa nearest neighbor (Vasanthanathan et al. 2023)	• Predict PROTAC permeability using binary classification models

1. Novel ligand generation: Generative models have the capability to produce chemical compounds that possess certain features, making them ideal candidates for the design of PROTAC ligands. These synthesized ligands can investigate chemical space that goes beyond what is typically investigated by medicinal chemists. This could potentially result in the identification of new PROTAC candidates.
2. Linker design: Generative AI can aid in designing novel linker molecules that connect the ligand targeting the disease-causing protein with the ligand targeting the E3 ligase.
3. Molecular optimization: By iteratively generating and evaluating molecular structures, generative AI can optimize PROTAC candidates for desired properties such as potency, selectivity, solubility, and metabolic stability.
4. De novo drug design: Generative models can be used to design PROTAC molecules from scratch, exploring chemical space to identify new scaffolds and linker chemistries that enhance PROTAC efficacy and safety.

5. Multi-objective optimization: Generative AI techniques can optimize PROTAC molecules for multiple objectives simultaneously, such as efficacy, selectivity, and pharmacokinetic properties, leading to the discovery of compounds with improved overall drug-like properties.
6. Target-specific PROTAC design: Generative models trained on specific disease-related protein structures can be used to generate PROTAC molecules tailored to degrade these specific targets, potentially increasing the precision and effectiveness of PROTAC therapy.
7. Predicting PROTAC efficacy: Generative AI models can predict the efficacy of created PROTAC compounds, allowing quick screening of huge chemical libraries for interesting experimental candidates.
8. Data augmentation: Generative AI can augment small datasets of experimentally validated PROTAC molecules by generating synthetic examples, thereby expanding the training data for predictive models, and improving their performance.

By harnessing the power of generative artificial intelligence, PROTAC drug discovery efforts can explore vast chemical space efficiently, accelerate the identification of novel candidates, and enhance the overall success rate of PROTAC development for various therapeutic applications.

14.5 Related Work

A recent significant study focusing on predicting the cell permeability of proteolysis targeting chimera aims to minimize the need for resource-intensive synthesis and testing of low-permeable PROTACs (Vasanthanathan et al. 2023). The authors present a thorough analysis of the capabilities and constraints of machine learning-based binary classification models. These models were created using 17 simple descriptors of a wide range of structurally diverse cerebelon (CRBN) and von Hippel–Lindau (VHL) PROTACs. The VHL PROTAC set was evaluated using kappa nearest neighbor and random-forest models. These models could predict the permeability of a blinded test set with an accuracy of over 80% ($k \geq 0.57$). The models were retrained by merging the initial training data with the blinded test set. These models demonstrated comparable performance when evaluated on the second blinded VHL set. However, the models developed for CRBN PROTACs had limited success, mostly attributed to the imbalanced nature of the CRBN datasets. All descriptors played a role in the models, however, size and lipophilicity emerged as the most significant factors. The researchers determined that well-trained machine learning models can be successfully used as efficient filters in the PROTAC design process.

Another recent study determined that computational technologies, such as deep learning and structure-based approaches, offer substantial prospects for tackling the obstacles in PROTAC development (Danishuddin et al. 2023). Deep learning algorithms have the capability to analyze extensive datasets, identify the most optimal linkers, predict binding affinities, and contribute to the enhancement of ternary complex structures. These technologies can expedite PROTAC design by narrowing the selection of candidates to those more likely to demonstrate efficacy and safety. In addition, they enable the virtual evaluation of potential PROTAC candidates against various E3 ligases, hence broadening the range of possible ligands. Through the integration of predictive capabilities and structural insights, researchers can make well-informed choices on PROTAC design, linker selection, and target engagement. With the increasing availability of crystal structures of ternary complexes, the accuracy of these computational techniques will potentially improve. This will further offer a clear direction for the future development of PROTACs.

A recent study discovered that DeepPROTACs, a deep learning-based targeted degradation algorithm for PROTACs, can assist in the creation of highly effective PROTAC molecules (Li et al. 2022). It can predict the ability of a suggested PROTAC molecule to degrade based on the structures of a provided target protein and E3 ligase. The experimental dataset was mostly sourced from PROTAC-DB. In the DeepPROTACs model, both the ligands and the ligand binding pockets are created and depicted as graphs. These graphs are then inputted into Graph Convolutional Networks to extract features. The SMILES representations of linkers are inputted into a Bidirectional Long Short-Term Memory layer to produce the features. Their investigations demonstrate that the DeepPROTACs model obtains an average prediction accuracy of 77.95% and ROC (under receiver operating characteristic curve) of 0.8470 on the test set.

High level network architecture of DeepPROTAC is illustrated in Fig. 14.12:

The entire network was constructed using the PyTorch and PyTorch Geometric frameworks. The encoding of linker SMILES was inputted into an embedding layer to acquire the distributed representation. The superiority of this procedure over one-hot encoding lies in the fact that the latter considers all characters as separate entities without any interconnection. Afterward, the result of this embedding layer was sent into a Bidirectional LSTM layer and a fully connected layer in sequence. The output of this linker network was combined with the outputs of pocket/ligand networks and then inputted into a Multi-Layer Perceptron (MLP) with two fully connected layers to produce the result.

Over a span of more than 20 years, consistent and dedicated efforts have resulted in the accumulation of a substantial amount of high-quality data for PROTACs. For example, the recently released PROTAC-DB, an online database, contains a total of 2258 PROTACs, 275 warheads (small compounds that bind to the protein of interest), 68 E3 ligands (small molecules that recruit E3 ligases), and 1099 linkers (Zhou 2021). PROTAC-DB provides information on the binding affinities, degradation efficiencies, and cellular activities of different PROTACs. More recently, the rapid advancement of deep Learning enhanced its capability to uncover the probable relationships between inputs and outcomes. This offers an excellent chance to combine deep learning with PROTACs.

PROTAC technology offers a significant advantage in its ability to effectively target proteins that were previously considered difficult or impossible to drug. Due to the effective binding of small- molecule ligands to target proteins, the majority of successful PROTACs now utilize small- molecule inhibitors (SMIs) as ligands to target druggable proteins. Furthermore, recent research demonstrates that the use of

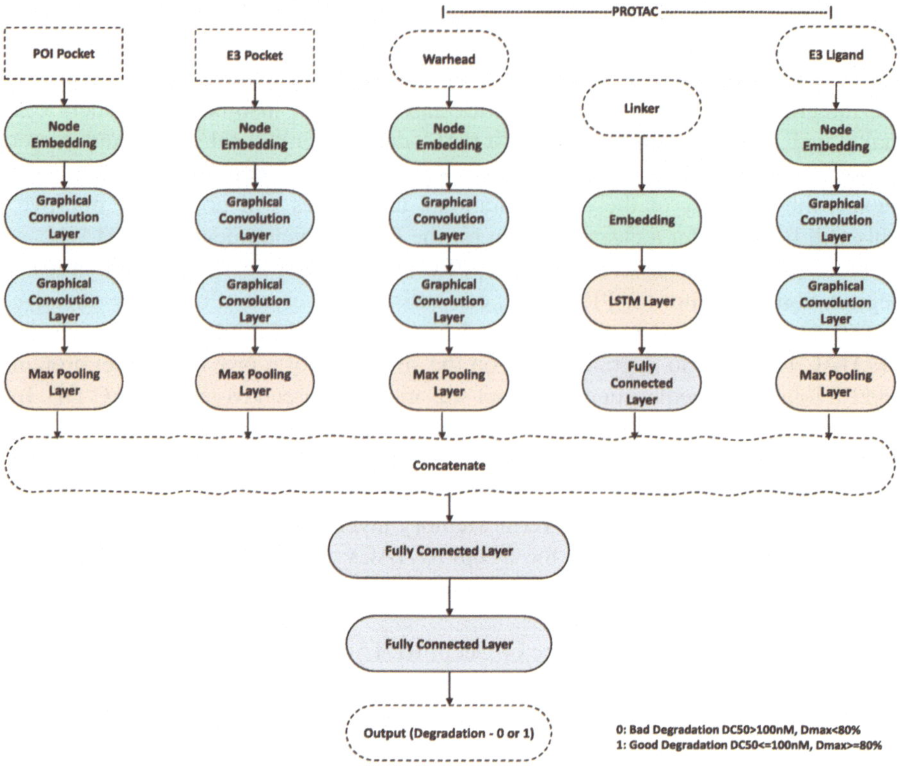

Fig. 14.12 The network architecture of DeepPROTACs (Li et al. 2022)

PROTAC in conjunction with kinase inhibitors might result in a synergistic impact on the suppression of tumors. This potentially means that the use of PROTAC in combination with targeted inhibitors or chemotherapy/antibody medicines could as well be a promising approach for cancer treatment (Qi et al. 2021).

A recent study found that drug- resistant targets exhibited a high degree of sensitivity toward PROTACs. Historically, chemotherapy has been the primary approach for treating cancer. However, the development of resistance to chemotherapeutic medications has impeded its effectiveness, leading to the recurrence of the disease. The subsequent development of kinase inhibitors and immunotherapy has also shown the issue of drug resistance. PROTACs, which impact both enzymatic and non-enzymatic functions of proteins, work by eliminating the entire target protein. As a result, this technique is anticipated to tackle the issue of drug resistance that is encountered by existing treatments. Also, PROTACs have the potential to be effective in targeting proteins that were previously considered difficult or impossible to drug. The efficacy of most small-molecule drugs or large-molecule antibodies relies on the presence of an active site on a binding enzyme or receptor. However, it is believed that more than 80% of proteins in human cells lack these sites.

Traditional small-molecule drugs commonly function by inhibiting the enzymatic activity of their targets. On the contrary, PROTACs can bind to target proteins by exploiting several binding sites. Furthermore, PROTACs can impact non-enzymatic activities. Multiple studies have demonstrated that PROTACs possess the ability to expand the range of targets that can be effectively treated with drugs and can also regulate both protease and non-enzyme functions. A study suggests that certain challenges associated with traditional small- molecule inhibitors can be surmounted by PROTACs (Yao et al. 2022). Despite all research, it is also true that today PROTAC medications are currently in their nascent phase when it comes to clinical application. Various challenges, including the slow advancement in the development of PROTACs, a low success rate, limited ability to enter cell membranes and be absorbed orally, and insufficient evidence from human clinical studies, together contribute to the barriers.

Artificial intelligence and machine learning are becoming integral to the process of drug discovery. However, it might be argued that the implementation of AI/ML techniques in this field has been very sluggish and frequently met with doubt, in contrast to their utilization in other industrial sectors. One reason for this is that, as independent tools, most AI/ML systems lack adequate evidence to substantiate their predictions. Instead, these systems are depicted as "black boxes" that take in an input (such as a molecule structure) and generate an output (such as a prediction of binding affinity), without offering any physical or mechanistic explanations. Enhancing the transparency of AI/ML models, for example, by emphasizing the structural components that influence a binding event, has been recognized as a crucial element in establishing trust in AI/ML (Miquel et al. 2023). Although the design of PROTACs is progressing, the de novo design of MGDs and other methods are currently more of a wish list that will require more study and implementation in the next years.

With the increasing accessibility of computing power, Generative AI has demonstrated significant potential in expediting and enhancing drug discovery procedures. The primary applications of generative AI have predominantly been in the field of de novo drug design, where it is utilized to create novel pharmacological compounds. However, there is also considerable promise for this technology to be employed in target identification and drug repurposing. Companies in diverse sectors have promptly acknowledged the promise of generative AI and have aggressively used its applications to acquire a competitive advantage and enhance their goods in the field of medication discovery. For Example, a drug discovery company, known for being at the forefront of utilizing generative artificial intelligence in drug development, has successfully created and verified an AI platform based on generative adversarial networks. This platform has been seamlessly incorporated into their commercially available Pharma.AI platform. The primary objective of this platform is to expedite the creation of innovative compounds, the discovery of disease targets, and the prediction of outcomes in clinical trials.

A recent work describes the utilization of a generative artificial intelligence (AI) approach for the design of PROTACs (Geoffrey Ben et al. 2022). Their methodology generates many PROTAC candidates and employs a docking-like scoring/

ranking system to evaluate their interactions with the pocket residues of FGFR1-MDM2. This method also assesses their capacity to facilitate the development of the ternary complex. Moreover, they employ computationally demanding techniques to estimate quantities measured in PROTAC assays, which serve as performance indicators for the PROTACs.

Developing PROTACs requires conducting thorough experiments and trials to investigate the wide range of chemical possibilities. To speed up this procedure, recent work has introduced an innovative deep generative model for the purpose of designing PROTACs in a resource-limited environment. This model is then directed to generate PROTACs with the most favorable pharmacokinetics using deep reinforcement learning (Zeng et al. 2022). By utilizing this approach on the target protein bromodomain-containing protein 4, a total of 5000 compounds were produced. These compounds underwent additional refinement using machine learning-based classifiers and physics-driven simulations. As a demonstration of feasibility, they found, created, and empirically evaluated six potential PROTACs that degrade bromodomain-containing protein 4. Out of these, three were confirmed using cell-based tests and western blot analysis. A particular candidate was subjected to additional testing and shown favorable pharmacokinetics in mice. Thus showing that integration of deep learning and molecular simulations can aid in the rational design and optimization of PROTACs.

Another instance involves a biotechnology startup based in Switzerland, which declared in March 2023 that it had developed a platform that employs generative artificial intelligence for protein engineering. The protein engineering platform employs generative algorithms, advanced robotics, microfluidics, and synthetic biology techniques to create and enhance protein sequences.

In June 2023, a Japanese specialist in cryo-electron microscopy (EM) formed a cooperation with the goal of utilizing AI and cryo-EM technology to accelerate the process of identifying and designing new preclinical drug candidates. Their generative AI system will leverage the knowledge acquired from cryo-EM research to optimize the design of molecular structures that enhance protein-ligand interactions.

The determination of pharmacodynamic and pharmacokinetic parameters of PROTACs has posed challenges to traditional approaches previously employed. The progress in the fields of "molecular modeling" and "crystallography" has facilitated a move toward the development of PROTACs. This has resulted in progress not only in basic research but also in the therapeutic space, with the incorporation of PROTACs into clinical trials. According to them, this endeavor has the potential to uncover novel therapeutics for orphan illnesses (Guedeney et al. 2023).

With the pharmaceutical industry's focus shifting toward the development of increasingly intricate and specific treatments, the importance of generative AI will grow even more significant. While the use of technology in drug discovery is increasing, there are still a few obstacles to its adoption. Obstacles to the implementation of generative AI encompass the need for extensive and top-notch datasets for training, a necessity for specialized proficiency in machine learning and data science, and complex ethical and legal deliberations.

14.6 Conclusion

The identification of effective PROTAC molecules is a time and resource- intensive activity, such as refinement of linker length and structure, etc. Currently, the primary emphasis in the design and optimization of PROTAC is on researching the structure-activity interactions of POI ligands and linkers. Of these factors, the linker plays a crucial role not only in the degrading activity of PROTACs, but also in determining the membrane permeability, metabolic stability, and drug availability. Hence, the crucial aspect of molecular design for PROTACs lies in properly designing and connecting POI and E3 ligands. Thus far, there is still a lack of comprehensive understanding regarding the principles that govern the design of linker molecules, particularly in terms of their length and composition. A well-defined outline of a comprehensive approach for developing highly effective PROTAC molecules is still needed.

One of the recent developments and potential for further research include photo-PROTAC, which is built using "photo control linkers," compared to typical medications. Photo-PRTOAC is anticipated to become a prominent method among PROTAC drugs. The exploration of new E3 ligases and their ligands lags significantly behind the study on PROTACs. Out of the numerous E3 ligases present in the human genome, only a small number are utilized in PROTACs. Currently, most PROTACs achieve the degradation of target proteins by enlisting E3 ligases such as CRBN, VHL, MDM2, and IAP. However, the investigation of PROTACs using solely these E3 ligases is currently insufficient. Hence, it is important to investigate additional new E3 ligases to expedite the advancement of PROTACs.

To summarize, Artificial Intelligence (AI) has a distinct and important role to play in PROTACs design. Rapidly developing computational technologies have made substantial contributions to PROTAC research in recent times. Researchers currently can employ AI-driven methods to investigate a wider variety of E3 ligases (the proteins accountable for ubiquitination) and enhance linker regions in the PROTAC molecules. These computational methodologies can improve our comprehension of the efficacy and safety of PROTACs in clinical environments.

Although there have been many positive outcomes, there are still some obstacles that remain in the development of PROTACs. These issues include the need to create strong ternary complexes, the occurrence of unintended effects on non-target molecules, poor cell permeability, and the need to optimize the binding affinity. AI-powered computational tools can assist in tackling these difficulties by offering insights into the relationships of PROTAC components and facilitating rational design and optimization.

References

Danishuddin JMS, Song KS, Lee KW, Kim JJ, Park YM (2023) Revolutionizing drug targeting strategies: integrating artificial intelligence and structure-based methods in PROTAC development. Pharmaceuticals 16(12):1649

Geoffrey Ben AS, Kulkarni NM, Agrawal D, Vetrivel R, Gurram K (2022) A new in-silico approach for PROTAC design and quantitative rationalization of PROTAC mediated ternary complex formation. Online bioRxiv.org

Guedeney N, Cornu M, Schwalen F, Kieffer C, Voisin-Chiret AS (2023) PROTAC technology: a new drug design for chemical biology with many challenges in drug discovery. Drug Discov Today 28(1):103395

Li F, Hu Q, Zhang X, Sun R, Liu Z, Wu S, Tian S, Ma X, Dai Z, Yang X, Gao S, Bai F (2022) DeepPROTACs is a deep learning-based target degradation predictor for PROTACS. Nat Commun 13(1):7133

Miquel DF, Cigler M, Winter GE (2023) Advancing targeted protein degradation via multiomics profiling and artificial intelligence. J Am Chem Soc 145(5):2711–2732

Nori D, Coley CW, Mercado R (2022) De novo PROTAC design using Graph-based deep generative models. Quantitative Biology

Qi SM, Dong J, Xu ZY, Cheng XD, Zhang WD, Qin JJ (2021) PROTAC: an effective targeted protein degradation strategy for cancer therapy. Front Pharmacol 12:692574

Vasanthanathan V, Kolling F, Giese A, Goller AH, Lehmann L, Meibom D, Kihlberg J (2023) Predictive modeling of PROTAC cell permeability with machine learning. ACS Omega 8(6):5901–5916

Yao T, Xiao H, Wang H, Xu X (2022) Recent advances in PROTACs for drug targeted protein research. Int J Mol Sci 23(18):10328

Zeng S, Tan Y, Wang Z, Li C, Zhang Z, Sang X, Chen H, Yang Y (2022) Accelerated rational PROTAC design via deep learning and molecular simulations. Nat Machine Intell 4:739–748

Zhang W, Burman SS, Roy CJ, Donovan KA, Yang C, Chelsea S, Boning Z, Zeng Zexian G, Shengqing ZY, Dian L, Fischer Eric S, Collin T, Shirley LX (2022) Machine learning modeling protein intrinsic features predicts tractability of targeted protein degradation. Genomics Proteomics Bioinformatics 20:882–898

Zhou Z-H (2021) Machine learning. Springer, Cham

PROTACs in Treatment of Glioma

15

Sachin Kumar and Vikas Maharshi

Abstract

Treatment for gliomas, especially glioblastoma (GBM), is extremely difficult because of the disease's poor prognosis and low 5-year survival rate (less than 5%). There is an urgent need for novel strategies because existing treatments have not been able to provide long-term solutions. Proteolysis-targeting chimeras (PROTACs) present a potential approach since they utilize the ubiquitin-proteasome system to degrade particular proteins that are essential for the growth and dissemination of gliomas. Targeting oncogenic proteins necessary for glioma survival, PROTACs interfere with critical mechanisms that promote tumor growth.

PROTAC treatment targets the epidermal growth factor receptor (EGFR), which is often dysregulated in gliomas. By destroying mutant EGFR variants, PROTACs may be able to overcome resistance that is frequently seen with conventional targeted therapies, therefore improving therapy efficacy as a whole. Furthermore, the complex and diverse biology of gliomas is addressed by PROTACs' capacity to target many proteins at once, which may delay the formation of resistance. Important information on PROTACs' translational potential is provided by ongoing clinical trials evaluating the safety and effectiveness of the treatment in glioma patients. In the end, glioma treatment that uses targeted protein degradation to interfere with neoplastic pathways is called PROTAC therapy.

S. Kumar (✉)
Department of Medical Laboratory Technology, School of Allied Health Sciences and Management, Delhi Pharmaceutical Sciences and Research University (DPSRU), New Delhi, India
e-mail: sachinkr84@dpsru.edu.in

V. Maharshi
Department of Pharmacology, All India Institute of Medical Sciences (AIIMS), Patna, India
e-mail: vikas11315@aiimspatna.org

M. Nandave, P. Jain (eds.), *PROTAC-Mediated Protein Degradation: A Paradigm Shift in Cancer Therapeutics*, https://doi.org/10.1007/978-981-97-5077-1_15

In the context of glioblastoma and its pivotal signaling pathways such as EGFR, TGF-b, and p53, this chapter will delve into the role of E3 ubiquitin ligases and deubiquitinating enzymes, offering fresh perspectives on their therapeutic utility for glioblastoma.

Keywords

Brain tumors · Cancer therapy · Glioma · Proteolysis-targeting chimeras · Targeted protein degradation · Tumor-specific protein degradation · Ubiquitin-proteasome system

Abbreviations

AMP	Adenosine monophosphate
AR	Androgen receptor
ATP	Adenosine triphosphate
BCL-XL	B-cell lymphoma-extra-large
CDK4/6	Cyclin-dependent kinase 4/6
CDKN2A/B	Cyclin-dependent kinase inhibitor 2A/B
CNS	Central nervous system
Cys	Cysteine
EGFR	Epidermal growth factor receptor
ER	Estrogenic receptor
GBM	Glioblastoma
GSCs	GB stem-like cells
IDH	Isocitrate dehydrogenase
NF1	Neurofibromatosis type 1
PDGFRα	Platelet-derived growth factor receptor alpha
PEG	Polyethylene glycol
POIs	Protein of interest
PROTAC	Proteolysis-targeting chimeras
PTEN	Phosphatase and tensin homolog
RIBOTACs	Ribonuclease targeting chimeras
SMIs	Small molecule inhibitors
TFs	Transcription factors
TGF-β	Transforming growth factor-β
TMZ	Temozolomide
UPS	Ubiquitin-proteasome system
WHO	World Health Organization

15.1 Introduction

Glioblastoma is a type of glioma and is considered one of the most common and most malignant CNS tumors in adults (Lospinoso Severini et al. 2022; Scholz et al. 2020). Presently the standard of care for glioblastoma includes maximal safe surgical resection followed by radiotherapy and chemotherapy (with Temozolomide) (Lospinoso Severini et al. 2022; Scholz et al. 2020). Despite this aggressive therapeutic strategy, overall prognosis is poor with a median survival of 15 months (Lospinoso Severini et al. 2022; Scholz et al. 2020). Glioblastoma is a rapidly progressing tumor invading surrounding brain tissue, also characterized by genetic and molecular intra- and inter-tumoral heterogeneity and presence of drug-resistant glioblastoma stem-like cells. All these features may contribute to its poor response to currently available treatment (Lospinoso Severini et al. 2022; Scholz et al. 2020). Various genetic alterations in molecular drivers involved in pathogenesis of glioblastoma have been identified. These molecular drivers include but are not limited to epidermal growth factor receptor (EGFR), cyclin-dependent kinase(s) 4/6 (CDK4/6), cyclin-dependent kinase inhibitor 2A/B (CDKN2A/B), neurofibromatosis 1 (NF1), platelet-derived growth factor receptor alpha (PDGFRα) and isocitrate dehydrogenase (IDH) (Lospinoso Severini et al. 2022). Identification of this aberration in molecular network led to exploration and testing of number of therapeutic strategies, however, none of those proved to be clinically beneficial (Lospinoso Severini et al. 2022). One of the recent emerging strategies in this field is 'proteolysis targeting chimera (PROTAC)' (Lospinoso Severini et al. 2022; Scholz et al. 2020; Ocaña and Pandiella 2020). PROTAC molecules are developed to target and facilitate the proteins involved in tumorigenesis (Lospinoso Severini et al. 2022; Scholz et al. 2020; Ocaña and Pandiella 2020).

Proteins are degraded in the cell through various processes including receptor internalization followed by enzymatic action in lysosomes and ubiquitin proteasome system (UPS) (Ocaña and Pandiella 2020). PROTAC technology exploits UPS of tumor cell for its action (Lospinoso Severini et al. 2022; Scholz et al. 2020; Ocaña and Pandiella 2020).

15.2 Ubiquitin-Proteasome System

Proteasome is a term given to a complex of protease(s) that degrade selected intracellular proteins (Tanaka 2009). UPS involves ubiquitination (also termed ubiquitylation) of target proteins, a step required for recognition of the target protein by proteasome (Ocaña and Pandiella 2020).

Before explaining the functioning of UPS as a whole, it is worth describing the structure of proteasome for better understanding of the former. Proteasome is a complex consisting of central part having catalytic activity and is known as core particle (CP) and peripheral part known as regulatory particle (RP). One CP may have one RP attached or a pair of RPs attached to either of the ends to form enzymatically active proteasome. The active proteasome may have sedimentation coefficient of 26S (1 CP and 1 RP) or 30S (1 CP and 2 RPs) and accordingly may be called 26S and 30S proteasome respectively. RP is known as 19S proteasome and CP as 20S proteasome (Tanaka 2009) (Fig. 15.1).

RP or 19S proteasome is composed of two subcomplexes *viz* lid (away from CP) and base (toward CP). Lid consists of nine subunits with no ATPase activity (Rpn). Rpn subunits of lid especially Rpn11 cause de-ubiquitination of target protein. Base subcomplex consists of two types of subunits, Rpn (total four in number including Rpn10 and Rpn 13) and those with triple ATPase activity (Rpt; total six in number). Rpn10 and 13 act as receptors for target proteins. Rpn10 and 13 recognize ubiquitinated protein and capture it. The other two functions of base subcomplex (mediated by Rpt subunits) are unfolding of target protein and opening the central pore present in CP (described ahead) (Tanaka 2009) (Fig. 15.1).

CP or 20S proteasome is composed of two outer α-subunit rings and two inner β-subunit rings. Outer rings have a narrow central pore for entry of target protein inside the CP which is closed by a gate. Three β-subunits in each inner ring have threonine hydrolase (catalytic) activity (so proteasome is a type of threonine protease) required for degradation of target protein (Tanaka 2009) (Fig. 15.1).

As said above, ubiquitination of target protein is required for its degradation by proteasome machinery. Ubiquitination is a multi-step (activation, conjugation, and ligation) process that involves action of multiple enzymes in sequence namely E1, E2 conjugase, and E3 ligase (Ocaña and Pandiella 2020). Activation process involves E1 enzyme catalyzed acyl-adenylation of C-(carboxy) terminus of ubiquitin in presence of ATP and transfer of this ubiquitin to E1 enzyme through a thioester link between C-terminus of ubiquitin and --SH group of cysteine moiety of E1 enzyme (Ocaña and Pandiella 2020). Conjugation step leads to transfer of ubiquitin from E1-thio-esterified ubiquitin to cysteine moiety of E2 (conjugating enzyme). In third step, E3 enzyme makes a bridge between E2-ubiquitin complex and the target protein and creates an isopeptide bond between lysine residue to target protein and C-terminal glycine of ubiquitin (Ligation) (Ocaña and Pandiella 2020) (Fig. 15.2).

E3 enzymes impart substrate specificity through diversity in their structure and function and tissue-specific distribution. Based on their structural and functional differences, E3 ligases (>700 have been identified) are broadly grouped into four

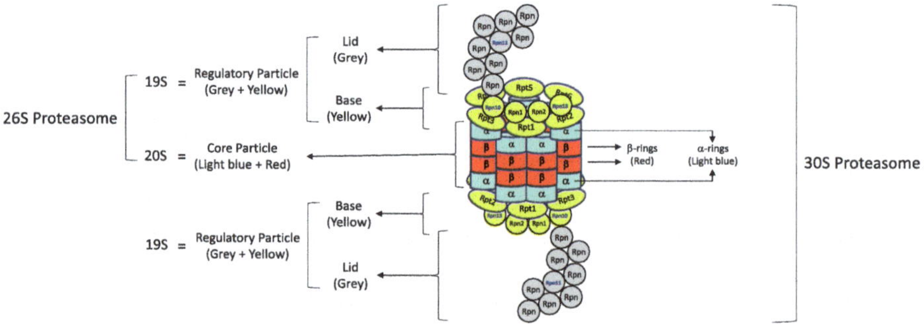

Fig. 15.1 Schematic diagram showing structure of proteasome machinery

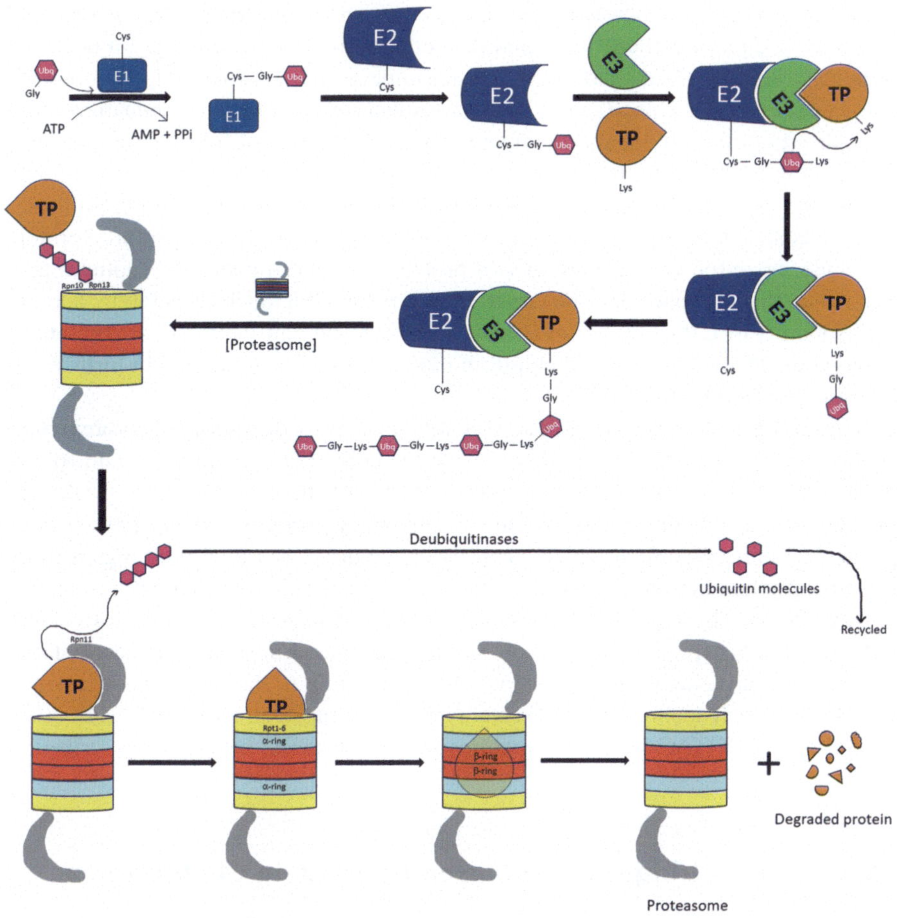

Fig. 15.2 Schematic diagram showing role of ubiquitin-proteasome system in protein degradation. *AMP* adenosine mono phosphate, *ATP* adenosine tri phosphate, *Cys* cysteine, *E1* enzyme involved in ubiquitin activation step, *E2* enzyme involved in conjugation step, *E3* E3 ligase enzyme involved in ligation step, *Gly* glycine, *Lys* lysine, *Rpn* regulatory particle subunit with no ATPase activity (Rpn10 and Rpn13 recognizes polyubiquitinated protein, Rpn11 has de-ubiquitinase activity), *Rpt* regulatory particle subunit with triple ATPase activity (Rpt1-6 unfold target protein, open the central pore in α-rings and facilitate entry of target protein into the chamber formed by β-rings which have catalytic/proteolytic activity), *TP* target protein, *Ubq* ubiquitin

families viz 'Really Interesting New Gene (RING), Homologous to E6-AP Carboxy Terminus (HECT), Ring-in-Between-Ring (RBR), and U box E3 ubiquitin ligases family (Scholz et al. 2020). A detailed description of these families is not in the scope of this chapter. Multiple (usually more than 4) molecules of ubiquitin should covalently bind to the target protein (polyubiquitylation) for its proteolysis by proteasome (Ocaña and Pandiella 2020; Tanaka 2009; Sakamoto 2010). For polyubiquitination to occur, next ubiquitin molecule binds at the lysine residue of the former ubiquitin molecule (which is bound to target protein), and the process is repeated (Qi et al. 2021). Specificity of signal depends on the length of ubiquitin chain on protein molecules (Maksoud 2021). Proteasome is attracted toward polyubiquitinated target protein. Rpn 10 and 13 (of base subcomplex) of proteasome machinery recognize polyubiquitinated protein and capture it. Rpn11 (of lid subcomplex) then causes de-ubiquitination to release the ubiquitin chain that may further be broken down into individual ubiquitin molecules by the action of other de-ubiquitinases (Scholz et al. 2020). Rpt subunits of (base subcomplex) unfold the captured protein molecule, open the central pore of 20S proteasome, and transfer the captured protein inside the core particle. The target protein is then degraded by catalytic activity of β-subunits of CP of proteasome machinery (Ocaña and Pandiella 2020; Tanaka 2009) (Fig. 15.2). The released ubiquitin molecules can be reutilized in further cellular processes (Qi et al. 2021).

Thus, UPS system helps maintain cell homeostasis by degrading the protein molecules. Number of proteins are there inside the cells that play an important role in pathogenesis of diseases including tumors. Many of these proteins (e.g. transcription factors, scaffolding proteins, and rat sarcoma proteins) do not have a deep pocket in their structure at which small molecule drugs can bind with high affinity for showing their action. Therefore, it has not been possible to develop drugs against these proteins. These proteins are known as undruggable proteins targets. Recently UPS has been exploited to develop small molecule drugs that can target these undruggable proteins despite their low affinity binding to these proteins (Liu et al. 2022). One such technology is 'Proteolysis Targeting Chimera (PROTAC)' (Liu et al. 2022). PROTAC-based therapeutics for glioblastoma have been described in this chapter.

15.3 Role of Ubiquitin-Proteasome System in Glioblastoma

15.3.1 Hyperactivity of Epidermal Growth Factor Receptor

Epidermal growth factor receptor (EGFR or ErbB1 or HER1) is an ErbB family receptor with tyrosine kinase activity and is involved in cell proliferation and survival. EGFR activity is increased in around 60% of the primary glioblastomas (Xu et al. 2017). Mechanisms of its increased activity could be multiple including mutations in gene encoding this protein. UPS also plays a role in hyperactivity of EGFR in glioblastoma. De-ubiquitinase CSN6 overexpression in glioblastoma leads to stabilization of EGFR (Scholz et al. 2020) through its de-ubiquitination and thus preventing its recognition by protease machinery. Moreover, AKT is also involved in the signal transduction pathway of EGFR to cell proliferation. PTEN is an inhibitor of AKT. Polyubiquitination of PTEN (leading to its rapid degradation) is also one of the mechanisms for EGFR hyperactivity. Likewise, autoubiquitination of E3 ligase specific for EGFR has also been reported in glioblastoma leading to stabilization of EGFR and its higher activity (Scholz et al. 2020).

15.3.2 Uncontrolled TGF-β Signaling

Transforming growth factor-β (TGF-β) acts on its receptor TβR. This binding, signals to cause cell proliferation and survival through different cascades. TGF-β signaling is also controlled by ubiquitin system. SMAD7 protein binds to TβR and also acts as docking site for an E3 ligase (SMURF2) and a de-ubiquitinase (USP15). SMURF2 facilitates polyubiquitination and in turn degradation of TβR whereas UPS15 functions in opposite direction. These and a few other ubiquitin system mechanisms are dysregulated in glioblastoma to net increase in TGF-β signaling (Scholz et al. 2020).

15.3.3 Reduction in p53-mediated Tumor Suppressor Function

p53 acts as a tumor suppressor molecule. Levels and/or activity of this molecule are also controlled by ubiquitin pathway. A RING E3 ligase (MDM2) facilitates its ubiquitination and in turn its degradation. Overexpression of MDM2 in glioblastoma thus decreases the levels of p53 inside the tumor cell. One de-ubiquitinase (USP4) that stabilizes another HECT E3 ligase (HUWE1/Mule), is also overexpressed and thus in-turn hastens the degradation of p53 protein inside the tumor cells (Scholz et al. 2020).

15.4 Role of PROTAC Technology in Glioblastoma

A novel strategy for medication development, PROTAC holds great promise for treating a variety of malignancies, including GBM, one of the deadliest and most aggressive types of brain cancer. As scientists look for novel approaches to tackle the intricate problems presented by GBM, they have focused their attention on the distinct mode of action of PROTACs, which are intended to cause targeted degradation of particular proteins inside cells.

Heterobifunctional molecules known as PROTACs are made up of two ligands joined by a linker. First ligand interacts with protein of interest (POIs), then another ligand attaches to an E3, and a linker joins them (Fig. 15.3) (An and Fu 2018). The UPS catalyzes the POI's degradation and promotes ubiquitylation when the E3 and POI are close to each other, as mediated by PROTAC.

Over two decades have passed since the development of PROTAC compounds (Sakamoto et al. 2001), and throughout that time, numerous attempts have been made to increase their efficacy. For instance, to improve cell permeability, peptide ligands in the PROTAC structure have been altered in small molecules (Schneekloth et al. 2008).

PROTACs provide a number of benefits over conventional small molecule inhibitors (SMIs), but they also have certain drawbacks. In contrast to SMIs, a PROTAC molecule can catalyze the breakdown of several POI compounds and has a much lower pharmacological impact, which lowers toxicity. Notably, proteins deemed

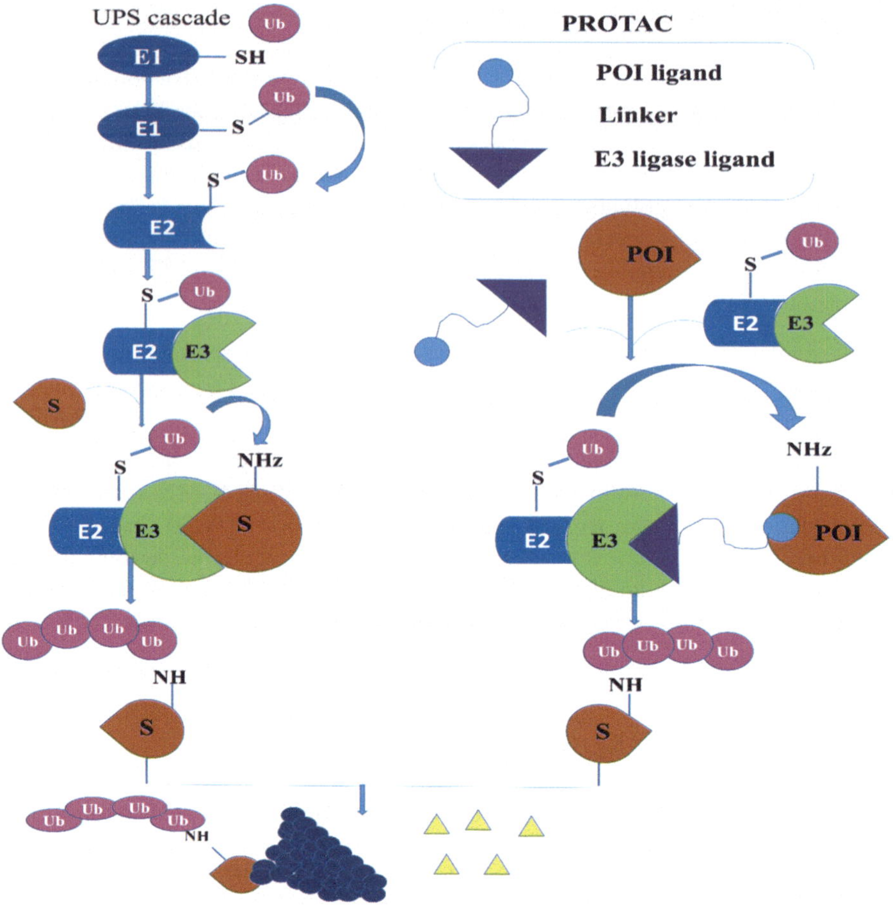

Fig. 15.3 PROTAC and ubiquitin-proteasome systems. Diagrammatic illustration of the Ubiquin-Proteasome System's (UPS cascade) enzymatic cascade (left side). The E1 activating enzyme's ATP-dependent activation of ubiquitin causes ubiquitylation. Subsequently, an E3-Ub ligase (E3) transfers the attached ubiquitin (Ub) to a Lys residue on a substrate protein (S) after it has been bound to the E2 Ub-conjugating enzyme. The proteasome may break down the substrate when a poly-Ub chain comprising more than four Ub moieties forms. The elements of PROTAC and how they work (PROTAC; right side). A ligand specific for the POI and another ligand for E3 are combined to form PROTACs, which are heterobifunctional small molecules joined by a linker. When an E3 ligase is brought close to a particular POI, it can be tagged with Ub and broken down by a proteasome, which is how PROTACs function

"undruggable" may be the target of PROTAC activity. This is especially important because transcription factors (TFs) have been linked to the development of many cancers (Bai et al. 2019; Zhou et al. 2019). For instance, it has been reported that neuroblastoma, breast cancer, and prostate cancer, respectively, are associated with genetic changes in c-MYC, FOXO1, or the androgen receptor (AR) (Bushweller 2019; Yu et al. 2019). One potential therapeutic approach for these human malignancies is to inhibit their expression by means of protein breakdown. Consistent with the findings in this field, two phase I studies have introduced two PROTACs targeting the androgen receptor (AR) and the estrogenic receptor (ER) into clinical practice for the treatment of prostate cancer and ER-positive breast cancer, respectively (Mullard 2019).

Additionally, PROTACs can overcome SMIs resistance by concentrating on mutant POIs and the resistance caused by POIs overexpression (Burslem et al. 2018; Zhao et al. 2019; Gonzalez et al. 2020; Kregel et al. 2020). When promoting PROTACs' adoption into clinical practice, it is necessary to take into account a few safety considerations. PROTACs' toxicities, both on and off target, are the main factor limiting their utilization. The on-target toxicity has to do with how POI acts physiologically. In addition to their enzymatic functions, some proteins, such as kinases, act as scaffolds and are therefore required for normal cell activity. SMIs only reduce the enzymatic activity of POI, but the entire degradation of PROTACs affects both the scaffolding and the enzymatic functions of POI, resulting in undesirable results (Cromm et al. 2018; Nunes et al. 2019). Furthermore, an effective PROTAC can exhaust its POIs entirely, in contrast to SMI, which can only partially impede the functioning of their targets.

If POIs are essential for cell survival, then partial inhibition brought on by SMIs therapy may be tolerated, whereas degradation brought on by PROTAC may be harmful (Winter et al. 2015). According to several studies (Chan et al. 2018; Cromm et al. 2018; Olson et al. 2018; Testa et al. 2018; Smith et al. 2019), the degree of cellular damage is determined by the rate of depleted protein resynthesis. Furthermore, in contrast to PROTAC-mediated persistent depletion, SMI-mediated suppression of POIs is temporary. In this instance, the benefits or drawbacks of PROTACs depend on the cellular/tissue context and the target properties. A POI cannot be fatal to cells if it has redundant activity in healthy tissues and is gradually degraded (Mason et al. 2007; Eichhorn et al. 2014; Khan et al. 2019). Targeting a POI that is necessary for physiological cellular processes, on the other hand, may result in on-target toxicity. A light-induced switch on PROTACs, which is thereafter referred to as an opto-PROTAC, facilitates the spatiotemporal degradation of protein targets. We showed light-inducible protein degradation by incorporating two additional PROTACs, dBET1, and dALK, with pomalidomide, the original chemical, to create a photolabile caging group. When exposed to ultraviolet A radiation, these opto-PROTACs can induce controlled deterioration at a defined time and pace, but they do not exhibit any activity in the dark. Our approach allows PROTAC to be a precision medicine and offers a broadly applicable platform for the creation of light-controlled PROTACs (Liu et al. 2020).

The "unintentional" breakdown of proteins is a common cause of off-target toxicity. This phenomenon could manifest when the non-target protein is in proximity to or in complex with the POI rather than directly attached to the PROTAC (Hsu et al. 2020). A phenomenon called the "Hook effect" may occur because PROTACs create a ternary complex between POI and E3. More specifically, large concentrations of PROTACs prevent the formation of ternary complexes, leading to an excess of binary binds such as PROTAC-POI or PROTAC-E3, which invalidates target degradation (Pettersson and Crews 2019). Moreover, the production of binary complexes consisting of PROTAC-E3 has the ability to trigger the degradation of non-targeted proteins with lesser affinity (Moreau et al. 2020). The buildup of off-target ubiquitylated proteins that saturate the UPS and dysregulate proteostasis could result from this occurrence, or it could impact nutrients necessary for cellular homeostasis (Schmitt et al. 2002).

15.5 PROTAC Optimization Strategies

PROTAC is a relatively new field of research that is developing quickly, but it still needs a lot of work to optimize. It is possible to adjust the technology's biological and physical-chemical characteristics. One important structural component that can be enhanced is the linker length. A steric conflict brought on by excessively short linkers can destabilize the ternary complex and impede PROTAC function. On the other hand, too lengthy linkers may increase the motility of two PROTAC heads, altering the stability of the molecule. Additionally, a PROTAC's cell permeability is decreased and its molecular weight is increased by an excessive linker length.

Initially, flexible linkers like polyethylene glycol (PEG) or polymethylene chains were utilized in PROTAC design to enhance water solubility (Bai et al. 2019; Khan et al. 2019). The synthesis of PROTAC has recently made use of "click chemistry," which is based on the Diels-Alder (DA) reaction and copper-catalyzed azide-alkyne cycloaddition (CuAAC) (Wang et al. 2020). According to Lebraud et al. (2016), the resultant PROTACs can self-assemble as active molecules in living cells and can be more quickly evaluated for their capacity for destruction.

Another significant factor influencing the pharmacokinetic characteristics and oral bioavailability of PROTACs is the linker's stiffness (Farnaby et al. 2019; Testa et al. 2020). Nevertheless, if the ternary complex's cocrystal structure is unknown, designing the ideal stiff linker could be challenging.

Of the almost 600 E3s that the human genome encodes, only one percent have been investigated for substrate degradation (Khan et al. 2020). As target specificity is defined by E3, this characteristic may help the PROTACs work more effectively and be less harmful. One approach to PROTAC optimization, for instance, relies on E3 specific expression in tissues (e.g., F-box and leucine-rich repeat protein 16, FBXL16, is expressed specifically in cerebral cortex (Clifford et al. 1999) and/or cellular compartments (e.g., DDB1- and CUL4-associated factor 16, DCAF16, localizes only in the nucleus (Robb et al. 2017).

15.6 Role of PROTACs in Human Cancer Therapy

As a result of genetic and epigenomic changes, cancer is a multi-step process marked by aberrant cellular proliferation and dissemination (Hanahan and Weinberg 2011). Finding the molecular changes responsible for the carcinogenic characteristics has grown in interest as a means of creating new treatments (Ocaña et al. 2018).

Protease inhibitors have been used clinically in oncology to show how an imbalance in protein homeostasis in certain cancers indicates an oncogenic vulnerability (Inobe and Matouschek 2014; Schapira et al. 2019). Undoubtedly, precise proteostasis is essential for cells with high rates of protein turnover, including tumor cells, which require rapid and highly effective protein synthesis and degradation (Bard et al. 2019; Pohl and Dikic 2019).

In the past 20 years, many PROTACs have been created, but sadly, very few of them are tumor cell selective. Numerous PROTACs enlist E3 ligases, which are widely expressed in both tumor and normal tissues, resulting in toxicities that are directed toward the intended target. There are several approaches that can be used to get tumor-specific POIs to be selectively degraded by PROTACs.

It is feasible to target the POI with any E3 that is expressed in the tumor tissues if it is tumor-specific (Burslem et al. 2019). Alternatively, it is possible to optimize PROTACs by utilizing any accessible tissue-specific E3 if the POI is typical of a tumor-derived tissue (Schapira et al. 2019; Sun et al. 2019). Additionally, it is possible that a tumor-associated POI is expressed in healthy tissues and contributes to physiological cell processes, but that its expression is elevated in cancerous tissues. Utilizing tumor-specific E3s, which are abundantly expressed in tumor cells but expressed poorly or not at all in normal tissues, may provide a greater advantage in killing cancer cells specifically while reducing damage to normal tissues. One recent example is the creation of a B-cell lymphoma-extra-large (BCL-XL) PROTAC (Chung et al. 2020; Kolb et al. 2021).

Target protein degradation by PROTACs can now be accomplished selectively and specifically in tumors thanks to the identification of tissue-selective E3s made possible by the availability of public-omics data (Consortium 2015; Melé et al. 2015). The ability of light-controllable photo-PROTACs to cause tumor-specific degradation of photooxidants under visible or UVA light has attracted the attention of several research groups in recent years (Pfaff et al. 2019; Xue et al. 2019; Liu et al. 2020; Reynders et al. 2020). Only a few forms of cancer can be treated using this tactic in a clinical setting with photodynamic treatment.

Numerous preclinical investigations have proven the effectiveness of PROTACs (Bai et al. 2019; Khan et al. 2019; Li et al. 2019). Notably, it has also been demonstrated that PROTAC technology can activate an immune response to cancer by causing peptides produced from POI breakdown to be presented to antigen-presenting cells (Moser et al. 2017; Jensen et al. 2018). Additionally, according to Lai et al. (2018), PROTAC can be utilized to produce novel MHC-I peptides on the cell surface that will favor the development of a new immunopeptidome that is "targetable" by T-cell-based therapies (Lai et al. 2018).

PROTAC incorporation into nanoparticles that can be encapsulated with antibodies can help to specifically reach the tumoral environment and malignant cells (Beck et al. 2017; Niza et al. 2019; Pillow et al. 2020). PROTAC strategy can be used to exploit E3s having tumor suppressor natural substrates (Hines et al. 2019). Protease-like approaches have recently been developed to recruit nucleases and cause the breakdown of RNAs, or carcinogenic micro-RNAs. These compounds are referred to as ribonuclease targeting chimeras (RIBOTACs), and they represent novel potential anticancer treatments (Costales et al. 2020a, b). Overall, PROTACs and related technologies represent a potentially beneficial class of biological medicines for cancer treatment.

15.7 PROTAC as Therapeutic Strategies for Glioblastoma

The tumor entities that comprise central nervous system (CNS) cancers are diverse in terms of their etiology, clinical behavior, site of onset, and molecular biology (Kristensen et al. 2019; Lospinoso Severini et al. 2020). The most deadly and malignant of them all in adults is glioblastoma (Louis et al. 2016). The World Health Organization (WHO) has classified grade IV diffuse glioma, which includes about 54% of gliomas and has a median survival of approximately 15 months (Ostrom et al. 2014; Louis et al. 2016). The current standard of care for newly diagnosed GB patients is the use of an oral alkylating drug called temozolomide (TMZ), together with radiation and chemotherapy after maximal surgical resection (Stupp et al. 2005; Stupp et al. 2009). Although this treatment approach is vigorous, it is not very effective, rendering GB an incurable tumor that frequently recurs (Lieberman 2017). The primary characteristics of this cancer that impede treatment are its quick progression, the invasiveness of cancer cells in the brain's surrounding tissue, genetic and molecular heterogeneity both within and between tumors, and the existence of drug-resistant GB stem-like cells (GSCs), which promote tumor recurrence (Brennan et al. 2012; Meyer et al. 2015; Gangoso et al. 2021).

The identification of genetic change patterns affecting molecular drivers of GB tumorigenesis, such as epidermal growth factor receptor (EGFR), phosphatase and tensin homolog (PTEN), cyclin-dependent kinase 4/6 (CDK4/6) and cyclin-dependent kinase inhibitor 2A/B (CDKN2A/B), neurofibromatosis type 1 (NF1), platelet-derived growth factor receptor alpha (PDGFRα), isocitrate dehydrogenase (IDH) genes, and cyclin-dependent kinase inhibitor 2A/B (CDKN2A/B), has been made possible through transcriptome and genomic profiling (Verhaak et al. 2010; Dunn et al. 2012; Stoyanov and Dzhenkov 2018).

The identification of the faulty molecular networks responsible for the cancerous appearance of GB has brought attention to important mechanisms that may be used therapeutically. To date, a number of targeted therapeutics have been investigated for GB; the majority of these medications target growth factor receptors, such as EGFR, and downstream pathways that are frequently changed in GB, such as MAPK/ERK and PI3K/AKT/mTOR (Le Rhun et al. 2019). Nevertheless, none of these methods have received formal validation for their efficacy in clinical trials; this is probably because of toxicity, poor target coverage, or molecular compensatory mechanism (Touat et al. 2017; Le Rhun et al. 2019). Studies on several immunotherapeutic strategies have also been conducted to treat GB; however, the benefits of these strategies are limited by the tumor immunosuppressive milieu (Bufalieri et al. 2020a; Weenink et al. 2020; Bufalieri et al. 2021; Medikonda et al. 2021).

PROTACs in particular offer an intriguing targeted therapy for the treatment of this deadly tumor (Bufalieri et al. 2020b; Scholz et al. 2020; Maksoud 2021; Farrell and Jarome 2021). The UPS is showing promise as a source for the development of new therapeutic options for GB.

In GB cells, two distinct PROTAC approaches that can cause the degradation of CDK4 and/or CDK6 have been tried. Important for controlling the cell cycle, CDK4 and CDK6 are also intriguing targets for treating a variety of malignancies, including GB, which is often characterized by deregulation of the CDK4/6 pathway (Network 2008; Brennan et al. 2013; Bronner et al. 2019). In 2019, Zhao and Burgess investigated the efficacy of PROTACs in GB cell lines and breast cancer using two specific CDK4/6 inhibitors: ribociclib (Kisqali®, Novartis, Basel, Switzerland) and palbociclib (Ibrance®, Pfizer, New York, NY) (Zhao and Burgess 2019). The US Food and Drug Administration (FDA) has approved these medications for use in combination therapy for advanced breast cancer that is ER-positive but HER2-negative. These medications are presently being used in clinical trials, some of which are for the treatment of GB (NCT03158389; NCT02345824; NCT02933736; NCT03834740; NCT03355794; NCT03355794) (Lospinoso Severini et al. 2022). Palbociclib and ribociclib's PROTACs, also known as pal-pom and rib-pom, respectively, involve conjugating these two medications to pomalidomide (pom), a cereblon (CRBN) E3 ligand, through the process of cycloadding a known azide produced from pomalidomide to N-propargyl derivatives of either drug.

The decrease of CDK4 protein levels in U87 GB cells treated with pal-pom and rib-pom at 20–200 nM is considerable, indicating the efficacy of these PROTACs in countering the abnormal overexpression of this kinase in GB (Zhao and Burgess 2019). Su and colleagues also created a PROTAC and tested its effectiveness in GB cells by combining the CDK6 inhibitor palbociclib with the E3 CRBN recruiter pom (Su et al. 2019). Additionally, as recruitment moiety for the E3 ligases Mdm2 and VHL, as well as inhibitor of apoptosis (cIAP), nutlin-3b, VH032, and bestatin were employed in this investigation. The scientists discovered an interesting fact: only in U251 GB cells did CDK4 and CDK6 undergo degradation when PROTAC recruited CRBN, not the other E3s. Additionally, the degradation of CDK4 was observed to be less dramatic than that of CDK6. According to Su et al. (2019), CDK6 degraders with shorter linkers also had a better capacity for degradation, which favored the recruitment of CRBN toward CDK6. These observations indicate the possible applicability of PROTAC technology for the targeted CDK4/6 degradation for the treatment of GB, even if further research on the biological effect and antitumor potential of these PROTACs is still required.

In a recent study, the scientists utilized J22352, a highly selective histone deacetylase inhibitor, to reduce the growth of GB tumors. This offered the first in vivo proof of PROTACs' potential as anticancer treatments for GB (Liu et al. 2019). Since TMZ resistance and proliferation in GB are linked to HDAC6 upregulation, inhibiting this enzyme presents a viable approach for GB therapeutic interventions (Wang et al. 2016). J22352 exhibits PROTAC-like characteristics, which cause HDAC6 to be ubiquitylated and then degraded by proteasomes. As a result, both in vitro and in vivo, the reduction of HDAC6 expression level dramatically suppresses the formation of GB tumors in U87MG cells by promoting autophagic cancer cell death and stimulating the antitumor immune response (Liu et al. 2019).

PROTACs have shown great promise for treating GB, as evidenced by these groundbreaking studies on their effects in GB cells and the proof that they have already been developed against oncoproteins relevant to the tumor's progression, such as EGFR (Zhang et al. 2020; Zhao et al. 2020), mitogen-activated MAP-kinases (MAPs) (Pandey et al. 2016; Trauner and Shemet 2019), and bromodomain and extraterminal (BET) protein BRD4 (Xu et al. 2018; Yang et al. 2019; Hu and Crews 2021; Yang et al. 2021). Table 15.1 provides the pros and cons of applications of PROTACs in treatment of brain cancer.

15.8 Conclusion

GBM is still the worst malignancy with the fewest available treatment options. New findings that are beginning to identify its heterogeneity suggest that, like with other cancers, tailored therapy will be the way of the future. The widespread phenomenon of protein breakdown is crucial for preserving cellular homeostasis. The tiny protein modifier ubiquitin is involved in the regulation of protein fate and function, which affects the majority of signal transduction pathways and cellular functions.

PROTACs have demonstrated some promising possibilities in the little over 20 years since the initial reports of UPS's discovery. This has given rise to renewed optimism over the successful targeting of the numerous oncogenic proteins that have been found to be disease drivers, especially those that were given the moniker "undruggable." Nevertheless, it has taken nearly an additional two decades to advance targeted protein degradation and PROTACs to the forefront of drug discovery. There's no doubt that the initial clinical trials, along with recent advancements in chemical biology and synthetic biology, will hasten the introduction of novel treatments.

In fact, it will be necessary to determine the tissue-specific expression of the E3 ligases employed in PROTACs as well as the potential effects on the system of drafting an endogenous E3 ligase for therapeutic assistance. Combining optogenetics with protein degrader techniques like opto-PROTACs is another promising development that may offer more control over the timing and triggering of protein breakdown.

Table 15.1 Pros and Cons of PROTAC technology in brain tumor

PROS	CONS
• Targeted protein mutagenesis and undruggability may be addressed	• Due to the high molecular weight of PROTAC, there is poor blood-brain barrier permeability
• Less harmful than pharmacological options	• Varying E3 ligase expression in different parts of the brain
• Administered at extremely low concentrations because of the enzymatic activity that is sub-stoichiometric	• Breakdown of non-targeted proteins

References

An S, Fu L (2018) Small-molecule PROTACs: an emerging and promising approach for the development of targeted therapy drugs. EBioMedicine 36:553–562. https://doi.org/10.1016/j.ebiom.2018.09.005

Bai L, Zhou H, Xu R, Zhao Y, Chinnaswamy K, McEachern D et al (2019) A potent and selective small-molecule degrader of stat3 achieves complete tumor regression in vivo. Cancer Cell 36:498–511. https://doi.org/10.1016/j.ccell.2019.10.002

Bard JAM, Bashore C, Dong KC, Martin A (2019) The 26S proteasome utilizes a kinetic gateway to prioritize substrate degradation. Cell 177(286–298):e215. https://doi.org/10.1016/j.cell.2019.02.031

Beck A, Goetsch L, Dumontet C, Corvaïa N (2017) Strategies and challenges for the next generation of antibody-drug conjugates. Nat Rev Drug Discov 16:315–337. https://doi.org/10.1038/nrd.2016.268

Brennan D, Chen X, Cheng L, Mahoney M, Riobo NA (2012) Noncanonical hedgehog signaling. Vitam Horm 88:55–72. https://doi.org/10.1016/b978-0-12-394622-5.00003-1

Brennan CW, Verhaak RG, McKenna A, Campos B, Noushmehr H, Salama SR et al (2013) The somatic genomic landscape of glioblastoma. Cell 155:462–477. https://doi.org/10.1016/j.cell.2013.09.034

Bronner SM, Merrick KA, Murray J, Salphati L, Moffat JG, Pang J et al (2019) Design of a brain-penetrant CDK4/6 inhibitor for glioblastoma. Bioorg Med Chem Lett 29(29):2294–2301. https://doi.org/10.1016/j.bmcl.2019.06.021

Bufalieri F, Caimano M, Lospinoso Severini L, Basili I, Paglia F, Sampirisi L et al (2020a) The RNA-binding ubiquitin ligase MEX3A affects glioblastoma tumorigenesis by inducing ubiquitylation and degradation of RIG-I. Cancer 12:321. https://doi.org/10.3390/cancers12020321

Bufalieri F, Lospinoso Severini L, Caimano M, Infante P, Di Marcotullio L (2020b) DUBs activating the hedgehog signaling pathway: a promising therapeutic target in cancer. Cancer 12:1518. https://doi.org/10.3390/cancers12061518

Bufalieri F, Basili I, Di Marcotullio L, Infante P (2021) Harnessing the activation of RIG-I like receptors to inhibit glioblastoma tumorigenesis. Front Mol Neurosci 14:710171. https://doi.org/10.3389/fnmol.2021.710171

Burslem GM, Smith BE, Lai AC, Jaime-Figueroa S, McQuaid DC, Bondeson DP et al (2018) The advantages of targeted protein degradation over inhibition: an RTK case study. Cell Chem Biol 25:67. https://doi.org/10.1016/j.chembiol.2017.09.009

Burslem GM, Schultz AR, Bondeson DP, Eide CA, Savage Stevens SL, Druker BJ et al (2019) Targeting BCR-ABL1 in chronic myeloid leukemia by PROTAC-mediated targeted protein degradation. Cancer Res 79:4744–4753. https://doi.org/10.1158/0008-5472.CAN-19-1236

Bushweller JH (2019) Targeting transcription factors in cancer - from undruggable to reality. Nat Rev Cancer 19:611–624. https://doi.org/10.1038/s41568-019-0196-7

Chan KH, Zengerle M, Testa A, Ciulli A (2018) Impact of target warhead and linkage vector on inducing protein degradation: comparison of bromodomain and extra-terminal (BET) degraders derived from triazolodiazepine (JQ1) and tetrahydroquinoline (I-Bet726) BET inhibitor scaffolds. J Med Chem 61:504–513. https://doi.org/10.1021/acs.jmedchem.6b01912

Chung CW, Dai H, Fernandez E, Tinworth CP, Churcher I, Cryan J et al (2020) Structural insights into PROTAC-mediated degradation of Bcl-xL. ACS Chem Biol 15:2316–2323. https://doi.org/10.1021/acschembio.0c00266

Clifford SC, Walsh S, Hewson K, Green EK, Brinke A, Green PM et al (1999) Genomic organization and chromosomal localization of the humanCUL2 gene and the role of von Hippel-Lindau tumor suppressorbinding protein (CUL2 and VBP1) mutation and loss in renal-cell carcinoma development. Genes Chromosom Cancer 26:20–28. https://doi.org/10.1002/(sici)1098-2264(199909)26:13.0.co;2-6

Consortium (2015) Human genomics. The genotype-tissue expression (GTEx) pilot analysis: multitissue gene regulation in humans. Science 348:648–660. https://doi.org/10.1126/science.1262110

Costales MG, Aikawa H, Li Y, Childs-Disney JL, Abegg D, Hoch DG et al (2020a) Small-molecule targeted recruitment of a nuclease to cleave an oncogenic RNA in a mouse model of metastatic cancer. Proc Natl Acad Sci USA 117:2406–2411. https://doi.org/10.1073/pnas.1914286117

Costales MG, Childs-Disney JL, Haniff HS, Disney MD (2020b) How we think about targeting RNA with small molecules. J Med Chem 63:8880–8900. https://doi.org/10.1021/acs.jmedchem.9b01927

Cromm PM, Samarasinghe KTG, Hines J, Crews CM (2018) Addressing kinase-independent functions of Fak via PROTAC-mediated degradation. J Am Chem Soc 140:17019–17026. https://doi.org/10.1021/jacs.8b08008

Dunn GP, Rinne ML, Wykosky J, Genovese G, Quayle SN, Dunn IF et al (2012) Emerging insights into the molecular and cellular basis of glioblastoma. Genes Dev 26(26):756–784. https://doi.org/10.1101/gad.187922.112

Eichhorn JM, Alford SE, Sakurikar N, Chambers TC (2014) Molecular analysis of functional redundancy among anti-apoptotic Bcl-2 proteins and its role in cancer cell survival. Exp Cell Res 322:415–424. https://doi.org/10.1016/j.yexcr.2014.02.010

Farnaby W, Koegl M, Roy MJ, Whitworth C, Diers E, Trainor N et al (2019) BAF complex vulnerabilities in cancer demonstrated via structure based PROTAC design. Nat Chem Biol 15:672–680. https://doi.org/10.1038/s41589-019-0294-6

Farrell K, Jarome TJ (2021) Is PROTAC technology really a game changer for central nervous system drug discovery. Expert Opin Drug Discov 16:833–840. https://doi.org/10.1080/17460441.2021.1915979

Gangoso E, Southgate B, Bradley L, Rus S, Galvez-Cancino F, McGivern N et al (2021) Glioblastomas acquire myeloid-affiliated transcriptional programs via epigenetic immunoediting to elicit immune evasion. Cell 29(184):2454–2470. https://doi.org/10.1016/j.cell.2021.03.023

Gonzalez TL, Hancock M, Sun S, Gersch CL, Larios JM, David W et al (2020) Targeted degradation of activating estrogen receptor α ligand binding domain mutations in human breast cancer. Breast Cancer Res Treat 180:611–622. https://doi.org/10.1007/s10549-020-05564-y

Hanahan D, Weinberg RA (2011) Hallmarks of cancer: the next generation. Cell 144:646–674. https://doi.org/10.1016/j.cell.2011.02.013

Hines J, Lartigue S, Dong H, Qian Y, Crews CM (2019) MDM2-recruiting PROTAC offers superior, synergistic antiproliferative activity via simultaneous degradation of BRD4 and stabilization of P53. Cancer Res 79:251–262. https://doi.org/10.1158/0008-5472.can-18-2918

Hsu JH, Rasmusson T, Robinson J, Pachl F, Read J, Kawatkar S et al (2020) EED-targeted PROTACs degrade EED, EZH2, and SUZ12 in the PRC2 complex. Cell Chem Biol 27:41–e17. https://doi.org/10.1016/j.chembiol.2019.11.004

Hu Z, Crews CM (2021) Recent developments in PROTAC-mediated protein degradation: from bench to clinic. Chembiochem 23:e202100270. https://doi.org/10.1002/cbic.202100270

Inobe T, Matouschek A (2014) Paradigms of protein degradation by the proteasome. Curr Opin Struct Biol 24:156–164. https://doi.org/10.1016/j.sbi.2014.02.002

Jensen SM, Potts GK, Ready DB, Patterson MJ (2018) Specific MHC-I peptides are induced using PROTACs. Front Immunol 9:2697. https://doi.org/10.3389/fimmu.2018.02697

Khan S, Zhang X, Lv D, Zhang Q, He Y, Zhang P et al (2019) A selective BCL-XL PROTAC degrader achieves safe and potent antitumor activity. Nat Med 25:1938–1947. https://doi.org/10.1038/s41591-019-0668-z

Khan S, He Y, Zhang X, Yuan Y, Pu S, Kong Q et al (2020) Proteolysis targeting chimeras (PROTACs) as emerging anticancer therapeutics. Oncogene 639:4909–4924. https://doi.org/10.1038/s41388-020-1336-y

Kolb R, De U, Khan S, Luo Y, Kim MC, Yu H et al (2021) Proteolysistargeting chimera against BCL-X. Nat Commun 24(12):1281. https://doi.org/10.1038/s41467-021-21573-x

Kregel S, Wang C, Han X, Xiao L, Fernandez-Salas E, Bawa P et al (2020) Androgen receptor degraders overcome common resistance mechanisms developed during prostate cancer treatment. Neoplasia 22:111–119. https://doi.org/10.1016/j.neo.2019.12.003
Kristensen BW, Priesterbach-Ackley LP, Petersen JK, Wesseling P (2019) Molecular pathology of tumors of the central nervous system. Ann Oncol 30:1265–1278. https://doi.org/10.1093/annonc/mdz164
Lai J, Wang Y, Wu S-S, Ding D, Sun Z-Y, Zhang Y et al (2018) Elimination of melanoma by sortase A-generated TCR-like antibody-drug conjugates (TL-ADCs) targeting intracellular melanoma antigen MART-1. Biomaterials 178:158–169. https://doi.org/10.1016/j.biomaterials.2018.06.017
Le Rhun E, Preusser M, Roth P, Reardon DA, van den Bent M, Wen P et al (2019) Molecular targeted therapy of glioblastoma. Cancer Treat Rev 80:101896. https://doi.org/10.1016/j.ctrv.2019.101896
Lebraud H, Wright DJ, Johnson CN, Heightman TD (2016) Protein degradation by in-cell self-assembly of proteolysis targeting chimeras. ACS Cent Sci 2(2):927–934. https://doi.org/10.1021/acscentsci.6b00280
Li Y, Yang J, Aguilar A, McEachern D, Przybranowski S, Liu L et al (2019) Discovery of MD-224 as a first-in-class, highly potent, and efficacious proteolysis targeting chimera murine double minute 2 degrader capable of achieving complete and durable tumor regression. J Med Chem 62:448–466. https://doi.org/10.1021/acs.jmedchem.8b00909
Lieberman F (2017) Glioblastoma update: molecular biology, diagnosis, treatment, response assessment, and translational clinical trials. F1000Res 6:1892. https://doi.org/10.12688/f1000research.11493.1
Liu J-R, Yu C-W, Hung P-Y, Hsin L-W, Chern J-W (2019) Highselective HDAC6 inhibitor promotes HDAC6 degradation following autophagy modulation and enhanced antitumor immunity in glioblastoma. Biochem Pharmacol 163:458–471. https://doi.org/10.1016/j.bcp.2019.03.023
Liu J, Chen H, Ma L, He Z, Wang D, Liu Y et al (2020) Light-induced control of protein destruction by opto-PROTAC. Sci Adv 6:eaay5154. https://doi.org/10.1126/sciadv.aay5154
Liu Z, Hu M, Yang Y, Du C, Zhou H, Liu C et al (2022) An overview of PROTACs: a promising drug discovery paradigm. Mol Biomed 3(1):46. https://doi.org/10.1186/s43556-022-00112-0
Lospinoso Severini L, Ghirga F, Bufalieri F, Quaglio D, Infante P, Di Marcotullio L (2020) The SHH/GLI signaling pathway: a therapeutic target for medulloblastoma. Expert Opin Ther Targets 24:1159–1181. https://doi.org/10.1080/14728222.2020.1823967
Lospinoso Severini L, Bufalieri F, Infante P, Di Marcotullio L (2022) Proteolysis-targeting chimera (PROTAC): is the technology looking at the treatment of brain tumors? Front Cell Dev Biol 10:854352. https://doi.org/10.3389/fcell.2022.854352
Louis DN, Perry A, Reifenberger G, von Deimling A, Figarella-Branger D, Cavenee WK et al (2016) The 2016 World Health Organization classification of tumors of the central nervous system: a summary. Acta Neuropathol 6131:803–820. https://doi.org/10.1007/s00401-016-1545-1
Maksoud S (2021) The role of the ubiquitin proteasome system in glioma: analysis emphasizing the main molecular players and therapeutic strategies identified in glioblastoma multiforme. Mol Neurobiol 58(7):3252–3269. https://doi.org/10.1007/s12035-021-02339-4
Mason KD, Carpinelli MR, Fletcher JI, Collinge JE, Hilton AA, Ellis S et al (2007) Programmed anuclear cell death delimits platelet life span. Cell 128:1173–1186. https://doi.org/10.1016/j.cell.2007.01.037
Medikonda R, Dunn G, Rahman M, Fecci P, Lim M (2021) A review of glioblastoma immunotherapy. J Neuro-Oncol 151:41–53. https://doi.org/10.1007/s11060-020-03448-1
Melé M, Ferreira PG, Reverter F, DeLuca DS, Monlong J, Sammeth M et al (2015) The human transcriptome across tissues and individuals. Science 348:660–665. https://doi.org/10.1126/science.aaa0355
Meyer M, Reimand J, Lan X, Head R, Zhu X, Kushida M et al (2015) Single cell-derived clonal analysis of human glioblastoma links functional and genomic heterogeneity. Proc Natl Acad Sci USA 112:851–856. https://doi.org/10.1073/pnas.1320611111

Moreau K, Coen M, Zhang AX, Pachl F, Castaldi MP, Dahl G et al (2020) Proteolysis-targeting chimeras in drug development: a safety perspective. Br J Pharmacol 04177:1709–1718. https://doi.org/10.1111/bph.15014

Moser SC, Voerman JSA, Buckley DL, Winter GE, Schliehe C (2017) Acute pharmacologic degradation of a stable antigen enhances its direct presentation on MHC class I molecules. Front Immunol 8:1920. https://doi.org/10.3389/fimmu.2017.01920

Mullard A (2019) Arvinas's PROTACs pass first safety and PK analysis. Nat Rev Drug Discov 1118:895. https://doi.org/10.1038/d41573-019-00188-4

Network CGAR (2008) Comprehensive genomic characterization defines human glioblastoma genes and core pathways. Nature 455:1061–1068. https://doi.org/10.1038/nature07385

Niza E, Noblejas-López MDM, Bravo I, Nieto-Jiménez C, Castro-Osma JA, Canales-Vázquez J et al (2019) Trastuzumab-targeted biodegradable nanoparticles for enhanced delivery of dasatinib in HER2+ metastasic breast cancer. Nanomaterials 9:9. https://doi.org/10.3390/nano9121793

Nunes J, McGonagle GA, Eden J, Kiritharan G, Touzet M, Lewell X et al (2019) Targeting IRAK4 for degradation with PROTACs. ACS Med Chem Lett 10:1081–1085. https://doi.org/10.1021/acsmedchemlett.9b00219

Ocaña A, Pandiella A (2020) Proteolysis targeting chimeras (PROTACs) in cancer therapy. J Exp Clin Cancer Res 39(1):189. https://doi.org/10.1186/s13046-020-01672-1

Ocaña A, García-Alonso S, Amir E, Pandiella A (2018) Refining early antitumoral drug development. Trends Pharmacol Sci 39:922–925. https://doi.org/10.1016/j.tips.2018.09.003

Olson CM, Jiang B, Erb MA, Liang Y, Doctor ZM, Zhang Z et al (2018) Pharmacological perturbation of CDK9 using selective CDK9 inhibition or degradation. Nat Chem Biol 14:163–170. https://doi.org/10.1038/nchembio.2538

Ostrom QT, Gittleman H, Liao P, Rouse C, Chen Y, Dowling J et al (2014) CBTRUS statistical report: primary brain and central nervous system tumors diagnosed in the United States in 2007-2011. Neuro-Oncology 16(4):1–63. https://doi.org/10.1093/neuonc/nou223

Pandey V, Bhaskara VK, Babu PP (2016) Implications of mitogen activated protein kinase signaling in glioma. J Neurosci Res 94:114–127. https://doi.org/10.1002/jnr.23687

Pettersson M, Crews CM (2019) Proteolysis targeting chimeras (PROTACs) - past, present and future. Drug Discov Today Technol 31:15–27. https://doi.org/10.1016/j.ddtec.2019.01.002

Pfaff P, Samarasinghe KTG, Crews CM, Carreira EM (2019) Reversible spatiotemporal control of induced protein degradation by bistable photoPROTACs. ACS Cent Sci 5:1682–1690. https://doi.org/10.1021/acscentsci.9b00713

Pillow TH, Adhikari P, Blake RA, Chen J, Del Rosario G, Deshmukh G et al (2020) Antibody conjugation of a chimeric BET degrader enables in vivo activity. ChemMedChem 15:17–25. https://doi.org/10.1002/cmdc.201900497

Pohl C, Dikic I (2019) Cellular quality control by the ubiquitinproteasome system and autophagy. Science 366(15366):818–822. https://doi.org/10.1126/science.aax3769

Qi SM, Dong J, Xu ZY, Cheng XD, Zhang WD, Qin JJ (2021) PROTAC: an effective targeted protein degradation strategy for cancer therapy. Front Pharmacol 12:692574. https://doi.org/10.3389/fphar.2021.692574

Reynders M, Matsuura BS, Bérouti M, Simoneschi D, Marzio A, Pagano M et al (2020) PHOTACs enable optical control of protein degradation. Sci Adv 6:eaay5064. https://doi.org/10.1126/sciadv.aay5064

Robb CM, Contreras JI, Kour S, Taylor MA, Abid M, Sonawane YA et al (2017) Chemically induced degradation of CDK9 by a proteolysis targeting chimera (PROTAC). Chem Commun 53:7577–7580. https://doi.org/10.1039/c7cc03879h

Sakamoto KM (2010) Protacs for treatment of cancer. Pediatr Res 67(5):505–508. https://doi.org/10.1203/PDR.0b013e3181d35017

Sakamoto KM, Kim KB, Kumagai A, Mercurio F, Crews CM, Deshaies RJ (2001) Protacs: chimeric molecules that target proteins to the Skp1-Cullin-F box complex for ubiquitination and degradation. Proc Natl Acad Sci USA 98:8554–8559. https://doi.org/10.1073/pnas.141230798

Schapira M, Calabrese MF, Bullock AN, Crews CM (2019) Targeted protein degradation: expanding the toolbox. Nat Rev Drug Discov 18:949–963. https://doi.org/10.1038/s41573-019-0047-y

Schmitt C, Tonnelle C, Dalloul A, Chabannon C, Debré P, Rebollo A (2002) Aiolos and ikaros: regulators of lymphocyte development, homeostasis and lymphoproliferation. Apoptosis 7:277–284. https://doi.org/10.1023/a:1015372322419

Schneekloth AR, Pucheault M, Tae HS, Crews CM (2008) Targeted intracellular protein degradation induced by a small molecule: en route to chemical proteomics. Bioorg Med Chem Lett 18:5904–5908. https://doi.org/10.1016/j.bmcl.2008.07.114

Scholz N, Kurian KM, Siebzehnrubl FA, Licchesi JDF (2020) Targeting the ubiquitin system in glioblastoma. Front Oncol 10:574011. https://doi.org/10.3389/fonc.2020.574011

Smith BE, Wang SL, Jaime-Figueroa S, Harbin A, Wang J, Hamman BD et al (2019) Differential PROTAC substrate specificity dictated by orientation of recruited E3 ligase. Nat Commun 10:131. https://doi.org/10.1038/s41467-018-08027-7

Stoyanov GS, Dzhenkov DL (2018) On the concepts and history of glioblastoma multiforme - morphology, genetics and epigenetics. Folia Med 60:48–66. https://doi.org/10.1515/folmed-2017-0069

Stupp R, Mason WP, van den Bent MJ, Weller M, Fisher B, Taphoorn MJ et al (2005) Radiotherapy plus concomitant and adjuvant temozolomide for glioblastoma. N Engl J Med 352:987–996. https://doi.org/10.1056/NEJMoa043330

Stupp R, Hegi ME, Mason WP, van den Bent MJ, Taphoorn MJ, Janzer RC et al (2009) Effects of radiotherapy with concomitant and adjuvant temozolomide versus radiotherapy alone on survival in glioblastoma in a randomised phase III study: 5-year analysis of the EORTC-NCIC trial. Lancet Oncol 10:459–466. https://doi.org/10.1016/s1470-2045(09)70025-7

Su S, Yang Z, Gao H, Yang H, Zhu S, An Z et al (2019) Potent and preferential degradation of CDK6 via proteolysis targeting chimera degraders. J Med Chem 62(62):7575–7582. https://doi.org/10.1021/acs.jmedchem.9b00871

Sun Y, Ding N, Song Y, Yang Z, Liu W, Zhu J et al (2019) Degradation of Bruton's tyrosine kinase mutants by PROTACs for potential treatment of ibrutinib-resistant non-Hodgkin lymphomas. Leukemia 833:2105–2110. https://doi.org/10.1038/s41375-019-0440-x

Tanaka K (2009) The proteasome: overview of structure and functions. Proc Jpn Acad Ser B Phys Biol Sci 85(1):12–36. https://doi.org/10.2183/pjab.85.12

Testa A, Lucas X, Castro GV, Chan KH, Wright JE, Runcie AC et al (2018) 3-Fluoro-4-hydroxyprolines: synthesis, conformational analysis, and stereoselective recognition by the VHL E3 ubiquitin ligase for targeted protein degradation. J Am Chem Soc 140:9299–9313. https://doi.org/10.1021/jacs.8b05807

Testa A, Hughes SJ, Lucas X, Wright JE, Ciulli A (2020) Structure based design of a macrocyclic PROTAC. Angew Chem Int Ed Engl 59:1727–1734. https://doi.org/10.1002/anie.201914396

Touat M, Idbaih A, Sanson M, Ligon KL (2017) Glioblastoma targeted therapy: updated approaches from recent biological insights. Ann Oncol 28:1457–1472. https://doi.org/10.1093/annonc/mdx106

Trauner D, Shemet A (2019) Selective targeting of P38 MAPK isoforms using PROTACs. Synfacts 15:0438. https://doi.org/10.1038/s41467-018-08027-7

Verhaak RGW, Hoadley KA, Purdom E, Wang V, Qi Y, Wilkerson MD et al (2010) Integrated genomic analysis identifies clinically relevant subtypes of glioblastoma characterized by abnormalities in PDGFRA, IDH1, EGFR, and NF1. Cancer Cell 17:98–110. https://doi.org/10.1016/j.ccr.2009.12.020

Wang Z, Hu P, Tang F, Lian H, Chen X, Zhang Y et al (2016) HDAC6 promotes cell proliferation and confers resistance to temozolomide in glioblastoma. Cancer Lett 379:134–142. https://doi.org/10.1016/j.canlet.2016.06.001

Wang Y, Jiang X, Feng F, Liu W, Sun H (2020) Degradation of proteins by PROTACs and other strategies. Acta Pharm Sin B 10:207–238. https://doi.org/10.1016/j.apsb.2019.08.001

Weenink B, French PJ, Sillevis Smitt PAE, Debets R, Geurts M (2020) Immunotherapy in glioblastoma: current shortcomings and future perspectives. Cancer 12:751. https://doi.org/10.3390/cancers12030751

Winter GE, Buckley DL, Paulk J, Roberts JM, Souza A, DhePaganon S et al (2015) Phthalimide conjugation as a strategy for in vivo target protein degradation. Science 348:1376–1381. https://doi.org/10.1126/science.aab1433
Xu H, Zong H, Ma C, Ming X, Shang M, Li K et al (2017) Epidermal growth factor receptor in glioblastoma (review). Oncol Lett 14:512–516
Xu L, Chen Y, Mayakonda A, Koh L, Chong YK, Buckley DL et al (2018) Targetable BET proteins- and E2F1-dependent transcriptional program maintains the malignancy of glioblastoma. Proc Natl Acad Sci USA 115:E5086–E5095. https://doi.org/10.1073/pnas.1712363115
Xue G, Wang K, Zhou D, Zhong H, Pan Z (2019) Light-induced protein degradation with photocaged PROTACs. J Am Chem Soc 141:14118370–14118374. https://doi.org/10.1021/jacs.9b06422
Yang C-Y, Qin C, Bai L, Wang S (2019) Small-molecule PROTAC degraders of the bromodomain and extra terminal (BET) proteins - a review. Drug Discov Today Technol 31:43–51. https://doi.org/10.1016/j.ddtec.2019.04.001
Yang H, Wei L, Xun Y, Yang A, You H (2021) BRD4: an emerging prospective therapeutic target in glioma. Mol Ther Oncol 21:1–14. https://doi.org/10.1016/j.omto.2021.03.005
Yu JM, Sun W, Wang ZH, Liang X, Hua F, Li K et al (2019) TRIB3 supports breast cancer stemness by suppressing FOXO1 degradation and enhancing SOX2 transcription. Nat Commun 10:5720. https://doi.org/10.1038/s41467-019-13700-6
Zhang H, Zhao H-Y, Xi X-X, Liu Y-J, Xin M, Mao S et al (2020) Discovery of potent epidermal growth factor receptor (EGFR) degraders by proteolysis targeting chimera (PROTAC). Eur J Med Chem 189:112061. https://doi.org/10.1016/j.ejmech.2020.112061
Zhao B, Burgess K (2019) PROTACs suppression of CDK4/6, crucial kinases for cell cycle regulation in cancer. Chem Commun 55:2704–2707. https://doi.org/10.1039/c9cc00163h
Zhao Q, Ren C, Liu L, Chen J, Shao Y, Sun N et al (2019) Discovery of SIAIS178 as an effective BCR-ABL degrader by recruiting Von Hippel-Lindau (VHL) E3 ubiquitin ligase. J Med Chem 62:9281–9298. https://doi.org/10.1021/acs.jmedchem.9b01264
Zhao HY, Yang XY, Lei H, Xi XX, Lu SM, Zhang JJ et al (2020) Discovery of potent small molecule PROTACs targeting mutant EGFR. Eur J Med Chem 208:112781. https://doi.org/10.1016/j.ejmech.2020.112781
Zhou H, Bai L, Xu R, Zhao Y, Chen J, McEachern D et al (2019) Structure based discovery of SD-36 as a potent, selective, and efficacious PROTAC degrader of STAT3 protein. J Med Chem 62:11280–11300. https://doi.org/10.1021/acs.jmedchem.9b01530

Proteolysis Targeting Chimera (PROTACs): An Attractive Technology in CVD Therapeutics—Opportunities and Challenges

16

Dhruv Sanjay Jaya Gupta, Nidhi Khedpande, and Kalyani Barve

Abstract

Cardiovascular diseases (CVDs) affect the heart and blood vessels. Stroke and heart attack are the leading cause of death globally, responsible for 17.9 million annual deaths. Both these conditions need prompt management, surgical intervention, and drug therapy thereafter. Other CVDs also require chronic therapy. Most of the currently available drugs are small molecules, which face issues like drug resistance, side effects, and reduced activity against many proteins. PROTAC technology, which utilizes ATP-dependent ubiquitin-proteasome system (UPS) pathway to degrade the protein of interest, can be a prospective solution. PROTAC is a bifunctional molecule containing a ligand for engaging E3 ligase, a linker, and another ligand to connect to the target protein for degradation. E3 ligase labels the protein with ubiquitin, which then undergoes degradation. Several proteins like IRAK4, TRIM 24, and ion channel proteins are involved in the pathogenesis of CVDs. PROTAC technology can not only target the dysfunctional proteins but also overcome the resistance to drugs like statins which on prolonged use cause protein overexpression. However, it faces disadvantages including low bioavailability and cell specificity. Another pressing issue is the availability and synthesis of specific PROTACs for research. To cater to this need, several service providers offer support to drug discovery teams for PROTAC design and synthesis. Translation of this technology to clinical trials has begun for cancer, however, for CVDs no suitable candidate has been developed. This chapter mainly focuses on the development of PROTACs for the treat-

D. S. J. Gupta · N. Khedpande · K. Barve (✉)
Department of Pharmacology, Shobhaben Pratapbhai Patel School of Pharmacy & Technology Management, SVKM's Narsee Monjee Institute of Management Studies (NMIMS) Deemed-to-University, Mumbai, India
e-mail: kalyani.barve@nmims.edu

M. Nandave, P. Jain (eds.), *PROTAC-Mediated Protein Degradation: A Paradigm Shift in Cancer Therapeutics*, https://doi.org/10.1007/978-981-97-5077-1_16

ment of various CVDs, their advantages, limitations, and future scope in clinical practice.

Keywords

Atherosclerosis · Coagulation · Hyperlipidemia · Myocardial infarction · Stroke · Thrombosis · Ubiquitin-proteasome system

Abbreviations

AD	Alzheimer's disease
APOA2	Apolipoprotein A-II precursor
AR	Androgen receptor
B2M	Beta-2-microglobulin
BET	Bromodomain and extra-terminal domain
Brd4	Bromodomain-containing protein 4
CAMs	Cell adhesion molecules
CDKs	Cyclin-dependent kinases
CMA	Chaperone-mediated autophagy
COX-1/2	Cyclooxygenase 1/2
CRBN	Cereblon
CRP	C-reactive protein
CUL2	Cullin-2
CVD	Cardiovascular disease
DAPK1	Protein kinase 1
ER	Estrogen receptor
ERα	Estrogen receptor alpha
HER2	Human epidermal growth factor receptor 2
HIF-1α	Hypoxia-inducible factor 1α
hs-CRP	High-sensitivity C-reactive protein
HTS	High-throughput screening
IDO	Indoleamine 2,3-dioxygenase
IGFALS	Insulin-like growth factor binding protein, acid labile subunit
IGFBP4	Insulin-like growth factor binding protein 4
MCL-1	Myeloid leukemia 1
MetAP-2	Methionine aminopeptidase 2
MI	Myocardial infarction
MMPs	Matrix metalloproteinases
NAPT	Nicotinamide phosphoribosyltransferase
NGAL	Neutrophil gelatinase-associated lipocalin
ORM1	Alpha-1-acid glycoprotein 1
PPIA	Peptidyl-prolyl isomerase A
PROTAC	Proteolysis targeting chimera

ROS	Reactive oxygen species
SHP2	Src homology 2 domain-containing phosphatase 2
SMIs	Small-molecule inhibitors
TAMs	Tumor-associated macrophages
TPD	Targeted protein degradation
VHL1	Von-Hippel-Lindau ubiquitin ligase-1

16.1 Introduction

A relatively novel and emerging technology, proteolysis targeting chimera (PROTAC) has been explored to mediate the induction of protein degradation, facilitated by a targeting molecule (Li et al. 2022). In a quest to design targeted, site-specific management strategies for metabolic disorders such as cardiovascular disease (CVD) and cancer, research has been focused on designing delivery systems with minimal off-target activity and undesired effects (Zou et al. 2019). While the roadmap of this pursuit started with an initial screening of small molecules displaying protein binding and successive inhibition, drug resistance was a common challenge associated with such therapies. In addition to the onset of resistance, other challenges such as protein accumulation and lack of identification by targeted proteins due to the structure of such small moieties were significant concerns (Békés et al. 2022). This has prompted further brainstorming and focused efforts in the domain of targeting proteins, and PROTAC has taken the stage as a promising technology to facilitate protein degradation.

Utilizing the ubiquitin-protease system to enable robust protein targeting and subsequent degradation, PROTAC makes use of the intrinsic protein degradation mechanism of cells to clear out denatured and harmful proteins, that may contribute to pathogenesis and disease progression (Wang et al. 2022a). This technology has been tapped into for the targeting of a variety of enzymes, transcription factors, and signaling proteins, with benefits ranging from malignancies to complex neurological conditions (Kargbo 2019). Research has indicated that the degradation of proteins may prove to be a more ideal strategy as opposed to simple inhibition, paving the way for PROTAC molecules tailored and designed for specific disease conditions (Tamatam and Shin 2023).

The starting point for cultivating an understanding of the benefits of protein degradation in disease management may be dated back to the late 1990s, with several research groups observed the protein degradation induced by geldanamycin, an antibiotic. While its activity was initially understood to be binding to HSP90, it was seen to bring about the degradation of several proteins such as human epidermal growth factor receptor 2 (HER2, a tyrosine kinase commonly expressed in breast malignancies), Raf-1 (a proto-oncogene), p53 (a transcription factor responsible for the maintenance of stability of gene sequences) and Bcr-Abl (linked with an overproduction of tyrosine kinase, chiefly in leukemic conditions). However, in an attempt to lower the toxicity and prevent the degradation of off-target proteins, it was linked with steroid hormones such as estradiol and testosterone, to enhance affinity to selected receptors (Zou et al. 2019). This laid the foundation for the exploitation of chimeric proteins from ubiquitin ligase complexes, with Zhou et al. being credited with the designing of first SCF E3 ubiquitin ligase complex, for targeting osteosarcoma cells (Zhou et al. 2000). While these attempts were regarded essentially as predecessors to the development of PROTAC, their contribution is important to consider. Finally, the starting point of engineering of PROTAC as we know it today was initiated by Kathleen M. Sakamoto and her research group, who hypothesizes a system utilizing the linkage of an E3 ubiquitin ligase with target proteins, to trigger degradation (Sakamoto 2010). These efforts further resulted in the designing of chimeric molecules utilizing ovalicin, an inhibitor of angiogenesis, and linkage with phosphopeptide proteins. This complex, recognized by the protein methionine aminopeptidase 2 (MetAP-2), further brought about its degradation. While such chimeric complexes were initially harnessed to target endocrine receptors such as estrogen receptor (ER) and androgen receptor (AR) through peptide-based PROTACs, focused research efforts paved the way for more specific applications (Sakamoto et al. 2001). Some notable examples include the work undertaken by Montrose et al. for the targeting of cancer-forming X-protein in patients with an underlying viral basis of hepatocellular carcinoma (Montrose and Krissansen 2014), Kim et al. engaging in PROTAC engineering centered around an octapeptide associated with HIF-1α (hypoxia-inducible factor 1α) (Zhang et al. 2004), and Fan et al. utilizing chaperone-mediated autophagy (CMA) to design a pentapeptide system, with specificity for three proteins, namely protein kinase 1 (DAPK1), scaffolding protein PSD-95, and a-synuclein (Fan et al. 2014). In summary, advancements in the designing of PROTACs have indicated possibilities for knockdown of multiple proteins, confirmed by the demonstration of efficacy in preclinical disease models. Figure 16.1 offers a holistic insight into the synthesis of PROTACs, as well as the current and emerging potentials for clinical applications.

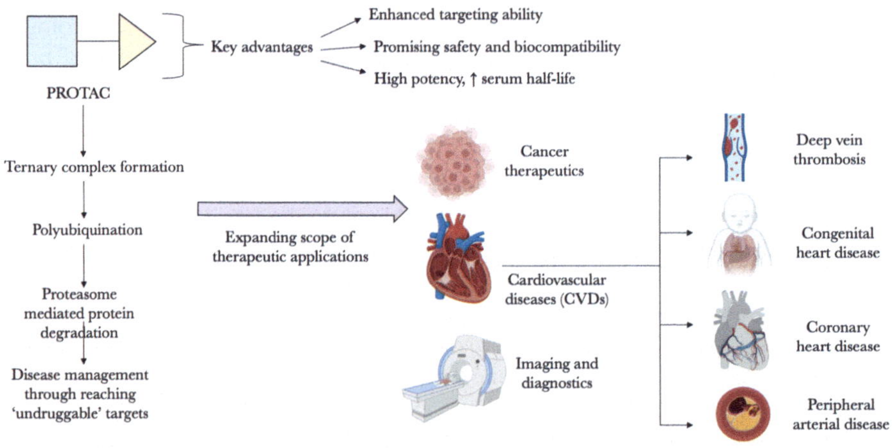

Fig. 16.1 A comprehensive overview of the synthesis of PROTACs, as well as the current and emerging scope of therapeutic applications

As these initial hypotheses had a basis in short-chain peptides as the focal point facilitating the recognition of E3 ubiquitin ligases, they were dubbed 'peptide-based PROTACs' (Jin et al. 2020). The mechanism of action of such peptide-based moieties may be understood to be an initial binding of the peptide chain to E3 ubiquitin ligase, with a small molecule that is also a part of the linker chain undergoing binding with the protein of interest, resulting in polyubiquitination and subsequent proteolysis, mediated by proteasomes (Wang et al. 2024a).

However, these models were marred by several challenges, including low permeability across cellular membranes; thereby requiring specialized delivery strategies, exertion of activity exclusively at a micromolar range, as well as the large size of the chimeric molecule, which in turn resulted in immune recognition and recruitment of antibodies, thereby nullifying activity in biological systems (Au et al. 2020). Advancements in research and development have now led us to the usage of small molecules in PROTAC engineering, ushering in a new era of targeted interventions.

To overcome the drawbacks of peptide-based PROTACs, small molecule-based PROTAC strategies have been explored. The starting point for this shift was the usage of small molecules as E3 ubiquitin ligase moieties, to bring about its recognition (Jin et al. 2022). Potential advantages of such a strategy include easier absorption and transport across biological membranes, owing to a smaller size. As for the mechanism of action, it was initially proposed that these small molecules bring about protein denaturation, followed by subsequent degradation via a ubiquitin-proteasome or autophagy (An and Fu 2018). It has been observed that research initiatives have largely focused on the development of small-molecule PROTACs for prudently targeting the bromodomain and extra-terminal domain (BET) family of proteins, by a linkage with BRD4 inhibitors, primarily for cancer management (Simonetta et al. 2019).

Through focused research efforts, over 30 proteins playing a key role in human diseases and pathogenesis have been targeted. While most of these endeavors have been focused on cancer management, major targets may be grouped into three categories: nuclear receptors (such as ER and AR), protein kinases (such as BCR-Abl, cyclin-dependent kinases), regulatory and neurodegeneration-linked proteins (such as X-protein, Tau) and metabolic enzymes (Chirnomas et al. 2023).

In addition to this, it is essential to develop an understanding of the various components involved in the targeting process. Starting with target proteins (as previously summarized), we direct our attention to ligands for the chosen targets. Popular ligands include natural agents (estradiol, flavonoids such as apigenin), inhibitory agents (such as thiazolidinedione, and ceitinib), as well as peptides (Schneider et al. 2021). The ligands that have been explored for E3 ubiquitin ligase so far, attached to a linker, chiefly encompass agents such as peptides, thalidomide, and von-Hippel-Lindau ubiquitin ligase-1 (VHL1). Finally, this is linked with E3 ubiquitin ligase (example: Cereblon) (Maneiro et al. 2021). Figure 16.2 offers an overview of the mechanism of action, and the sequence of events leading from initial target recognition and ternary complex formation to protein degradation.

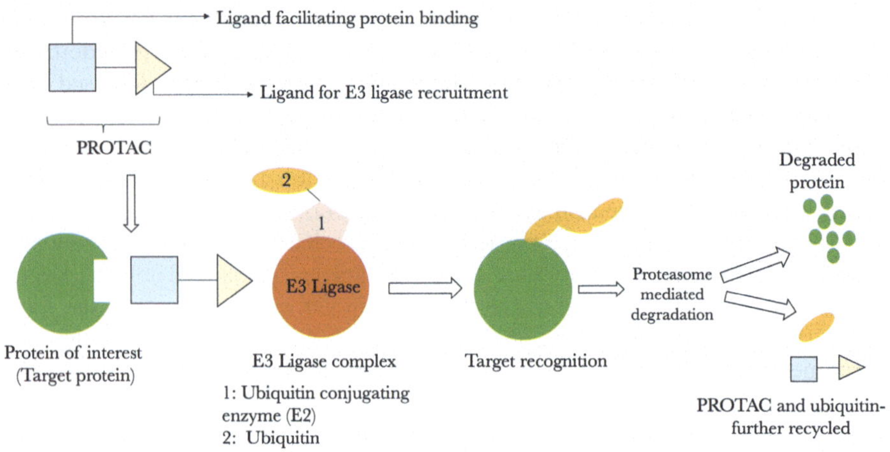

Fig. 16.2 Bird's-eye view of the mechanism of action of PROTAC, from initiation to proteolysis

16.2 Proteins in the Pathogenesis of CVDs

Cardiovascular diseases (CVDs) broadly encompass a group of disorders of the heart and associated blood vessels. As per the classification provided by the World Health Organization, CVDs may be understood to include- coronary heart disease, cerebrovascular disease, peripheral arterial disease, rheumatic heart disease, congenital heart disease, and deep vein thrombosis, also including pulmonary embolism (Goldsborough et al. 2022). Through this section, the authors have aimed to provide a basic understanding of each, as well as key proteins involved in the pathogenesis of these ailments.

In the simplest terms, coronary heart disease may be understood to be a disease of blood vessels directly supplying the heart muscle. Coronary heart disease is characterized by the blockade of blood supply to the heart due to lipid buildup in the arteries, through a process referred to as atherosclerosis (Malakar et al. 2019). The fatty deposits, known as atheromas, further trigger cardiac ischemia owing to an insufficient supply of oxygenated blood. Coronary heart disease may be attributed to many lifestyle factors, such as high-fat diets, a sedentary lifestyle, as well as smoking and drinking (Steg and Ducrocq 2016). MRI scans and coronary angiography are most commonly employed for its detection. As for the pathogenesis, a misfolding of light chain proteins and transthyretin proteins results in the formation of amyloid fibrils, in a process known as amyloidosis (Raval et al. 2023). In addition, an increase in oxidative stress causes reduced nitric oxide production and an upregulation of cytokines. This, in turn, elevates the levels of C-reactive protein (CRP). Oxidized LDL lipid accumulation, adhesion of monocytes and T-lymphocytes, and platelet aggregation give rise to extracellular lipid accumulation, recruitment of foam cells, and smooth muscle proliferation (Libby et al. 2019). Major contributing proteins include collagen, elastin, and proteoglycans, resulting in the formation of atheromatous plaque. Based on the type of lesion, clinical manifestations may be observed- stenotic lesions are fewer and fibrotic, and result in ischemia, while non-stenotic lesions are more widespread and lipid-rich and result in myocardial infarction (MI) (Dolmaci et al. 2021). Exosomes from erythrocytes, lymphocytes, and platelets result in a reduction of nitric-oxide-mediated vasodilation, resulting in endothelial dysfunction and calcification of vasculature (Gao et al. 2019). In response to stress stimuli such as cytokines, there is an upregulation of p38 mitogen-activated protein kinases (Hua et al. 2022). Some other proteins include Beta-2-microglobulin (B2M), alpha-1-acid glycoprotein 1 (ORM1), and insulin-like growth factor binding protein, acid labile subunit (IGFALS), showing a strong correlation with the risk for coronary heart disease. Other prominent biomarkers in coronary heart disease include apolipoprotein A-II precursor (APOA2), peptidyl-prolyl isomerase A (PPIA), and IGFBP4 insulin-like growth factor binding protein 4 (IGFBP4) (Prentice et al. 2010).

Peripheral artery disease (PAD) is characterized clinically by the narrowing or blockage of blood vessels to the extremities, primarily occurring due to atherosclerosis (Shamaki et al. 2022). Inflammatory biomarkers associated with PAD include CRP and interleukins, chiefly IL-6. In addition to this, markers of endothelial dysfunction include ADMA and soluble cell adhesion molecules (CAMs) (Morley et al. 2018). Oxidized LDL and lipoprotein A are presented as lipoproteins, with hepatocyte growth factor and VEGF acting as modulators of angiogenesis. In addition to this, various coagulation proteins such as fibrinogen thrombomodulin and PAI-1 play a crucial role in the pathogenesis of PAD (Feng et al. 2022). These biomarkers, however, essentially lack specificity and are elevated in other CVDs as well. Other plasma biomarkers showing a strong clinical correlation with PAD include Beta-2-microglobulin (B2M) and cystatin C, a non-glycosylated biomarker that is also used to assess kidney function (Cooke and Wilson 2010). This has prompted the quest for biomarkers specific to PAD, and Galectin-3 and hs-CRP (high-sensitivity C-reactive protein) have been identified as two prominent protein biomarkers associated with peripheral artery disease and resulting ischemia (Ding et al. 2020).

Rheumatic heart disease is an immune condition, that is observed to occur as a complication resulting from rheumatic fever, chiefly caused by streptococcal bacterial infections. It is characterized by single or repeated attacks of rheumatic fever, resulting in the deformity of heart valves, and shortening of the chordae tendinea, with the mitral valve bearing the greatest impact (Leal et al. 2019). Concerning protein biomarkers, those suffering from this ailment display elevated levels of serum tenascin C, also utilized as a prominent biomarker for its diagnosis (Karatas et al. 2013). In addition to this, another potential biomarker for risk assessment is NT-proBNP (Khare and Dwivedi 2016). Adding to this, the lipoteichoic acids, N-Acetyl-Beta-D-Glucosamine, and M-protein sub-units from the causative organism trigger a lowering of Ficolin-2 levels, and an elevation of MBL levels. In addition to complement activation as well as phagocytosis triggering, genetic polymorphisms include the MBL2 and FCN2 genes (Beltrame et al. 2015).

Congenital heart disease encompasses a range of defects in the working of the heart, existing since birth (Scott and Neal 2021). They may be further understood under two categories- acyanotic defects (arising from volume overload and pressure overload- including outflow and inflow obstruction), and cyanotic defects (characterized by a decrease or increase in pulmonary blood flow) (Rohit and Shrivastava 2018). Some common types of congenital heart defects include atrial and ventricular septal defects, aortic coarctation (narrowing of the aorta, resulting in lowered blood flow), and tetrad falls (aortal narrowing, thickening of heart walls, narrowing of pulmonary artery valve and septal defect) (Rohit and Rajan 2020). In terms of biomarkers, some key markers of oxidative stress include uric acid and myeloperoxidase, while neuroendocrine markers such as copeptin and endothelin-1 may also be upregulated (Özlü et al. 2023). In addition to this, natriuretic peptides result in a myocardial stretch, eventually culminating in heart failure. Inflammation is chiefly regulated by IL-6, TNF-α, and procalcitonin, while markers of renal injury such as creatinine and Neutrophil gelatinase-associated lipocalin (NGAL) may also be expressed (Houska and Schwartz 2020).

In summation, the biomarkers of CVD may broadly be understood under the following categories: those associated with renal dysfunction (such as NAG, Cystatin C), those linked with neurohumoral activation (Endothelin-1, renin, urocortin), inflammatory biomarkers (Cytokines, CRP, TNF- α), those associated with oxidative stress (oxidized LDLs, plasma malondialdehyde) and finally, biomarkers linked with matrix remodeling and myocyte injury (MMPs, galectin-3, ILs, heart shock proteins) (Ibrahim and Januzzi 2018). These biomarkers can be targets in developing PROTACs for CVDs.

16.3 PROTACs in the Preclinical/Clinical Phase

This section offers a bird's-eye view of the advancements in PROTAC applications, through the lens of results from pre-clinical testing. While the authors have initiated this section with a broad overview of the applications of PROTACs in cancer management, the scope of discussion subsequently expands to other conditions such as neurodegeneration, as well as CVD. In addition to this, as a conclusion to this section, challenges and opportunities for clinical translation have also been briefly touched upon.

A study undertaken by Wang et al. explored the benefits offered by PROTAC in a model of liver cancer. Utilizing a FOXM1-focused PROTAC, reduced cellular migration and in turn, lowered tumor proliferation, was observed in vivo. In addition to this, the glucose metabolism of cancerous cells was modulated, through the downregulation of GLUT1 (Wang et al. 2022b). GLUT1 knockdown has been associated with cardiac hypertrophy and myocardial infarction, and its suppression using PROTAC-systems may have pronounced clinical benefits (Chen et al. 2021; Kashihara et al. 2022). Yang et al. published an interesting study, illustrating the dual-pronged benefit offered by the degradation of cyclin-dependent kinases (CDKs), using PROTAC, for breast cancer management (Yang et al. 2022). In addition to being a prominent biomarker in malignancies, CDKs have also shown benefits in cardio-oncology and the suppression of cardiac hypertrophy and may be explored further (Kohoutek and Blazek 2012).

Moving toward other indications, PROTAC strategies have also been employed in the management of neurodegenerative conditions, such as Alzheimer's disease (AD). Through ubiquitination-proteasome-governed protein degradation, a promising tau clearance was observed, with an improvement in cognitive functioning (Wang et al. 2021). This indicates a positive, promising shift in focus to other pathologies and potentials for wider-scale applications. PROTAC technologies have also been utilized in the development of attenuated vaccine candidates, through the degradation of viral proteins in the cells of the host. This, in turn, showcased strong cellular and humoral immunity responses, which may be extended to many conditions requiring immunomodulation (Si et al. 2022).

Recent research initiatives have focused on the engineering of PROTACs sensitive and responsive to certain stimuli, for controlled proteolysis. Various conditions of the cellular environment, such as the concentration of ROS, enzymes such as phosphatase and hypoxia as well as external environmental triggers, have been employed to activate PROTACs, and the way forward could include antibody-therapeutic agent complexes as well as smart PROTACs, highly responsive to such triggers (An et al. 2023).

Directing our focus to more specific applications rooted in the domain of cardiology, several recently undertaken studies have been detailed in the following paragraph. A study by Wang et al. focused on the targeting of Cullin-2 (CUL2)-RING ubiquitin ligases, with a marked role in calcium ion homeostasis and obesity-linked heart dysfunction (Diaz et al. 2022). PROTAC-mediated downregulation of the CAND1 gene was observed, allowing selective targeting of proteins belonging to that family, and bringing about a modification of the cellular microenvironment, to prevent the acceleration of pathologies (Wang et al. 2024b). Recently, several patents have been approved by the FDA for the usage of the BCL-2 family of proteins, as well as Myeloid leukemia 1(MCL-1) inhibitors. While these proteins have a foundation in cancer pathogenesis, studies have indicated the involvement of Bcl-2 in cardiac hypertrophy, mediated by the PI3K-Akt signaling pathway (Meng et al. 2021). Therefore, the positively evolving patent landscape is promising, concerning CVD management using PROTACs. In addition to this, the usage of such technologies would help to overcome the adverse effects and toxicity posed by conventional inhibitors of these targets, such as platelet toxicity (Pal et al. 2022). Cereblon (CRBN), which is an integral part of E3 ubiquitin ligase complexes used for ion channel targeting, plays a crucial role in the regulation of heart failure and a lowering of ejection fraction. Park et al. proposed PROTAC usage to bring about a knockdown of this protein selectively, thereby exerting a cardioprotective effect (Park et al. 2022). In continuation with the discussion on MCL-1 inhibitors, PROTAC interventions have been harnessed to overcome the cardiotoxicity posed by these agents. This, in turn, has led to the expansion of their therapeutic benefits, as well as expedited clinical translation, that further aids the designing of personalized treatment strategies (Liu et al. 2023a). Lastly, another interesting advancement has been the targeting of Bromodomain-containing protein 4 (Brd4), playing a pivotal role in the upregulation of inflammatory markers and as a result, acute heart failure. In addition to this, Brd4 plays a role in a variety of other CVDs, including pulmonary arterial hypertension and hypertension (Lin and Du 2020). The usage of PROTACs on this frontier has emerged as an upcoming frontier in drug discovery and clinical translation, with several benefits coming to light, in addition to proteolysis (Duan et al. 2018).

Our discussion would remain incomplete without offering a glimpse of the current clinical status of PROTAC strategies. Despite the exploration of several clinical targets supported by human E3 ligases, a vast majority of PROTAC interventions do not make it past the stage of preclinical development, impeding equitable market access and availability to patients in need. There is a need to explore several targeting strategies (nano-strategies have been discussed in subsequent sections) to improve target protein specificity and reduce off-target adverse effects (Chen et al. 2022). In 2021, a small-molecule PROTAC, ARV-110, intended for oral delivery, entered phase I of clinical evaluation. With AR specificity, this engineered formulation was intended for the management of prostate cancer, expanding the scope of discussion for the lysis of various proteins through the ubiquitin-proteasome pathway (Zeng et al. 2021). Additionally, another avenue for clinical interest is the combination of immunomodulators with PROTACs, for targeted protein degradation (TPD), for benefits concerning draggability (Fang et al. 2023). The rational engineering and designing of these agents is crucial, as minor variations may have a pronounced shift in their safety and biocompatibility, as well as selectivity concerning protein degradation (Chen et al. 2023). It is, therefore, essential to supplement clinical research with comprehensive data from preclinical screening, including pharmacokinetic and pharmacodynamic profiling.

Another major challenge impeding clinical translation is the availability of a small population of E3 ligase recruiters, despite the possibility of the human genome encoding over 600 ligases of this nature (Han and Sun 2023). Before engaging in full-fledged clinical testing, there is a pressing need to identify a greater number of ligands, to work in conjunction with PROTAC interventions.

16.4 Role of PROTACs in Diagnosis and Imaging

This section delineates the applications of PROTAC systems in diagnostics and imaging. An attempt has been made to highlight recent, relevant research from several disciplines, to offer readers a holistic overview of the progress that has been made.

A study undertaken by Shi et al. explored the benefits offered by BRD4-focused PROTACs in the studying of biomolecular condensates, through high-throughput screening (HTS). Tying in cellular imaging, a quantitative method was established, for the mapping of cellular components of eukaryotic cells, to further understand signaling and responses to DNA damage (Shi et al. 2023). In a separate study by Trapotsi et al, visualization of cellular morphology was carried out using a technology referred to as 'cell painting' to gauge the phenotype of PROTACs. Based on cellular cluster formation and HTS, the toxicity profiles of these agents were evaluated, which could have future implications for drug design (Trapotsi et al. 2022). Additionally, PROTACs have been utilized for the imaging of CAR T cell therapy. This, in turn, facilitates quantitative imaging helmed by cellular activity, and possibilities for modulation and regulation (Lee et al. 2023). Cheng et al. proposed the development of theranostic PROTACs, to map the degradation of nicotinamide phosphoribosyltransferase (NAPT), a cytokine expressed in lymphocytes. Using an ovarian cancer A2780 cell line, PROTAC fulfilled a theranostic role and downregulated the expression of nicotinamide adenine dinucleotide (NAD^+), thereby exerting a robust antiproliferative effect (Cheng et al. 2022). There are currently a limited number of studies to support the usage of PROTAC in theranostics and imaging, and there is a need for greater lab-scale testing and development to put forth contemporary models that boost the role of these agents in disease screening and eventually, disease management.

16.5 Nanotechnological and Targeted-Delivery Interventions

In recent times, nanotechnological interventions have been employed to overcome the drawbacks associated with conventional PROTAC therapy. While a majority of these studies remain restricted to the domain of oncology, it would be interesting to see how some of these strategies may be proposed for the management of CVDs as well.

Zhang et al. proposed the development of Nano-PROTAC, designed by glutathione-responsive PLGA, to enable camouflaging with alveolar membranes in the lung. This nano strategy, in turn, offered a double-pronged benefit- the targeting of cancerous cells as well as tumor-associated macrophages (TAMs). In response to the tumor microenvironmental pH, condition-specific PROTAC release was also facilitated (Zhang et al. 2022a). In addition to this, nano-mediated PROTACs have been combined with phototherapy and immunotherapy (condensed as photo-immunotherapy). A Src homology 2 domain-containing phosphatase 2 (SHP2)-targeting PROTAC peptide engineered by Zhang et al. silenced immunosuppressive signaling pathways, thereby activating T-cells and providing a robust immune response (Zhang et al. 2023a). Immunotherapy is a recent emerging frontier in CVD management, primarily as a means to reduce inflammation, stabilize plaque formation, and restrict the production of inflammatory markers (Jyotsna et al. 2023). It would be interesting to see PROTAC usage in combination with immunotherapy, for CVD amelioration. A noteworthy advancement has been PROTAC delivery through alternative dosage forms, such as hydrogels, in combination with conventional chemotherapeutic agents such as paclitaxel. As protein degradation continues, a sustained release pattern for the drug moiety is also established (Wu et al. 2023). Such temperature-responsive hydrogel systems have been used in the delivery of stem cells to the heart (Lu et al. 2009), as well as the management of congenital heart defects (Garven et al. 2022), and may harness PROTACs to modulate angiogenesis and the targeting of matrix metalloproteinases (MMPs) (Shaik et al. 2023).

A prominent challenge impeding the exertion of benefits offered by PROTAC is the onset of a 'hook effect', characterized by the production of undesirable binary complexes, at a high concentration of PROTAC (Pettersson and Crews 2019). Another study reported the development of nano-PROTACs, assembled in-situ through hydrogen bonding, was observed to form E3 ligases-led degradation complexes at various binding sites, improved triggering of degradation as well as a greater in vitro potency (Zhang et al. 2023b). This, in turn, may be utilized for the development of potent PROTAC platforms, for precision medicine in the management of CVDs, as well as subsequent clinical translation.

Liu et al. explored the development of nanoparticle-based PROTACs with glutathione scavenging activity, for the amplification of anticancer effects through oncogene suppression and degradation of proteins associated with ferroptosis. This formulation showcased promising safety and biocompatibility (Liu et al. 2023b). Given the implications of glutathione in the induction and acceleration of ferroptosis in cardiomyocytes and the exertion of a protective effect against reactive oxygen species (ROS) (Ichihara et al. 2023), advancements moving forward could extend to the management of cardiac ischemia.

An intriguing application has been microenvironmental modulation through the usage of smart, nanotechnology-driven PROTAC systems. On combing the polymeric skeleton structure with a cyclooxygenase 1/2 (COX-1/2)-focused PROTAC, proteolysis was mediated through the ubiquitin-proteasome system. This, in turn, downregulated prostaglandin E2 and improved immuno-sensitivity (Zhang et al. 2022b). Another recent study by Lin et al. explored nanofibril-facilitated PROTAC delivery, for stronger binding with proteins in proximity, and the formation of a greater number of plenary complexes to trigger disintegration (Lin et al. 2024). More recently, semiconductor-based PROTAC systems have been evaluated for the elicitation of immunogenic responses, as well as the degradation of proteins responsible for suppressed immune functioning, such as indoleamine 2,3-dioxygenase (IDO) (Zhang et al. 2021). This enzyme has been linked with undesirable cardiac remodeling after myocardial infarction (MI) (Melhem et al. 2021) and may be evaluated in that domain.

Owing to challenges associated with solubility, PROTACs have also been delivered through microneedle patches, to promote tissue-specific activity. The patches formulated were observed to demonstrate enhanced tumor permeation and retention of chemotherapeutic and immunotherapeutic agents, as well as a more robust degradation of estrogen receptor alpha (ERα) (Cheng et al. 2023). Finally, biomimetic nanoparticles, with a desirable size and ability to camouflage with cell membranes, have been used for the controlled delivery of PROTACs and to enhance their biocompatibility. Such a smart system mediates the downregulation and silencing of RAS signaling (Fan et al. 2022), which has also been seen to play a role in the pathogenesis of cardiac hypertrophy as well as remodeling, indicating potential for translational benefits (Ramos-Kuri et al. 2021). As previously discussed, a majority of these benefits have been realized in anticancer applications, however, given several common underlying pathogenic mechanisms, there is a promising scope for their applications in the management of CVDs as well. This, however, would require focused research efforts starting with in silico testing, and further scaling up of technologies to in vivo and clinical-scale assessment.

16.6 Future Scope and Challenges of PROTACs

Over the past 20 years, the PROTAC technology, which is superior to conventional small-molecule compounds, has grown to be an important biological and therapeutic tool in both industry and academia. The field of targeted treatments has been greatly expanded by the extraordinary development of PROTAC technology. The idea of creating more potent PROTACs as possible treatments is becoming more feasible as additional POI-PROTAC-E3 ternary complex structures are being found. The ability of PROTAC technology to target "undruggable" proteins is one of its most significant benefits. Qi et al. report that the majority of effective PROTACs now use small-molecule inhibitors (SMIs) as ligands to target druggable proteins because small-molecule ligands are capable of binding to the target proteins with high affinity (Qi et al. 2021). The application of PROTACs in oncology, degradation of proteins involved in neurodegenerative disorders, and cardiovascular diseases supported by some recent preclinical and clinical evidence promote their further advancements in developing safer more effective and targeted therapeutic strategies. Nanotechnology interventions are marking a relevant step to overcome the limitations associated with the dosing and delivery of PROTACs. Even though PROTAC technology was developed in 2001 and has come a long way since then, there are still certain challenges with its design and implementation. For instance, the length and chemical makeup of the linkers that link the ligands determines how efficient PROTAC is in addition to the ligands of POIs and E3 ligases. Furthermore, ligand binding strength, cell permeability, spatial orientation, and other variables will significantly affect how effective PROTACs are. Consequently, one of the main scientific issues that needs to be resolved is how these elements interact to produce the greatest efficiency. Off-target effect is yet another challenge wherein PROTAC cannot locate the target protein. Tissue-specific E3 protein can help achieve high selectivity since different tissues and times express E3 ligase differently. The ligands of the POIs and the E3 ligases must eventually be located into an acceptable spatial configuration to form ternary degradation complexes, which requires multiple rounds of tests after the theoretical design is finished. Rational design of PROTACs requires a significant investment of time and people, which is why new design techniques and technologies, like AI and CADD are crucial. Targeted protein degradation is an exciting approach that we think could provide patients with novel therapeutic choices for a variety of diseases. It is expected that the discoveries in basic biology and chemistry will further enhance this modality in the next decades.

16.7 Conclusion

Over the last few decades, there has been tremendous progress in the domain of CVDs and their management. With a growing focus on targeting proteins to check disease progression, several targets including but not limited to IRAK4, TRIM 24, ion channel proteins, BRD4, UCP 1, BTK, Akt have been explored for their therapeutic benefits in controlling development and progression of CVDs. As briefly touched upon, conventional treatment modalities come with their own drawbacks, the most prominent of which is drug resistance. As PROTACs emerge as an attractive strategy, they don't come without their own pitfalls, some of which include poor bioavailability and a low rate of bench to bedside translation. These issues are being tackled through continuous research effort and innovation. Some forward-looking attempts include the design and synthesis of antibody-PROTAC, oligonucleotide PROTAC and transcription factor PROTAC. Owing to increasing emphasis on site-specific delivery, various approaches such as targeted delivery of TRIM 24 by PLGA nanoparticles coated with M2 macrophage membrane have been assessed. Pharmaceutical companies are steadily investing in PROTAC technologies to develop sustainable products that will solve problems with existing therapeutics. These agents may also be of value as diagnostic agents especially when they are designed to track protein degradation such as theranostic PROTACs. While the data available on theranostic /diagnostic PROTACs in CVDs is very limited, the authors though have attempted to shed light on the latest PROTAC diagnostics. In conclusion, while there has been promising progress in recent times, there is a need to further validate the safety and efficacy profiles of these agents, especially through clinical testing. This would, in turn, permit therapeutic evaluation in diverse populations, highlighting newer applications and previously missed challenges. Additionally, there is a pressing need to institute and refine regulatory guidelines for such modalities, to permit incorporate into treatment regimens and improve market access. This, in turn, would facilitate health for all, and reduce the growing burden on healthcare systems worldwide.

References

An S, Fu L (2018) Small-molecule PROTACs: an emerging and promising approach for the development of targeted therapy drugs. EBioMedicine 36:553–562. https://doi.org/10.1016/j.ebiom.2018.09.005

An K, Deng X, Chi H, Zhang Y, Li Y, Cheng M, Ni Z, Yang Z, Wang C, Chen J, Bai J, Ran C, Wei Y, Li J, Zhang P, Xu F, Tan W (2023) Stimuli-responsive PROTACs for controlled protein degradation. Angew Chem Int Ed 62(39):e202306824. https://doi.org/10.1002/anie.202306824

Au YZ, Wang T, Sigua LH, Qi J (2020) Peptide-based PROTAC: the predator of pathological proteins. Cell Chem Biol 27(6):637–639. https://doi.org/10.1016/j.chembiol.2020.06.002

Békés M, Langley DR, Crews CM (2022) PROTAC targeted protein degraders: the past is prologue. Nat Rev Drug Discov 21(3):181–200. https://doi.org/10.1038/s41573-021-00371-6

Beltrame MH, Catarino SJ, Goeldner I, Boldt ABW, De Messias-Reason IJ (2015) The lectin pathway of complement and rheumatic heart disease. Front Pediatr 2:148. https://doi.org/10.3389/fped.2014.00148

Chen Z, Dudek J, Maack C, Hofmann U (2021) Pharmacological inhibition of GLUT1 as a new immunotherapeutic approach after myocardial infarction. Biochem Pharmacol 190:114597. https://doi.org/10.1016/j.bcp.2021.114597

Chen Y, Tandon I, Heelan W, Wang Y, Tang W, Hu Q (2022) Proteolysis-targeting chimera (PROTAC) delivery system: advancing protein degraders towards clinical translation. Chem Soc Rev 51(13):5330–5350. https://doi.org/10.1039/D1CS00762A

Chen S, Zheng Y, Liang B, Yin Y, Yao J, Wang Q, Liu Y, Neamati N (2023) The application of PROTAC in HDAC. Eur J Med Chem 260:115746. https://doi.org/10.1016/j.ejmech.2023.115746

Cheng J, He S, Xu J, Huang M, Dong G, Sheng C (2022) Making protein degradation visible: discovery of theranostic PROTACs for detecting and degrading NAMPT. J Med Chem 65(23):15725–15737. https://doi.org/10.1021/acs.jmedchem.2c01243

Cheng X, Hu S, Cheng K (2023) Microneedle patch delivery of PROTACs for anti-cancer therapy. ACS Nano 17(12):11855–11868. https://doi.org/10.1021/acsnano.3c03166

Chirnomas D, Hornberger KR, Crews CM (2023) Protein degraders enter the clinic—a new approach to cancer therapy. Nat Rev Clin Oncol 20(4):265–278. https://doi.org/10.1038/s41571-023-00736-3

Cooke JP, Wilson AM (2010) Biomarkers of peripheral arterial disease. J Am Coll Cardiol 55(19):2017–2023. https://doi.org/10.1016/j.jacc.2009.08.090

Diaz S, Wang K, Sjögren B, Liu X (2022) Roles of cullin-RING ubiquitin ligases in cardiovascular diseases. Biomolecules 12(3):416. https://doi.org/10.3390/biom12030416

Ding N, Yang C, Ballew SH, Kalbaugh CA, McEvoy JW, Salameh M, Aguilar D, Hoogeveen RC, Nambi V, Selvin E, Folsom AR, Heiss G, Coresh J, Ballantyne CM, Matsushita K (2020) Fibrosis and inflammatory markers and long-term risk of peripheral artery disease: the ARIC study. Arterioscler Thromb Vasc Biol 40(9):2322–2331. https://doi.org/10.1161/ATVBAHA.120.314824

Dolmaci OB, Legué J, Lindeman JHN, Driessen AHG, Klautz RJM, Van Brakel TJ, Siebelink HJ, Mertens BJA, Poelmann RE, Gittenberger-de Groot AC, Grewal N (2021) Extent of coronary artery disease in patients with stenotic bicuspid versus tricuspid aortic valves. J Am Heart Assoc 10(12):e020080. https://doi.org/10.1161/JAHA.120.020080

Duan Y, Guan Y, Qin W, Zhai X, Yu B, Liu H (2018) Targeting Brd4 for cancer therapy: inhibitors and degraders. Med Chem Commun 9(11):1779–1802. https://doi.org/10.1039/C8MD00198G

Fan X, Jin WY, Lu J, Wang J, Wang YT (2014) Rapid and reversible knockdown of endogenous proteins by peptide-directed lysosomal degradation. Nat Neurosci 17(3):471–480. https://doi.org/10.1038/nn.3637

Fan R, He S, Wang Y, Qiao J, Liu H, Galstyan L, Ghazaryan A, Cai H, Feng S, Ni P, Dong G, Li H (2022) Targeted delivery of a PROTAC induced PDEδ degrader by a biomimetic drug delivery system for enhanced cytotoxicity against pancreatic cancer cells. Am J Cancer Res 12(3):1027–1041

Fang Y, Wang S, Han S, Zhao Y, Yu C, Liu H, Li N (2023) Targeted protein degrader development for cancer: advances, challenges, and opportunities. Trends Pharmacol Sci 44(5):303–317. https://doi.org/10.1016/j.tips.2023.03.003

Feng X, Zhang Y, Du M, Li S, Ding J, Wang J, Wang Y, Liu P (2022) Identification of diagnostic biomarkers and therapeutic targets in peripheral immune landscape from coronary artery disease. J Transl Med 20(1):399. https://doi.org/10.1186/s12967-022-03614-1

Gao X-F, Wang Z-M, Wang F, Gu Y, Zhang J-J, Chen S-L (2019) Exosomes in coronary artery disease. Int J Biol Sci 15(11):2461–2470. https://doi.org/10.7150/ijbs.36427

Garven E, Rodell CB, Shema K, Govender K, Cassel SE, Ferrick B, Kupsho G, Kung E, Spiller KL, Stevens R, Throckmorton AL (2022) Tunable blood shunt for neonates with complex congenital heart defects. Front Bioeng Biotechnol 9:734310. https://doi.org/10.3389/fbioe.2021.734310

Goldsborough E, Osuji N, Blaha MJ (2022) Assessment of cardiovascular disease risk. Endocrinol Metab Clin N Am 51(3):483–509. https://doi.org/10.1016/j.ecl.2022.02.005

Han X, Sun Y (2023) PROTACs: a novel strategy for cancer drug discovery and development. Media Commun 4(3):e290. https://doi.org/10.1002/mco2.290

Houska NM, Schwartz LI (2020) The year in review: anesthesia for congenital heart disease 2019. Semin Cardiothorac Vasc Anesth 24(2):175–186. https://doi.org/10.1177/1089253220920476

Hua B, Liu Q, Gao S, Li W, Li H (2022) Protective role of activating PPARγ in advanced glycation end products-induced impairment of coronary artery vasodilation via inhibiting p38 phosphorylation and reactive oxygen species production. Biomed Pharmacother 147:112641. https://doi.org/10.1016/j.biopha.2022.112641

Ibrahim NE, Januzzi JL (2018) Established and emerging roles of biomarkers in heart failure. Circ Res 123(5):614–629. https://doi.org/10.1161/CIRCRESAHA.118.312706

Ichihara G, Katsumata Y, Sugiura Y, Matsuoka Y, Maeda R, Endo J, Anzai A, Shirakawa K, Moriyama H, Kitakata H, Hiraide T, Goto S, Ko S, Iwasawa Y, Sugai K, Daigo K, Goto S, Sato K, Yamada K, Sano M (2023) MRP1-dependent extracellular release of glutathione induces cardiomyocyte ferroptosis after ischemia-reperfusion. Circ Res 133(10):861–876. https://doi.org/10.1161/CIRCRESAHA.123.323517

Jin J, Wu Y, Chen J, Shen Y, Zhang L, Zhang H, Chen L, Yuan H, Chen H, Zhang W, Luan X (2020) The peptide PROTAC modality: a novel strategy for targeted protein ubiquitination. Theranostics 10(22):10141–10153. https://doi.org/10.7150/thno.46985

Jin J, Wu Y, Zhao Z, Wu Y, Zhou Y, Liu S, Sun Q, Yang G, Lin J, Nagle DG, Qin J, Zhang Z, Chen H, Zhang W, Sun S, Luan X (2022) Small-molecule PROTAC mediates targeted protein degradation to treat STAT3-dependent epithelial cancer. JCI Insight 7(22):e160606. https://doi.org/10.1172/jci.insight.160606

Jyotsna F, Ikram J, Nageeta F, Komal F, Anjlee F, Patel H, Nassri T, Kumari M, Kumar R, Shah SU, Kashif M, Varrassi G, Kumar S, Patel T (2023) Unlocking the potential of immunotherapy in cardiovascular disease: a comprehensive review of applications and future directions. Cureus. https://doi.org/10.7759/cureus.42790

Karatas Z, Baysal T, Alp H, Toker A (2013) Serum tenascin-C: a novel biomarker for diagnosis and predicting prognosis of rheumatic carditis? J Trop Pediatr 59(6):476–482. https://doi.org/10.1093/tropej/fmt058

Kargbo RB (2019) PROTAC degradation of IRAK4 for the treatment of neurodegenerative and cardiovascular diseases. ACS Med Chem Lett 10(9):1251–1252. https://doi.org/10.1021/acsmedchemlett.9b00385

Kashihara T, Mukai R, Oka S, Zhai P, Nakada Y, Yang Z, Mizushima W, Nakahara T, Warren JS, Abdellatif M, Sadoshima J (2022) YAP mediates compensatory cardiac hypertrophy through aerobic glycolysis in response to pressure overload. J Clin Invest 132(6):e150595. https://doi.org/10.1172/JCI150595

Khare R, Dwivedi S (2016) NT-ProBNP as a potential marker of left atrial dysfunction in rheumatic mitral stenosis: correlation with left atrial function after PBMV. J Heart Valve Dis 25(5):613–618

Kohoutek J, Blazek D (2012) Cyclin K goes with Cdk12 and Cdk13. Cell Div 7(1):12. https://doi.org/10.1186/1747-1028-7-12

Leal MTBC, Passos LSA, Guarçoni FV, Aguiar JMDS, Silva RBRD, Paula TMND, Santos RFD, Nassif MCL, Gomes NFA, Tan TC, Nunes MCP (2019) Rheumatic heart disease in the modern era: recent developments and current challenges. Rev Soc Bras Med Trop 52:e20180041. https://doi.org/10.1590/0037-8682-0041-2019

Lee IK, Sharma N, Noguera-Ortega E, Liousia M, Baroja ML, Etersque JM, Pham J, Sarkar S, Carreno BM, Linette GP, Puré E, Albelda SM, Sellmyer MA (2023) A genetically encoded protein tag for control and quantitative imaging of CAR T cell therapy. Mol Ther 31(12):3564–3578. https://doi.org/10.1016/j.ymthe.2023.10.020

Li D, Yu D, Li Y, Yang R (2022) A bibliometric analysis of PROTAC from 2001 to 2021. Eur J Med Chem 244:114838. https://doi.org/10.1016/j.ejmech.2022.114838

Libby P, Pasterkamp G, Crea F, Jang I-K (2019) Reassessing the mechanisms of acute coronary syndromes: the "vulnerable plaque" and superficial erosion. Circ Res 124(1):150–160. https://doi.org/10.1161/CIRCRESAHA.118.311098

Lin S, Du L (2020) The therapeutic potential of BRD4 in cardiovascular disease. Hypertens Res 43(10):1006–1014. https://doi.org/10.1038/s41440-020-0459-4

Lin Z, Garcia BA, Lv D (2024) Bifunctional peptide nanofibrils for targeted protein degradation. Angew Chem Int Ed 63(3):e202316581. https://doi.org/10.1002/anie.202316581

Liu H, Chen W, Wu G, Zhou J, Liu C, Tang Z, Huang X, Gao J, Xiao Y, Kong N, Joshi N, Cao Y, Abdi R, Tao W (2023a) Glutathione-scavenging nanoparticle-mediated PROTACs delivery for targeted protein degradation and amplified antitumor effects. Adv Sci 10(16):2207439. https://doi.org/10.1002/advs.202207439

Liu J, Yin F, Wang Z, Song T, Zhang Z (2023b) An updated patent review of Mcl-1 inhibitors (2020–2022). Expert Opin Ther Pat 33(5):371–383. https://doi.org/10.1080/13543776.2023.2219394

Lu W-N, Lü S-H, Wang H-B, Li D-X, Duan C-M, Liu Z-Q, Hao T, He W-J, Xu B, Fu Q, Song YC, Xie X-H, Wang C-Y (2009) Functional improvement of infarcted heart by co-injection of embryonic stem cells with temperature-responsive chitosan hydrogel. Tissue Eng A 15(6):1437–1447. https://doi.org/10.1089/ten.tea.2008.0143

Malakar AK, Choudhury D, Halder B, Paul P, Uddin A, Chakraborty S (2019) A review on coronary artery disease, its risk factors, and therapeutics. J Cell Physiol 234(10):16812–16823. https://doi.org/10.1002/jcp.28350

Maneiro M, De Vita E, Conole D, Kounde CS, Zhang Q, Tate EW (2021) PROTACs, molecular glues and bifunctionals from bench to bedside: Unlocking the clinical potential of catalytic drugs. In: Progress in medicinal chemistry, vol 60. Elsevier, Amsterdam, pp 67–190. https://doi.org/10.1016/bs.pmch.2021.01.002

Melhem NJ, Chajadine M, Gomez I, Howangyin K-Y, Bouvet M, Knosp C, Sun Y, Rouanet M, Laurans L, Cazorla O, Lemitre M, Vilar J, Mallat Z, Tedgui A, Ait-Oufella H, Hulot J-S, Callebert J, Launay J-M, Fauconnier J, Taleb S (2021) Endothelial cell indoleamine 2,3-dioxygenase 1 alters cardiac function after myocardial infarction through kynurenine. Circulation 143(6):566–580. https://doi.org/10.1161/CIRCULATIONAHA.120.050301

Meng X, Cui J, He G (2021) Bcl-2 Is involved in cardiac hypertrophy through PI3K-Akt pathway. Biomed Res Int 2021:1–8. https://doi.org/10.1155/2021/6615502

Montrose K, Krissansen GW (2014) Design of a PROTAC that antagonizes and destroys the cancer-forming X-protein of the hepatitis B virus. Biochem Biophys Res Commun 453(4):735–740. https://doi.org/10.1016/j.bbrc.2014.10.006

Morley RL, Sharma A, Horsch AD, Hinchliffe RJ (2018) Peripheral artery disease. BMJ 360:j5842. https://doi.org/10.1136/bmj.j5842

Özlü F, Demiray G, Güneş D, Yıldızdaş H, Yıldız Ş (2023) Assessment of antioxidant system status before and after operation in neonates with congenital heart disease. Niger J Clin Pract 26(10):1557–1562. https://doi.org/10.4103/njcp.njcp_252_23

Pal P, Zhang P, Poddar SK, Zheng G (2022) Patent landscape of inhibitors and PROTACs of the anti-apoptotic BCL-2 family proteins. Expert Opin Ther Pat 32(9):1003–1026. https://doi.org/10.1080/13543776.2022.2116311

Park N, Marquez J, Pham TK, Ko TH, Youm JB, Kim M, Choi SH, Moon J, Flores J, Ko KS, Rhee BD, Shimizu I, Minamino T, Ha JD, Hwang JY, Yang SJ, Park C-S, Kim HK, Han J (2022) Cereblon contributes to cardiac dysfunction by degrading Cav1.2α. Eur Heart J 43(20):1973–1989. https://doi.org/10.1093/eurheartj/ehac072

Pettersson M, Crews CM (2019) Proteolysis targeting chimeras (PROTACs)—past, present and future. Drug Discov Today Technol 31:15–27. https://doi.org/10.1016/j.ddtec.2019.01.002

Prentice RL, Paczesny S, Aragaki A, Amon LM, Chen L, Pitteri SJ, McIntosh M, Wang P, Buson Busald T, Hsia J, Jackson RD, Rossouw JE, Manson JE, Johnson K, Eaton C, Hanash SM (2010) Novel proteins associated with risk for coronary heart disease or stroke among postmenopausal women identified by in-depth plasma proteome profiling. Genome Med 2(7):48. https://doi.org/10.1186/gm169

Qi S-M, Dong J, Xu Z-Y, Cheng X-D, Zhang W-D, Qin J-J (2021) PROTAC: an effective targeted protein degradation strategy for cancer therapy. Front Pharmacol 12:692574. https://doi.org/10.3389/fphar.2021.692574

Ramos-Kuri M, Meka SH, Salamanca-Buentello F, Hajjar RJ, Lipskaia L, Chemaly ER (2021) Molecules linked to Ras signaling as therapeutic targets in cardiac pathologies. Biol Res 54(1):23. https://doi.org/10.1186/s40659-021-00342-6

Raval M, Siddiq S, Sharma K, Sanghvi L, Jain A, Patel S, Trivedi J, Uttam Chandani K, Patel D, Desai R (2023) A review of recent advances in the diagnosis of cardiac amyloidosis, treatment of its cardiac complications, and disease-modifying therapies. F1000Research 12:192. https://doi.org/10.12688/f1000research.130285.1

Rohit M, Rajan P (2020) Approach to cyanotic congenital heart disease in children. Indian J Pediatr 87(5):372–380. https://doi.org/10.1007/s12098-020-03274-3

Rohit M, Shrivastava S (2018) Acyanotic and cyanotic congenital heart diseases. Indian J Pediatr 85(6):454–460. https://doi.org/10.1007/s12098-017-2454-6

Sakamoto KM (2010) Protacs for treatment of cancer. Pediatr Res 67(5):505–508. https://doi.org/10.1203/PDR.0b013e3181d35017

Sakamoto KM, Kim KB, Kumagai A, Mercurio F, Crews CM, Deshaies RJ (2001) Protacs: chimeric molecules that target proteins to the Skp1–Cullin–F box complex for ubiquitination and degradation. Proc Natl Acad Sci 98(15):8554–8559. https://doi.org/10.1073/pnas.141230798

Schneider M, Radoux CJ, Hercules A, Ochoa D, Dunham I, Zalmas L-P, Hessler G, Ruf S, Shanmugasundaram V, Hann MM, Thomas PJ, Queisser MA, Benowitz AB, Brown K, Leach AR (2021) The PROTACtable genome. Nat Rev Drug Discov 20(10):789–797. https://doi.org/10.1038/s41573-021-00245-x

Scott M, Neal AE (2021) Congenital heart disease. Prim Care 48(3):351–366. https://doi.org/10.1016/j.pop.2021.04.005

Shaik R, Xu J, Wang Y, Hong Y, Zhang G (2023) Fibrin-enriched cardiac extracellular matrix hydrogel promotes in vitro angiogenesis. ACS Biomater Sci Eng 9(2):877–888. https://doi.org/10.1021/acsbiomaterials.2c01148

Shamaki GR, Markson F, Soji-Ayoade D, Agwuegbo CC, Bamgbose MO, Tamunoinemi B-M (2022) Peripheral artery disease: a comprehensive updated review. Curr Probl Cardiol 47(11):101082. https://doi.org/10.1016/j.cpcardiol.2021.101082

Shi Y, Liao Y, Liu Q, Ni Z, Zhang Z, Shi M, Li P, Li H, Rao Y (2023) BRD4-targeting PROTAC as a unique tool to study biomolecular condensates. Cell Discov 9(1):47. https://doi.org/10.1038/s41421-023-00544-0

Si L, Shen Q, Li J, Chen L, Shen J, Xiao X, Bai H, Feng T, Ye AY, Li L, Zhang C, Li Z, Wang P, Oh CY, Nurani A, Niu S, Zhang C, Wei X, Yuan W, Plebani R (2022) Generation of a live attenuated influenza A vaccine by proteolysis targeting. Nat Biotechnol 40(9):1370–1377. https://doi.org/10.1038/s41587-022-01381-4

Simonetta KR, Taygerly J, Boyle K, Basham SE, Padovani C, Lou Y, Cummins TJ, Yung SL, Von Soly SK, Kayser F, Kuriyan J, Rape M, Cardozo M, Gallop MA, Bence NF, Barsanti PA, Saha A (2019) Prospective discovery of small molecule enhancers of an E3 ligase-substrate interaction. Nat Commun 10(1):1402. https://doi.org/10.1038/s41467-019-09358-9

Steg PG, Ducrocq G (2016) Future of the prevention and treatment of coronary artery disease. Circ J 80(5):1067–1072. https://doi.org/10.1253/circj.CJ-16-0266

Tamatam R, Shin D (2023) Emerging strategies in proteolysis-targeting chimeras (PROTACs): highlights from 2022. Int J Mol Sci 24(6):5190. https://doi.org/10.3390/ijms24065190

Trapotsi M-A, Mouchet E, Williams G, Monteverde T, Juhani K, Turkki R, Miljković F, Martinsson A, Mervin L, Pryde KR, Müllers E, Barrett I, Engkvist O, Bender A, Moreau K (2022) Cell morphological profiling enables high-throughput screening for proteolysis targeting chimera (PROTAC) phenotypic signature. ACS Chem Biol 17(7):1733–1744. https://doi.org/10.1021/acschembio.2c00076

Wang W, Zhou Q, Jiang T, Li S, Ye J, Zheng J, Wang X, Liu Y, Deng M, Ke D, Wang Q, Wang Y, Wang J-Z (2021) A novel small-molecule PROTAC selectively promotes tau clearance to improve cognitive functions in Alzheimer-like models. Theranostics 11(11):5279–5295. https://doi.org/10.7150/thno.55680

Wang K, Dai X, Yu A, Feng C, Liu K, Huang L (2022a) Peptide-based PROTAC degrader of FOXM1 suppresses cancer and decreases GLUT1 and PD-L1 expression. J Exp Clin Cancer Res 41(1):289. https://doi.org/10.1186/s13046-022-02483-2

Wang C, Zheng C, Wang H, Zhang L, Liu Z, Xu P (2022b) The state of the art of PROTAC technologies for drug discovery. Eur J Med Chem 235:114290. https://doi.org/10.1016/j.ejmech.2022.114290

Wang H, Chen M, Zhang X, Xie S, Qin J, Li J (2024a) Peptide-based PROTACs: current challenges and future perspectives. Curr Med Chem 31(2):208–222. https://doi.org/10.2174/0929867330666230130121822

Wang K, Diaz S, Li L, Lohman JR, Liu X (2024b) CAND1 inhibits cullin-2-RING ubiquitin ligases for enhanced substrate specificity. Nat Struct Mol Biol. https://doi.org/10.1038/s41594-023-01167-5

Wu Y, Chang X, Yang G, Chen L, Wu Q, Gao J, Tian R, Mu W, Gooding JJ, Chen X, Sun S (2023) A physiologically responsive nanocomposite hydrogel for treatment of head and neck squamous cell carcinoma via proteolysis-targeting chimeras enhanced immunotherapy. Adv Mater 35(12):2210787. https://doi.org/10.1002/adma.202210787

Yang J, Chang Y, Tien JC-Y, Wang Z, Zhou Y, Zhang P, Huang W, Vo J, Apel IJ, Wang C, Zeng VZ, Cheng Y, Li S, Wang GX, Chinnaiyan AM, Ding K (2022) Discovery of a highly potent and selective dual PROTAC degrader of CDK12 and CDK13. J Med Chem 65(16):11066–11083. https://doi.org/10.1021/acs.jmedchem.2c00384

Zeng S, Huang W, Zheng X, Cheng L, Zhang Z, Wang J, Shen Z (2021) Proteolysis targeting chimera (PROTAC) in drug discovery paradigm: recent progress and future challenges. Eur J Med Chem 210:112981. https://doi.org/10.1016/j.ejmech.2020.112981

Zhang D, Baek S-H, Ho A, Kim K (2004) Degradation of target protein in living cells by small-molecule proteolysis inducer. Bioorg Med Chem Lett 14(3):645–648. https://doi.org/10.1016/j.bmcl.2003.11.042

Zhang C, Zeng Z, Cui D, He S, Jiang Y, Li J, Huang J, Pu K (2021) Semiconducting polymer nano-PROTACs for activatable photo-immunometabolic cancer therapy. Nat Commun 12(1):2934. https://doi.org/10.1038/s41467-021-23194-w

Zhang C, He S, Zeng Z, Cheng P, Pu K (2022a) Smart nano-PROTACs reprogram tumor microenvironment for activatable photo-metabolic cancer immunotherapy. Angew Chem Int Ed 61(8):e202114957. https://doi.org/10.1002/anie.202114957

Zhang H, Peng R, Chen S, Shen A, Zhao L, Tang W, Wang X, Li Z, Zha Z, Yi M, Zhang L (2022b) Versatile nano-PROTAC-induced epigenetic reader degradation for efficient lung cancer therapy. Adv Sci 9(29):2202039. https://doi.org/10.1002/advs.202202039

Zhang N, Hou D, Hu X, Liang J, Wang M, Song Z, Yi L, Wang Z, An H, Xu W, Wang H (2023a) Nano proteolysis targeting chimeras (PROTACs) with anti-hook effect for tumor therapy. Angew Chem Int Ed 62(37):e202308049. https://doi.org/10.1002/anie.202308049

Zhang C, Xu M, He S, Huang J, Xu C, Pu K (2023b) Checkpoint nano-PROTACs for activatable cancer photo-immunotherapy. Adv Mater 35(6):2208553. https://doi.org/10.1002/adma.202208553

Zhou P, Bogacki R, McReynolds L, Howley PM (2000) Harnessing the ubiquitination machinery to target the degradation of specific cellular proteins. Mol Cell 6(3):751–756. https://doi.org/10.1016/S1097-2765(00)00074-5

Zou Y, Ma D, Wang Y (2019) The PROTAC technology in drug development. Cell Biochem Funct 37(1):21–30. https://doi.org/10.1002/cbf.3369

Delivery Systems: Miniaturised PROTAC, Nano PROTAC, and Aptamer-Based RNA PROTAC

17

Manish P. Patel, Hard K. Kalyani, Kashyap M. Patel, Bharat R. Patel, Dipti H. Patel, and Jayvadan K. Patel

Abstract

Proteolysis Targeting Chimeras (PROTACs) have emerged as a promising approach in drug discovery, enabling the targeted degradation of specific proteins implicated in various diseases. The development history of PROTACs dates to the early 2000s, marked by significant milestones leading to their current state of advancement. Proteolysis targeting chimeras (PROTACs) are hetero-bifunctional small molecules with three chemical elements: a ligand binding to a target protein, a ligand binding to E3 ubiquitin ligase, and a linker for conjugating these two ligands. Despite their therapeutic potential, PROTACs face challenges such as selectivity, stability, and delivery. The future of PROTACs lies in overcoming these hurdles and enhancing their efficacy and safety profiles. Nano PROTAC delivery systems have garnered attention for their unique characteristics and functionality, facilitating targeted protein degradation with improved pharmacokinetics and biodistribution. Aptamer-based RNA PROTAC delivery systems offer a promising avenue for targeted drug delivery, leveraging the specificity of aptamers for precise targeting and integration with PROTAC technology. In comparative analysis, Various drug delivery systems like liposomes,

M. P. Patel (✉) · H. K. Kalyani · K. M. Patel
Department of Pharmaceutics, L. M. College of Pharmacy, Ahmedabad, Gujarat, India
e-mail: manish.patel@lmcp.ac.in

B. R. Patel
Department of Pharmacology, L. M. College of Pharmacy, Ahmedabad, Gujarat, India
e-mail: bharat.patel@lmcp.ac.in

D. H. Patel
Faculty of Pharmacy, Department of Pharmaceutics, Institute of Pharmaceutical Sciences, Parul University, Vadodara, India

J. K. Patel
Aavis Pharmaceuticals, Hoschton, GA, USA

M. Nandave, P. Jain (eds.), *PROTAC-Mediated Protein Degradation: A Paradigm Shift in Cancer Therapeutics*, https://doi.org/10.1007/978-981-97-5077-1_17

nanotechnology and co-crystallization play crucial roles in enhancing the efficacy and selectivity of PROTACs, presenting distinct advantages and challenges. Looking ahead, the future of targeted protein degradation through PROTACs holds immense promise, with potential advancements and applications across diverse disease areas like cancer, targeted protein degradation and overcoming common resistance mechanisms.

Keywords

PROTAC · E3 ubiquitin ligase · Aptamer-based RNA PROTAC · Nano PROTAC · Cereblon · Von Hippel-Lindau (VHL)

Abbreviations

Ab-TAC	Antibody-based PROTAC
ALK	Anaplastic lymphoma kinase
APC	Aptamer-PROTAC conjugates
AR	Androgen receptor
BCL-XLB	Cell lymphoma extra-large
BRAF	v-Raf murine sarcoma viral oncogene homolog B1
CRABP1	Cellular retinoic acid-binding protein 1
CRBN	Cereblon
eGFR	Estimated glomerular filtration rate
EPR	Enhanced permeability and retention effect
ER	Estrogen receptor
HNSCC	Head and neck squamous cell carcinoma
IAPs	Specific inhibitors of apoptosis proteins
KRAS	Kirsten rat sarcoma viral oncogene homologue
LYTAC	Lysosome-targeting chimaera
mABs	Monoclonal antibodies
MDM2	Mouse double minute 2 homolog
NSCLC	Non-small cell lung cancer
POI	Protein of interest
PROTAC	Proteolysis targeting chimeras
RING	Really interesting new gene
RTKs	Receptor tyrosine kinases
SARS-CoV2	Severe acute respiratory syndrome coronavirus 2
SBD	Substrate binding domain
SELEX	Systemic evolution of ligands by exponential enrichment
siRNA	Small interfering RNA
SMIs	Small molecular inhibitors
TC	Ternary complex
TP	Target proteins

TPD	Targeted protein degradation
UPS	Ubiquitin-proteosome system
VHL	Von-Hippel-Lindau

17.1 Development History of PROTAC

Professors Crews and Deshaies first introduced the idea of "PROTAC" in a paper that was published in 2001. They created protac-1, the first PROTAC molecule, a chimeric chemical, to demonstrate the idea. Normal maintenance of cellular function, including proliferation, differentiation, and cell death, depends on the degradation of cellular proteins. Ubiquitin-dependent proteolysis is a key mechanism for post-translational regulation of proteins (Sakamoto et al. 2001).

17.1.1 Overview of PROTAC Technology

Many diseases are treated primarily by targeting proteins. As a result, numerous methods have been created to disrupt target proteins (TP) associated with these disorders. Small interfering RNA (siRNA), monoclonal antibodies (mABs), and small molecular inhibitors (SMIs) can all be used to block, inhibit, or silence target proteins. However, such traditional approaches can only target 20–25% of the known TP at this time (Saraswat et al. 2023).

Proteolysis targeting chimaeras (PROTACs) are hetero-bifunctional small molecules with three chemical elements: a ligand binding to a target protein, a ligand binding to E3 ubiquitin ligase, and a linker for conjugating these two ligands. Using the ubiquitin-proteasome system, the target protein is chemically knocked down using this method. Crews and associates introduced the idea of PROTACs in 2001 (An and Fu 2018). PROTAC is a novel therapeutic system that uses the ubiquitin-proteasome system which degrades the target proteins due to its endogenous decomposition system, which is different from competitive inhibition or irreversible inhibition system (Tashima 2023). The ubiquitin-proteasome system, an important protein degradation system present in all eukaryotes. PRO-TACs have opened a new chapter for the development of new drugs and novel drug delivery system assuming role of tool for industries and academics (Gao et al. 2020).

PROTACs consist of two ligands; an "anchor" to bind to an E3 ubiquitin ligase and a "warhead" to bind to a protein of interest, connected by a chemical linker. It is becoming more and more obvious that the linker's length and structure have a major effect on the physical and chemical properties and bioactivity of PROTACs. A "warhead" ligand hooks itself to the particular protein of interest (POI) which is to be targeted, whereas a "anchor" ligand hooks itself onto the substrate binding domain (SBD) of an E3 ubiquitin (Ub) ligase. The POI is transformed into a ternary complex (TC) with the E3 ligase by the PROTAC, which attaches to both proteins in cells.

Most of PROTACs now used are in research and aimed at cancer, neurologic illnesses, and inflammation diseases. The most commonly E3-ligases used for the developing the PROTACs are VHL (von-Hippel-Lindau) (Ivan and Kaelin 2001) and CRBN (Cereblon). Conversely, prolines and amides are two of the E3 binding types that are employed in research. E3 ligases have an important function in substrate selection and cascade efficiency control, which makes them a vital element for protein ubiquitination. They are classified in to three types on the basis of their mechanism of ubiquitin transfer. CRBN stands for Cereblon is a 442-amino acid protein that makes complex and interact with the adaptor protein damaged DNA-binding protein1 (Bricelj et al. 2021). The von Hippel-Lindau (VHL) protein is the substrate receptor subunit of the Cullin2, really interesting new gene (RING)-VHL ($CRL2^{VHL}$) multi-subunit E3 ligase (Diehl and Ciulli 2022). At the moment, the majority of PROTACs used as examples in the literature use either cereblon, inhibitor of apoptosis proteins (IAP), von Hippel-Lindau tumour suppressor (VHL), or Mouse double minute 2 homolog (MDM2) as their E3 ligase; however, this list continues to be developing (Pike et al. 2020).

17.2 PROTACs as a Therapeutic Approach

Chemical biology and drug development are entering an exciting new age with targeted protein degradation, or TPD. The Proteolysis Targeting Chimera (PROTAC) technology co-opts the ubiquitin-proteasome system to identify and eliminate cellular proteins (Nalawansha and Crews 2020). By targeting proteins, PROTACs technology demonstrates potential in mitigating the side effects of established medications. These medications have shown to be successful in a variety of therapeutic applications and conditions. By taking over the Ubiquitin-proteosome system (UPS), the Proteolysis targeting Chimera (PROTAC) technology was created to deliberately cause the breakdown of a specific protein of interest (POI).

17.2.1 Advantages of PROTACs

17.2.1.1 Potent and Profound Degradation of Target Proteins

Protein function is altered by small molecule inhibitors, which stoichiometrically occupy the active regions of target proteins (Bondeson et al. 2015). Target proteins were effectively destroyed by PROTACs that used an LCL161 derivative as the IAP binding ligand at low nanomolar or low micromolar concentrations (Shibata et al. 2018).

17.2.1.2 Overcome Drug Resistance Due to Mutations

One of the major obstacles today's clinical treatment is drug resistance. In order to combat medication resistance, PROTAC technology has been successful in targeting several oncogenic targets, including AR in instances involving enzalutamide-resistant prostate cancer (Kregel et al. 2020), targeting ER for drug-resistant breast cancer (Hu et al. 2019). Resistance to small molecule inhibitors can be brought on by non-synonymous mutations at active sites. Their subsequent signalling steps are also promptly and persistently blocked in addition to the target proteins (An and Fu 2018; Zhang et al. 2018). Burslem et al. generated small molecules that act as PROTACs to break down transmembrane receptor tyrosine kinases (RTKs). They discovered that downstream signalling was quickly suppressed as a result of the deterioration caused by PROTAC (Burslem et al. 2018). In addition to cancer treatments, PROTAC technology has been applied in other fields. For instance, small molecule protease degraders of the hepatitis C virus have been shown to lessen the vulnerability to resistance mutations. The first PROTAC molecule to successfully breakdown viral protein was DGY-08-097 (de Wispelaere et al. 2019). In contrast to the conventional approaches, PROTAC seeks to eliminate proteins instead of preventing POI from acting. Thus, in theory, PROTAC molecules may be utilized to overcome the acquired resistance brought on by the POI mutation.

17.2.1.3 Enhanced Target Selectivity

Based on proteome-wide approaches, several studies have discovered that small-molecule PROTACs displayed high selectivity for target proteins. Some research examining the proteome as a whole as well as a number of other research revealed that PROTAC selectivity could exceed the intrinsic binding specificity of target binding ligands. PROTAC molecules have proven the benefits of increasing target protein selectivity, which may be attributed to the synergistic actions of E3 ubiquitin ligase, POI ligands, and linkers, including the ternary complex's strict conformational requirements.

17.2.1.4 Modifying the Potency and Selectivity

Enhancing binding selectivity and specificity properties is a primary goal for all drugs during the drug development process. Recent studies have shown that PROTAC can show the path for increasing the selectivity and specificity (Zeng et al. 2021). When developing PROTACs, the length and make-up of the linkers can be changed to alter the potency and selectivity of the PROTACs towards specific target proteins. CRABP-I, related to Alzheimer's disease14 and CRABP-II is connected with neuroblastoma (Hu et al. 2019) Wilms tumour, head and neck squamous cell carcinoma (HNSCC). These constitute all target proteins that can be used in conjunction with PROTAC to treat various illnesses.

17.2.1.5 PROTAC for Targeting the "Undruggable" Proteome

Proteases that are more appropriately difficult to target with drugs are referred to be "undruggable" proteomes. Over 80% of human proteins are targets that can be "undruggable". Because of the special qualities of PROTAC technology, it is feasible to create compounds that can control difficult, unconventional therapeutic targets (Zeng et al. 2021).

17.2.1.6 PROTAC for Achieving Tissue Specificity

For the treatment of diseases, achieving tissue specificity is a recurring theme, particularly for cancer therapies that can significantly minimize on-target toxicity regardless of target. In human tissue, not every protein, enzyme, or receptor is expressed in the same way; this also applies to E3 ubiquitin ligases. Evidence suggests that various tissue expressions vary E3 ubiquitin ligases. Photo-PROTAC, a hybrid of photo-pharmacology and PROTAC technology, is a wonderful tool for precisely controlling POI deterioration in a spatiotemporal manner (Zeng et al. 2021).

17.2.2 Few Applications of PROTACs

17.2.2.1 PROTAC-Based Antiviral Strategies

Throughout human history, viruses have been responsible for multiple worldwide pandemics that resulted in a significant number of fatalities. The severe acute respiratory syndrome coronavirus 2 (SARS-CoV-2) pandemic in recent times has forced technology to bow before the forces of nature. Newer technologies are being developed every day, PROTAC being one of them. The two forms of antivirals currently in use are vaccines that help trigger an immune response against a viral disease and small molecule inhibitors that target viral proteins. The disadvantages of conventional techniques include low immunogenic response and insufficient suppression of viral proteins (Pica and Palese 2013). Currently, the majority of treatments and prevention strategies for human viral infections involve an assortment of drugs and vaccinations. Recently, Si et al. outlined a unique PROTAC-based method for creating influenza virus strains that have been engineered to be attenuated for vaccination. They significantly reduced viral replication by using PROTAC technology to specifically target and degrade certain influenza virus viral proteins by the host ubiquitin-proteasome system. PROTAC viruses could trigger strong immune responses all while exhibiting replication difficulties (Ahmad et al. 2023).

17.2.3 PROTAC in Development for Lung Cancer Treatment

In recent years, PROTACs have been rapidly established as novel treatments for lung cancer and as innovative ways to combat drug resistance. Multiple PROTAC molecules have been under development for validated therapeutic targets in NSCLC such as EGFR, KRAS, ALK, BRAF, and BCL-XL, with promising compounds showing promising antitumor efficacy in cell models and other preclinical tumour models. PROTACs have the probable likelihood to be developed as new lung cancer therapeutics. Currently, two PROTACs have gone into phase II clinical trials: ARV-110, which targets the androgen receptor (AR), and ARV-471, which also targets the estrogenic receptor (ER) (Li et al. 2023; Paiva and Crews 2019).

17.3 Mechanism of Action and Components

PROTACs generally work by taking advantage of the UPS to ubiquitinate the target protein and then eliminate it. The ubiquitin-proteasome system is made of Ubiquitin-activating enzymes (E1), ubiquitin conjugates (E2), ubiquitin ligases (E3), ubiquitin, and the 26S proteasome which are the principal component playing a crucial role in intracellular protein (Tashima 2023).

PROTACs induce the proximity of E3 ligase and the target POI to form a ternary complex. Following the formation of the ternary complex, the target protein is ubiquitinated by E3 ligase, initiating the degradation process. The endogenous 26S proteasome subsequently recognizes and breaks down the ubiquitinated target protein, ultimately leading to the proteasomal breakdown of the target proteins in cells (Saraswat et al. 2023). The process consists of two distinct steps: (1) the protein to be broken down is specifically marked with a polyubiquitin chain and (2) the marked substrate is subsequently broken down by the multi-catalytic protease complex known as the 26S proteasome. First, an enzyme called ubiquitin-activating enzyme (E1) activates the carboxyl group at the C-terminus of ubiquitin. Following ubiquitin's attachment to the E1 enzyme, a thioester is formed which is then transferred to ubiquitin-conjugating enzyme (E2) via transacylation step. Finally, ubiquitin is transferred to a lysine (or, less commonly, the amino terminus) of the protein substrate that is specifically bound by an ubiquitin ligase (E3) (Schneekloth et al. 2004).

The process of breaking down a protein via this pathway begins with the specific alteration of internal lysine residues' εNH2 groups through ubiquitination. This amino group has been the only known target for ubiquitination up until this point (Breitschopf et al. 1998).

The possible benefit of PROTAC technology can offset the drawbacks of conventional medication therapy, such as SMIs (small molecular inhibitors) (Guo et al. 2023), mAbs (monoclonal antibodies) (Hu and Crews 2022) and nucleic acid therapeutics, encouraging its rapid development for clinical translation. Ub-activating enzyme (E1), Ub-conjugating enzyme (E2), and Ub ligase (E3) are a cascade of three enzymes that covalently tag proteins with ubiquitin (Ub) before the proteasome needs to break them down (Nalawansha and Crews 2020). PROTACs use ligase activity to enlist an E3 to a POI, which ubiquitinates the POI and allows the 26S proteasome to degrade it. Because the PROTAC is not broken down during this process, it can encourage the ubiquitination and breakdown of many POI counterparts, functioning in a sub-stoichiometric manner.

17.4 Challenges of PROTAC

To date, the vast substantial of PROTAC molecules are outside the conventional Lipinski's "rule-of-five" space (Edmondson et al. 2019). PROTAC has unique challenges like the process by which PROTAC enters the cell membrane is currently unclear, and further research and development are required to clarify the substance's absorption, distribution, metabolism, excretion, and toxicity. Getting molecules with the correct physicochemical qualities and boosting cell uptake and bioavailability to sustain the critical concentration of PROTAC for pharmacological activity remain extremely difficult tasks.

An essential element in the proteasome-mediated degradation of proteins is the E3 ligase. Less than 1% of the known E3 ligases in the human body are found in the body's more than 600 members (Konstantinidou et al. 2019). At the moment, the most prevalent E3 ligase is used to create more than 90% of documented PROTACs (Schapira et al. 2019). Extending the E3 ubiquitin ligase that can be utilized in PROTAC technology is one of the main issues facing the technology. Only when a stable "target protein-PROTAC-E3 ubiquitin ligase" ternary complex is formed can PROTAC effectively and selectively ubiquitinate a target protein (Gadd et al. 2017), an intermediary species essential to the degrader molecule's biological action. However, it is challenging to capture and recognize the intricate system of ternary crystal formations. PROTAC's linker design is also very important.

Lastly, medicinal chemists face a great deal of difficulty due to the complexity of PROTAC molecules. The connection location, connector length, and conjugation vector composition all play a major role in the creation of productive PROTAC and have proven to be difficult to address methodically (Zeng et al. 2021).

17.5 Future of PROTAC

Regardless of the mechanical understanding gained, PROTAC technology's full potential is still unrealized. Given the observed differences in POI degradation depending on whether E3 ligase is recruited, it is essential that we investigate the "PROTAC-ability" of alternative E3 ligases in addition to investigating other protein types. Methods that facilitate the categorization and suppression of active E2/E3 pairings could potentially lead to the discovery of new E3 modulators (Paiva and Crews 2019).

Understanding the molecular pathways by which ubiquitin E3 ligases catalyse substrate ubiquitination has advanced significantly in the last several years. However, more research is needed to understand how E3 ligases specifically target the N termini of proteins or certain substrate lysine's for modification.

Chain linkage type in polyubiquitin-chain assembly is determined by the final thioester in the ubiquitination cascade; therefore, in polyubiquitination catalysed by RING E3s, the E2 mostly controls chain linkage (Berndsen and Wolberger 2014). Research on targeted protein degradation with PROTACs has generated a lot of interest in both academia and business. As our understanding of the mechanisms underlying this process grows, new investigations have shown even more intriguing possibilities for precisely controlling these bifunctional molecules (Hughes et al. 2021). Only a small number of the more than 600 E3s predicted in the human genome have been confirmed or used to further PROTAC development. In order to propel PROTAC technology forward, more work needs to be done to find useful E3s (Pettersson and Crews 2019). While the amount of biologics—such as insulin replacement proteins or monoclonal antibodies—that are making their way into clinics has surged recently, small molecule-based pharmaceuticals will still account for the majority of the drug market (Campbell et al. 2018). Since PROTAC was able to successfully degrade harmful proteins that were previously thought to be "undruggable," individuals have started using PROTACs in clinics to treat conditions like tumours, autoimmune disorders, neurodegenerative diseases, alopecia, acne, and asthma (Xiao et al. 2022).

17.6 Nano PROTAC Delivery Systems

17.6.1 Characteristics and Functionality

Nano PROTAC delivery systems are an innovative approach that combines nanotechnology with PROTAC (Proteolysis-Targeting Chimeras) technology to enhance the delivery and efficacy of PROTACs. Here's an overview of their characteristics and functionality:

17.6.1.1 Characteristics of Nano PROTAC Delivery Systems

Size: Nano PROTACs are typically designed to be within the nanometre scale, which facilitates their entry into cells and allows for more precise targeting of intracellular proteins.

Solubility and stability: Nanotechnology can improve the solubility and stability of PROTAC molecules, which is crucial for their bioavailability and effectiveness in biological systems.

Targeted delivery: By leveraging the unique properties of nanoparticles, nano PROTACs can be directed to specific tissues or cell types, reducing off-target effects, and improving therapeutic outcomes.

Controlled release: Nano carriers can be engineered to release PROTACs in a controlled manner, ensuring sustained degradation of target proteins over time.

17.6.1.2 Functionality of Nano PROTAC Delivery Systems

Enhanced permeability: The small size of nano PROTACs enhances their ability to permeate biological membranes, increasing their access to target proteins within cells.

Reduced immunogenicity: Nanoparticles can be designed to evade the immune system, reducing the potential for immunogenic responses, and allowing for repeated administration if necessary.

Combinatorial therapeutic effects: Nano PROTACs can be combined with other therapeutic agents to achieve synergistic effects, such as simultaneous drug delivery and protein degradation.

17.7 Aptamer-Based RNA PROTAC Delivery Systems

Aptamer-based RNA PROTAC delivery systems utilize genetically encoded RNA scaffolds combined with aptamers to target specific proteins for degradation. These systems offer precise targeting and degradation of disease-related proteins, showing promise in therapeutic applications. By conjugating aptamers with PROTACs, researchers have enhanced their in vivo potency and tumour-specific targeting capabilities, thereby improving therapeutic efficacy. Aptamers, as single-stranded DNA or RNA molecules, provide high specificity and affinity for their target proteins, making them ideal for guiding PROTACs to the desired cellular locations.

17.7.1 Explanation of Aptamers in Drug Delivery

Aptamers are short, single-stranded oligonucleotides (DNA or RNA) which can fold into special three-dimensional structures that allow them to bind to their target molecules—which can be proteins, tiny compounds, or even cells—with a high degree of specificity and affinity. Usually, an iterative screening procedure known as Systemic Evolution of Ligands by Exponential Enrichment (SELEX) is used to create them from complicated nucleic acid libraries (Gao et al. 2020).

The ability of aptamers to connect with a wide range of target molecules, from tiny molecules to whole organisms, is unique. Often dubbed as "chemical antibodies," they have generated a great deal of interest in a variety of clinical trials because of their benefits, which include high biostability, adaptability in chemical modification, minimal immunogenicity, and rapid tissue penetration (Gao et al. 2020).

Here's a detailed explanation of how aptamers are used in drug delivery:

1. Target recognition: Aptamers are selected through a process called SELEX (Systematic Evolution of Ligands by Exponential Enrichment), which identifies aptamers with the highest affinity for a specific target. Once identified, these aptamers can be synthesized and used to recognize and bind to the target molecule or cell type in the body.
2. Drug conjugation: Aptamers can be conjugated to therapeutic agents, either directly or through a linker molecule. This conjugation allows the aptamer to carry the drug directly to the target site, reducing off-target effects and improving the therapeutic index of the drug.
3. Controlled release: Aptamers can be engineered to change their structure in response to external stimuli, such as pH or temperature changes. This property can be exploited to design controlled-release systems where the aptamer releases the drug only when it reaches the target environment.
4. Reduced immunogenicity: Compared to antibodies, aptamers typically elicit a lower immune response, making them suitable for repeated administration. Their small size and lack of Fc region also reduce the risk of immune complex formation.
5. Nanocarrier targeting: Aptamers can be attached to the surface of nanoparticles, liposomes, or other nanocarriers. These aptamer-functionalized nanocarriers can

deliver a wide range of therapeutic agents, including small-molecule drugs, proteins, and nucleic acids, to the target cells or tissues.
6. Clinical applications: Aptamers have been explored for the targeted delivery of chemotherapeutics, gene therapy vectors, and siRNA molecules. They have shown promise in treating various diseases, including cancer, where targeted delivery is crucial for maximizing therapeutic efficacy while minimizing side effects.
7. Advantages over antibodies: Aptamers offer several advantages over traditional antibodies, including easier synthesis, better tissue penetration due to their smaller size, and the ability to undergo chemical modifications to enhance their stability and binding properties.
8. Challenges: Despite their potential, the clinical translation of aptamer-based drug delivery systems faces challenges such as stability in the bloodstream, efficient escape from the endosomal compartment after cellular uptake, and the cost of large-scale synthesis.

17.7.2 Integration with PROTAC Technology

The integration of aptamers with PROTAC (Proteolysis-Targeting) technology represents a novel strategy for targeted protein degradation, especially in cancer therapy. Here's how aptamers are being integrated with (Table 17.1)

Table 17.1 Example of PROTAC integration (Zhang et al. 2018)

Aspect	Description
Aptamer-PROTAC conjugates (APCs)	Researchers have developed aptamer-PROTAC conjugates where an aptamer is used as a targeting ligand. This approach enhances the selectivity of PROTAC molecules, allowing them to selectively degrade target proteins within tumour cells while sparing normal cells
Tumour-specific targeting	By using aptamers that bind to proteins overexpressed on the surface of cancer cells, APCs can deliver PROTACs directly to the tumour site. This targeted delivery system improves the in vivo antitumor potency of PROTACs and reduces toxicity, potentially leading to better therapeutic outcomes
Enhanced solubility and stability	Aptamers can improve the solubility and stability of PROTACs in aqueous solutions, which is crucial for their bioavailability and effectiveness in biological systems. The use of nucleic acid aptamers as targeting ligands has shown promise in overcoming the solubility challenges associated with traditional PROTACs
Selective degradation	The selective breakdown of proteins like nucleolin, which is overexpressed in some cancer types, is made possible by the creation of novel PROTACs employing nucleic acid aptamers. The target protein can be ubiquitinated and degraded by these aptamer-based PROTACs when they form a ternary complex with it and an E3 ligase
Broadening the PROTAC toolbox	The use of nucleic acid aptamers expands the toolbox for constructing PROTACs, providing a promising strategy for the development of tumour selective PROTACs. This approach could lead to the creation of a new class of PROTACs with improved properties for clinical applications

17.8 Comparative Analysis

Various drug delivery systems play crucial roles in enhancing the efficacy and selectivity of PROTACs, presenting distinct advantages and challenges:

1. *Liposomes*: Liposomes nano-PROTACs are the lipid-based NPs with greatest potential for a quick clinical translation. Efficient oral delivery of PROTACs via liposomes poses challenges, although it's a promising route. They will be able to enhance biodistribution and prevent undesired PROTAC metabolization, just like any other nanocarrier. It has been demonstrated that liposome drug carriers enhance the pharmacokinetic characteristics and therapeutic efficacy of small molecules (Bitounis et al. 2012). Since lipophilic materials like polymers and liposomes can carry PROTACs to specific intracellular locations that are necessary for a therapeutic response, these nanoparticle carriers have been studied as potential delivery methods (Cimas et al. 2020; Fu et al. 2020; Kou et al. 2023) (Fig. 17.1).
2. Nanoparticles: Nanoparticles offer diverse options for drug delivery, including inorganic and polymeric nanoparticles. Nanoparticles have been successfully used for improving the PK and low water solubility of drugs (Moon et al. 2023). They improve the action of PROTACs, especially in cancer treatment (Juan et al. 2021). These NPs work through the enhanced permeability and retention effect (EPR), which facilitates the delivery of the material to the target site through convection and diffusion mechanism. PROTAC molecules have demonstrated

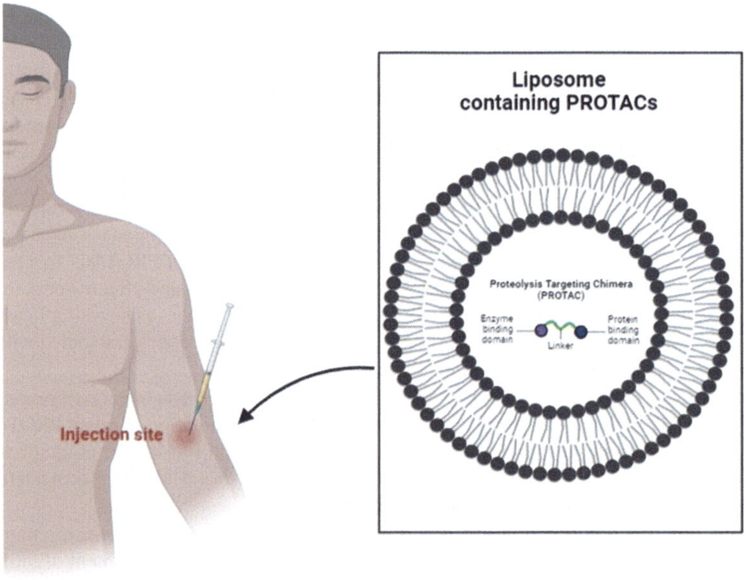

Fig. 17.1 Liposomes containing PROTAC

the ability to directly conjugate antibodies or aptamers to both effectively modulate the unfavourable PK behaviour of PROTACs and enhance their tumour-selective accumulation (Moon et al. 2023). The distinct benefit of nanotechnology is that it significantly increases the bioavailability of active pharmaceutical compounds that are encapsulated (Zhang and Pu 2020). Lately, efforts have been undertaken to design nanoplatforms in order to improve PROTACs' bioavailability for the treatment of cancer (Saraswat et al. 2020) (Fig. 17.2).

3. Co-crystallization: While not directly mentioned in the provided search results, co-crystallization is a drug delivery technique with its own advantages, potentially applicable to PROTACs. It involves the formation of crystalline structures to enhance drug stability and solubility.
4. Other delivery systems: Though not specifically outlined in the search results, various other delivery systems like nanoparticles and bioconjugates are under investigation for PROTAC delivery. Each system presents unique advantages and challenges, influencing their suitability for specific application (Chen et al. 2022).

Compared to other drug delivery systems:

Unlike nanoparticle-based systems, PROTACs are small molecules that do not face challenges related to stability and scalability.

Unlike cocrystallization and solid dispersions, which aim to improve drug solubility and dissolution, PROTACs target the actual protein of interest, potentially offering a more direct and efficient therapeutic approach.

While local drug delivery systems provide targeted delivery, they may have limited applicability based on the target site. PROTACs can potentially target proteins in various tissues and organs.

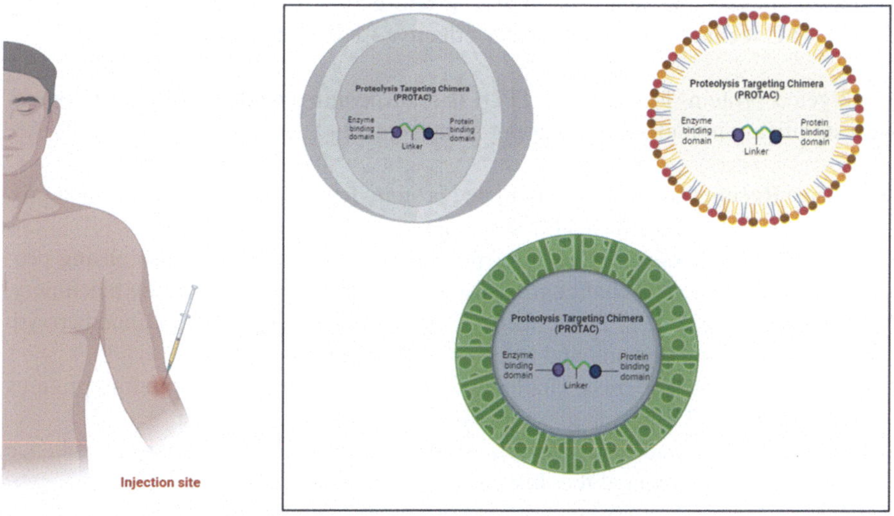

Fig. 17.2 Nanoproducts with PROTAC

However, the PROTAC system also faces challenges, such as potential off-target effects, limited bioavailability, and the need for optimization to ensure efficient target protein degradation.

17.9 Future Perspectives and Challenges

17.9.1 Outlook on Targeted Protein Degradation

The PROTAC (proteolysis targeting chimeras) technology has opened a new chapter for novel drug development and has garnered the interest of academic establishments, major pharmaceutical companies, and biotechnology firms (Gao et al. 2020).

PROTAC technology has opened new avenues for drug development, yet it faces several challenges and holds promising future perspectives:

Opportunities: PROTACs represent a novel approach for targeting previously undruggable proteins, offering potential therapeutic solutions for various diseases.

Challenges: Despite its potential, PROTAC technology encounters hurdles such as limited bioavailability, tissue penetration, and off-target effects. Overcoming these challenges requires further research and innovation (Gao et al. 2020).

Future directions: Advances in delivery systems, target engagement, and degradation kinetics hold promise for improving PROTAC efficacy and specificity.

Technological innovations: Continued research into new delivery systems and PROTAC structures may address current limitations, expanding the application of PROTACs in diverse therapeutic areas.

Collaborative efforts: Collaboration among researchers, pharmaceutical companies, and regulatory agencies is vital for addressing challenges and translating PROTAC technology into clinically effective treatments (Gao et al. 2020).

Targeted protein degradation (TPD) is a burgeoning field in drug development with significant potential and ongoing advancements:

Advances: Recent research has elucidated protein degradation mechanisms and propelled the development of TPD strategies.

Draggability Perspective: TPDs offer promise in eliminating disease-causing proteins, even previously challenging targets, through harnessing cellular machinery.

Market Expansion: The TPD therapeutics market is forecasted for continued growth due to increasing demand and research in the coming years.

Challenges: Despite progress, challenges such as degrading DNA-binding proteins and optimizing TPD strategies remain.

Future Prospects: Continued research efforts aim to overcome challenges, enhance draggability, and expand the therapeutic potential of TPD.

17.9.2 Potential Advancements and Applications

The field of targeted protein degradation (TPD) has seen significant advancements and has a wide range of potential applications. Here are some key points:

17.9.2.1 Advancements

Utilizing the destruction machinery found in cells, proteolysis targeting chimeras (PROTAC) technology has emerged as one of the most intriguing techniques for eliminating particular disease-associated proteins. Apart from PROTAC, numerous other targeted protein degradation (TPD) techniques are being developed, such as molecular glue, LYTAC (lysosome-targeting chimaera), and AbTAC (antibody-based PROTAC).

17.9.2.2 PROTACs with TPD

PROTACs (proteolysis-targeting chimeras) are innovative molecules designed for targeted protein degradation. A major process that breaks down intracellular proteins as part of regular cellular maintenance is ubiquitin-dependent proteolysis. They induce the degradation of specific proteins by utilizing the ubiquitin-proteasome system (UPS). This process involves the recruitment of E3 ligase, leading to ubiquitination and subsequent proteasomal degradation of the target protein. PROTACs consist of three key components: a ligand that binds to the target protein, a linker, and a ligand that recruits an E3 ubiquitin ligase. This innovative approach allows for the selective degradation of disease-causing proteins, offering potential therapeutic benefits across various diseases, including cancer and neurodegenerative disorders.

17.9.2.3 PROTACs with Lysosome-Targeting Chimaera

In 2020 Banik et al. created a novel approach to specifically target extracellular proteins in general and IMPs in particular for destruction. This novel approach utilizes lysosome-targeting chimeras (LYTACs), which are small molecule–antibody conjugates that bind a target protein's extracellular domain and a cell-surface lysosome-shuttling receptor at the same time. Lysosome-Targeting Chimaeras focus on degrading proteins via lysosomal pathways. Here's a brief comparison:

17.9.2.4 Location

PROTACs act in the cytosol, tagging proteins for proteasomal degradation and lysosome-targeting chimaeras function within lysosomes to induce protein degradation.

17.9.2.5 Mechanism

PROTACs recruit E3 ligases to label proteins for proteasomal degradation and Lysosome-Targeting Chimaeras direct proteins for degradation through lysosomal pathways.

17.9.2.6 Antibody-Based PROTAC

Antibody-based PROTACs (AbTACs) are an innovative form of targeted protein degradation (TPD) that employ antibodies to specifically bind to protein targets. IgGs can be quickly produced via phage display as high-affinity binders to target proteins and have lengthy serum half-lives. Bispecific IgGs have the ability to bind and co-localize two proteins at the same time. Our hypothesis was that this biological construct might function similarly to PROTACs by attracting membrane-bound E3 ligases to target proteins and causing their destruction. These antibodies-based PROTACs are known as AbTACs. They work especially well in causing extracellular and membrane proteins to break down. AbTACs offer a dual mechanism of action, engaging in both proteasomal and lysosomal degradation pathways.

These technologies have significantly broadened the application of TPD and offered new perspectives on drug development.

17.9.2.7 Applications (Zhang et al. 2018)

TPD has potential applications in various diseases, including cancer, viral infection, immune disorders, and neurodegenerative diseases.

It is particularly useful for targeting "undruggable" proteins that are not amenable to traditional small molecule inhibitors.

TPD can also be used to regulate signalling pathways associated with disease.

17.9.2.8 Future Perspectives

The field of TPD is vibrant and rapidly evolving, with ongoing research into new degradation strategies and potential therapeutic applications (Table 17.2).

Table 17.2 Advancements in TPD strategies

Advancements in TPD strategies	Description
PROTACs	Utilizes bifunctional molecules to induce target protein degradation
Molecular glues	Facilitates the degradation of target proteins by recruiting E3 ligase
Degradation tags (dTAGs)	Allows for the degradation of specific proteins of interest
Trim away	Utilizes enzymatic trimming to degrade target proteins
Specific inhibitors of apoptosis proteins (IAPs)	Targets IAPs to promote protein degradation
Small molecule drug discovery	Shifts from static structures to dynamic protein degradation for drug development
Therapeutic modality	Emerges as a promising approach with applications in various diseases
Widespread development	Catalyses the creation of innovative therapies over the past two decades
Emerging therapeutic strategy	Utilizes small-molecule ligands for targeting dysregulated proteins across therapeutic areas

17.10 Conclusion

The development and application of PROTAC technology have revolutionized the field of targeted protein degradation. Its unique mechanism of action and components have opened new therapeutic avenues, despite the challenges it faces. The future of PROTAC looks promising, with continuous advancements being made to overcome these challenges. The advantage of Nano PROTAC Delivery Systems and Aptamer-Based RNA PROTAC Delivery Systems has further broadened the scope of this technology. These systems, with their distinct characteristics and functionalities, have enhanced the delivery and efficacy of PROTACs. Comparative analysis of different delivery systems has provided valuable insights into their advantages and disadvantages, guiding the selection of the most suitable system for specific applications. Looking forward, the field of targeted protein degradation holds immense potential. With ongoing research and development, we can anticipate significant advancements and novel applications of PROTAC technology. However, it's important to address the existing challenges to fully harness the potential of this technology. Overall, PROTAC technology represents a significant breakthrough in the field of medicine, offering a promising approach to the treatment of various diseases. As we continue to explore and understand this technology, we can look forward to more innovative solutions in the future.

References

Ahmad H, Zia B, Husain H, Husain A (2023) Recent advances in PROTAC-based antiviral strategies. Vaccine 11(2):2. https://doi.org/10.3390/vaccines11020270

An S, Fu L (2018) Small-molecule PROTACs: an emerging and promising approach for the development of targeted therapy drugs. EBioMedicine 36:553–562. https://doi.org/10.1016/j.ebiom.2018.09.005

Berndsen CE, Wolberger C (2014) New insights into ubiquitin E3 ligase mechanism. Nat Struct Mol Biol 21(4):301–307. https://doi.org/10.1038/nsmb.2780

Bitounis D, Fanciullino R, Iliadis A, Ciccolini J (2012) Optimizing druggability through liposomal formulations: new approaches to an old concept. Int Sch Res Notices 2012:e738432. https://doi.org/10.5402/2012/738432

Bondeson DP et al (2015) Catalytic in vivo protein knockdown by small-molecule PROTACs. Nat Chem Biol 11:611–617

Breitschopf K, Bengal E, Ziv T, Admon A, Ciechanover A (1998) A novel site for ubiquitination: the N-terminal residue, and not internal lysines of MyoD, is essential for conjugation and degradation of the protein. EMBO J 17(20):5964–5973. https://doi.org/10.1093/emboj/17.20.5964

Bricelj A, Steinebach C, Kuchta R, Gütschow M, Sosič I (2021) E3 ligase ligands in successful PROTACs: an overview of syntheses and linker attachment points. Front Chem 9:707317. https://doi.org/10.3389/fchem.2021.707317

Burslem GM, Smith BE, Lai AC, Jaime-Figueroa S, McQuaid DC, Bondeson DP, Toure M, Dong H, Qian Y, Wang J, Crew AP, Hines J, Crews CM (2018) The advantages of targeted protein degradation over inhibition: an RTK case study. Cell Chem Biol 25(1):67–77. https://doi.org/10.1016/j.chembiol.2017.09.009

Campbell IB, Macdonald SJF, Procopiou PA (2018) Medicinal chemistry in drug discovery in big pharma: past, present and future. Drug Discov Today 23(2):219–234. https://doi.org/10.1016/j.drudis.2017.10.007

Chen W, Schilperoort M, Cao Y, Shi J, Tabas I, Tao W (2022) Macrophage-targeted nanomedicine for the diagnosis and treatment of atherosclerosis. Nat Rev Cardiol 19(4):228–249

Cimas FJ, Niza E, Juan A, Noblejas-López M, Bravo I, Lara-Sanchez A, Alonso-Moreno C, Ocaña A (2020) Controlled delivery of BET-PROTACs: in vitro evaluation of MZ1-loaded polymeric antibody conjugated nanoparticles in breast cancer. Pharmaceutics 12:10. https://doi.org/10.3390/pharmaceutics12100986

de Wispelaere M, Du G, Donovan KA, Zhang T, Eleuteri NA, Yuan JC, Kalabathula J, Nowak RP, Fischer ES, Gray NS, Yang PL (2019) Small molecule degraders of the hepatitis C virus protease reduce susceptibility to resistance mutations. Nat Commun 10(1):3468. https://doi.org/10.1038/s41467-019-11429-w

Diehl CJ, Ciulli A (2022) Discovery of small molecule ligands for the von Hippel-Lindau (VHL) E3 ligase and their use as inhibitors and PROTAC degraders. Chem Soc Rev 51(19):8216–8257. https://doi.org/10.1039/d2cs00387b

Edmondson SD, Yang B, Fallan C (2019) Proteolysis targeting chimeras (PROTACs) in "beyond rule-of-five" chemical space: recent progress and future challenges. Bioorg Med Chem Lett 29(13):1555–1564. https://doi.org/10.1016/j.bmcl.2019.04.030

Fu Y, Rathod D, Patel K (2020) Protein kinase C inhibitor anchored BRD4 PROTAC PEGylated nanoliposomes for the treatment of vemurafenib-resistant melanoma. Exp Cell Res 396(1):112275. https://doi.org/10.1016/j.yexcr.2020.112275

Gadd MS, Testa A, Lucas X, Chan K-H, Chen W, Lamont DJ, Zengerle M, Ciulli A (2017) Structural basis of PROTAC cooperative recognition for selective protein degradation. Nat Chem Biol 13(5):514–521. https://doi.org/10.1038/nchembio.2329

Gao H, Sun X, Rao Y (2020) PROTAC technology: opportunities and challenges. ACS Med Chem Lett 11(3):237–240. https://doi.org/10.1021/acsmedchemlett.9b00597

Guo Y, Bao Q, Ping H, Shi J (2023) Nanomedicine-based co-delivery of a calcium channel inhibitor and a small molecule targeting CD47 for lung cancer immunotherapy. Nat Commun 14:7306

Hu Z, Crews CM (2022) Recent developments in PROTAC-mediated protein degradation: from bench to clinic. ChemBioChem 23(2):e202100270. https://doi.org/10.1002/cbic.202100270

Hu J, Hu B, Wang M, Xu F, Miao B, Yang C-Y, Wang M, Liu Z, Hayes DF, Chinnaswamy K, Delproposto J, Stuckey J, Wang S (2019) Discovery of ERD-308 as a highly potent proteolysis targeting chimera (PROTAC) degrader of estrogen receptor (ER). J Med Chem 62(3):1420–1442. https://doi.org/10.1021/acs.jmedchem.8b01572

Hughes GR, Dudey AP, Hemmings AM, Chantry A (2021) Frontiers in PROTACs. Drug Discov Today 26(10):2377–2383. https://doi.org/10.1016/j.drudis.2021.04.010

Ivan M, Kaelin WG (2001) The von Hippel–Lindau tumor suppressor protein. Curr Opin Genet Dev 11(1):27–34. https://doi.org/10.1016/S0959-437X(00)00152-0

Juan A, del Mar Noblejas-López M, Arenas-Moreira M, Alonso-Moreno C, Ocaña A (2021) Frontiers options to improve the action of PROTACs in cancer: development of controlled delivery nanoparticles. Front Cell Dev Biol 9:805336

Konstantinidou M, Li J, Zhang B, Wang Z, Shaabani S, Ter Brake F, Essa K, Dömling A (2019) PROTACs- a game-changing technology. Expert Opin Drug Discov 14(12):1255–1268. https://doi.org/10.1080/17460441.2019.1659242

Kou P, Levy ES, Nguyen AD, Zhang D, Chen S, Cui Y, Zhang X, Broccatelli F, Pizzano J, Cantley J, Bortolon E, Rousseau E, Berlin M, Dragovich P, Sethuraman V (2023) Development of liposome systems for enhancing the PK properties of bivalent PROTACs. Pharmaceutics 15(8):8. https://doi.org/10.3390/pharmaceutics15082098

Kregel S, Wang C, Han X, Xiao L, Fernandez-Salas E, Bawa P, McCollum BL, Wilder-Romans K, Apel IJ, Cao X, Speers C, Wang S, Chinnaiyan AM (2020) Androgen receptor degraders overcome common resistance mechanisms developed during prostate cancer treatment. Neoplasia 22(2):111–119. https://doi.org/10.1016/j.neo.2019.12.003

Li JW, Zheng G, Kaye FJ, Wu L (2023) PROTAC therapy as a new targeted therapy for lung cancer. Mol Ther 31(3):647–656. https://doi.org/10.1016/j.ymthe.2022.11.011

Moon Y, Jeon SI, Shim MK, Kim K (2023) Cancer-specific delivery of proteolysis-targeting chimeras (PROTACs) and their application to cancer immunotherapy. Pharmaceutics 15(2):2. https://doi.org/10.3390/pharmaceutics15020411

Nalawansha DA, Crews CM (2020) PROTACs: an emerging therapeutic modality in precision medicine. Cell Chem Biol 27(8):998–1014. https://doi.org/10.1016/j.chembiol.2020.07.020

Paiva S-L, Crews CM (2019) Targeted protein degradation: elements of PROTAC design. Curr Opin Chem Biol 50:111–119. https://doi.org/10.1016/j.cbpa.2019.02.022

Pettersson M, Crews CM (2019) Proteolysis targeting chimeras (PROTACs)—past, present and future. Drug Discov Today Technol 31:15–27. https://doi.org/10.1016/j.ddtec.2019.01.002

Pica N, Palese P (2013) Toward a universal influenza virus vaccine: prospects and challenges. Annu Rev Med 64(1):189–202. https://doi.org/10.1146/annurev-med-120611-145115

Pike A, Williamson B, Harlfinger S, Martin S, McGinnity DF (2020) Optimising proteolysis-targeting chimeras (PROTACs) for oral drug delivery: a drug metabolism and pharmacokinetics perspective. Drug Discov Today 25(10):1793–1800. https://doi.org/10.1016/j.drudis.2020.07.013

Sakamoto KM, Kim KB, Kumagai A, Mercurio F, Crews CM, Deshaies RJ (2001) Protacs: chimeric molecules that target proteins to the Skp1–Cullin–F box complex for ubiquitination and degradation. Proc Natl Acad Sci 98(15):8554–8559. https://doi.org/10.1073/pnas.141230798

Saraswat A, Patki M, Fu Y, Barot S, Dukhande VV, Patel K (2020) Nanoformulation of proteolysis targeting chimera targeting "undruggable" c-Myc for the treatment of pancreatic cancer. Nanomedicine 15(18):1761–1777. https://doi.org/10.2217/nnm-2020-0156

Saraswat AL, Vartak R, Hegazy R, Patel A, Patel K (2023) Drug delivery challenges and formulation aspects of proteolysis targeting chimera (PROTACs). Drug Discov Today 28(1):103387. https://doi.org/10.1016/j.drudis.2022.103387

Schapira M, Calabrese MF, Bullock AN, Crews CM (2019) Targeted protein degradation: expanding the toolbox. Nat Rev Drug Discov 18:949–963

Schneekloth JS, Fonseca FN, Koldobskiy M, Mandal A, Deshaies R, Sakamoto K, Crews CM (2004) Chemical genetic control of protein levels: selective in vivo targeted degradation. J Am Chem Soc 126(12):3748–3754. https://doi.org/10.1021/ja039025z

Shibata N, Nagai K, Morita Y, Ujikawa O, Ohoka N, Hattori T, Koyama R, Sano O, Imaeda Y, Nara H, Cho N, Naito M (2018) Development of protein degradation inducers of androgen receptor by conjugation of androgen receptor ligands and inhibitor of apoptosis protein ligands. J Med Chem 61(2):543–575. https://doi.org/10.1021/acs.jmedchem.7b00168

Tashima T (2023) Proteolysis-targeting chimera (PROTAC) delivery into the brain across the blood-brain barrier. Antibodies 12(3):3. https://doi.org/10.3390/antib12030043

Xiao M, Zhao J, Wang Q, Liu J, Ma L (2022) Recent advances of degradation technologies based on PROTAC mechanism. Biomolecules 12(9):9. https://doi.org/10.3390/biom12091257

Zeng S, Huang W, Zheng X, Cheng L, Zhang Z, Wang J, Shen Z (2021) Proteolysis targeting chimera (PROTAC) in drug discovery paradigm: recent progress and future challenges. Eur J Med Chem 210:112981. https://doi.org/10.1016/j.ejmech.2020.112981

Zhang C, Pu K (2020) Molecular and nanoengineering approaches towards activatable cancer immunotherapy. Chem Soc Rev 49(13):4234–4253. https://doi.org/10.1039/C9CS00773C

Zhang C, Han X-R, Yang X, Jiang B, Liu J, Xiong Y, Jin J (2018) Proteolysis targeting chimeras (PROTACs) of anaplastic lymphoma kinase (ALK). Eur J Med Chem 151:304–314. https://doi.org/10.1016/j.ejmech.2018.03.071

The manufacturer's authorised representative in the EU is Springer Nature Customer Service Centre GmbH, Europaplatz 3, 69115 Heidelberg, Germany. If you have any concerns regarding our products, please contact ProductSafety@springernature.com

Printed and bound by CPI Group (UK) Ltd, Croydon, CR0 4YY
01/10/2026
02225859-0014